# Universitext

For further volumes:
http://www.springer.com/series/223

Wolfgang Rautenberg

# A Concise Introduction
# to Mathematical Logic

## Third Edition

 Springer

Prof. Dr. Wolfgang Rautenberg
Fachbereich Mathematik und Informatik
14195 Berlin
Germany
raut@math.fu-berlin.de

*Cover illustration*: Photographer unknown. Courtesy of The Shelby White and Leon Levy
Archives Center, Institute for Advanced Study, Princeton, NJ, USA.

ISBN 978-1-4419-1220-6     e-ISBN 978-1-4419-1221-3
DOI 10.1007/978-1-4419-1221-3
Springer New York Dordrecht Heidelberg London

Library of Congress Control Number: 2009932782

Mathematics Subject Classification (2000): 03-XX, 68N 17

Printed on acid-free paper

Springer is part of Springer Science+Business Media (www.springer.com)

# Foreword

by Lev Beklemishev, Moscow

The field of mathematical logic—evolving around the notions of logical validity, provability, and computation—was created in the first half of the previous century by a cohort of brilliant mathematicians and philosophers such as Frege, Hilbert, Gödel, Turing, Tarski, Malcev, Gentzen, and some others. The development of this discipline is arguably among the highest achievements of science in the twentieth century: it expanded mathematics into a novel area of applications, subjected logical reasoning and computability to rigorous analysis, and eventually led to the creation of computers.

The textbook by Professor Wolfgang Rautenberg is a well-written introduction to this beautiful and coherent subject. It contains classical material such as logical calculi, beginnings of model theory, and Gödel's incompleteness theorems, as well as some topics motivated by applications, such as a chapter on logic programming. The author has taken great care to make the exposition readable and concise; each section is accompanied by a good selection of exercises.

A special word of praise is due for the author's presentation of Gödel's second incompleteness theorem, in which the author has succeeded in giving an accurate and simple proof of the derivability conditions and the provable $\Sigma_1$-completeness, a technically difficult point that is usually omitted in textbooks of comparable level. This work can be recommended to all students who want to learn the foundations of mathematical logic.

# Preface

The third edition differs from the second mainly in that parts of the text have been elaborated upon in more detail. Moreover, some new sections have been added, for instance a separate section on Horn formulas in Chapter **4**, particularly interesting for logic programming. The book is aimed at students of mathematics, computer science, and linguistics. It may also be of interest to students of philosophy (with an adequate mathematical background) because of the epistemological applications of Gödel's incompleteness theorems, which are discussed in detail.

Although the book is primarily designed to accompany lectures on a graduate level, most of the first three chapters are also readable by undergraduates. The first hundred twenty pages cover sufficient material for an undergraduate course on mathematical logic, combined with a due portion of set theory. Only that part of set theory is included that is closely related to mathematical logic. Some sections of Chapter **3** are partly descriptive, providing a perspective on decision problems, on automated theorem proving, and on nonstandard models.

Using this book for independent and individual study depends less on the reader's mathematical background than on his (or her) ambition to master the technical details. Suitable examples accompany the theorems and new notions throughout. We always try to portray simple things simply and concisely and to avoid excessive notation, which could divert the reader's mind from the essentials. Line breaks in formulas have been avoided. To aid the student, the indexes have been prepared very carefully. Solution hints to most exercises are provided in an extra file ready for download from Springer's or the author's website.

Starting from Chapter **4**, the demands on the reader begin to grow. The challenge can best be met by attempting to solve the exercises without recourse to the hints. The density of information in the text is rather high; a newcomer may need one hour for one page. Make sure to have paper and pencil at hand when reading the text. Apart from sufficient training in logical (or mathematical) deduction, additional prerequisites are assumed only for parts of Chapter **5**, namely some knowledge of classical algebra, and at the very end of the last chapter some acquaintance with models of axiomatic set theory.

On top of the material for a one-semester lecture course on mathematical logic, basic material for a course in logic for computer scientists is included in Chapter **4** on logic programming. An effort has been made to capture some of the interesting aspects of this discipline's logical foundations. The resolution theorem is proved constructively. Since all recursive functions are computable in PROLOG, it is not hard to deduce the undecidability of the existence problem for successful resolutions.

Chapter **5** concerns applications of mathematical logic in mathematics itself. It presents various methods of model construction and contains the basic material for an introductory course on model theory. It contains in particular a model-theoretic proof of quantifier eliminability in the theory of real closed fields, which has a broad range of applications.

A special aspect of the book is the thorough treatment of Gödel's incompleteness theorems in Chapters **6** and **7**. Chapters **4** and **5** are not needed here. **6.1**[1] starts with basic recursion theory needed for the arithmetization of syntax in **6.2** as well as in solving questions about decidability and undecidability in **6.5**. Defining formulas for arithmetical predicates are classified early, to elucidate the close relationship between logic and recursion theory. Along these lines, in **6.5** we obtain in one sweep Gödel's first incompleteness theorem, the undecidability of the tautology problem by Church, and Tarski's result on the nondefinability of truth, all of which are based on certain diagonalization arguments. **6.6** includes among other things a sketch of the solution to Hilbert's tenth problem.

Chapter **7** is devoted mainly to Gödel's second incompleteness theorem and some of its generalizations. Of particular interest thereby is the fact that questions about self-referential arithmetical statements are algorithmically decidable due to Solovay's completeness theorem. Here and elsewhere, Peano arithmetic (PA) plays a key role, a basic theory for the foundations of mathematics and computer science, introduced already in **3.3**. The chapter includes some of the latest results in the area of self-reference not yet covered by other textbooks.

Remarks in small print refer occasionally to notions that are undefined and direct the reader to the bibliography, or will be introduced later. The bibliography can represent an incomplete selection only. It lists most

---

[1] This is to mean Section **6.1**, more precisely, Section **1** in Chapter **6**. All other boldface labels are to be read accordingly throughout the book.

English textbooks on mathematical logic and, in addition, some original papers mainly for historical reasons. It also contains some titles treating biographical, historical, and philosophical aspects of mathematical logic in more detail than this can be done in the limited size of our book. Some brief historical remarks are also made in the *Introduction*. Bibliographical entries are sorted alphabetically by author names. This order may slightly diverge from the alphabetic order of their citation labels.

The material contained in this book will remain with high probability the subject of lectures on mathematical logic in the future. Its streamlined presentation has allowed us to cover many different topics. Nonetheless, the book provides only a selection of results and can at most accentuate certain topics. This concerns above all Chapters **4**, **5**, **6**, and **7**, which go a step beyond the elementary. Philosophical and foundational problems of mathematics are not systematically discussed within the constraints of this book, but are to some extent considered when appropriate.

The seven chapters of the book consist of numbered sections. A reference like Theorem 5.4 is to mean Theorem 4 in Section **5** of a given chapter. In cross-referencing from another chapter, the chapter number will be adjoined. For instance, Theorem 6.5.4 means Theorem 5.4 in Chapter **6**. You may find additional information about the book or contact me on my website www.math.fu-berlin.de/~raut. Please contact me if you propose improved solutions to the exercises, which may afterward be included in the separate file *Solution Hints to the Exercises*.

I would like to thank the colleagues who offered me helpful criticism along the way. Useful for Chapter **7** were hints from Lev Beklemishev and Wilfried Buchholz. Thanks also to Peter Agricola for his help in parts of the contents and in technical matters, and to Michael Knoop and David Kramer for their thorough reading of the manuscript and finding a number of mistakes.

Wolfgang Rautenberg, June 2009

# Contents

# Introduction

Traditional logic as a part of philosophy is one of the oldest scientific disciplines. It can be traced back to the Stoics and to Aristotle[2] and is the root of what is nowadays called philosophical logic. Mathematical logic, however, is a relatively young discipline, having arisen from the endeavors of Peano, Frege, and Russell to reduce mathematics entirely to logic. It steadily developed during the twentieth century into a broad discipline with several subareas and numerous applications in mathematics, computer science, linguistics, and philosophy.

One feature of modern logic is a clear distinction between object language and metalanguage. The first is formalized or at least formalizable. The latter is, like the language of this book, a kind of a colloquial language that differs from author to author and depends also on the audience the author has in mind. It is mixed up with semiformal elements, most of which have their origin in set theory. The amount of set theory involved depends on one's objectives. Traditional semantics and model theory as essential parts of mathematical logic use stronger set-theoretic tools than does proof theory. In some model-theoretic investigations these are often the strongest possible ones. But on average, little more is assumed than knowledge of the most common set-theoretic terminology, presented in almost every mathematical course or textbook for beginners. Much of it is used only as a *façon de parler*.

The language of this book is similar to that common to almost all mathematical disciplines. There is one essential difference though. In mathematics, metalanguage and object language strongly interact with each other, and the latter is semiformalized in the best of cases. This method has proved successful. Separating object language and metalanguage is relevant only in special context, for example in axiomatic set theory, where formalization is needed to specify what certain axioms look like. Strictly formal languages are met more often in computer science. In analyzing complex software or a programming language, as in logic, formal linguistic entities are the central objects of consideration.

---

[2] The Aristotelian syllogisms are easy but useful examples for inferences in a first-order language with unary predicate symbols. One of these syllogisms serves as an example in Section **4.6** on logic programming.

The way of arguing about formal languages and theories is traditionally called the *metatheory*. An important task of a metatheoretic analysis is to specify procedures of logical inference by so-called *logical calculi*, which operate purely syntactically. There are many different logical calculi. The choice may depend on the formalized language, on the logical basis, and on certain aims of the formalization. Basic metatheoretic tools are in any case the naive natural numbers and inductive proof procedures. We will sometimes call them proofs by *metainduction*, in particular when talking about formalized object theories that speak about natural numbers. Induction can likewise be carried out on certain sets of strings over a fixed alphabet, or on the system of rules of a logical calculus.

The logical means of the metatheory are sometimes allowed or even explicitly required to be different from those of the object language. But in this book the logic of object languages, as well as that of the metalanguage, are classical, two-valued logic. There are good reasons to argue that classical logic is the logic of common sense. Mathematicians, computer scientists, linguists, philosophers, physicists, and others are using it as a common platform for communication.

It should be noticed that logic used in the sciences differs essentially from logic used in everyday language, where logic is more an art than a serious task of saying what follows from what. In everyday life, nearly every utterance depends on the context. In most cases logical relations are only alluded to and rarely explicitly expressed. Some basic assumptions of two-valued logic mostly fail, in particular, a context-free use of the logical connectives. Problems of this type are not dealt with here. To some extent, many-valued logic or Kripke semantics can help to clarify the situation, and sometimes intrinsic mathematical methods must be used in order to solve such problems. We shall use Kripke semantics here for a different goal, though, the analysis of self-referential sentences in Chapter **7**.

Let us add some historical remarks, which, of course, a newcomer may find easier to understand *after* and not *before* reading at least parts of this book. In the relatively short period of development of modern mathematical logic in the twentieth century, some highlights may be distinguished, of which we mention just a few. Many details on this development can be found in the excellent biographies [Daw] and [FF] on Gödel and Tarski, the leading logicians in the last century.

The first was the axiomatization of set theory in various ways. The most important approaches are those of Zermelo (improved by Fraenkel and von Neumann) and the theory of types by Whitehead and Russell. The latter was to become the sole remnant of Frege's attempt to reduce mathematics to logic. Instead it turned out that mathematics can be based entirely on set theory as a first-order theory. Actually, this became more salient after the rest of the hidden assumptions by Russell and others were removed from axiomatic set theory around 1915; see [Hei]. For instance, the notion of an ordered pair, crucial for reducing the notion of a function to set theory, is indeed a set-theoretic and not a logical one.

Right after these axiomatizations were completed, Skolem discovered that there are countable models of the set-theoretic axioms, a drawback to the hope for an axiomatic characterization of a set. Just then, two distinguished mathematicians, Hilbert and Brouwer, entered the scene and started their famous quarrel on the foundations of mathematics. It is described in a comprehensive manner for instance in [Kl2, Chapter IV] and need therefore not be repeated here.

As a next highlight, Gödel proved the completeness of Hilbert's rules for predicate logic, presented in the first modern textbook on mathematical logic, [HA]. Thus, to some extent, a dream of Leibniz became real, namely to create an *ars inveniendi* for mathematical truth. Meanwhile, Hilbert had developed his view on a foundation of mathematics into a program. It aimed at proving the consistency of arithmetic and perhaps the whole of mathematics including its nonfinitistic set-theoretic methods by finitary means. But Gödel showed by his incompleteness theorems in 1931 that Hilbert's original program fails or at least needs thorough revision.

Many logicians consider these theorems to be the top highlights of mathematical logic in the twentieth century. A consequence of these theorems is the existence of consistent extensions of Peano arithmetic in which true and false sentences live in peaceful coexistence with each other, called "dream theories" in **7.3**. It is an intellectual adventure of holistic beauty to see wisdom from number theory known for ages, such as the Chinese remainder theorem, simple properties of prime numbers, and Euclid's characterization of coprimeness (page 249), unexpectedly assuming pivotal positions within the architecture of Gödel's proofs. Gödel's methods were also basic for the creation of recursion theory around 1936.

Church's proof of the undecidability of the tautology problem marks another distinctive achievement. After having collected sufficient evidence by his own investigations and by those of Turing, Kleene, and some others, Church formulated his famous thesis (see **6.1**), although in 1936 no computers in the modern sense existed nor was it foreseeable that computability would ever play the basic role it does today.

Another highlight of mathematical logic has its roots in the work of Tarski, who proved first the undefinability of truth in formalized languages as explained in **6.5**, and soon thereafter started his fundamental work on decision problems in algebra and geometry and on model theory, which ties logic and mathematics closely together. See Chapter **5**.

As already mentioned, Hilbert's program had to be revised. A decisive step was undertaken by Gentzen, considered to be another groundbreaking achievement of mathematical logic and the starting point of contemporary proof theory. The logical calculi in **1.4** and **3.1** are akin to Gentzen's calculi of natural deduction.

We further mention Gödel's discovery that it is not the axiom of choice (AC) that creates the consistency problem in set theory. Set theory with AC and the continuum hypothesis (CH) is consistent, provided set theory without AC and CH is. This is a basic result of mathematical logic that would not have been obtained without the use of strictly formal methods. The same applies to the independence proof of AC and CH from the axioms of set theory by Cohen in 1963.

The above indicates that mathematical logic is closely connected with the aim of giving mathematics a solid foundation. Nonetheless, we confine ourself to logic and its fascinating interaction with mathematics, which characterizes mathematical logic. History shows that it is impossible to establish a programmatic view on the foundations of mathematics that pleases everybody in the mathematical community. Mathematical logic is the right tool for treating the technical problems of the foundations of mathematics, but it cannot solve its epistemological problems.

# Notation

We assume that the reader is familiar with the most basic mathematical terminology and notation, in particular with the *union*, *intersection*, and *complementation* of sets, denoted by ∪, ∩, and \, respectively. Here we summarize only some notation that may differ slightly from author to author or is specific for this book. $\mathbb{N}$, $\mathbb{Z}$, $\mathbb{Q}$, $\mathbb{R}$ denote the sets of natural numbers including 0, integers, rational, and real numbers, respectively, and $\mathbb{N}_+$, $\mathbb{Q}_+$, $\mathbb{R}_+$ the sets of positive members of the corresponding sets. $n, m, i, j, k$ always denote natural numbers unless stated otherwise. Hence, extended notation like $n \in \mathbb{N}$ is mostly omitted.

In the following, $M, N$ denote sets, $M \subseteq N$ denotes inclusion, while $M \subset N$ means proper inclusion (i.e., $M \subseteq N$ and $M \neq N$). As a rule, we write $M \subset N$ only if the circumstance $M \neq N$ has to be emphasized. If $M$ is fixed in a consideration and $N$ varies over subsets of $M$, then $M \setminus N$ may also be symbolized by $\setminus N$ or $\neg N$.

$\emptyset$ denotes the *empty set*, and $\mathfrak{P}M$ the *power set* (= set of all subsets) of $M$. If one wants to emphasize that all elements of a set $S$ are sets, $S$ is also called a *system* or *family* of sets. $\bigcup S$ denotes the union of $S$, that is, the set of elements belonging to at least one $M \in S$, and $\bigcap S$ stands for the intersection of a nonempty system $S$, the set of elements belonging to all $M \in S$. If $S = \{M_i \mid i \in I\}$ then $\bigcup S$ and $\bigcap S$ are mostly denoted by $\bigcup_{i \in I} M_i$ and $\bigcap_{i \in I} M_i$, respectively.

A *relation* between $M$ and $N$ is a subset of $M \times N$, the set of ordered pairs $(a, b)$ with $a \in M$ and $b \in N$. A precise definition of $(a, b)$ is given on page 114. Such a relation, $f$ say, is said to be a *function* or *mapping from $M$ to $N$* if for each $a \in M$ there is precisely one $b \in N$ with $(a, b) \in f$. This $b$ is denoted by $f(a)$ or $fa$ or $a^f$ and called the *value of $f$ at $a$*. We denote a function $f$ from $M$ to $N$ also by $f : M \to N$, or by $f : x \mapsto t(x)$, provided $f(x) = t(x)$ for some term $t$ (see **2.2**). $\operatorname{ran} f = \{fx \mid x \in M\}$ is called the *range* of $f$, and $\operatorname{dom} f = M$ its *domain*. $\operatorname{id}_M$ denotes the *identical function* on $M$, that is, $\operatorname{id}_M(x) = x$ for all $x \in M$.

$f : M \to N$ is *injective* if $fx = fy \Rightarrow x = y$, for all $x, y \in M$, *surjective* if $\operatorname{ran} f = N$, and *bijective* if $f$ is both injective and surjective. The reader should basically be familiar with this terminology. The phrase "let $f$ be a function from $M$ to $N$" is sometimes shortened to "let $f : M \to N$."

The set of all functions from a set $I$ to a set $M$ is denoted by $M^I$. If $f, g$ are functions with $\operatorname{ran} g \subseteq \operatorname{dom} f$ then $h \colon x \mapsto f(g(x))$ is called their *composition* (or *product*). It will preferably be written as $h = f \circ g$.

Let $I$ and $M$ be sets, $f \colon I \to M$, and call $I$ the *index set*. Then $f$ will often be denoted by $(a_i)_{i \in I}$ and is named, depending on the context, an (indexed) *family*, an *I-tuple*, or a *sequence*. If 0 is identified with $\emptyset$ and $n > 0$ with $\{0, 1, \ldots, n-1\}$, as is common in set theory, then $M^n$ can be understood as the set of $n$-tuples $(a_i)_{i < n} = (a_0, \ldots, a_{n-1})$ *of length* $n$ whose members belong to $M$. In particular, $M^0 = \{\emptyset\}$. Also the set of sequences $(a_1, \ldots, a_n)$ with $a_i \in M$ will frequently be denoted by $M^n$. In concatenating finite sequences, which has an obvious meaning, the *empty sequence* (i.e., $\emptyset$), plays the role of a neutral element. $(a_1, \ldots, a_n)$ will mostly be denoted by $\vec{a}$. Note that this is the empty sequence for $n = 0$, similar to $\{a_1, \ldots, a_n\}$ for $n = 0$ always being the empty set. $f\vec{a}$ means $f(a_1, \ldots, a_n)$ throughout.

If $A$ is an *alphabet*, i.e., if the elements $\mathsf{s} \in A$ are symbols or at least named symbols, then the sequence $(\mathsf{s}_1, \ldots, \mathsf{s}_n) \in A^n$ is written as $\mathsf{s}_1 \cdots \mathsf{s}_n$ and called a *string* or a *word* over $A$. The empty sequence is called in this context the *empty string*. A string consisting of a single symbol $\mathsf{s}$ is termed an *atomic string*. It will likewise be denoted by $\mathsf{s}$, since it will be clear from the context whether $\mathsf{s}$ means a symbol or an atomic string.

Let $\xi \eta$ denote the concatenation of the strings $\xi$ and $\eta$. If $\xi = \xi_1 \eta \xi_2$ for some strings $\xi_1, \xi_2$ and $\eta \neq \emptyset$ then $\eta$ is called a *segment* (or *substring*) of $\xi$, termed a *proper* segment in case $\eta \neq \xi$. If $\xi_1 = \emptyset$ then $\eta$ is called an *initial*, if $\xi_2 = \emptyset$, a *terminal* segment of $\xi$.

Subsets $P, Q, R, \ldots \subseteq M^n$ are called *n-ary predicates of $M$* or $n$-ary *relations*. A unary predicate will be identified with the corresponding subset of $M$. We may write $P\vec{a}$ for $\vec{a} \in P$, and $\neg P\vec{a}$ for $\vec{a} \notin P$. Metatheoretical predicates (or properties) cast in words will often be distinguished from the surrounding text by single quotes, for instance, if we speak of the syntactic predicate 'The variable $x$ occurs in the formula $\alpha$'. We can do so since quotes inside quotes will not occur in this book. Single-quoted properties are often used in induction principles or reflected in a theory, while ordinary ("double") quotes have a stylistic function only.

An $n$-ary *operation of $M$* is a function $f \colon M^n \to M$. Since $M^0 = \{\emptyset\}$, a 0-ary operation of $M$ is of the form $\{(\emptyset, c)\}$, with $c \in M$; it is denoted by

$c$ for short and called a *constant*. Each operation $f: M^n \to M$ is uniquely described by the *graph of $f$*, defined as

$$\text{graph } f := \{(a_1, \ldots, a_{n+1}) \in M^{n+1} \mid f(a_1, \ldots, a_n) = a_{n+1}\}.[1]$$

Both $f$ and graph $f$ are essentially the same, but in most situations it is more convenient to distinguish between them.

The most important operations are binary ones. The corresponding symbols are mostly written between the arguments, as in the following listing of properties of a binary operation $\circ$ on a set $A$. $\circ: A^2 \to A$ is

| | | |
|---|---|---|
| *commutative* | if | $a \circ b = b \circ a$ for all $a, b \in A$, |
| *associative* | if | $a \circ (b \circ c) = (a \circ b) \circ c$ for all $a, b, c \in A$, |
| *idempotent* | if | $a \circ a = a$ for all $a \in A$, |
| *invertible* | if | for all $a, b \in A$ there are $x, y \in A$ |
| | | with $a \circ x = b$ and $y \circ a = b$. |

If $H, \Theta$ (read *eta, theta*) are expressions of our metalanguage, $H \Leftrightarrow \Theta$ stands for '$H$ *iff* $\Theta$' which abbreviates '$H$ *if and only if* $\Theta$'. Similarly, $H \Rightarrow \Theta$ and $H \, \& \, \Theta$ mean '*if $H$ then $\Theta$*' and '$H$ *and* $\Theta$', respectively, and $H \vee \Theta$ is to mean '$H$ *or* $\Theta$.' This notation does not aim at formalizing the metalanguage but serves improved organization of metatheoretic statements. We agree that $\Rightarrow, \Leftrightarrow, \ldots$ separate stronger than linguistic binding particles such as "there is" or "for all." Therefore, in the statement

$$\text{`}X \vdash \alpha \Leftrightarrow X \vDash \alpha, \text{ for all } X \text{ and all } \alpha\text{'} \quad \text{(Theorem 1.4.6)}$$

the comma should not be dropped; otherwise, some serious misunderstanding may arise: '$X \vDash \alpha$ *for all $X$ and all $\alpha$*' is simply false.

$H :\Leftrightarrow \Theta$ means that the expression $H$ is defined by $\Theta$. When integrating formulas in the colloquial metalanguage, one may use certain abbreviating notation. For instance, '$\alpha \equiv \beta$ and $\beta \equiv \gamma$' is occasionally shortened to $\alpha \equiv \beta \equiv \gamma$. ('the formulas $\alpha, \beta$, and $\beta, \gamma$ are equivalent'). This is allowed, since in this book the symbol $\equiv$ will never belong to the formal language from which the formulas $\alpha, \beta, \gamma$ are taken. W.l.o.g. or w.l.o.g. is a colloquial shorthand of "without loss of generality" used in mathematics.

---

[1] This means that the left-hand term graph $f$ is *defined* by the right-hand term. A corresponding meaning has $:=$ throughout, except in programs and flow diagrams, where $x := t$ means the allocation of the value of the term $t$ to the variable $x$.

# Chapter 1

# Propositional Logic

Propositional logic, by which we here mean two-valued propositional logic, arises from analyzing connections of given sentences $A, B$, such as

$$A \text{ and } B, \quad A \text{ or } B, \quad \text{not } A, \quad \text{if } A \text{ then } B.$$

These connection operations can be approximately described by two-valued logic. There are other connections that have temporal or local features, for instance, *first A then B* or *here A there B*, as well as unary modal operators like *it is necessarily true that*, whose analysis goes beyond the scope of two-valued logic. These operators are the subject of temporal, modal, or other subdisciplines of many-valued or nonclassical logic. Furthermore, the connections that we began with may have a meaning in other versions of logic that two-valued logic only incompletely captures. This pertains in particular to their meaning in natural or everyday language, where meaning may strongly depend on context.

In two-valued propositional logic such phenomena are set aside. This approach not only considerably simplifies matters, but has the advantage of presenting many concepts, for instance those of consequence, rule induction, or resolution, on a simpler and more perspicuous level. This will in turn save a lot of writing in Chapter **2** when we consider the corresponding concepts in the framework of predicate logic.

We will not consider everything that would make sense in two-valued propositional logic, such as two-valued fragments and problems of definability and interpolation. The reader is referred instead to [KK] or [Ra1]. We will concentrate our attention more on propositional calculi. While there exists a multitude of applications of propositional logic, we will not consider technical applications such as the designing of Boolean circuits

W. Rautenberg, *A Concise Introduction to Mathematical Logic*,
Universitext, DOI 10.1007/978-1-4419-1221-3_1,
© Springer Science+Business Media, LLC 2010

and problems of optimization. These topics have meanwhile been integrated into computer science. Rather, some useful applications of the propositional compactness theorem are described comprehensively.

## 1.1   Boolean Functions and Formulas

Two-valued logic is based on two foundational principles: the *principle of bivalence*, which allows only two truth values, namely *true* and *false*, and the *principle of extensionality*, according to which the truth value of a connected sentence depends only on the truth values of its parts, not on their meaning. Clearly, these principles form only an idealization of the actual relationships.

Questions regarding degrees of truth or the sense-content of sentences are ignored in two-valued logic. Despite this simplification, or indeed because of it, such a method is scientifically successful. One does not even have to know exactly what the truth values *true* and *false* actually are. Indeed, in what follows we will identify them with the two symbols 1 and 0. Of course, one could have chosen any other apt symbols such as $\top$ and $\bot$ or $\mathsf{t}$ and $\mathsf{f}$. The advantage here is that all conceivable interpretations of *true* and *false* remain open, including those of a purely technical nature, for instance the two states of a gate in a Boolean circuit.

According to the meaning of the word *and*, the conjunction $A$ *and* $B$ of sentences $A, B$, in formalized languages written as $A \wedge B$ or $A \& B$, is true if and only if $A, B$ are both true and is false otherwise. So conjunction corresponds to a binary function or operation over the set $\{0,1\}$ of truth values, named the $\wedge$-*function* and denoted by $\wedge$. It is given by its *value matrix* $\begin{pmatrix} 1 & 0 \\ 0 & 0 \end{pmatrix}$, where, in general, $\begin{pmatrix} 1 \circ 1 & 1 \circ 0 \\ 0 \circ 1 & 0 \circ 0 \end{pmatrix}$ represents the value matrix or *truth table* of a binary function $\circ$ with arguments and values in $\{0,1\}$. The delimiters of these small matrices will usually be omitted.

A function $f \colon \{0,1\}^n \to \{0,1\}$ is called an $n$-ary *Boolean function* or *truth function*. Since there are $2^n$ $n$-tuples of $0, 1$, it is easy to see that the number of $n$-ary Boolean functions is $2^{2^n}$. We denote their totality by $\boldsymbol{B}_n$. While $\boldsymbol{B}_2$ has $2^4 = 16$ members, there are only four unary Boolean functions. One of these is *negation*, denoted by $\neg$ and defined by $\neg 1 = 0$ and $\neg 0 = 1$. $\boldsymbol{B}_0$ consists just of the constants 0 and 1.

The first column of the table below contains the common binary connections with examples of their instantiation in English. The second column lists some of its traditional symbols, which also denote the corresponding truth function, and the third its truth table. *Disjunction* is the *inclusive or* and is to be distinguished from the *exclusive disjunction*. The latter corresponds to addition modulo 2 and is therefore given the symbol $+$. In Boolean circuits the functions $+, \downarrow, \uparrow$ are often denoted by *xor, nor,* and *nand*; the latter is also known as the *Sheffer function*. Recall our agreement in the section *Notation* that the symbols $\&, \vee, \Rightarrow$, and $\Leftrightarrow$ will be used only on the metatheoretic level.

A connected sentence and its corresponding truth function need not be denoted by the same symbol; for example, one might take $\wedge$ for conjunction and *et* as the corresponding truth function. But in doing so one would only be creating extra notation, but no new insights. The meaning of a symbol will always be clear from the context: if $\alpha, \beta$ are sentences of a formal language, then $\alpha \wedge \beta$ denotes their conjunction; if $a, b$ are truth values, then $a \wedge b$ just denotes a truth value. Occasionally, we may want to refer to the symbols $\wedge, \vee, \neg, \dots$ themselves, setting their meaning temporarily aside. Then we talk of the *connectives* or *truth functors* $\wedge, \vee, \neg, \dots$

| compound sentence | symbol | truth table |
|---|---|---|
| conjunction<br>*A and B; A as well as B* | $\wedge$, $\&$ | 1 0<br>0 0 |
| disjunction<br>*A or B* | $\vee$, $\mathsf{V}$ | 1 1<br>1 0 |
| implication<br>*if A then B; B provided A* | $\rightarrow$, $\Rightarrow$ | 1 0<br>1 1 |
| equivalence<br>*A if and only if B; A iff B* | $\leftrightarrow$, $\Leftrightarrow$ | 1 0<br>0 1 |
| exclusive disjunction<br>*either A or B but not both* | $+$ | 0 1<br>1 0 |
| nihilation<br>*neither A nor B* | $\downarrow$ | 0 0<br>0 1 |
| incompatibility<br>*not at once A and B* | $\uparrow$ | 0 1<br>1 1 |

Sentences formed using connectives given in the table are said to be logically equivalent if their corresponding truth tables are identical. This is the case, for example, for the sentences *A provided B* and *A or not B*, which represent the *converse implication*, denoted by $A \leftarrow B$.[1] It does not appear in the table, since it arises by swapping $A, B$ in the implication. This and similar reasons explain why only a few of the sixteen binary Boolean functions require notation. Another example of logical equivalent sentences are *if A and B then C*, and *if B then C provided A*.

In order to recognize and describe logical equivalence of compound sentences it is useful to create a suitable formalism or a formal language. The idea is basically the same as in arithmetic, where general statements are more clearly expressed by means of certain formulas. As with arithmetical terms, we consider propositional formulas as strings of signs built in given ways from basic symbols. Among these basic symbols are variables, for our purposes called *propositional variables*, the set of which is denoted by *PV*. Traditionally, these are symbolized by $p_0, p_1, \ldots$ However, our numbering of the variables below begins with $p_1$ rather than with $p_0$, enabling us later on to represent Boolean functions more conveniently. Further, we use certain logical signs such as $\wedge, \vee, \neg, \ldots$, similar to the signs $+, \cdot, \ldots$ of arithmetic. Finally, parentheses ( , ) will serve as technical aids, although these are dispensable, as will be seen later on.

Each time a propositional language is in question, the set of its logical symbols, called the *logical signature*, and the set of its variables must be given in advance. For instance, it is crucial in some applications of propositional logic in Section **1.5** for *PV* to be an arbitrary set, and not a countably infinite one as indicated previously. Put concretely, we define a propositional language $\mathcal{F}$ of formulas built up from the symbols $( , ), \wedge, \vee, \neg, p_1, p_2, \ldots$ inductively as follows:

(F1) The atomic strings $p_1, p_2, \ldots$ are formulas, called *prime formulas*, also called *atomic* formulas, or simply *prime*.

(F2) If the strings $\alpha, \beta$ are formulas, then so too are the strings $(\alpha \wedge \beta)$, $(\alpha \vee \beta)$, and $\neg \alpha$.

This is a recursive (somewhat sloppily also called inductive) definition in the set of strings on the alphabet of the mentioned symbols, that is,

---

[1] Converse implication is used in the programming language PROLOG, see **4.6**.

only those strings gained using (F1) or (F2) are in this context formulas. Stated set-theoretically, $\mathcal{F}$ is the smallest (i.e., the intersection) of all sets of strings $S$ built from the aforementioned symbols with the properties

(f1) $p_1, p_2, \ldots \in S$,  (f2) $\alpha, \beta \in S \Rightarrow (\alpha \wedge \beta), (\alpha \vee \beta), \neg \alpha \in S$.

**Example.** $(p_1 \wedge (p_2 \vee \neg p_1))$ is a formula. On the other hand, its initial segment $(p_1 \wedge (p_2 \vee \neg p_1)$ is not, because a closing parenthesis is missing. It is intuitively clear and will rigorously be proved on the next page that the number of left parentheses occurring in a formula coincides with the number of its right parentheses.

**Remark 1.** (f1) and (f2) are set-theoretic translations of (F1) and (F2). Some authors like to add a third condition to (F1), (F2), namely (F3): *No other strings than those obtained by* (F1) *and* (F2) *are formulas in this context*. But this at most underlines that (F1), (F2) are the only formula-building rules; (F3) *follows* from our definition, as its set-theoretic translation by (f1), (f2) indicates. Note that we do not strictly distinguish between the symbol $p_i$ and the prime formula or atomic string $p_i$. Note also that in the formula definition parentheses are needed only for binary connectives, not if a formula starts with $\neg$. By a slightly more involved definition at least the outermost parentheses in formulas of the form $(\alpha \circ \beta)$ with a binary connective $\circ$ could be saved. Howsoever propositional formulas are defined, what counts is their unique readability, see page 7.

The formulas defined by (F1), (F2) are called *Boolean formulas*, because they are obtained using the *Boolean signature* $\{\wedge, \vee, \neg\}$. Should further connectives belong to the logical signature, for example $\rightarrow$ or $\leftrightarrow$, (F2) of the above definition must be augmented accordingly. But unless stated otherwise, $(\alpha \rightarrow \beta)$ and $(\alpha \leftrightarrow \beta)$ are here just abbreviations; the first is $\neg(\alpha \wedge \neg \beta)$, the second is $((\alpha \rightarrow \beta) \wedge (\beta \rightarrow \alpha))$.

Occasionally, it is useful to have symbols in the logical signature for always false and always true, $\bot$ and $\top$ respectively, say, called *falsum* and *verum* and sometimes also denoted by 0 and 1. These are to be regarded as supplementary prime formulas, and clause (F1) should be altered accordingly. However, we prefer to treat $\bot$ and $\top$ as abbreviations: $\bot := (p_1 \wedge \neg p_1)$ and $\top := \neg \bot$.

For the time being we let $\mathcal{F}$ be the set of all Boolean formulas, although everything said about $\mathcal{F}$ holds correspondingly for any propositional language. Propositional variables will henceforth be denoted by $p, q, \ldots$, formulas by $\alpha, \beta, \gamma, \delta, \varphi, \ldots$, prime formulas also by $\pi$, and sets of propositional formulas by $X, Y, Z$, where these letters may also be indexed.

For the reason of parenthesis economy in formulas, we set some conventions similar to those used in writing arithmetical terms.

1. The outermost parentheses in a formula may be omitted (if there are any). For example, $(p \vee q) \wedge \neg p$ may be written in place of $((p \vee q) \wedge \neg p)$. Note that $(p \vee q) \wedge \neg p$ is not itself a formula but *denotes* the formula $((p \vee q) \wedge \neg p)$.

2. In the order $\neg, \wedge, \vee, \rightarrow, \leftrightarrow$, each connective binds more strongly than those following it. Thus, one may write $p \vee q \wedge \neg p$ instead of $p \vee (q \wedge \neg p)$, which means $(p \vee (q \wedge \neg p))$ by convention 1.

3. By the multiple use of $\rightarrow$ we *associate to the right*. So $p \rightarrow q \rightarrow p$ is to mean $p \rightarrow (q \rightarrow p)$. Multiple occurrences of other binary connectives are associated to the left, for instance, $p \wedge q \wedge \neg p$ means $(p \wedge q) \wedge \neg p$. In place of $\alpha_0 \wedge \cdots \wedge \alpha_n$ and $\alpha_0 \vee \cdots \vee \alpha_n$ we may write $\bigwedge_{i \leqslant n} \alpha_i$ and $\bigvee_{i \leqslant n} \alpha_i$, respectively.

Also, in arithmetic, one normally associates to the left. An exception is the term $x^{y^z}$, where traditionally association to the right is used, that is, $x^{y^z}$ equals $x^{(y^z)}$. Association to the right has some advantages in writing tautologies in which $\rightarrow$ occurs several times; for instance in the examples of tautologies listed in **1.3** on page 18.

The above conventions are based on a reliable syntax in the framework of which intuitively clear facts, such as the identical number of left and right parentheses in a formula, are rigorously provable. These proofs are generally carried out using induction on the construction of a formula. To make this clear we denote by $\mathcal{E}\varphi$ that a property $\mathcal{E}$ holds for a string $\varphi$. For example, let $\mathcal{E}$ mean the property '$\varphi$ is a formula that has equally many right- and left-hand parentheses'. $\mathcal{E}$ is trivially valid for prime formulas, and if $\mathcal{E}\alpha$, $\mathcal{E}\beta$ then clearly also $\mathcal{E}(\alpha \wedge \beta)$, $\mathcal{E}(\alpha \vee \beta)$, and $\mathcal{E}\neg\alpha$. From this we may conclude that $\mathcal{E}$ applies to all formulas, our reasoning being a particularly simple instance of the following

**Principle of formula induction.** *Let $\mathcal{E}$ be a property of strings that satisfies the conditions*

(o) *$\mathcal{E}\pi$ for all prime formulas $\pi$,*

(s) *$\mathcal{E}\alpha, \mathcal{E}\beta \Rightarrow \mathcal{E}(\alpha \wedge \beta), \mathcal{E}(\alpha \vee \beta), \mathcal{E}\neg\alpha$, for all $\alpha, \beta \in \mathcal{F}$.*

*Then $\mathcal{E}\varphi$ holds for all formulas $\varphi$.*

The justification of this principle is straightforward. The set $S$ of all strings with property $\mathcal{E}$ has, thanks to (o) and (s), the properties (f1) and (f2) on page 5. But $\mathcal{F}$ is the smallest such set. Therefore, $\mathcal{F} \subseteq S$. In words, $\mathcal{E}$ applies to all formulas $\varphi$. Clearly, if other connectives are involved, condition (s) must accordingly be modified.

It is intuitively clear and easily confirmed inductively on $\varphi$ that a compound Boolean formula $\varphi$ (i.e., $\varphi$ is not prime) is of the form $\varphi = \neg \alpha$ or $\varphi = (\alpha \wedge \beta)$ or $\varphi = (\alpha \vee \beta)$ for suitable $\alpha, \beta \in \mathcal{F}$. Moreover, this decomposition is unique. For instance, $(\alpha \wedge \beta)$ cannot at the same time be written $(\alpha' \vee \beta')$ with perhaps different formulas $\alpha', \beta'$. Thus, compound formulas have the unique readability property, more precisely, the

**Unique formula reconstruction property.** *Each compound formula $\varphi \in \mathcal{F}$ is either of the form $\neg \alpha$ or $(\alpha \circ \beta)$ for some uniquely determined formulas $\alpha, \beta \in \mathcal{F}$, where $\circ$ is either $\wedge$ or $\vee$.*

This property is less obvious than it might seem. Nonetheless, the proof is left as an exercise (Exercise 4) in order to maintain the flow of things. It may be a surprise to the novice that for the unique formula reconstruction, parentheses are dispensable throughout. Indeed, propositional formulas, like arithmetical terms, can be written without any parentheses; this is realized in *Polish notation* ($=$ PN), also called *prefix notation,* once widely used in the logic literature. The idea consists in altering (F2) as follows: *if $\alpha, \beta$ are formulas then so too are $\wedge \alpha \beta$, $\vee \alpha \beta$, and $\neg \alpha$.* Similar to PN is RPN (*reverse Polish notation*), still used in some programming languages like PostScript. RPN differs from PN only in that a connective is placed *after* the arguments. For instance, $(p \wedge (q \vee \neg p))$ is written in RPN as $pqp\neg\vee\wedge$. Reading PN or RPN requires more effort due to the high density of information; but by the same token it can be processed very fast by a computer or a high-tech printer getting its job as a PostScript program. The only advantage of the parenthesized version is that its decoding is somewhat easier for our eye through the dilution of information.

Intuitively it is clear what a *subformula* of a formula $\varphi$ is; for example, $(q \wedge \neg p)$ is a subformula of $(p \vee (q \wedge \neg p))$. All the same, for some purposes it is convenient to characterize the set $\mathrm{Sf}\,\varphi$ of all subformulas of $\varphi$ inductively:

$$\mathrm{Sf}\,\pi = \{\pi\} \text{ for prime formulas } \pi; \quad \mathrm{Sf}\,\neg\alpha = \mathrm{Sf}\,\alpha \cup \{\neg\alpha\},$$
$$\mathrm{Sf}(\alpha \circ \beta) = \mathrm{Sf}\,\alpha \cup \mathrm{Sf}\,\beta \cup \{(\alpha \circ \beta)\} \text{ for a binary connective } \circ.$$

Thus, a formula is always regarded as a subformula of itself. The above is a typical example of a *recursive definition on the construction of formulas*. Another example of such a definition is the *rank* $\mathrm{rk}\,\varphi$ of a formula $\varphi$, which provides a sometimes more convenient measure of the complexity of $\varphi$ than its length as a string and occasionally simplifies inductive arguments. Intuitively, $\mathrm{rk}\,\varphi$ is the highest number of nested connectives in $\varphi$. Let $\mathrm{rk}\,\pi = 0$ for prime formulas $\pi$, and if $\mathrm{rk}\,\alpha$ and $\mathrm{rk}\,\beta$ are already defined, then $\mathrm{rk}\,\neg\alpha = \mathrm{rk}\,\alpha + 1$ and $\mathrm{rk}(\alpha \circ \beta) = \max\{\mathrm{rk}\,\alpha, \mathrm{rk}\,\beta\} + 1$. Here $\circ$ denotes any binary connective. We will not give here a general formulation of this definition procedure because it is very intuitive and similar to the well-known procedure of recursive definitions on $\mathbb{N}$. It has been made sufficiently clear by the preceding examples. Its justification is based on the unique reconstruction property and insofar not quite trivial, in contrast to the proof procedure by induction on formulas that immediately follows from the definition of propositional formulas.

If a property is to be proved by induction on the construction of formulas $\varphi$, we will say that it is a *proof by induction on $\varphi$*. Similarly, the recursive construction of a function $f$ on $\mathcal{F}$ will generally be referred to as *defining $f$ by recursion on $\varphi$*, often somewhat sloppily paraphrased as *defining $f$ by induction on $\varphi$*. Examples are Sf and rk. Others will follow.

Since the truth value of a connected sentence depends only on the truth values of its constituent parts, we may assign to every propositional variable of $\alpha$ a truth value rather than a sentence, thereby evaluating $\alpha$, i.e., calculating a truth value. Similarly, terms are evaluated in, say, the arithmetic of real numbers, whose value is then a real (= real number). An arithmetical term $t$ in the variables $x_1, \ldots, x_n$ describes an $n$-ary function whose arguments and values are reals, while a formula $\varphi$ in $p_1, \ldots, p_n$ describes an $n$-ary Boolean function. To be precise, a propositional *valuation*, or alternatively, a (propositional) *model*, is a mapping $w \colon PV \to \{0,1\}$ that can also be understood as a mapping from the set of prime formulas to $\{0,1\}$. We can extend this to a mapping from the whole of $\mathcal{F}$ to $\{0,1\}$ (likewise denoted by $w$) according to the stipulations

$(*)\quad w(\alpha \wedge \beta) = w\alpha \wedge w\beta; \quad w(\alpha \vee \beta) = w\alpha \vee w\beta; \quad w\neg\alpha = \neg w\alpha.^{2}$

By *the value $w\varphi$ of a formula $\varphi$* under a valuation $w \colon PV \to \{0,1\}$

---

[2] We often use $(*)$ or $(\star)$ as a temporary label for a condition (or property) that we refer back to in the text following the labeled condition.

we mean the value given by this extension. We could denote the extended mapping by $\hat{w}$, say, but it is in fact not necessary to distinguish it symbolically from $w \colon PV \to \{0,1\}$ because the latter determines the extension uniquely. Similarly, we keep the same symbol if an operation in $\mathbb{N}$ extends to a larger domain. If the logical signature contains further connectives, for example $\to$, then $(*)$ must be supplemented accordingly, with $w(\alpha \to \beta) = w\alpha \to w\beta$ in the example. However, if $\to$ is defined as in the Boolean case, then this equation must be provable. Indeed, it is provable, because from our definition of $\alpha \to \beta$ it follows that

$$w(\alpha \to \beta) = w\neg(\alpha \wedge \neg\beta) = \neg w(\alpha \wedge \neg\beta) = \neg(w\alpha \wedge \neg w\beta) = w\alpha \to w\beta,$$

for any $w$. A corresponding remark could be made with respect to $\leftrightarrow$ and to $\top$ and $\bot$. Always $w\top = 1$ and $w\bot = 0$ by our definition of $\top, \bot$, in accordance with the meaning of these symbols. However, if these or similar symbols belong to the logical signature, then suitable equations must be added to the definition of $w$.

Let $\mathcal{F}_n$ denote the set of all formulas of $\mathcal{F}$ in which at most the variables $p_1, \ldots, p_n$ occur $(n > 0)$. Then it can easily be seen that $w\alpha$ for the formula $\alpha \in \mathcal{F}_n$ depends only on the truth values of $p_1, \ldots, p_n$. In other words, $\alpha \in \mathcal{F}_n$ satisfies for all valuations $w, w'$,

$$(\star) \quad w\alpha = w'\alpha \text{ whenever } wp_i = w'p_i \text{ for } i = 1, \ldots, n.$$

The simple proof of $(\star)$ follows from induction on the construction of formulas in $\mathcal{F}_n$, observing that these are closed under the operations $\neg, \wedge, \vee$. Clearly, $(\star)$ holds for $p \in \mathcal{F}_n$, and if $(\star)$ is valid for $\alpha, \beta \in \mathcal{F}_n$, then also for $\neg\alpha$, $\alpha \wedge \beta$, and $\alpha \vee \beta$. It is then clear that each $\alpha \in \mathcal{F}_n$ defines or represents an $n$-ary Boolean function according to the following

**Definition.** $\alpha \in \mathcal{F}_n$ *represents* the function $f \in \boldsymbol{B}_n$ (or $f$ is represented by $\alpha$) whenever $w\alpha = f w\vec{p} \ (:= f(wp_1, \ldots, wp_n))$ for all valuations $w$.

Because $w\alpha$ for $\alpha \in \mathcal{F}_n$ is uniquely determined by $wp_1, \ldots, wp_n$, $\alpha$ represents precisely one function $f \in \boldsymbol{B}_n$, sometimes written as $\alpha^{(n)}$. For instance, both $p_1 \wedge p_2$ and $\neg(\neg p_1 \vee \neg p_2)$ represent the $\wedge$-function, as can easily be illustrated using a table. Similarly, $\neg p_1 \vee p_2$ and $\neg(p_1 \wedge \neg p_2)$ represent the $\to$-function, and $p_1 \vee p_2$, $\neg(\neg p_1 \wedge \neg p_2)$, $(p_1 \to p_2) \to p_2$ all represent the $\vee$-function. Incidentally, the last formula shows that the $\vee$-connective can be expressed using implication alone.

There is a caveat though: since $\alpha = p_1 \vee p_2$, for instance, belongs not only to $\mathcal{F}_2$ but to $\mathcal{F}_3$ as well, $\alpha$ also represents the Boolean function $f : (x_1, x_2, x_3) \mapsto x_1 \vee x_2$. However, the third argument is only "fictional," or put another way, the function $f$ is not essentially ternary.

In general we say that an operation $f : M^n \to M$ is *essentially n-ary* if $f$ has no fictional arguments, where the $i$th argument of $f$ is called *fictional* whenever for all $x_1, \ldots, x_i, \ldots x_n \in M$ and all $x_i' \in M$,

$$ f(x_1, \ldots, x_i, \ldots, x_n) = f(x_1, \ldots, x_i', \ldots, x_n). $$

Identity and the $\neg$-function are the essentially unary Boolean functions, and out of the sixteen binary functions, only ten are essentially binary, as is seen in scrutinizing the possible truth tables.

**Remark 2.** If $a_n$ denotes temporarily the number of all $n$-ary Boolean functions and $e_n$ the number of all essentially $n$-ary Boolean functions, it is not particularly difficult to prove that $a_n = \sum_{i \leqslant n} \binom{n}{i} e_i$. Solving for $e_n$ results in $e_n = \sum_{i \leqslant n} (-1)^{n-i} \binom{n}{i} a_i$. However, we will not make use of these equations. These become important only in a more specialized study of Boolean functions.

### Exercises

1. $f \in \boldsymbol{B}_n$ is called *linear* if $f(x_1, \ldots, x_n) = a_0 + a_1 x_1 + \cdots + a_n x_n$ for suitable coefficients $a_0, \ldots, a_n \in \{0, 1\}$. Here $+$ denotes exclusive disjunction (addition modulo 2) and the not written multiplication is conjunction (i.e., $a_i x_i = x_i$ for $a_i = 1$ and $a_i x_i = 0$ for $a_i = 0$). (a) Show that the above representation of a linear function $f$ is unique. (b) Determine the number of $n$-ary linear Boolean functions. (c) Prove that each formula $\alpha$ in $\neg, +$ (i.e., $\alpha$ is a formula of the logical signature $\{\neg, +\}$) represents a linear Boolean function.

2. Verify that a compound Boolean formula $\varphi$ is either of the form $\varphi = \neg\alpha$ or else $\varphi = (\alpha \wedge \beta)$ or $\varphi = (\alpha \vee \beta)$ for suitable formulas $\alpha, \beta$ (this is the easy part of the unique reconstruction property).

3. Prove that a proper initial segment of a formula $\varphi$ is never a formula. Equivalently: If $\alpha\xi = \beta\eta$ with $\alpha, \beta \in \mathcal{F}$ and arbitrary strings $\xi, \eta$, then $\alpha = \beta$. The same holds for formulas in PN, but not in RPN.

4. Prove (with Exercise 3) the second more difficult part of the unique reconstruction property, the claim of uniqueness.

## 1.2   Semantic Equivalence and Normal Forms

Throughout this chapter $w$ will always denote a propositional valuation. Formulas $\alpha, \beta$ are called (logically or semantically) *equivalent,* and we write $\alpha \equiv \beta$, when $w\alpha = w\beta$ for all valuations $w$. For example $\alpha \equiv \neg\neg\alpha$. Obviously, $\alpha \equiv \beta$ iff for any $n$ such that $\alpha, \beta \in \mathcal{F}_n$, both formulas represent the same $n$-ary Boolean function. It follows that at most $2^{2^n}$ formulas in $\mathcal{F}_n$ can be pairwise inequivalent, since there are no more than $2^{2^n}$ $n$-ary Boolean functions.

   In arithmetic one writes simply $s = t$ to express that the terms $s, t$ represent the same function. For example, $(x+y)^2 = x^2 + 2xy + y^2$ expresses the equality of values of the left- and right-hand terms for all $x, y \in \mathbb{R}$. This way of writing is permissible because formal syntax plays a minor role in arithmetic. In formal logic, however, as is always the case when syntactic considerations are to the fore, one uses the equality sign in messages like $\alpha = \beta$ only for the syntactic identity of the strings $\alpha$ and $\beta$. Therefore, the equivalence of formulas must be denoted differently. Clearly, for all formulas $\alpha, \beta, \gamma$ the following equivalences hold:

$$
\begin{aligned}
\alpha \wedge (\beta \wedge \gamma) &\equiv \alpha \wedge \beta \wedge \gamma, & \alpha \vee (\beta \vee \gamma) &\equiv \alpha \vee \beta \vee \gamma & &\text{(associativity)}; \\
\alpha \wedge \beta &\equiv \beta \wedge \alpha, & \alpha \vee \beta &\equiv \beta \vee \alpha & &\text{(commutativity)}; \\
\alpha \wedge \alpha &\equiv \alpha, & \alpha \vee \alpha &\equiv \alpha & &\text{(idempotency)}; \\
\alpha \wedge (\alpha \vee \beta) &\equiv \alpha, & \alpha \vee \alpha \wedge \beta &\equiv \alpha & &\text{(absorption)}; \\
\alpha \wedge (\beta \vee \gamma) &\equiv \alpha \wedge \beta \vee \alpha \wedge \gamma, & & & &\text{($\wedge$-distributivity)}; \\
\alpha \vee \beta \wedge \gamma &\equiv (\alpha{\vee}\beta) \wedge (\alpha{\vee}\gamma) & & & &\text{($\vee$-distributivity)}; \\
\neg(\alpha \wedge \beta) &\equiv \neg\alpha \vee \neg\beta, & \neg(\alpha \vee \beta) &\equiv \neg\alpha \wedge \neg\beta & &\text{(de Morgan rules)}.
\end{aligned}
$$

Furthermore, $\alpha \vee \neg\alpha \equiv \top$, $\alpha \wedge \neg\alpha \equiv \bot$, and $\alpha \wedge \top \equiv \alpha \vee \bot \equiv \alpha$. It is also useful to list certain equivalences for formulas containing $\rightarrow$, for example the frequently used $\alpha \rightarrow \beta \equiv \neg\alpha \vee \beta$ $(\equiv \neg(\alpha \wedge \neg\beta))$, and the important

$$
\alpha \rightarrow \beta \rightarrow \gamma \equiv \alpha \wedge \beta \rightarrow \gamma \equiv \beta \rightarrow \alpha \rightarrow \gamma.
$$

To generalize: $\alpha_1 \rightarrow \cdots \rightarrow \alpha_n \equiv \alpha_1 \wedge \cdots \wedge \alpha_{n-1} \rightarrow \alpha_n$. Further, we mention the "left distributivity" of implication with respect to $\wedge$ and $\vee$, namely

$$
\alpha \rightarrow \beta \wedge \gamma \equiv (\alpha \rightarrow \beta) \wedge (\alpha \rightarrow \gamma); \quad \alpha \rightarrow \beta \vee \gamma \equiv (\alpha \rightarrow \beta) \vee (\alpha \rightarrow \gamma).
$$

Should the symbol $\rightarrow$ lie to the right then the following are valid:

$$
\alpha \wedge \beta \rightarrow \gamma \equiv (\alpha \rightarrow \gamma) \vee (\beta \rightarrow \gamma); \quad \alpha \vee \beta \rightarrow \gamma \equiv (\alpha \rightarrow \gamma) \wedge (\beta \rightarrow \gamma).
$$

**Remark 1.** These last two logical equivalences are responsible for a curious phenomenon in everyday language. For example, the two sentences

   A: *Students and pensioners pay half price,*

   B: *Students or pensioners pay half price*

evidently have the same meaning. How to explain this? Let *student* and *pensioner* be abbreviated by $S$, $P$, and *pay half price* by $H$. Then

$$\alpha: \ (S \to H) \wedge (P \to H), \qquad \beta: \ (S \vee P) \to H$$

express somewhat more precisely the factual content of $A$ and $B$, respectively. Now, according to our truth tables, the formulas $\alpha$ and $\beta$ are simply logically equivalent. The everyday-language statements $A$ and $B$ of $\alpha$ and $\beta$ obscure the structural difference of $\alpha$ and $\beta$ through an apparently synonymous use of the words *and* and *or*.

Obviously, $\equiv$ is an equivalence relation, that is,

$$\alpha \equiv \alpha \qquad \text{(reflexivity)},$$
$$\alpha \equiv \beta \ \Rightarrow \ \beta \equiv \alpha \qquad \text{(symmetry)},$$
$$\alpha \equiv \beta, \beta \equiv \gamma \ \Rightarrow \ \alpha \equiv \gamma \quad \text{(transitivity)}.$$

Moreover, $\equiv$ is a *congruence relation* on $\mathcal{F}$,[3] i.e., for all $\alpha, \alpha', \beta, \beta'$,

$$\alpha \equiv \alpha', \beta \equiv \beta' \ \Rightarrow \ \alpha \circ \beta \equiv \alpha' \circ \beta', \neg \alpha \equiv \neg \alpha' \qquad (\circ \in \{\wedge, \vee\}).$$

For this reason the *replacement theorem* holds: $\alpha \equiv \alpha' \ \Rightarrow \ \varphi \equiv \varphi'$, where $\varphi'$ is obtained from $\varphi$ by replacing one or several of the possible occurrences of the subformula $\alpha$ in $\varphi$ by $\alpha'$. For instance, by replacing the subformula $\neg p \vee \neg q$ by the equivalent formula $\neg(p \wedge q)$ in $\varphi = (\neg p \vee \neg q) \wedge (p \vee q)$ we obtain $\varphi' = \neg(p \wedge q) \wedge (p \vee q)$, which is equivalent to $\varphi$. A similar replacement theorem also holds for arithmetical terms and is constantly used in their manipulation. This mostly goes unnoticed, because $=$ is written instead of $\equiv$, and the replacement for $=$ is usually correctly applied. The simple inductive proof of the replacement theorem will be given in a somewhat broader context in **2.4**.

Furnished with the equivalences $\neg\neg\alpha \equiv \alpha$, $\neg(\alpha \wedge \beta) \equiv \neg\alpha \vee \neg\beta$, and $\neg(\alpha \vee \beta) \equiv \neg\alpha \wedge \neg\beta$, and using replacement it is easy to construct for each formula an equivalent formula in which $\neg$ stands only in front of variables. For example, $\neg(p \wedge q \vee r) \equiv \neg(p \wedge q) \wedge \neg r \equiv (\neg p \vee \neg q) \wedge \neg r$ is obtained in this way. This observation follows also from Theorem 2.1.

---

[3] This concept, stemming originally from geometry, is meaningfully defined in every algebraic structure and is one of the most important and most general mathematical concepts; see **2.1**. The definition is equivalent to the condition
$$\alpha \equiv \alpha' \ \Rightarrow \ \alpha \circ \beta \equiv \alpha' \circ \beta, \beta \circ \alpha \equiv \beta \circ \alpha', \neg \alpha \equiv \neg \alpha', \text{ for all } \alpha, \alpha', \beta.$$

It is always something of a surprise to the newcomer that independent of its arity, *every* Boolean function can be represented by a Boolean formula. While this can be proved in various ways, we take the opportunity to introduce certain normal forms and therefore begin with the following

**Definition.** Prime formulas and negations of prime formulas are called *literals*. A disjunction $\alpha_1 \vee \cdots \vee \alpha_n$, where each $\alpha_i$ is a conjunction of literals, is called a *disjunctive normal form*, a DNF for short. A conjunction $\beta_1 \wedge \cdots \wedge \beta_n$, where every $\beta_i$ is a disjunction of literals, is called a *conjunctive normal form*, a CNF for short.

**Example 1.** The formula $p \vee (q \wedge \neg p)$ is a DNF; $p \vee q$ is at once a DNF and a CNF; $p \vee \neg(q \wedge \neg p)$ is neither a DNF nor a CNF.

Theorem 2.1 states that every Boolean function is represented by a Boolean formula, indeed by a DNF, and also by a CNF. It would suffice to show that for given $n$ there are at least $2^{2^n}$ pairwise inequivalent DNFs (resp. CNFs). However, we present instead a constructive proof whereby for a Boolean function given in tabular form a representing DNF (resp. CNF) can explicitly be written down. In Theorem 2.1 we temporarily use the following notation: $p^1 := p$ and $p^0 := \neg p$. With this stipulation, $w(p_1^{x_1} \wedge p_2^{x_2}) = 1$ iff $wp_1 = x_1$ and $wp_2 = x_2$. More generally, induction on $n \geqslant 1$ easily shows that for all $x_1, \ldots, x_n \in \{0, 1\}$,

$(*)$   $w(p_1^{x_1} \wedge \cdots \wedge p_n^{x_n}) = 1 \Leftrightarrow w\vec{p} = \vec{x}$  (i.e., $wp_1 = x_1, \ldots, wp_n = x_n$).

**Theorem 2.1.** *Every Boolean function $f$ with $f \in \boldsymbol{B}_n$ $(n > 0)$ is representable by a DNF, namely by*

$$\alpha_f := \bigvee_{f\vec{x}=1} p_1^{x_1} \wedge \cdots \wedge p_n^{x_n}.^4$$

*At the same time, $f$ is representable by the CNF*

$$\beta_f := \bigwedge_{f\vec{x}=0} p_1^{\neg x_1} \vee \cdots \vee p_n^{\neg x_n}.$$

**Proof.** By the definition of $\alpha_f$, the following equivalences hold for an arbitrary valuation $w$:

---

[4] The disjuncts of $\alpha_f$ can be arranged, for instance, according to the lexicographical order of the $n$-tuples $(x_1, \ldots, x_n) \in \{0, 1\}^n$. If the disjunction is empty (that is, if $f$ does not take the value 1) let $\alpha_f$ be $\bot$ $(= p_1 \wedge \neg p_1)$. Thus, the empty disjunction is $\bot$. Similarly, the empty conjunction equals $\top$ $(= \neg\bot)$. These conventions correspond to those in arithmetic, where the empty sum is 0 and the empty product is 1.

$$
\begin{aligned}
w\alpha_f = 1 \quad &\Leftrightarrow \quad \text{there is an } \vec{x} \text{ with } f\vec{x} = 1 \text{ and } w(p_1^{x_1} \wedge \cdots \wedge p_n^{x_n}) = 1 \\
&\Leftrightarrow \quad \text{there is an } \vec{x} \text{ with } f\vec{x} = 1 \text{ and } w\vec{p} = \vec{x} \quad \big(\text{by } (*)\big) \\
&\Leftrightarrow \quad f w\vec{p} = 1 \quad (\text{replace } \vec{x} \text{ by } w\vec{p}).
\end{aligned}
$$

Thus, $w\alpha_f = 1 \Leftrightarrow f w\vec{p} = 1$. From this equivalence, and because there are only two truth values, $w\alpha_f = f w\vec{p}$ follows immediately. The representability proof of $f$ by $\beta_f$ runs analogously; alternatively, Theorem 2.4 below may be used. $\square$

**Example 2.** For the *exclusive-or function* $+$, the construction of $\alpha_f$ in Theorem 2.1 gives the representing DNF $p_1 \wedge \neg p_2 \vee \neg p_1 \wedge p_2$, because $(1,0), (0,1)$ are the only pairs for which $+$ has the value 1. The CNF given by the theorem, on the other hand, is $(p_1 \vee p_2) \wedge (\neg p_1 \vee \neg p_2)$; the equivalent formula $(p_1 \vee p_2) \wedge \neg(p_1 \wedge p_2)$ makes the meaning of the exclusive-or compound particularly intuitive.

$p_1 \wedge p_2 \vee \neg p_1 \wedge p_2 \vee \neg p_1 \wedge \neg p_2$ is the DNF given by Theorem 2.1 for the Boolean function $\rightarrow$. It is longer than the formula $\neg p_1 \vee p_2$, which is also a representing DNF. But the former is distinctive in that each of its disjuncts contains each variable occurring in the formula exactly once. A DNF of $n$ variables with the analogous property is called *canonical*. The notion of canonical CNF is correspondingly explained. For instance, the function $\leftrightarrow$ is represented by the canonical CNF $(\neg p_1 \vee p_2) \wedge (p_1 \vee \neg p_2)$ according to Theorem 2.1, which always provides canonical normal forms as representing formulas.

Since each formula represents a certain Boolean function, Theorem 2.1 immediately implies the following fact, which has also a (more lengthy) syntactical proof with the replacement theorem mentioned on page 12.

**Corollary 2.2.** *Each $\varphi \in \mathcal{F}$ is equivalent to a DNF and to a CNF.*

**Functional completeness.** A logical signature is called *functional complete* if every Boolean function is representable by a formula in this signature. Theorem 2.1 shows that $\{\neg, \wedge, \vee\}$ is functional complete. Because of $p \vee q \equiv \neg(\neg p \wedge \neg q)$ and $p \wedge q \equiv \neg(\neg p \vee \neg q)$, one can further leave aside $\vee$, or alternatively $\wedge$. This observation is the content of

**Corollary 2.3.** *Both $\{\neg, \wedge\}$ and $\{\neg, \vee\}$ are functional complete.*

Therefore, to show that a logical signature $L$ is functional complete, it is enough to represent $\neg, \wedge$ or else $\neg, \vee$ by formulas in $L$. For example, because $\neg p \equiv p \to 0$ and $p \wedge q \equiv \neg(p \to \neg q)$, the signature $\{\to, 0\}$ is functional complete. On the other hand, $\{\to, \wedge, \vee\}$, and a fortiori $\{\to\}$, are not. Indeed, $w\varphi = 1$ for any formula $\varphi$ in $\to, \wedge, \vee$ and any valuation $w$ such that $wp = 1$ for all $p$. This can readily be confirmed by induction on $\varphi$. Thus, never $\neg p \equiv \varphi$ for any such formula $\varphi$.

It is noteworthy that the signature containing only $\downarrow$ is functional complete: from the truth table for $\downarrow$ we get $\neg p \equiv p \downarrow p$ as well as $p \wedge q \equiv \neg p \downarrow \neg q$. Likewise for $\{\uparrow\}$, because $\neg p \equiv p \uparrow p$ and $p \vee q \equiv \neg p \uparrow \neg q$. That $\{\uparrow\}$ must necessarily be functional complete once we know that $\{\downarrow\}$ is will become obvious in the discussion of the duality theorem below. Even up to term equivalence, there still exist infinitely many signatures. Here signatures are called *term equivalent* if the formulas of these signatures represent the same Boolean functions as in Exercise 2, for instance.

Define inductively on the formulas from $\mathcal{F}$ a mapping $\delta : \mathcal{F} \to \mathcal{F}$ by

$$p^\delta = p, \quad (\neg\alpha)^\delta = \neg\alpha^\delta, \quad (\alpha \wedge \beta)^\delta = \alpha^\delta \vee \beta^\delta, \quad (\alpha \vee \beta)^\delta = \alpha^\delta \wedge \beta^\delta.$$

$\alpha^\delta$ is called the *dual formula* of $\alpha$ and is obtained from $\alpha$ simply by interchanging $\wedge$ and $\vee$. Obviously, for a DNF $\alpha$, $\alpha^\delta$ is a CNF, and vice versa. Define the *dual* of $f \in \boldsymbol{B}_n$ by $f^\delta \vec{x} := \neg f \neg \vec{x}$ with $\neg \vec{x} := (\neg x_1, \ldots, \neg x_n)$. Clearly $f^{\delta^2} := (f^\delta)^\delta = f$ since $(f^\delta)^\delta \vec{x} = \neg\neg f \neg\neg \vec{x} = f\vec{x}$. Note that $\wedge^\delta = \vee$, $\vee^\delta = \wedge$, $\leftrightarrow^\delta = +$, $\downarrow^\delta = \uparrow$, but $\neg^\delta = \neg$. In other words, $\neg$ is *self-dual*. One may check by going through all truth tables that essentially binary self-dual Boolean functions do not exist. But it was Dedekind who discovered the interesting ternary self-dual function

$$d_3 : (x_1, x_2, x_3) \mapsto x_1 \wedge x_2 \vee x_1 \wedge x_3 \vee x_2 \wedge x_3.$$

The above notions of duality are combined in the following

**Theorem 2.4 (The duality principle for two-valued logic).** *If $\alpha$ represents the function $f$ then $\alpha^\delta$ represents the dual function $f^\delta$.*

**Proof** by induction on $\alpha$. Trivial for $\alpha = p$. Let $\alpha, \beta$ represent $f_1, f_2$, respectively. Then $\alpha \wedge \beta$ represents $f : \vec{x} \mapsto f_1\vec{x} \wedge f_2\vec{x}$, and in view of the induction hypothesis, $(\alpha \wedge \beta)^\delta = \alpha^\delta \vee \beta^\delta$ represents $g : \vec{x} \mapsto f_1^\delta \vec{x} \vee f_2^\delta \vec{x}$. This function is just the dual of $f$ because

$$f^\delta \vec{x} = \neg f \neg \vec{x} = \neg(f_1 \neg\vec{x} \wedge f_2 \neg\vec{x}) = \neg f_1 \neg\vec{x} \vee \neg f_2 \neg\vec{x} = f_1^\delta \vec{x} \vee f_2^\delta \vec{x} = g\vec{x}.$$

The induction step for $\vee$ is similar. Now let $\alpha$ represent $f$. Then $\neg\alpha$ represents $\neg f : \vec{x} \mapsto \neg f\vec{x}$. By the induction hypothesis, $\alpha^\delta$ represents $f^\delta$.

Thus $(\neg\alpha)^\delta = \neg\alpha^\delta$ represents $\neg f^\delta$, which coincides with $(\neg f)^\delta$ because of $(\neg f)^\delta \vec{x} = (\neg\neg f\vec{x})\neg\vec{x} = \neg(\neg f\neg\vec{x}) = \neg(f^\delta\vec{x})$. $\blacksquare$

For example, we know that $\leftrightarrow$ is represented by $p \wedge q \vee \neg p \wedge \neg q$. Hence, by Theorem 2.4, $+ (= \leftrightarrow^\delta)$ is represented by $(p \vee q) \wedge (\neg p \vee \neg q)$. More generally, if a canonical DNF $\alpha$ represents $f \in \boldsymbol{B}_n$, then the canonical CNF $\alpha^\delta$ represents $f^\delta$. Thus, if every $f \in \boldsymbol{B}_n$ is representable by a DNF then every $f$ must necessarily be representable by a CNF, since $f \mapsto f^\delta$ maps $\boldsymbol{B}_n$ bijectively onto itself as follows from $f^{\delta^2} = f$. Note also that Dedekind's just defined ternary self-dual function $d_3$ shows in view of Theorem 2.4 that $p \wedge q \vee p \wedge r \vee q \wedge r \equiv (p \vee q) \wedge (p \vee r) \wedge (q \vee r)$.

**Remark 2.** $\{\wedge, \vee, 0, 1\}$ is *maximally functional incomplete*, that is, if $f$ is any Boolean function not representable by a formula in $\wedge, \vee, 0, 1$, then $\{\wedge, \vee, 0, 1, f\}$ is functional complete (Exercise 4). As was shown by E. Post (1920), there are up to term equivalence only five maximally functional incomplete logical signatures: besides $\{\wedge, \vee, 0, 1\}$ only $\{\rightarrow, \wedge\}$, the dual of this, $\{\leftrightarrow, \neg\}$, and $\{d_3, \neg\}$. The formulas of the last one represent just the self-dual Boolean functions. Since $\neg p \equiv 1 + p$, the signature $\{0, 1, +, \cdot\}$ is functional complete, where $\cdot$ is written in place of $\wedge$. The deeper reason is that $\{0, 1, +, \cdot\}$ is at the same time the extralogical signature of fields (see **2.1**). Functional completeness in the two-valued case just derives from the fact that for a finite field, each operation on its domain is represented by a suitable polynomial. We mention also that for any finite set $M$ of truth values considered in many-valued logics there is a generalized two-argument Sheffer function, by which every operation on $M$ can be obtained, similarly to $\uparrow$ in the two-valued case.

## Exercises

1. Verify the logical equivalences
$$(p \rightarrow q_1) \wedge (\neg p \rightarrow q_2) \equiv p \wedge q_1 \vee \neg p \wedge q_2,$$
$$p_1 \wedge q_1 \rightarrow p_2 \vee q_2 \equiv (p_1 \rightarrow p_2) \vee (q_1 \rightarrow q_2).$$

2. Show that the signatures $\{+, 1\}$, $\{+, \neg\}$, $\{\leftrightarrow, 0\}$, and $\{\leftrightarrow, \neg\}$ are all term equivalent. The formulas of each of these signatures represent precisely the linear Boolean functions.

3. Show that the formulas in $\wedge, \vee, 0, 1$ represent exactly the *monotonic* Boolean functions. These are the constants from $\boldsymbol{B}_0$, and for $n > 0$ the $f \in \boldsymbol{B}_n$ such that for all $i$ with $1 \leqslant i \leqslant n$,
$$f(x_1, \ldots, x_{i-1}, 0, x_{i+1}, \ldots, x_n) \leqslant f(x_1, \ldots, x_{i-1}, 1, x_{i+1}, \ldots, x_n).$$

4. Show that the logical signature $\{\wedge, \vee, 0, 1\}$ is maximally functional incomplete.

5. If one wants to prove Corollary 2.2 syntactically with the properties of $\equiv$ (page 11) one needs generalizations of the distributivity, e.g., $\bigvee_{i \leqslant n} \alpha_i \wedge \bigvee_{j \leqslant m} \beta_j \equiv \bigvee_{i \leqslant n, j \leqslant m} (\alpha_i \wedge \beta_j)$. Verify the latter.

## 1.3   Tautologies and Logical Consequence

Instead of $w\alpha = 1$ we prefer from now on to write $w \vDash \alpha$ and read this $w$ *satisfies* $\alpha$. Further, if $X$ is a set of formulas, we write $w \vDash X$ if $w \vDash \alpha$ for all $\alpha \in X$ and say that $w$ is a (propositional) *model for $X$*. A given $\alpha$ (resp. $X$) is called *satisfiable* if there is some $w$ with $w \vDash \alpha$ (resp. $w \vDash X$). $\vDash$, called the *satisfiability relation*, evidently has the following properties:

$$w \vDash p \quad \Leftrightarrow \quad wp = 1 \quad (p \in PV); \qquad w \vDash \neg\alpha \quad \Leftrightarrow \quad w \nvDash \alpha;$$

$$w \vDash \alpha \wedge \beta \quad \Leftrightarrow \quad w \vDash \alpha \text{ and } w \vDash \beta; \qquad w \vDash \alpha \vee \beta \quad \Leftrightarrow \quad w \vDash \alpha \text{ or } w \vDash \beta.$$

One can define the satisfiability relation $w \vDash \alpha$ for a given $w \colon PV \to \{0, 1\}$ also inductively on $\alpha$, according to the clauses just given. This approach is particularly useful for extending the satisfiability conditions in **2.3**.

It is obvious that $w \colon PV \to \{0, 1\}$ will be uniquely determined by setting down in advance for which variables $w \vDash p$ should be valid. Likewise the notation $w \vDash \alpha$ for $\alpha \in \mathcal{F}_n$ is already meaningful when $w$ is defined only for $p_1, \ldots, p_n$. One could extend such a $w$ to a global valuation by setting, for instance, $wp = 0$ for all unmentioned variables $p$.

For formulas containing other connectives the satisfaction conditions are to be formulated accordingly. For example, we expect

$$(*) \quad w \vDash \alpha \to \beta \quad \Leftrightarrow \quad \text{if } w \vDash \alpha \text{ then } w \vDash \beta.$$

If $\to$ is taken to be a primitive connective, $(*)$ is required. However, we defined $\to$ in such a way that $(*)$ is provable.

**Definition.** $\alpha$ is called *logically valid* or a (two-valued) *tautology*, in short $\vDash \alpha$, whenever $w \vDash \alpha$ for all valuations $w$. A formula not satisfiable at all, i.e. $w \nvDash \alpha$ for all $w$, is called a *contradiction*.

**Examples.** $p \vee \neg p$ is a tautology and so is $\alpha \vee \neg\alpha$ for every formula $\alpha$, the so-called *law of the excluded middle* or the *tertium non datur*. On the other hand, $\alpha \wedge \neg\alpha$ and $\alpha \leftrightarrow \neg\alpha$ are always contradictions. The following tautologies in $\rightarrow$ are mentioned in many textbooks on logic. Remember our agreement about association to the right in formulas in which $\rightarrow$ repeatedly occurs.

$$p \rightarrow p \qquad \qquad \text{(self-implication)},$$
$$(p \rightarrow q) \rightarrow (q \rightarrow r) \rightarrow (p \rightarrow r) \qquad \text{(chain rule)},$$
$$(p \rightarrow q \rightarrow r) \rightarrow (q \rightarrow p \rightarrow r) \qquad \text{(exchange of premises)},$$
$$p \rightarrow q \rightarrow p \qquad \qquad \text{(premise charge)},$$
$$(p \rightarrow q \rightarrow r) \rightarrow (p \rightarrow q) \rightarrow (p \rightarrow r) \qquad \text{(Frege's formula)},$$
$$((p \rightarrow q) \rightarrow p) \rightarrow p \qquad \text{(Peirce's formula)}.$$

It will later turn out that all tautologies in $\rightarrow$ alone are derivable (in a sense still to be explained) from the last three formulas.

Clearly, it is decidable whether a formula $\alpha$ is a tautology, in that one tries out the valuations of the variables of $\alpha$. Unfortunately, no essentially more efficient method is known; such a method exists only for formulas of a certain form. We will have a somewhat closer look at this problem in **4.3**. Various questions such as checking the equivalence of formulas can be reduced to a decision about whether a formula is a tautology. For notice the obvious equivalence of $\alpha \equiv \beta$ and $\vDash \alpha \leftrightarrow \beta$.

Basic in propositional logic is the following

**Definition.** $\alpha$ is a *logical consequence* of $X$, written $X \vDash \alpha$, if $w \vDash \alpha$ for every model $w$ of $X$. In short, $w \vDash X \Rightarrow w \vDash \alpha$, for all valuations $w$.

While we use $\vDash$ both as the symbol for logical consequence (which is a relation between sets of formulas $X$ and formulas $\alpha$) and the satisfiability property, it will always be clear from the context what $\vDash$ actually means. Evidently, $\alpha$ is a tautology iff $\emptyset \vDash \alpha$, so that $\vDash \alpha$ can be regarded as an abbreviation for $\emptyset \vDash \alpha$.

In this book, $X \vDash \alpha, \beta$ will always mean '$X \vDash \alpha$ and $X \vDash \beta$'. More generally, $X \vDash Y$ is always to mean '$X \vDash \beta$ for all $\beta \in Y$'. We also write throughout $\alpha_1, \ldots, \alpha_n \vDash \beta$ in place of $\{\alpha_1, \ldots, \alpha_n\} \vDash \beta$, and more briefly, $X, \alpha \vDash \beta$ in place of $X \cup \{\alpha\} \vDash \beta$.

**Examples of logical consequence.** (a) $\alpha, \beta \vDash \alpha \wedge \beta$ and $\alpha \wedge \beta \vDash \alpha, \beta$. This is evident from the truth table of $\wedge$. (b) $\alpha, \alpha \rightarrow \beta \vDash \beta$, because $1 \rightarrow x = 1 \Rightarrow x = 1$ according to the truth table of $\rightarrow$.

(c) $X \vDash \bot \Rightarrow X \vDash \alpha$ for each $\alpha$. Indeed, $X \vDash \bot = p_1 \wedge \neg p_1$ obviously means that $X$ is unsatisfiable (has no model), as e.g. $X = \{p_2, \neg p_2\}$.
(d) $X, \alpha \vDash \beta$ & $X, \neg \alpha \vDash \beta \Rightarrow X \vDash \beta$. In order to see this let $w \vDash X$. If $w \vDash \alpha$ then $X, \alpha \vDash \beta$ and hence $w \vDash \beta$, and if $w \nvDash \alpha$ (i.e., $w \vDash \neg \alpha$) then $w \vDash \beta$ clearly follows from $X, \neg \alpha \vDash \beta$. Note that (d) reflects our case distinction made in the naive metatheory while proving (d).

Example (a) could also be stated as $X \vDash \alpha, \beta \Leftrightarrow X \vDash \alpha \wedge \beta$. The property exemplified by (b) is called the *modus ponens* when formulated as a rule of inference, as will be done in **1.6**. Example (d) is another formulation of the often-used procedure of proof by cases: In order to conclude a sentence $\beta$ from a set of premises $X$ it suffices to show it to be a logical consequence both under an additional supposition and under its negation. This is generalized in Exercise 3.

Important are the following general and obvious properties of $\vDash$:

$$
\begin{array}{lll}
\text{(R)} & \alpha \in X \Rightarrow X \vDash \alpha & \textit{(reflexivity)}, \\
\text{(M)} & X \vDash \alpha \ \& \ X \subseteq X' \Rightarrow X' \vDash \alpha & \textit{(monotonicity)}, \\
\text{(T)} & X \vDash Y \ \& \ Y \vDash \alpha \Rightarrow X \vDash \alpha & \textit{(transitivity)}.
\end{array}
$$

Useful for many purposes is also the closure of the logical consequence relation under substitution, which generalizes the fact that from $p \vee \neg p$ all tautologies of the form $\alpha \vee \neg \alpha$ arise from substituting $\alpha$ for $p$.

**Definition.** A (propositional) *substitution* is a mapping $\sigma : PV \to \mathcal{F}$ that is extended in a natural way to a mapping $\sigma : \mathcal{F} \to \mathcal{F}$ as follows:

$$
(\alpha \wedge \beta)^\sigma = \alpha^\sigma \wedge \beta^\sigma, \quad (\alpha \vee \beta)^\sigma = \alpha^\sigma \vee \beta^\sigma, \quad (\neg \alpha)^\sigma = \neg \alpha^\sigma.
$$

Thus, like valuations, substitutions are considered as operations on the whole of $\mathcal{F}$. For example, if $p^\sigma = \alpha$ for some fixed $p$ and $q^\sigma = q$ otherwise, then $\varphi^\sigma$ arises from $\varphi$ by substituting $\alpha$ for $p$ at all occurrences of $p$ in $\varphi$. From $p \vee \neg p$ arises in this way the schema $\alpha \vee \neg \alpha$. For $X \subseteq \mathcal{F}$ let $X^\sigma := \{\varphi^\sigma \mid \varphi \in X\}$. The observation $\vDash \varphi \Rightarrow \vDash \varphi^\sigma$ turns out to be the special instance $X = \emptyset$ of the useful property

$$
\text{(S)} \quad X \vDash \alpha \Rightarrow X^\sigma \vDash \alpha^\sigma \quad \textit{(substitution invariance)}.
$$

In order to verify (S), define $w^\sigma$ for a given valuation $w$ in such a way that $w^\sigma p = wp^\sigma$. We first prove by induction on $\alpha$ that

$$
(*) \quad w \vDash \alpha^\sigma \Leftrightarrow w^\sigma \vDash \alpha.
$$

If $\alpha$ is prime, $(*)$ certainly holds. As regards the induction step, note that

$$w \vDash (\alpha \wedge \beta)^\sigma \Leftrightarrow w \vDash \alpha^\sigma \wedge \beta^\sigma \Leftrightarrow w \vDash \alpha^\sigma, \beta^\sigma$$

$$\Leftrightarrow w^\sigma \vDash \alpha, \beta \quad \text{(induction hypothesis)}$$

$$\Leftrightarrow w^\sigma \vDash \alpha \wedge \beta.$$

The reasoning for $\vee$ and $\neg$ is analogous and so $(*)$ holds. Now let $X \vDash \alpha$ and $w \vDash X^\sigma$. By $(*)$, we get $w^\sigma \vDash X$. Thus $w^\sigma \vDash \alpha$, and again by $(*)$, $w \vDash \alpha^\sigma$. This confirms (S). Another important property of $\vDash$ that is not so easily obtained will be proved in **1.4**, namely

(F)    $X \vDash \alpha \Rightarrow X_0 \vDash \alpha$ for some finite subset $X_0 \subseteq X$.

$\vDash$ shares the properties (R), (M), (T), and (S) with almost all classical and nonclassical (many-valued) propositional *consequence relations*. This is to mean a relation $\vdash$ between sets of formulas and formulas of an arbitrary propositional language $\mathcal{F}$ that has the properties corresponding to (R), (M), (T), and (S). These properties are the starting point for a general and strong theory of logical systems created by Tarski, which underpins nearly all logical systems considered in the literature. Should $\vdash$ satisfy the property corresponding to (F) then $\vdash$ is called *finitary*.

**Remark 1.** Sometimes (S) is not demanded in defining a consequence relation, and if (S) holds, one speaks of a *structural* consequence relation. We omit this refinement. Notions such as tautology, consistency, maximal consistency, and so on can be used with reference to *any* consequence relation $\vdash$ in an arbitrary propositional language $\mathcal{F}$. For instance, a set of formulas $X$ is called *consistent* in $\vdash$ whenever $X \nvdash \alpha$ for some $\alpha$, and *maximally consistent* if $X$ is consistent but has no proper consistent extension. $\vdash$ itself is called consistent if $X \nvdash \alpha$ for some $X$ and $\alpha$ (this is equivalent to not $\vdash \alpha$ for all $\alpha$). Here as always, $\vdash \alpha$ stands for $\emptyset \vdash \alpha$. If $\mathcal{F}$ contains $\neg$ then the consistency of $X$ is often defined by $X \vdash \alpha, \neg\alpha$ for no $\alpha$. But the aforementioned definition has the advantage of being completely independent of any assumption concerning the occurring connectives. Another example of a general definition is this: A formula set $X$ is called *deductively closed in* $\vdash$ provided $X \vdash \alpha \Rightarrow \alpha \in X$, for all $\alpha \in \mathcal{F}$. Because of (R), this condition can be replaced by $X \vdash \alpha \Leftrightarrow \alpha \in X$. Examples in $\vDash$ are the set of all tautologies and the whole of $\mathcal{F}$. The intersection of a family of deductively closed sets is again deductively closed. Hence, each $X \subseteq \mathcal{F}$ is contained in a smallest deductively closed set, called the *deductive closure* of $X$ in $\vdash$. It equals $\{\alpha \in \mathcal{F} \mid X \vdash \alpha\}$, as is easily seen. The notion of a consequence relation can also be defined in terms of properties of the deductive closure. We mention that (F) holds not just for our relation $\vDash$ that is given by a two-valued matrix, but for the consequence relation of *any* finite logical matrix in *any* propositional language. This is stated and at once essentially generalized in Exercise 3 in **5.7** as an application of the ultraproduct theorem.

A special property of the consequence relation $\vDash$, easily provable, is

(D)    $X, \alpha \vDash \beta \;\Rightarrow\; X \vDash \alpha \rightarrow \beta,$

called the (semantic) *deduction theorem* for propositional logic. To see this suppose $X, \alpha \vDash \beta$ and let $w$ be a model for $X$. If $w \vDash \alpha$ then by the supposition, $w \vDash \beta$, hence $w \vDash \alpha \rightarrow \beta$. If $w \nvDash \alpha$ then $w \vDash \alpha \rightarrow \beta$ as well. Hence $X \vDash \alpha \rightarrow \beta$ in any case. This proves (D). As is immediately seen, the converse of (D) holds as well, that is, one may replace $\Rightarrow$ in (D) by $\Leftrightarrow$. Iterated application of this simple observation yields

$$\alpha_1, \ldots, \alpha_n \vDash \beta \Leftrightarrow \vDash \alpha_1 \rightarrow \alpha_2 \rightarrow \cdots \rightarrow \alpha_n \rightarrow \beta \Leftrightarrow \vDash \alpha_1 \wedge \alpha_2 \wedge \cdots \wedge \alpha_n \rightarrow \beta.$$

In this way, $\beta$'s being a logical consequence of a finite set of premises is transformed into a tautology. Using (D) it is easy to obtain tautologies. For instance, to prove $\vDash p \rightarrow q \rightarrow p$, it is enough to verify $p \vDash q \rightarrow p$, for which it in turn suffices to show that $p, q \vDash p$, and this is trivial.

**Remark 2.** By some simple applications of (D) each of the tautologies in the examples on page 18 can be obtained, except the formula of Peirce. As we shall see in Chapter **2**, all properties of $\vDash$ derived above and in the exercises will carry over to the consequence relation of a first-order language.

### Exercises

1. Use the deduction theorem as in the text in order to prove

   (a)    $\vDash (p \rightarrow q \rightarrow r) \rightarrow (p \rightarrow q) \rightarrow (p \rightarrow r),$
   (b)    $\vDash (p \rightarrow q) \rightarrow (q \rightarrow r) \rightarrow (p \rightarrow r).$

2. Suppose that $X \vDash \alpha \rightarrow \beta$. Prove that $X \vDash (\gamma \rightarrow \alpha) \rightarrow (\gamma \rightarrow \beta)$.

3. Verify the (rule of) *disjunctive case distinction*: if $X, \alpha \vDash \gamma$ and $X, \beta \vDash \gamma$ then $X, \alpha \vee \beta \vDash \gamma$. This implication is traditionally written more suggestively as
$$\frac{X, \alpha \vDash \gamma \mid X, \beta \vDash \gamma}{X, \alpha \vee \beta \vDash \gamma}.$$

4. Verify the *rules of contraposition* (notation as in Exercise 3):
$$\frac{X, \alpha \vDash \beta}{X, \neg\beta \vDash \neg\alpha} \quad ; \quad \frac{X, \neg\beta \vDash \neg\alpha}{X, \alpha \vDash \beta}.$$

5. Let $\vdash$ be a consequence relation and let $X$ be maximally consistent in $\vdash$ (see Remark 1). Show that $X$ is deductively closed in $\vdash$.

## 1.4    A Calculus of Natural Deduction

We will now define a derivability relation $\vdash$ by means of a calculus operating solely with some structural rules. $\vdash$ turns out to be identical to the consequence relation $\vDash$. The calculus $\vdash$ is of the so-called Gentzen type and its rules are given with respect to pairs $(X, \alpha)$ of formulas $X$ and formulas $\alpha$. Another calculus for $\vDash$, of the Hilbert type, will be considered in **1.6**. In distinction to [Ge], we do not require that $X$ be finite; our particular goals here make such a restriction dispensable. If $\vdash$ applies to the pair $(X, \alpha)$ then we write $X \vdash \alpha$ and say that $\alpha$ is *derivable* or *provable* from $X$ (made precise below); otherwise we write $X \nvdash \alpha$.

Following [Kl1], Gentzen's name for $(X, \alpha)$, *Sequenz*, is translated as *sequent*. The calculus is formulated in terms of $\wedge, \neg$ and encompasses the six rules below, called the *basic rules*. How to operate with these rules will be explained afterwards. The choice of $\{\wedge, \neg\}$ as the logical signature is a matter of convenience and justified by its functional completeness. The other standard connectives are introduced by the definitions

$$\alpha \vee \beta := \neg(\neg\alpha \wedge \neg\beta), \ \alpha \to \beta := \neg(\alpha \wedge \neg\beta), \ \alpha \leftrightarrow \beta := (\alpha \to \beta) \wedge (\beta \to \alpha).$$

$\top, \bot$ are defined as on page 5. Of course, one could choose any other functional complete signature and adapt the basic rules correspondingly. But it should be observed that a complete calculus in $\neg, \wedge, \vee, \to$, say, must also include basic rules concerning $\vee$ and $\to$, which makes induction arguments on the basic rules of the calculus more lengthy.

Each of the basic rules below has certain premises and a conclusion. Only (IS) has no premises. It allows the derivation of all sequents $\alpha \vdash \alpha$. These are called the *initial* sequents, because each derivation must start with these. (MR), the *monotonicity rule*, could be weakened. It becomes even provable if all pairs $(X, \alpha)$ with $\alpha \in X$ are called initial sequents.

$$
\begin{array}{ll}
\text{(IS)} \quad \dfrac{}{\alpha \vdash \alpha} \ \text{(initial sequent)} & \text{(MR)} \quad \dfrac{X \vdash \alpha}{X' \vdash \alpha} \ (X' \supseteq X), \\[3ex]
(\wedge 1) \quad \dfrac{X \vdash \alpha, \beta}{X \vdash \alpha \wedge \beta} & (\wedge 2) \quad \dfrac{X \vdash \alpha \wedge \beta}{X \vdash \alpha, \beta} \\[3ex]
(\neg 1) \quad \dfrac{X \vdash \alpha, \neg\alpha}{X \vdash \beta} & (\neg 2) \quad \dfrac{X, \alpha \vdash \beta \ \mid \ X, \neg\alpha \vdash \beta}{X \vdash \beta}
\end{array}
$$

Here and in the following $X \vdash \alpha, \beta$ is to mean $X \vdash \alpha$ and $X \vdash \beta$. This convention is important, since $X \vdash \alpha, \beta$ has another meaning in Gentzen calculi that operate with pairs of sets of formulas. The rules $(\wedge 1)$ and $(\neg 1)$ actually have two premises, just like $(\neg 2)$. Note further that $(\wedge 2)$ really consists of two subrules corresponding to the conclusions $X \vdash \alpha$ and $X \vdash \beta$. In $(\neg 2)$, $X, \alpha$ means $X \cup \{\alpha\}$, and this abbreviated form will always be used when there is no risk of misunderstanding.

$\alpha_1, \ldots, \alpha_n \vdash \beta$ stands for $\{\alpha_1, \ldots, \alpha_n\} \vdash \beta$; in particular, $\alpha \vdash \beta$ for $\{\alpha\} \vdash \beta$, and $\vdash \alpha$ for $\emptyset \vdash \alpha$, just as with $\vDash$.

$X \vdash \alpha$ (read from "$X$ is *provable* or *derivable* $\alpha$") is to mean that the sequent $(X, \alpha)$ can be obtained after a stepwise application of the basic rules. We can make this idea of "stepwise application" of the basic rules rigorous and formally precise (intelligible to a computer, so to speak) in the following way: a *derivation* is to mean a finite sequence $(S_0; \ldots; S_n)$ of sequents such that every $S_i$ is either an initial sequent or is obtained through the application of some basic rule to preceding elements in the sequence. Thus, *from $X$ is derivable $\alpha$* if there is a derivation $(S_0; \ldots; S_n)$ with $S_n = (X, \alpha)$. A simple example with the end sequent $\alpha, \beta \vdash \alpha \wedge \beta$, or minutely $(\{\alpha, \beta\}, \alpha \wedge \beta)$, is the derivation

$$(\alpha \vdash \alpha \; ; \; \alpha, \beta \vdash \alpha \; ; \; \beta \vdash \beta \; ; \; \alpha, \beta \vdash \beta \; ; \; \alpha, \beta \vdash \alpha \wedge \beta).$$

Here (MR) was applied twice, followed by an application of $(\wedge 1)$. Not shorter would be complete derivation of the sequent $(\emptyset, \top)$, i.e., a proof of $\vdash \top$. In this example both $(\neg 1)$ and $(\neg 2)$ are essentially involved.

Useful for shortening lengthy derivations is the derivation of additional rules, which will be illustrated with the examples to follow. The second example, a generalization of the first, is the often-used proof method *reductio ad absurdum*: $\alpha$ is proved from $X$ by showing that the assumption $\neg \alpha$ leads to a contradiction. The other examples are given with respect to the defined $\rightarrow$-connective. Hence, for instance, the $\rightarrow$-elimination mentioned below runs in the original language $\dfrac{X \vdash \neg(\alpha \wedge \neg\beta)}{X, \alpha \vdash \beta}$.

**Examples of derivable rules**

$$\frac{X, \neg\alpha \vdash \alpha}{X \vdash \alpha}$$

| | | *proof* | *applied* |
|---|---|---|---|
| ($\neg$-elimination) | 1 | $X, \alpha \vdash \alpha$ | (IS), (MR) |
| | 2 | $X, \neg\alpha \vdash \alpha$ | supposition |
| | 3 | $X \vdash \alpha$ | $(\neg 2)$ |

$$\frac{X, \neg\alpha \vdash \beta, \neg\beta}{X \vdash \alpha}$$

(reductio ad absurdum)

*proof*

| | | *applied* |
|---|---|---|
| 1 | $X, \neg\alpha \vdash \beta, \neg\beta$ | supposition |
| 2 | $X, \neg\alpha \vdash \alpha$ | $(\neg 1)$ |
| 3 | $X \vdash \alpha$ | $\neg$-elimination |

$$\frac{X \vdash \alpha \to \beta}{X, \alpha \vdash \beta}$$

( $\to$-elimination)

| 1 | $X, \alpha, \neg\beta \vdash \alpha, \neg\beta$ | (IS), (MR) |
|---|---|---|
| 2 | $X, \alpha, \neg\beta \vdash \alpha \wedge \neg\beta$ | $(\wedge 1)$ |
| 3 | $X \vdash \neg(\alpha \wedge \neg\beta)$ | supposition |
| 4 | $X, \alpha, \neg\beta \vdash \neg(\alpha \wedge \neg\beta)$ | (MR) |
| 5 | $X, \alpha, \neg\beta \vdash \beta$ | $(\neg 1)$ on 2 and 4 |
| 6 | $X, \alpha \vdash \beta$ | $\neg$-elimination |

$$\frac{X \vdash \alpha \mid X, \alpha \vdash \beta}{X \vdash \beta}$$

(cut rule)

| 1 | $X, \neg\alpha \vdash \alpha$ | supposition, (MR) |
|---|---|---|
| 2 | $X, \neg\alpha \vdash \neg\alpha$ | (IS), (MR) |
| 3 | $X, \neg\alpha \vdash \beta$ | $(\neg 1)$ |
| 4 | $X, \alpha \vdash \beta$ | supposition |
| 5 | $X \vdash \beta$ | $(\neg 2)$ on 4 and 3 |

$$\frac{X, \alpha \vdash \beta}{X \vdash \alpha \to \beta}$$

( $\to$-introduction)

| 1 | $X, \alpha \wedge \neg\beta, \alpha \vdash \beta$ | supposition, (MR) |
|---|---|---|
| 2 | $X, \alpha \wedge \neg\beta \vdash \alpha$ | (IS), (MR), $(\wedge 2)$ |
| 3 | $X, \alpha \wedge \neg\beta \vdash \beta$ | cut rule |
| 4 | $X, \alpha \wedge \neg\beta \vdash \neg\beta$ | (IS), (MR), $(\wedge 2)$ |
| 5 | $X, \alpha \wedge \neg\beta \vdash \alpha \to \beta$ | $(\neg 1)$ |
| 6 | $X, \neg(\alpha \wedge \neg\beta) \vdash \alpha \to \beta$ | (IS), (MR) |
| 7 | $X \vdash \alpha \to \beta$ | $(\neg 2)$ on 5 and 6 |

**Remark 1.** The example of $\to$-introduction is nothing other than the syntactic form of the deduction theorem that was semantically formulated in the previous section. The deduction theorem also holds for intuitionistic logic. However, it is not in general true for all logical systems dealing with implication, thus indicating that the deduction theorem is not an inherent property of every meaningful conception of implication. For instance, the deduction theorem does not hold for certain formal systems of relevance logic that attempt to model implication as a cause-and-effect relation.

A simple application of the $\to$-elimination and the cut rule is a proof of the *detachment rule*

$$\frac{X \vdash \alpha, \alpha \to \beta}{X \vdash \beta}.$$

Indeed, the premise $X \vdash \alpha \to \beta$ yields $X, \alpha \vdash \beta$ by $\to$-elimination, and since $X \vdash \alpha$, it follows $X \vdash \beta$ by the cut rule. Applying detachment on $X = \{\alpha, \alpha \to \beta\}$, we obtain $\alpha, \alpha \to \beta \vdash \beta$. This collection of sequents is known as *modus ponens*. It will be more closely considered in **1.6**.

Many properties of $\vdash$ are proved through rule induction, which we describe after introducing some convenient terminology. We identify a property $\mathcal{E}$ of sequents with the set of all pairs $(X, \alpha)$ to which $\mathcal{E}$ applies. In this sense the logical consequence relation $\vDash$ is the property that applies to all pairs $(X, \alpha)$ with $X \vDash \alpha$.

All the rules considered here are of the form

$$R: \frac{X_1 \vdash \alpha_1 \mid \quad \cdots \quad \mid X_n \vdash \alpha_n}{X \vdash \alpha}$$

and are referred to as Gentzen-style rules. We say that $\mathcal{E}$ is *closed under* $R$ when $\mathcal{E}(X_1, \alpha_1), \ldots, \mathcal{E}(X_n, \alpha_n)$ implies $\mathcal{E}(X, \alpha)$. For a rule without premises, i.e., $n = 0$, this is just to mean $\mathcal{E}(X, \alpha)$. For instance, consider the above already mentioned property $\mathcal{E}: X \vDash \alpha$. This property is closed under each basic rule of $\vdash$. In detail this means

$$\alpha \vDash \alpha, \quad X \vDash \alpha \Rightarrow X' \vDash \alpha \text{ for } X' \supseteq X, \quad X \vDash \alpha, \beta \Rightarrow X \vDash \alpha \wedge \beta, \text{ etc.}$$

From the latter we may conclude that $\mathcal{E}$ applies to all provable sequents; in other words, $\vdash$ is (semantically) *sound*. What we need here to verify this conclusion is the following easily justifiable

**Principle of rule induction.** *Let $\mathcal{E}$ ($\subseteq \mathcal{P}\mathcal{F} \times \mathcal{F}$) be a property closed under all basic rules of $\vdash$. Then $X \vdash \alpha$ implies $\mathcal{E}(X, \alpha)$.*

**Proof** by induction on the length of a derivation of $S = (X, \alpha)$. If the length is 1, $\mathcal{E}S$ holds since $S$ must be an initial sequent. Now let $(S_0; \ldots; S_n)$ be a derivation of the sequent $S := S_n$. By the induction hypothesis we have $\mathcal{E}S_i$ for all $i < n$. If $S$ is an initial sequent then $\mathcal{E}S$ holds by assumption. Otherwise $S$ has been obtained by the application of a basic rule on some of the $S_i$ for $i < n$. But then $\mathcal{E}S$ holds, because $\mathcal{E}$ is closed under all basic rules. $\blacksquare$

As already remarked, the property $X \vDash \alpha$ is closed under all basic rules. Therefore, the principle of rule induction immediately yields the *soundness* of the calculus, that is, $\vdash \subseteq \vDash$. More explicitly,

$$X \vdash \alpha \ \Rightarrow \ X \vDash \alpha, \text{ for all } X, \alpha.$$

There are several equivalent definitions of $\vdash$. A purely set-theoretic one is the following: $\vdash$ is the smallest of all relations $\subseteq \mathfrak{P}\mathcal{F} \times \mathcal{F}$ that are closed under all basic rules. $\vdash$ is equally the smallest consequence relation closed under the rules $(\wedge 1)$ through $(\neg 2)$. The equivalence proofs of such definitions are wordy but not particularly contentful. We therefore do not elaborate further, because we henceforth use only rule induction. Using rule induction one can also prove $X \vdash \alpha \Rightarrow X^\sigma \vdash \alpha^\sigma$, and in particular the following theorem, for which the soundness of $\vdash$ is irrelevant.

**Theorem 4.1 (Finiteness theorem for $\vdash$).** *If $X \vdash \alpha$ then there is a finite subset $X_0 \subseteq X$ with $X_0 \vdash \alpha$.*

**Proof.** Let $\mathcal{E}(X, \alpha)$ be the property '$X_0 \vdash \alpha$ for some finite $X_0 \subseteq X$'. We will show that $\mathcal{E}$ is closed under all basic rules. Certainly, $\mathcal{E}(X, \alpha)$ holds for $X = \{\alpha\}$, with $X_0 = X$ so that $\mathcal{E}$ is closed under (MI). If $X$ has a finite subset $X_0$ such that $X_0 \vdash \alpha$, then so too does every set $X'$ such that $X' \supseteq X$. Hence $\mathcal{E}$ is closed under (MR). Let $\mathcal{E}(X, \alpha)$, $\mathcal{E}(X, \beta)$, with, say, $X_1 \vdash \alpha$, $X_2 \vdash \beta$ for finite $X_1, X_2 \subseteq X$. Then we also have $X_0 \vdash \alpha, \beta$ for $X_0 = X_1 \cup X_2$ by (MR). Hence $X_0 \vdash \alpha \wedge \beta$ by $(\wedge 1)$. Thus $\mathcal{E}(X, \alpha \wedge \beta)$ holds, and $\mathcal{E}$ is closed under $(\wedge 1)$. Analogously one shows the same for all remaining basic rules of $\vdash$ so that rule induction can be applied.   ◻

Of great significance is the notion of formal consistency. It fully determines the derivability relation, as the lemma to come shows. It will turn out that *consistent* formalizes adequately the notion *satisfiable*. The proof of this adequacy is the clue to the completeness problem.

**Definition.** $X \subseteq \mathcal{F}$ is called *inconsistent* (in our calculus $\vdash$) if $X \vdash \alpha$ for all $\alpha \in \mathcal{F}$, and otherwise *consistent*. $X$ is called *maximally consistent* if $X$ is consistent but each $Y \supset X$ is inconsistent.

The inconsistency of $X$ can be identified by the derivability of a single formula, namely $\bot \ (= p_1 \wedge \neg p_1)$, because $X \vdash \bot$ implies $X \vdash p_1, \neg p_1$ by $(\wedge 2)$, hence $X \vdash \alpha$ for all $\alpha$ by $(\neg 1)$. Conversely, when $X$ is inconsistent then in particular $X \vdash \bot$. Thus, $X \vdash \bot$ may be read as '$X$ is inconsistent', and $X \nvdash \bot$ as '$X$ is consistent'. From this it easily follows that $X$ is maximally consistent iff either $\alpha \in X$ or $\neg\alpha \in X$ for each $\alpha$. The latter is necessary, for if $\alpha, \neg\alpha \notin X$ then both $X, \alpha \vdash \bot$ and $X, \neg\alpha \vdash \bot$, hence $X \vdash \bot$ by ($\neg 2$). This contradicts the consistency of $X$. Sufficiency is obvious. Most important is the following lemma, in which the properties $\mathsf{C}^+$ and $\mathsf{C}^-$ can also be understood each as a pair of provable rules.

**Lemma 4.2.** *The derivability relation $\vdash$ has the properties*

$$\mathsf{C}^+: \quad X \vdash \alpha \Leftrightarrow X, \neg\alpha \vdash \bot, \qquad \mathsf{C}^-: \quad X \vdash \neg\alpha \Leftrightarrow X, \alpha \vdash \bot.$$

**Proof.** Suppose that $X \vdash \alpha$. Then clearly $X, \neg\alpha \vdash \alpha$ and since certainly $X, \neg\alpha \vdash \neg\alpha$, we have $X, \neg\alpha \vdash \beta$ for all $\beta$ by ($\neg 1$), in particular $X, \neg\alpha \vdash \bot$. Conversely, let $X, \neg\alpha \vdash \bot$ be the case, so that in particular $X, \neg\alpha \vdash \alpha$, and thus $X \vdash \alpha$ by $\neg$-elimination on page 23. Property $\mathsf{C}^-$ is proved completely analogously. ◻

The claim $\vDash \subseteq \vdash$, not yet proved, is equivalent to $X \nvDash \alpha \Rightarrow X \nvdash \alpha$, for all $X$ and $\alpha$. But so formulated it becomes apparent what needs to be done to obtain the proof. Since $X \nvdash \alpha$ is by $\mathsf{C}^+$ equivalent to the consistency of $X' := X \cup \{\neg\alpha\}$, and likewise $X \nvDash \alpha$ to the satisfiability of $X'$, we need only show that consistent sets are satisfiable. To this end we state the following lemma, whose proof, exceptionally, jumps ahead of matters in that it uses Zorn's lemma from **2.1** (page 46).

**Lemma 4.3 (Lindenbaum's theorem).** *Every consistent set $X \subseteq \mathcal{F}$ can be extended to a maximally consistent set $X' \supseteq X$.*

**Proof.** Let $H$ be the set of all consistent $Y \supseteq X$, partially ordered with respect to $\subseteq$. $H \neq \emptyset$, because $X \in H$. Let $K \subseteq H$ be a chain, i.e., $Y \subseteq Z$ or $Z \subseteq Y$, for all $Y, Z \in K$. Claim: $U := \bigcup K$ is an upper bound for $K$. Since $Y \in K \Rightarrow Y \subseteq U$, we have to show that $U$ is consistent. Assume that $U \vdash \bot$. Then $U_0 \vdash \bot$ for some finite $U_0 = \{\alpha_0, \ldots, \alpha_n\} \subseteq U$. If, say, $\alpha_i \in Y_i \in K$, and $Y$ is the biggest of the sets $Y_0, \ldots, Y_n$, then $\alpha_i \in Y$ for all $i \leqslant n$, hence also $Y \vdash \bot$ by (MR). This contradicts $Y \in H$ and confirms the claim. By Zorn's lemma, $H$ has a maximal element $X'$, which is necessarily a maximally consistent extension of $X$. ◻

**Remark 2.** The advantage of this proof is that it is free of assumptions regarding the cardinality of the language, while Lindenbaum's original construction was based on countable languages $\mathcal{F}$ and runs as follows: Let $X_0 := X \subseteq \mathcal{F}$ be consistent and let $\alpha_0, \alpha_1, \ldots$ be an enumeration of $\mathcal{F}$. Put $X_{n+1} = X_n \cup \{\alpha_n\}$ if this set is consistent and $X_{n+1} = X_n$ otherwise. Then $Y = \bigcup_{n \in \omega} X_n$ is a maximally consistent extension of $X$, as can be easily verified. In this proof, Zorn's lemma, which is equivalent to the axiom of choice, is not required.

**Lemma 4.4.** *A maximally consistent set $X \subseteq \mathcal{F}$ has the property*

$$[\neg] \quad X \vdash \neg\alpha \Leftrightarrow X \nvdash \alpha, \text{ for arbitrary } \alpha.$$

**Proof.** If $X \vdash \neg\alpha$, then $X \vdash \alpha$ cannot hold due to the consistency of $X$. If, on the other hand, $X \nvdash \alpha$, then $X, \neg\alpha$ is a consistent extension of $X$ according by $\mathsf{C}^+$. But then $\neg\alpha \in X$, because $X$ is maximally consistent. Consequently $X \vdash \neg\alpha$. $\blacksquare$

**Lemma 4.5.** *A maximally consistent set $X$ is satisfiable.*

**Proof.** Define $w$ by $w \vDash p \Leftrightarrow X \vdash p$. We will show that for all $\alpha$,

$$(*) \quad X \vdash \alpha \Leftrightarrow w \vDash \alpha.$$

For prime formulas this is trivial. Further,

$$
\begin{aligned}
X \vdash \alpha \wedge \beta \quad &\Leftrightarrow \quad X \vdash \alpha, \beta && \text{(rules } (\wedge 1), (\wedge 2) \text{ )} \\
&\Leftrightarrow \quad w \vDash \alpha, \beta && \text{(induction hypothesis)} \\
&\Leftrightarrow \quad w \vDash \alpha \wedge \beta && \text{(definition)} \\
X \vdash \neg\alpha \quad &\Leftrightarrow \quad X \nvdash \alpha && \text{(Lemma 4.4)} \\
&\Leftrightarrow \quad w \nvDash \alpha && \text{(induction hypothesis)} \\
&\Leftrightarrow \quad w \vDash \neg\alpha && \text{(definition)}.
\end{aligned}
$$

By $(*)$, $w$ is a model for $X$, thereby completing the proof. $\blacksquare$

Only the properties $[\wedge]$ $X \vdash \alpha \wedge \beta \Leftrightarrow X \vdash \alpha, \beta$ and $[\neg]$ from Lemma 4.4 are used in the simple model construction in Lemma 4.5, which reveals the requirements for propositional model construction in the base $\{\wedge, \neg\}$. Since maximally consistent sets $X$ are deductively closed (Exercise 5 in **1.3**), these requirements may also be stated as

$$(\wedge) \quad \alpha \wedge \beta \in X \Leftrightarrow \alpha, \beta \in X \quad ; \quad (\neg) \quad \neg\alpha \in X \Leftrightarrow \alpha \notin X.$$

Lemma 4.3 and Lemma 4.5 confirm the equivalence of the consistency and the satisfiability of a set of formulas. From this fact we easily obtain the main result of the present section.

**Theorem 4.6 (Completeness theorem).** *$X \vdash \alpha \Leftrightarrow X \vDash \alpha$, for all formula sets $X$ and formulas $\alpha$.*

**Proof.** The direction $\Rightarrow$ is the soundness of $\vdash$. Conversely, $X \nvdash \alpha$ implies that $X, \neg\alpha$ is consistent. Let $Y$ be a maximally consistent extension of $X, \neg\alpha$ according to Lemma 4.3. By Lemma 4.5, $Y$ is satisfiable, hence also $X, \neg\alpha$. Therefore $X \nvDash \alpha$. $\square$

An immediate consequence of Theorem 4.6 is the finiteness property (F) mentioned in **1.3**, which is almost trivial for $\vdash$ but not for $\vDash$:

**Theorem 4.7 (Finiteness theorem for $\vDash$).** *If $X \vDash \alpha$, then so too $X_0 \vDash \alpha$ for some finite subset $X_0$ of $X$.*

This is clear because the finiteness theorem holds for $\vdash$ (Theorem 4.1), hence also for $\vDash$. A further highly interesting consequence of the completeness theorem is

**Theorem 4.8 (Propositional compactness theorem).** *A set $X$ of propositional formulas is satisfiable if each finite subset of $X$ is satisfiable.*

This theorem holds because if $X$ is unsatisfiable, i.e., if $X \vDash \bot$, then, by Theorem 4.7, we also know that $X_0 \vDash \bot$ for some finite $X_0 \subseteq X$, thus proving the claim indirectly. Conversely, one easily obtains Theorem 4.7 from Theorem 4.8; both theorems are directly derivable from one another. Because Theorem 4.6 makes no assumptions regarding the cardinality of the set of variables, the compactness theorem following from it is likewise valid without the respective restrictions. This means that Theorem 4.8 has many useful applications, as the next section will illustrate.

Let us notice that there are direct proofs of Theorem 4.8 or appropriate reformulations that have nothing to do with a logical calculus. For example, the theorem is equivalent to

$$\bigcap_{\alpha \in X} \mathrm{Md}\,\alpha = \emptyset \;\Rightarrow\; \bigcap_{\alpha \in X_0} \mathrm{Md}\,\alpha = \emptyset \text{ for some finite } X_0 \subseteq X,$$

where $\mathrm{Md}\,\alpha$ denotes the set of all models of $\alpha$. In this formulation the compactness of a certain naturally arising topological space is claimed. The points of this space are the valuations of the variables, hence the name "compactness theorem." More on this can be found in [RS].

Another approach to completeness (probably the simplest one) is provided by Exercises 3 and 4. This approach makes some elegant use of substitutions, hence is called *the completeness proof by the substitution method*. This method is explained in the Solution Hints (and in more detail in [Ra3]). It yields the maximality of the derivability relation $\vdash$

(see Exercise 3), a much stronger result than its semantic completeness. This result yields not only the Theorems 4.6, 4.7, and 4.8 in one go, but also some further remarkable properties: Neither new tautologies nor new Hilbert style rules can consistently be adjoined to the calculus $\vdash$. These properties (discussed in detail, e.g., in [Ra1]) are known under the names *Post completeness* and *structural completeness* of $\vdash$, respectively.

### Exercises

1. Prove using Theorem 4.7: if $X \cup \{\neg\alpha \mid \alpha \in Y\}$ is inconsistent and $Y$ is nonempty, then there exist formulas $\alpha_0, \ldots, \alpha_n \in Y$ such that $X \vdash \alpha_0 \vee \cdots \vee \alpha_n$.

2. Augment the signature $\{\neg, \wedge\}$ by $\vee$ and prove the completeness of the calculus obtained by supplementing the basic rules used so far with the rules

   $$(\vee 1) \ \frac{X \vdash \alpha}{X \vdash \alpha \vee \beta, \beta \vee \alpha} \quad ; \quad (\vee 2) \ \frac{X, \alpha \vdash \gamma \mid X, \beta \vdash \gamma}{X, \alpha \vee \beta \vdash \gamma}.$$

3. Let $\vdash$ be a finitary consistent consequence relation in $\mathcal{F}\{\wedge, \neg\}$ with the properties $(\wedge 1)$ through $(\neg 2)$. Show that $\vdash$ is *maximal* (or *maximally consistent*). This means that each consequence relation $\vdash' \supset \vdash$ in $\mathcal{F}\{\wedge, \neg\}$ is inconsistent, i.e., $\vdash' \alpha$ for all $\alpha$.

4. Show by referring to Exercise 3: there is exactly one (consistent) consequence relation in $\mathcal{F}\{\wedge, \neg\}$ satisfying $(\wedge 1)$–$(\neg 2)$. This clearly entails the completeness of $\vdash$.

## 1.5   Applications of the Compactness Theorem

Theorem 4.8 is very useful in carrying over certain properties of finite structures to infinite ones. This section presents some typical examples. While these could also be treated with the compactness theorem of first-order logic in **3.3**, the examples demonstrate how the consistency of certain sets of first-order sentences can also be obtained in propositional logic. This approach to consistency is also useful also for Herbrand's theorem and related results concerning logic programming.

### 1. Every set $M$ can be (totally) ordered.[5]

This means that there is an irreflexive, transitive, and connex relation $<$ on $M$. For finite $M$ this follows easily by induction on the number of elements of $M$. The claim is obvious when $M = \emptyset$ or is a singleton. Let now $M = N \cup \{a\}$ with an $n$-element set $N$ and $a \notin N$, so that $M$ has $n + 1$ elements. Then we clearly get an order on $M$ from that for $N$ by "setting $a$ to the end," that is, defining $x < a$ for all $x \in N$.

Now let $M$ be any set. We consider for every pair $(a, b) \in M \times M$ a propositional variable $p_{ab}$. Let $X$ be the set consisting of the formulas

$$\neg p_{aa} \qquad (a \in M),$$
$$p_{ab} \wedge p_{bc} \rightarrow p_{ac} \quad (a, b, c \in M),$$
$$p_{ab} \vee p_{ba} \qquad (a \neq b).$$

From $w \vDash X$ we obtain an order $<$, simply by putting $a < b \Leftrightarrow w \vDash p_{ab}$. $w \vDash \neg p_{aa}$ says the same thing as $a \not< a$. Analogously, the remaining formulas of $X$ reflect transitivity and connexity. Thus, according to Theorem 4.8, it suffices to show that every finite subset $X_0 \subseteq X$ has a model. In $X_0$ only finitely many variables occur. Hence, there are finite sets $M_1 \subseteq M$ and $X_1 \supseteq X_0$, where $X_1$ is given exactly as $X$ except that $a, b, c$ now run through the finite set $M_1$ instead of $M$. But $X_1$ is satisfiable, because if $<$ orders the finite set $M_1$ and $w$ is defined by $w \vDash p_{ab}$ iff $a < b$, then $w$ is clearly a model for $X_1$, hence also for $X_0$.

### 2. The four-color theorem for infinite planar graphs.

A *simple graph* is a pair $(V, E)$ with an irreflexive symmetrical relation $E \subseteq V^2$. The elements of $V$ are called *points* or *vertices*. It is convenient to identify $E$ with the set of all unordered pairs $\{a, b\}$ such that $aEb$ and to call these pairs the *edges* of $(V, E)$. If $\{a, b\} \in E$ then we say that $a, b$ are *neighbors*. $(V, E)$ is said to be $k$-*colorable* if $V$ can be decomposed into $k$ *color classes* $C_1, \ldots, C_k \neq \emptyset$, $V = C_1 \cup \cdots \cup C_k$, with $C_i \cap C_j = \emptyset$ for $i \neq j$, such that neighboring points do not carry the same color; in other words, if $a, b \in C_i$ then $\{a, b\} \notin E$ for $i = 1, \ldots, k$.

---

[5] Unexplained notions are defined in **2.1**. Our first application is interesting because in set theory the compactness theorem is weaker than the axiom of choice (AC) which is equivalent to the statement that every set can be well-ordered. Thus, the ordering principle is weaker than AC since it follows from the compactness theorem.

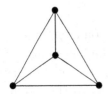

The figure shows the smallest four-colorable graph that is not three-colorable; all its points neighbor each other. We will show that a graph $(V, E)$ is $k$-colorable if every finite subgraph $(V_0, E_0)$ is $k$-colorable. $E_0$ consists of the edges $\{a, b\} \in E$ with $a, b \in V_0$. To prove our claim consider the following set $X$ of formulas built from the variables $p_{a,i}$ for $a \in V$ and $1 \leqslant i \leqslant k$:

$$p_{a,1} \vee \cdots \vee p_{a,k}, \quad \neg(p_{a,i} \wedge p_{a,j}) \quad (a \in V,\ 1 \leqslant i < j \leqslant k),$$
$$\neg(p_{a,i} \wedge p_{b,i}) \quad (\{a, b\} \in E,\ i = 1, \ldots, k).$$

The first formula states that every point belongs to at least one color class; the second ensures their disjointedness, and the third that no neighboring points have the same color. Once again it is enough to construct some $w \vDash X$. Defining then the $C_i$ by $a \in C_i \Leftrightarrow w \vDash p_{a,i}$ proves that $(V, E)$ is $k$-colorable. We must therefore satisfy each finite $X_0 \subseteq X$. Let $(V_0, E_0)$ be the finite subgraph of $(V, E)$ of all the points that occur as indices in the variables of $X_0$. The assumption on $(V_0, E_0)$ obviously ensures the satisfiability of $X_0$ for reasons analogous to those given in Example 1, and this is all we need to show. The *four-color theorem* says that every finite planar graph is four-colorable. Hence, the same holds for all graphs whose finite subgraphs are planar. These cover in particular all planar graphs embeddable in the real plane.

**3. König's tree lemma.** There are several versions of this lemma. For simplicity, ours refers to a *directed tree*. This is a pair $(V, \lhd)$ with an irreflexive relation $\lhd \subseteq V^2$ such that for a certain point $c$, the *root* of the tree, and any other point $a$ there is precisely one *path* connecting $c$ with $a$. This is a sequence $(a_i)_{i \leqslant n}$ with $a_0 = c$, $a_n = a$, and $a_i \lhd a_{i+1}$ for all $i < n$. From the uniqueness of a path connecting $c$ with any other point it follows that each $b \neq c$ has exactly one *predecessor* in $(V, \lhd)$, that is, there is precisely one $a$ with $a \lhd b$. Hence the name tree.

König's lemma then reads as follows: *If every $a \in V$ has only finitely many successors and $V$ contains arbitrarily long finite paths, then there is an infinite path through $V$ starting at $c$.* By such a path we mean a sequence $(c_i)_{i \in \mathbb{N}}$ such that $c_0 = c$ and $c_k \lhd c_{k+1}$ for each $k$. In order to prove the lemma we define the "layer" $S_k$ inductively by $S_0 = \{c\}$ and $S_{k+1} = \{b \in V \mid \text{there is some } a \in S_k \text{ with } a \lhd b\}$. Since every point has only finitely many successors, each $S_k$ is finite, and since there are arbitrarily long paths $c \lhd a_1 \lhd \cdots \lhd a_k$ and $a_k \in S_k$, no $S_k$ is empty. Now let $p_a$ for each $a \in V$ be a propositional variable, and let $X$ consist of the formulas

$$\text{(A)} \quad \bigvee_{a \in S_k} p_a, \quad \neg(p_a \wedge p_b) \quad (a, b \in S_k,\ a \neq b,\ k \in \mathbb{N}),$$
$$\text{(B)} \quad\quad\quad\quad p_b \to p_a \quad (a, b \in V,\ a \lhd b).$$

Suppose that $w \vDash X$. Then by the formulas under (A), for every $k$ there is precisely one $a \in S_k$ with $w \vDash p_a$, denoted by $c_k$. In particular, $c_0 = c$. Moreover, $c_k \lhd c_{k+1}$ for all $k$. Indeed, if $a$ is the predecessor of $b = c_{k+1}$, then $w \vDash p_a$ in view of (B), hence necessarily $a = c_k$. Thus, $(c_i)_{i \in \mathbb{N}}$ is a path of the type sought. Again, every finite subset $X_0 \subseteq X$ is satisfiable; for if $X_0$ contains variables with indices up to at most the layer $S_n$, then $X_0$ is a subset of a finite set of formulas $X_1$ that is defined as $X$, except that $k$ runs only up to $n$, and for this case the claim is obvious.

**4. The marriage problem (in linguistic guise).**

Let $N \neq \emptyset$ be a set of *words* or *names* (in speech) with meanings in a set $M$. A name $\nu \in N$ can be a *synonym* (i.e., it shares its meaning with other names in $N$), or a *homonym* (i.e., it can have several meanings), or even both. We proceed from the plausible assumption that each name $\nu$ has finitely many meanings only and that $k$ names have at least $k$ meanings. It is claimed that a *pairing-off* exists; that is, an injection $f : N \to M$ that associates to each $\nu$ one of its original meanings.

For finite $N$, the claim will be proved by induction on the number $n$ of elements of $N$. It is trivial for $n = 1$. Now let $n > 1$ and assume that the claim holds for all $k$-element sets of names whenever $0 < k < n$.

**Case 1**: For each $k$ $(0 < k < n)$: $k$ names in $N$ have at least $k + 1$ distinct meanings. Then to an arbitrarily chosen $\nu$ from $N$, assign one of its meanings $a$ to it so that from the names out of $N \setminus \{\nu\}$ any $k$ names still have at least $k$ meanings $\neq a$. By the induction hypothesis there is a pairing-off for $N \setminus \{\nu\}$ that together with the ordered pair $(\nu, a)$ yields a pairing-off for the whole of $N$.

**Case 2**: There is some $k$-element $K \subseteq N$ $(0 < k < n)$ such that the set $M_K$ of meanings of the $\nu \in K$ has only $k$ members. Every $\nu \in K$ can be assigned its meaning from $M_K$ by the induction hypothesis. From the names in $N \setminus K$ any $i$ names $(i \leqslant n - k)$ still have $i$ meanings not in $M_K$, as is not hard to see. By the induction hypothesis there is also a pairing-off for $N \setminus K$ with a set of values from $M \setminus M_K$. Joining the two obviously results in a pairing-off for the whole of $N$.

We will now prove the claim for arbitrary sets of names $N$: assign to each pair $(\nu, a) \in N \times M$ a variable $p_{\nu,a}$ and consider the set of formulas

$$X : \begin{cases} p_{\nu,a} \vee \cdots \vee p_{\nu,e} & (\nu \in N, \ a, \ldots, e \text{ the meanings of } \nu), \\ \neg(p_{\nu,x} \wedge p_{\nu,y}) & (\nu \in N, \ x, y \in M, \ x \neq y). \end{cases}$$

Assume that $w \vDash X$. Then to each $\nu$ there is exactly one $a_\nu$ with $w \vDash p_{\nu,a_\nu}$, so that $\{(\nu, \alpha_n) \mid \nu \in N\}$ is a pairing-off for $N$. Such a model $w$ exists by Theorem 4.8, for in a finite set $X_0 \subseteq X$ occur only finitely many names as indices and the case of finitely many names has just been treated.

## 5. The ultrafilter theorem.

This theorem is of fundamental significance in topology (from which it originally stems), model theory, set theory, and elsewhere. Let I be any nonempty set. A nonempty collection of sets $F \subseteq \mathfrak{P}I$ is called a *filter on* $I$ if for all $M, N \subseteq I$ hold the conditions

(a) $M, N \in F \Rightarrow M \cap N \in F$,    (b) $M \in F \ \& \ M \subseteq N \Rightarrow N \in F$.

Since $F \neq \emptyset$, (b) shows that always $I \in F$. As is easily verified, (a) and (b) together are equivalent to just a single condition, namely to

$$(\cap) \quad M \cap N \in F \ \Leftrightarrow \ M \in F \text{ and } N \in F.$$

For fixed $K \subseteq I$, $\{J \subseteq I \mid J \supseteq K\}$ is a filter, the *principal filter* generated by $K$. This is a *proper filter* provided $K \neq \emptyset$, which in general is to mean a filter with $\emptyset \notin F$. Another example on an infinite $I$ is the set of all *cofinite* subsets $M \subseteq I$, i.e., $\neg M \ (= I \setminus M)$ is finite. This holds because $M_1 \cap M_2$ is cofinite iff $M_1, M_2$ are both cofinite, so that $(\cap)$ is satisfied.

A filter $F$ is said to be an *ultrafilter on* $I$ provided it satisfies, in addition,

$$(\neg) \quad \neg M \in F \ \Leftrightarrow \ M \notin F.$$

Ultrafilters on an infinite set $I$ containing all cofinite subsets are called *nontrivial*. That such ultrafilters exist will be shown below. It is nearly impossible to describe them more closely. Roughly speaking, "we know they exist but we cannot see them." A trivial ultrafilter on $I$ contains at least one finite subset. $\{J \subseteq I \mid i_0 \in J\}$ is an example for each $i_0 \in I$. This is a *principal* ultrafilter. All trivial ultrafilters are of this form, Exercise 3. Thus, trivial and principal ultrafilters coincide. In particular, each ultrafilter on a finite set $I$ is trivial in this sense.

Each proper filter $F$ obviously satisfies the assumption of the following theorem and can thereby be extended to an ultrafilter.

**Theorem 5.1 (Ultrafilter theorem).** *Every subset $F \subseteq \mathfrak{P}I$ can be extended to an ultrafilter $U$ on a set $I$, provided $M_0 \cap \cdots \cap M_n \neq \emptyset$ for all $n$ and all $M_0, \ldots, M_n \in F$.*

**Proof.** Consider along with the propositional variables $p_J$ for $J \subseteq I$

$$X: \quad p_{M \cap N} \leftrightarrow p_M \wedge p_N, \quad p_{\neg M} \leftrightarrow \neg p_M, \quad p_J \qquad (M, N \subseteq I, \ J \in F).$$

Let $w \vDash X$. Then $(\cap), (\neg)$ are valid for $U := \{J \subseteq I \mid w \vDash p_J\}$; hence $U$ is an ultrafilter such that $F \subseteq U$. It therefore suffices to show that every finite subset of $X$ has a model, for which it is in turn enough to prove the ultrafilter theorem for finite $F$. But this is easy: Let $F = \{M_0, \ldots, M_n\}$, $D := M_0 \cap \cdots \cap M_n$, and $i_0 \in D$. Then $U = \{J \subseteq I \mid i_0 \in J\}$ is an ultrafilter containing $F$. ∎

### Exercises

1. Prove (using the compactness theorem) that every partial order $\leqslant_0$ on a set $M$ can be extended to a total order $\leqslant$ on $M$.

2. Let $F$ be a proper filter on $I$ ($\neq \emptyset$). Show that $F$ is an ultrafilter iff it satisfies $(\cup)$: $M \cup N \in F \Leftrightarrow M \in F$ or $N \in F$.

3. Let $I$ be an infinite set. Show that an ultrafilter $U$ on $I$ is trivial iff there is an $i_0 \in I$ such that $U = \{J \subseteq I \mid i_0 \in J\}$.

## 1.6 Hilbert Calculi

In a certain sense the simplest logical calculi are so-called *Hilbert calculi*. They are based on tautologies selected to play the role of *logical axioms*; this selection is, however, rather arbitrary and depends considerably on the logical signature. They use rules of inference such as, for example, modus ponens MP: $\alpha, \alpha \to \beta / \beta$.[6] An advantage of these calculi consists

---

[6] Putting it crudely, this notation should express the fact that $\beta$ is held to be proved from a formula set $X$ when $\alpha$ and $\alpha \to \beta$ are provable from $X$. Modus ponens is an example of a binary Hilbert-style rule; for a general definition of this type of rule see, for instance, [Ra1].

in the fact that formal proofs, defined below as certain finite sequences, are immediately rendered intuitive. This advantage will pay off above all in the arithmetization of proofs in **6.2**.

In the following we consider such a calculus with MP as the only rule of inference; we denote this calculus for the time being by $\vdash$, in order to distinguish it from the calculus $\vdash$ of **1.4**. The logical signature contains just $\neg$ and $\wedge$, the same as for $\vdash$. In the axioms of $\vdash$, however, we will also use implication defined by $\alpha \rightarrow \beta := \neg(\alpha \wedge \neg\beta)$, thus considerably shortening the writing down of the axioms.

The *logical axiom scheme* of our calculus consists of the set $\Lambda$ of all formulas of the following form (not forgetting the right association of parentheses in $\Lambda 1$, $\Lambda 2$, and $\Lambda 4$):

$\Lambda 1 \quad (\alpha \rightarrow \beta \rightarrow \gamma) \rightarrow (\alpha \rightarrow \beta) \rightarrow \alpha \rightarrow \gamma,$ $\qquad \Lambda 2 \quad \alpha \rightarrow \beta \rightarrow \alpha \wedge \beta,$

$\Lambda 3 \quad \alpha \wedge \beta \rightarrow \alpha, \quad \alpha \wedge \beta \rightarrow \beta,$ $\qquad\qquad \Lambda 4 \quad (\alpha \rightarrow \neg\beta) \rightarrow \beta \rightarrow \neg\alpha.$

$\Lambda$ consists only of tautologies. Moreover, all formulas derivable from $\Lambda$ using MP are tautologies as well, because $\vDash \alpha, \alpha \rightarrow \beta$ implies $\vDash \beta$. We will show that *all* 2-valued tautologies are provable from $\Lambda$ by means of MP. To this aim we first define the notion of a proof from $X \subseteq \mathcal{F}$ in $\vdash$.

**Definition.** A *proof* from $X$ (in $\vdash$) is a sequence $\Phi = (\varphi_0, \ldots, \varphi_n)$ such that for every $k \leqslant n$ either $\varphi_k \in X \cup \Lambda$ or there exist indices $i, j < k$ such that $\varphi_j = \varphi_i \rightarrow \varphi_k$ (i.e., $\varphi_k$ results from applying MP to terms of $\Phi$ preceding $\varphi_k$). A proof $(\varphi_0, \ldots, \varphi_n)$ with $\varphi_n = \alpha$ is called *a proof of $\alpha$ from $X$* of length $n+1$. Whenever such a proof exists we write $X \vdash \alpha$ and say that $\alpha$ is *provable* or *derivable from $X$*.

**Example.** $(p, q, p \rightarrow q \rightarrow p \wedge q, q \rightarrow p \wedge q, p \wedge q)$ is a proof of $p \wedge q$ from the set $X = \{p, q\}$. The last two terms in the proof sequence derive with MP from the previous ones, which are members of $X \cup \Lambda$.

Since a proof contains only finitely many formulas, the preceding definition leads immediately to the finiteness theorem for $\vdash$, formulated correspondingly to Theorem 4.1. Every proper initial segment of a proof is obviously a proof itself. Moreover, concatenating proofs of $\alpha$ and $\alpha \rightarrow \beta$ and tacking on $\beta$ to the resulting sequence will produce a proof for $\beta$, as is plain to see. This observation implies

$$(*) \quad X \vdash \alpha, \alpha \rightarrow \beta \;\Rightarrow\; X \vdash \beta.$$

In short, the set of all formulas derivable from $X$ is *closed under* MP. In applying the property (∗) we will often say "MP yields ..." It is easily seen that $X \vdash \alpha$ iff $\alpha$ belongs to the smallest set containing $X \cup \Lambda$ and closed under MP. For the arithmetization of proofs and for automated theorem proving, however, it is more appropriate to base derivability on the finitary notion of a proof that was given in the last definition. Fortunately, the following theorem relieves us of the necessity to verify a property of formulas $\alpha$ derivable from a given formula set $X$ each time by induction on the length of a proof of $\alpha$ from $X$.

**Theorem 6.1 (Induction principle for $\vdash$).** *Let $X$ be given and let $\mathcal{E}$ be a property of formulas. Then $\mathcal{E}$ holds for all $\alpha$ with $X \vdash \alpha$, provided*

(o) $\mathcal{E}$ *holds for all* $\alpha \in X \cup \Lambda$,

(s) $\mathcal{E}\alpha$ *and* $\mathcal{E}(\alpha \to \beta)$ *imply* $\mathcal{E}\beta$, *for all* $\alpha, \beta$.

**Proof** by induction on the length $n$ of a proof $\Phi$ of $\alpha$ from $X$. If $\alpha \in X \cup \Lambda$ then $\mathcal{E}\alpha$ holds by (o), which applies in particular if $n = 1$. If $\alpha \notin X \cup \Lambda$ then $n > 1$ and $\Phi$ contains members $\alpha_i$ and $\alpha_j = \alpha_i \to \alpha$ both having proofs of length $< n$. Hence, it holds $\mathcal{E}\alpha_i$ and $\mathcal{E}\alpha_j$ by the induction hypothesis, and so $\mathcal{E}\alpha$ according to (s). $\quad\blacksquare$

An application of Theorem 6.1 is the proof of $\vdash \ \subseteq \ \vDash$, or more explicitly,
$$X \vdash \alpha \ \Rightarrow \ X \vDash \alpha \qquad (soundness).$$
To see this let $\mathcal{E}\alpha$ be the property '$X \vDash \alpha$' for fixed $X$. Certainly, $X \vDash \alpha$ holds for $\alpha \in X$. The same is true for $\alpha \in \Lambda$. Thus, $\mathcal{E}\alpha$ for all $\alpha \in X \cup \Lambda$, and (o) is confirmed. Now let $X \vDash \alpha, \alpha \to \beta$; then so too $X \vDash \beta$, thus confirming the inductive step (s) in Theorem 6.1. Consequently, $\mathcal{E}\alpha$ (that is, $X \vDash \alpha$) holds for all $\alpha$ with $X \vdash \alpha$.

Unlike the proof of completeness for $\vdash$, the one for $\vdash$ requires a whole series of derivations to be undertaken. This is in accordance with the nature of things. To get Hilbert calculi up and running one must often begin with drawn-out derivations. In the derivations below we shall use without further comment the monotonicity (M) (page 19, with $\vdash$ for $\vDash$). (M) is obvious, for a proof in $\vdash$ from $X$ is also a proof from $X' \supseteq X$. Moreover, $\vdash$ is a consequence relation (as is every Hilbert calculus, based on Hilbert style rules). For example, if $X \vdash Y \vdash \alpha$, we construct a proof of $\alpha$ from $X$ by replacing each $\varphi \in Y$ occurring in a proof of $\alpha$ from $Y$ by a proof of $\varphi$ from $X$. This confirms the transitivity (T).

**Lemma 6.2.** (a) $X \vdash \alpha \rightarrow \neg \beta \ \Rightarrow \ X \vdash \beta \rightarrow \neg \alpha$,   (b) $\vdash \alpha \rightarrow \beta \rightarrow \alpha$,

$\qquad\qquad$ (c) $\vdash \alpha \rightarrow \alpha$,   (d) $\vdash \alpha \rightarrow \neg\neg\alpha$,   (e) $\vdash \beta \rightarrow \neg \beta \rightarrow \alpha$.

**Proof.** (a): Clearly $X \vdash (\alpha \rightarrow \neg\beta) \rightarrow \beta \rightarrow \neg\alpha$ by Axiom $\Lambda 4$. From this and from $X \vdash \alpha \rightarrow \neg\beta$ the claim is derived by MP. (b): By $\Lambda 3$, $\vdash \beta \wedge \neg\alpha \rightarrow \neg\alpha$, and so with (a), $\vdash \alpha \rightarrow \neg(\beta \wedge \neg\alpha) = \alpha \rightarrow \beta \rightarrow \alpha$.
(c): From $\gamma := \alpha$, $\beta := \alpha \rightarrow \alpha$ in $\Lambda 1$ we obtain

$$\vdash (\alpha \rightarrow (\alpha \rightarrow \alpha) \rightarrow \alpha) \rightarrow (\alpha \rightarrow \alpha \rightarrow \alpha) \rightarrow \alpha \rightarrow \alpha,$$

which yields the claim by applying (b) and MP twice; (d) then follows from (a) using $\vdash \neg\alpha \rightarrow \neg\alpha$. (e): Due to $\vdash \neg\beta \wedge \neg\alpha \rightarrow \neg\beta$ and (a), we get $\vdash \beta \rightarrow \neg(\neg\beta \wedge \neg\alpha) = \beta \rightarrow \neg\beta \rightarrow \alpha$.  ∎

Clearly, $\vdash$ satisfies the rules $(\wedge 1)$ and $(\wedge 2)$ of **1.4**, in view of $\Lambda 2, \Lambda 3$. Part (e) of Lemma 6.2 yields $X \vdash \beta, \neg\beta \Rightarrow X \vdash \alpha$, so that $\vdash$ satisfies also rule $(\neg 1)$. After some preparation we will show that rule $(\neg 2)$ holds for $\vdash$ as well, thereby obtaining the desired completeness result. A crucial step in this direction is

**Lemma 6.3 (Deduction theorem).** $X, \alpha \vdash \gamma$ *implies* $X \vdash \alpha \rightarrow \gamma$.

**Proof** by induction in $\vdash$ with a given set $X, \alpha$. Let $X, \alpha \vdash \gamma$, and let $\mathcal{E}\gamma$ now mean '$X \vdash \alpha \rightarrow \gamma$'. To prove (o) in Theorem 6.1, let $\gamma \in \Lambda \cup X \cup \{\alpha\}$. If $\gamma = \alpha$ then clearly $X \vdash \alpha \rightarrow \gamma$ by Lemma 6.2(c). If $\gamma \in X \cup \Lambda$ then certainly $X \vdash \gamma$. Because also $X \vdash \gamma \rightarrow \alpha \rightarrow \gamma$ by Lemma 6.2(b), MP yields $X \vdash \alpha \rightarrow \gamma$, thus proving (o). To show (s) let $X, \alpha \vdash \beta$ and $X, \alpha \vdash \beta \rightarrow \gamma$, so that $X \vdash \alpha \rightarrow \beta, \alpha \rightarrow \beta \rightarrow \gamma$ by the induction hypothesis. Applying MP to $\Lambda 1$ twice yields $X \vdash \alpha \rightarrow \gamma$, thus confirming (s). Therefore, by Theorem 6.1, $\mathcal{E}\gamma$ for all $\gamma$, which completes the proof.  ∎

**Lemma 6.4.** $\vdash \neg\neg\alpha \rightarrow \alpha$.

**Proof.** By $\Lambda 3$ and MP, $\neg\neg\alpha \wedge \neg\alpha \vdash \neg\alpha, \neg\neg\alpha$. Choose any $\tau$ with $\vdash \tau$. The already verified rule $(\neg 1)$ clearly yields $\neg\neg\alpha \wedge \neg\alpha \vdash \neg\tau$, and in view of Lemma 6.3, $\vdash \neg\neg\alpha \wedge \neg\alpha \rightarrow \neg\tau$. From Lemma 6.2(a) it follows that $\vdash \tau \rightarrow \neg(\neg\neg\alpha \wedge \neg\alpha)$. But $\vdash \tau$, hence using MP we obtain $\vdash \neg(\neg\neg\alpha \wedge \neg\alpha)$ and the latter formula is just $\neg\neg\alpha \rightarrow \alpha$.  ∎

Lemma 6.3 and Lemma 6.4 are preparations for the next lemma, which is decisive in proving the completeness of $\vdash$.

**Lemma 6.5.** $\vdash\!\!\!\sim$ *satisfies also rule* $(\neg 2)$ *of the calculus* $\vdash$.

**Proof.** Let $X, \beta \vdash\!\!\!\sim \alpha$ and $X, \neg\beta \vdash\!\!\!\sim \alpha$; then $X, \beta \vdash\!\!\!\sim \neg\neg\alpha$ and $X, \neg\beta \vdash\!\!\!\sim \neg\neg\alpha$ by Lemma 6.2(d). Hence, $X \vdash\!\!\!\sim \beta \to \neg\neg\alpha, \neg\beta \to \neg\neg\alpha$ (Lemma 6.3), and so $X \vdash\!\!\!\sim \neg\alpha \to \neg\beta$ and $X \vdash\!\!\!\sim \neg\alpha \to \neg\neg\beta$ by Lemma 6.2(a). Thus, MP yields $X, \neg\alpha \vdash\!\!\!\sim \neg\beta, \neg\neg\beta$, whence $X, \neg\alpha \vdash\!\!\!\sim \neg\tau$ by $(\neg 1)$, with $\tau$ as in Lemma 6.4. Therefore $X \vdash\!\!\!\sim \neg\alpha \to \neg\tau$, due to Lemma 6.3, and hence $X \vdash\!\!\!\sim \tau \to \neg\neg\alpha$ by Lemma 6.2(a). Since $X \vdash\!\!\!\sim \tau$ it follows that $X \vdash\!\!\!\sim \neg\neg\alpha$ and so eventually $X \vdash\!\!\!\sim \alpha$ by Lemma 6.4. $\square$

**Theorem 6.6 (Completeness theorem).** $\vdash\!\!\!\sim \; = \; \vDash$.

**Proof.** Clearly, $\vdash\!\!\!\sim \; \subseteq \; \vDash$. Now, by what was said already on page 38 and by the lemma above, $\vdash\!\!\!\sim$ satisfies all basic rules of $\vdash$. Therefore, $\vdash \; \subseteq \; \vdash\!\!\!\sim$. Since $\vdash \; = \; \vDash$ (Theorem 4.6), we obtain also $\vDash \; \subseteq \; \vdash\!\!\!\sim$. $\square$

This theorem implies in particular $\vdash\!\!\!\sim \varphi \Leftrightarrow \vDash \varphi$. In short, using MP one obtains from the axiom system $\Lambda$ exactly the two-valued tautologies.

**Remark 1.** It may be something of a surprise that $\Lambda 1$–$\Lambda 4$ are sufficient to obtain all propositional tautologies, because these axioms and all formulas derivable from them using MP are collectively valid in intuitionistic and minimal logic. That $\Lambda$ permits the derivation of all two-valued tautologies is based on the fact that $\to$ was *defined*. Had $\to$ been considered as a primitive connective, this would no longer have been the case. To see this, alter the interpretation of $\neg$ by setting $\neg 0 = \neg 1 = 1$. While one here indeed obtains the value 1 for every valuation of the axioms of $\Lambda$ and formulas derived from them using MP, one does not do so for $\neg\neg p \to p$, which therefore cannot be derived. Modifying the two-valued matrix or using many-valued logical matrices is a widely applied method to obtain independence results for logical axioms.

Thus, we have seen that there are very different calculi for deriving tautologies or to recover other properties of the semantic relation $\vDash$. We have studied here to some extend Gentzen-style and Hilbert-style calculi and this will be done also for first-order logic in Chapter **2**. In any case, logical calculi and their completeness proofs depend essentially on the logical signature, as can be seen, for example, from Exercise 1.

Besides Gentzen- and Hilbert-style calculi there are still other types of logical calculi, for example various tableau calculi, which are above all significant for their generalizations to nonclassical logical systems. Related to tableau calculi is the resolution calculus dealt with in **4.3**.

Using Hilbert-style calculi one can axiomatize 2-valued logic in other logical signatures and functional incomplete fragments. For instance, the fragment in $\wedge,\vee$, which, while having no tautologies, contains a lot of interesting Hilbert-style rules. Proving that this fragment is axiomatizable by finitely many such rules is less easy as might be expected. At least nine Hilbert rules are required. Easier is the axiomatization of the well-known $\rightarrow$-fragment in Exercise 3, less easy that of the $\vee$-fragment in Exercise 4. Each of the infinitely many fragments of two-valued logic with or without tautologies is axiomatizable by a calculus using only *finitely many* Hilbert-style rules of its respective language, as was shown in [HeR].

**Remark 2.** The calculus in Exercise 4 that treats the fragment in $\vee$ alone, is based solely on unary rules. This fact considerably simplifies the matter, but the completeness proof is nevertheless nontrivial. For instance, the indispensable rule $(\alpha\beta)\gamma/\alpha(\beta\gamma)$ is derivable in this calculus, since a tricky application of the rules (3) and (4) yields $(\alpha\beta)\gamma \vdash \gamma(\alpha\beta) \vdash (\gamma\alpha)\beta \vdash \beta(\gamma\alpha) \vdash (\beta\gamma)\alpha \vdash \alpha(\beta\gamma)$. Much easier would be a completeness proof of this fragment with respect to the Gentzen-style rules $(\vee1)$ and $(\vee2)$ from Exercise 2 in **1.4**.

### Exercises

1. Prove the completeness of the Hilbert calculus $\vdash$ in $\mathcal{F}\{\rightarrow,\bot\}$ with MP as the sole rule of inference, the definition $\neg\alpha := \alpha \rightarrow \bot$, and the axioms A1: $\alpha \rightarrow \beta \rightarrow \alpha$, A2: $(\alpha \rightarrow \beta \rightarrow \gamma) \rightarrow (\alpha \rightarrow \beta) \rightarrow \alpha \rightarrow \gamma$, and A3: $\neg\neg\alpha \rightarrow \alpha$.

2. Let $\vdash$ be a finitary consequence relation and let $X \nvdash \varphi$. Use Zorn's lemma to prove that there is a $\varphi$-*maximal* $Y \supseteq X$, that is, $Y \nvdash \varphi$ but $Y,\alpha \vdash \varphi$ whenever $\alpha \notin Y$. Such a $Y$ is deductively closed but need not be maximally consistent.

3. Let $\vdash$ denote the calculus in $\mathcal{F}\{\rightarrow\}$ with the rule of inference MP, the axioms A1, A2 from Exercise 1, and $((\alpha \rightarrow \beta) \rightarrow \alpha) \rightarrow \alpha$ (the Peirce axiom). Verify that (a) a $\varphi$-maximal set $X$ is maximally consistent, (b) $\vdash$ is a complete calculus in the propositional language $\mathcal{F}\{\rightarrow\}$.

4. Show the completeness of the calculus $\vdash$ in $\mathcal{F}\{\vee\}$ with the four unary Hilbert-style rules below. The writing of $\vee$ has been omitted:

$$(1)\ \alpha/\alpha\beta, \quad (2)\ \alpha\alpha/\alpha, \quad (3)\ \alpha\beta/\beta\alpha, \quad (4)\ \alpha(\beta\gamma)/(\alpha\beta)\gamma.$$

# Chapter 2

# First-Order Logic

Mathematics and some other disciplines such as computer science often consider domains of individuals in which certain relations and operations are singled out. When using the language of propositional logic, our ability to talk about the properties of such relations and operations is very limited. Thus, it is necessary to refine our linguistic means of expression, in order to procure new possibilities of description. To this end, one needs not only logical symbols but also variables for the individuals of the domain being considered, as well as a symbol for equality and symbols for the relations and operations in question. First-order logic, sometimes called also predicate logic, is the part of logic that subjects properties of such relations and operations to logical analysis.

Linguistic particles such as "for all" and "there exists" (called *quantifiers*) play a central role here, whose analysis should be based on a well prepared semantic background. Hence, we first consider mathematical structures and classes of structures. Some of these are relevant both to logic (in particular model theory) and to computer science. Neither the newcomer nor the advanced student needs to read all of **2.1**, with its mathematical flavor, at once. The first five pages should suffice. The reader may continue with **2.2** and later return to what is needed.

Next we home in on the most important class of formal languages, the *first-order* languages, also called *elementary languages*. Their main characteristic is a restriction of the quantification possibilities. We discuss in detail the semantics of these languages and arrive at a notion of *logical consequence* from arbitrary premises. In this context, the notion of a formalized theory is made more precise.

W. Rautenberg, *A Concise Introduction to Mathematical Logic*,
Universitext, DOI 10.1007/978-1-4419-1221-3_2,
© Springer Science+Business Media, LLC 2010

Finally, we treat the introduction of new notions by explicit definitions and other expansions of a language, for instance by Skolem functions. Not until Chapter **3** do we talk about methods of formal logical deduction. While a multitude of technical details have to be considered in this chapter, nothing is especially profound. Anyway, most of it is important for the undertakings of the subsequent chapters.

## 2.1   Mathematical Structures

By a *structure* $\mathcal{A}$ we understand a nonempty set $A$ together with certain distinguished relations and operations of $A$, as well as certain constants distinguished therein. The set $A$ is also termed the *domain* of $\mathcal{A}$, or its *universe*. The distinguished relations, operations, and constants are called the *(basic) relations, operations, and constants of $\mathcal{A}$*. A *finite structure* is one with a finite domain. An easy example is $(\{0,1\}, \wedge, \vee, \neg)$. Here $\wedge, \vee, \neg$ have their usual meanings on the domain $\{0,1\}$, and no distinguished relations or constants occur. An *infinite structure* has an infinite domain. $\mathcal{A} = (\mathbb{N}, <, +, \cdot, 0, 1)$ is an example with the domain $\mathbb{N}$; here $<, +, \cdot, 0, 1$ have again their ordinary meaning.

Without having to say so every time, for a structure $\mathcal{A}$ the corresponding letter $A$ will always denote the domain of $\mathcal{A}$; similarly $B$ denotes the domain of $\mathcal{B}$, etc. If $\mathcal{A}$ contains no operations or constants, then $\mathcal{A}$ is also called a *relational structure*. If $\mathcal{A}$ has no relations it is termed an *algebraic structure*, or simply an *algebra*. For example, $(\mathbb{Z}, <)$ is a relational structure, whereas $(\mathbb{Z}, +, 0)$ is an algebraic structure, the *additive group* $\mathbb{Z}$ (it is customary to use here the symbol $\mathbb{Z}$ as well). Also the set of propositional formulas from **1.1** can be understood as an algebra, equipped with the operations $(\alpha, \beta) \mapsto (\alpha \wedge \beta)$, $(\alpha, \beta) \mapsto (\alpha \vee \beta)$, and $\alpha \mapsto \neg\alpha$. Thus, one may speak of the *formula algebra* $\mathcal{F}$ whenever it is useful to do so.

Despite our interest in specific structures, whole classes of structures are also often considered, for instance the classes of groups, rings, fields, vector spaces, Boolean algebras, and so on. Even when initially just a single structure is viewed, call it the paradigm structure, one often needs to talk about similar structures in the same breath, in *one* language, so to speak. This can be achieved by setting aside the concrete meaning of the relation and operation symbols in the paradigm structure and considering the symbols in themselves, creating thereby a formal language that enables one to talk at once about all structures relevant to a topic. Thus, one distinguishes in this context clearly between denotation and what is denoted. To emphasize this distinction, for instance for $\mathcal{A} = (A, +, <, 0)$, it is better to write $\mathcal{A} = (A, +^{\mathcal{A}}, <^{\mathcal{A}}, 0^{\mathcal{A}})$, where $+^{\mathcal{A}}$, $<^{\mathcal{A}}$, and $0^{\mathcal{A}}$ mean the relation, operation, and constant denoted by $+$, $<$, and $0$ in $\mathcal{A}$. Only if it is clear from the context what these symbols denote may the superscripts be omitted. In this way we are free to talk on the one hand about the structure $\mathcal{A}$, and on the other hand about the symbols $+, <, 0$.

A finite or infinite set $L$ resulting in this way, consisting of relation, operation, and constant symbols of a given arity, is called an *extralogical signature*. For the class of all groups (see page 47), $L = \{\circ, e\}$ exemplifies a favored signature; that is, one often considers groups as structures of the form $(G, \circ, e)$, where $\circ$ denotes the group operation and $e$ the unit element. But one can also define groups as structures of the signature $\{\circ\}$, because $e$ is definable in terms of $\circ$, as we shall see later. Of course, instead of $\circ$, another operation symbol could be chosen such as $\cdot$, $*$, or $+$. The latter is mainly used in connection with commutative groups. In this sense, the actual appearance of a symbol is less important; what matters is its arity. $r \in L$ always means that $r$ is a relation symbol, and $f \in L$ that $f$ is an operation symbol, each time of some arity $n > 0$, which of course depends on the symbols $r$ and $f$, respectively.[1]

An *L-structure* is a pair $\mathcal{A} = (A, L^{\mathcal{A}})$, where $L^{\mathcal{A}}$ contains for every $r \in L$ a relation $r^{\mathcal{A}}$ on $A$ of the same arity as $r$, for every $f \in L$ an operation $f^{\mathcal{A}}$ on $A$ of the arity of $f$, and for every $c \in L$ a constant $c^{\mathcal{A}} \in A$. We may omit the superscripts, provided it is clear from the context which operation or relation on $A$ is meant. We occasionally shorten also the notation of structures. For instance, we sometimes speak of the ring $\mathbb{Z}$ or the field $\mathbb{R}$ provided there is no danger of misunderstanding.

Every structure is an $L$-structure for a certain signature, namely that consisting of the symbols for its relations, functions, and constants. But this does not make the name $L$-structure superfluous. Basic concepts,

---

[1] Here $r$ and $f$ represent the general case and look different in a concrete situation. Relation symbols are also called predicate symbols, in particular in the unary case, and operation symbols are sometimes called function symbols. In special contexts, we also admit $n = 0$, regarding constants as 0-ary operations.

such as isomorphism and substructure, each refer to structures of the same signature. From **2.2** on, once the first-order language $\mathcal{L}$ belonging to $L$ has been defined, $L$-structures will mostly be called $\mathcal{L}$-structures. We then also often say that $r$, $f$, or $c$ belongs to $\mathcal{L}$ instead of $L$.

If $A \subseteq B$ and $f$ is an $n$-ary operation on $B$ then $A$ is *closed* under $f$, briefly *$f$-closed*, if $f\vec{a} \in A$ for all $\vec{a} \in A^n$. If $n = 0$, i.e., if $f$ is a constant $c$, this simply means $c \in A$. The intersection of any nonempty family of $f$-closed subsets of $B$ is itself $f$-closed. Accordingly, we can talk of the smallest (the intersection) of all $f$-closed subsets of $B$ that contain a given subset $E \subseteq B$. All of this extends in a natural way if $f$ is here replaced by an arbitrary family of operations of $B$.

**Example.** For a given positive $m$, the set $m\mathbb{Z} := \{m \cdot n \mid n \in \mathbb{Z}\}$ of integers divisible by $m$ is closed in $\mathbb{Z}$ under $+$, $-$, and $\cdot$, and is in fact the smallest such subset of $\mathbb{Z}$ containing $m$.

The *restriction* of an $n$-ary relation $r^B \subseteq B^n$ to a subset $A \subseteq B$ is $r^A = r^B \cap A^n$. For instance, the restriction of the standard order of $\mathbb{R}$ to $\mathbb{N}$ is the standard order of $\mathbb{N}$. Only because of this fact can the same symbol be used to denote these relations. The restriction $f^A$ of an operation $f^B$ on $B$ to a set $A \subseteq B$ is defined analogously whenever $A$ is $f$-closed. Simply let $f^A\vec{a} = f^B\vec{a}$ for $\vec{a} \in A^n$. For instance, addition in $\mathbb{N}$ is the restriction of addition in $\mathbb{Z}$ to $\mathbb{N}$, or addition in $\mathbb{Z}$ is an extension of this operation in $\mathbb{N}$. Again, only this state of affairs allows us to denote the two operations by the same symbol.

Let $\mathcal{B}$ be an $L$-structure and let $A \subseteq B$ be nonempty and closed under all operations of $\mathcal{B}$; this will be taken to include $c^\mathcal{B} \in A$ for constant symbols $c \in L$. To such a subset $A$ corresponds in a natural way an $L$-structure $\mathcal{A} = (A, L^\mathcal{A})$, where $r^\mathcal{A}$ and $f^\mathcal{A}$ for $r, f \in L$ are the restrictions of $r^\mathcal{B}$ respectively $f^\mathcal{B}$ to $A$. Finally, let $c^\mathcal{A} = c^\mathcal{B}$ for $c \in L$. The structure $\mathcal{A}$ so defined is then called a *substructure* of $\mathcal{B}$, and $\mathcal{B}$ is called an *extension* of $\mathcal{A}$, in symbols $\mathcal{A} \subseteq \mathcal{B}$. This is a certain abuse of $\subseteq$ but it does not cause confusion, since the arguments indicate what is meant.

$\mathcal{A} \subseteq \mathcal{B}$ implies $A \subseteq B$ but not conversely, in general. For example, $\mathcal{A} = (\mathbb{N}, <, +, 0)$ is a substructure of $\mathcal{B} = (\mathbb{Z}, <, +, 0)$ since $\mathbb{N}$ is closed under addition in $\mathbb{Z}$ and $0$ has the same meaning in $\mathcal{A}$ and $\mathcal{B}$. Here we dropped the superscripts for $<$, $+$, and $0$ because there is no risk of misunderstanding.

A nonempty subset $G$ of the domain $B$ of a given $L$-structure $\mathcal{B}$ defines a smallest substructure $\mathcal{A}$ of $\mathcal{B}$ containing $G$. The domain of $\mathcal{A}$ is the smallest subset of $B$ containing $G$ and closed under all operations of $B$. $\mathcal{A}$ is called the substructure *generated from $G$ in $\mathcal{B}$*. For instance, $3\mathbb{N}$ $(= \{3n \mid n \in \mathbb{N}\})$ is the domain of the substructure generated from $\{3\}$ in $(\mathbb{N}, +, 0)$, since $3\mathbb{N}$ contains 0 and 3, is closed under $+$, and is clearly the smallest such subset of $\mathbb{N}$. A structure $\mathcal{A}$ is called *finitely generated* if for some finite $G \subseteq A$ the substructure generated from $G$ in $\mathcal{A}$ coincides with $\mathcal{A}$. For instance, $(\mathbb{Z}, +, -, 0)$ is finitely generated by $G = \{1\}$.

If $\mathcal{A}$ is an $L$-structure and $L_0 \subseteq L$ then the $L_0$-structure $\mathcal{A}_0$ with domain $A$ and where $\mathsf{s}^{\mathcal{A}_0} = \mathsf{s}^{\mathcal{A}}$ for all symbols $\mathsf{s} \in L_0$ is termed the *$L_0$-reduct of $\mathcal{A}$*, and $\mathcal{A}$ is called an *$L$-expansion* of $\mathcal{A}_0$. For instance, the group $(\mathbb{Z}, +, 0)$ is the $\{+, 0\}$-reduct of the ordered ring $(\mathbb{Z}, <, +, \cdot, 0)$. The notions reduct and substructure must clearly be distinguished. A reduct of $\mathcal{A}$ has always the same domain as $\mathcal{A}$, while the domain of a substructure of $\mathcal{A}$ is as a rule a proper subset of $A$.

Below we list some frequently cited properties of a binary relation $\triangleleft$ in a set $A$. It is convenient to write $a \triangleleft b$ instead of $(a, b) \in \triangleleft$, and $a \ntriangleleft b$ for $(a, b) \notin \triangleleft$. Just as $a < b < c$ often stands for $a < b \ \& \ b < c$, we write $a \triangleleft b \triangleleft c$ for $a \triangleleft b \ \& \ b \triangleleft c$. In the listing below, 'for all $a$' and 'there exists an $a$' respectively mean 'for all $a \in A$' and 'there exists some $a \in A$'. The relation $\triangleleft \subseteq A^2$ is called

| | | |
|---|---|---|
| *reflexive* | if | $a \triangleleft a$ for all $a$, |
| *irreflexive* | if | $a \ntriangleleft a$ for all $a$, |
| *symmetric* | if | $a \triangleleft b \Rightarrow b \triangleleft a$, for all $a, b$, |
| *antisymmetric* | if | $a \triangleleft b \triangleleft a \Rightarrow a = b$, for all $a, b$, |
| *transitive* | if | $a \triangleleft b \triangleleft c \Rightarrow a \triangleleft c$, for all $a, b, c$, |
| *connex* | if | $a = b$ or $a \triangleleft b$ or $b \triangleleft a$, for all $a, b$. |

Reflexive, transitive, and symmetric relations are also called *equivalence relations*. These are often denoted by $\sim$, $\approx$, $\equiv$, $\simeq$, or similar symbols. Such a relation generates a partition of its domain whose parts, consisting of mutually equivalent elements, are called *equivalence classes*.

We now present an overview of classes of structures to which we will later refer, mainly in Chapter **5**. Hence, for the time being, the beginner may skip the following and jump to **2.2**.

**1. Graphs, partial orders, and orders.** A relational structure $(A, \lhd)$ with some relation $\lhd \subseteq A^2$ is often termed a (directed) *graph*. If $\lhd$ is irreflexive and transitive we usually write $<$ for $\lhd$ and speak of a (strict) *partial order* or a *partially ordered set*, also called a *poset* for short. If we define $x \leqslant y$ by $x < y$ or $x = y$, then $\leqslant$ is reflexive, transitive, and antisymmetric, called a *reflexive partial order*, the one that belongs to $<$. If one starts with a reflexive partial order on $A$ and defines $x < y$ by $x \leqslant y \,\&\, x \neq y$, then $(A, <)$ is clearly a poset.

A connex partial order $\mathcal{A} = (A, <)$ is called a *total* or *linear* order, also termed an *ordered* or a *strictly ordered* set. $\mathbb{N}, \mathbb{Z}, \mathbb{Q}, \mathbb{R}$ are examples with respect to their standard orders. Here we follow the traditional habit of referring to ordered sets by their domains only.

Let $U$ be a nonempty subset of some ordered set $A$ such that for all $a, b \in A$, $a < b \in U \Rightarrow a \in U$. Such a $U$ is called an *initial segment* of $A$. In addition, let $V := A \backslash U \neq \emptyset$. Then the pair $(U, V)$ is called a *cut*. The cut is said to be a *gap* if $U$ has no largest and $V$ no smallest element. However, if $U$ *has* a largest element $a$, and $V$ a smallest element $b$, then $(U, V)$ is called a *jump*. $b$ is in this case called *the immediate successor* of $a$, and $a$ *the immediate predecessor* of $b$, because then there is no element from $A$ between $a$ and $b$. An infinite ordered set without gaps and jumps, like $\mathbb{R}$, is said to be *continuously ordered*. Such a set is easily seen to be *densely ordered*, i.e., between any two elements lies another one.

A totally ordered subset $K$ of a partially ordered set $H$ is called a *chain* in $H$. Such a $K$ is said to be *bounded* (to the above) if there is some $b \in H$ with $a \leqslant b$ for all $a \in K$. Call $c \in H$ *maximal* in $H$ if no $a \in H$ exists with $a > c$. An infinite partial order need not have a maximal element, nor need all chains be bounded, as is seen by the example $(\mathbb{N}, <)$. With these notions, a basic mathematical tool can now be stated:

**Zorn's lemma.** *If every chain in a nonempty poset $H$ is bounded then $H$ has a maximal element.*

A (totally) ordered set $A$ is *well-ordered* if every nonempty subset of $A$ has a smallest element; equivalently, there are no infinite decreasing sequences $a_0 > a_1 > \cdots$ of elements from $A$. Clearly, every finite ordered set is well-ordered. The simplest example of an infinite well-ordered set is $\mathbb{N}$ together with its standard order.

**2. Groupoids, semigroups, and groups.** Algebras $\mathcal{A} = (A, \circ)$ with an operation $\circ \colon A^2 \to A$ are termed *groupoids*. If $\circ$ is associative then $\mathcal{A}$ is called a *semigroup*, and if $\circ$ is additionally invertible, then $\mathcal{A}$ is said to be a *group*. It is provable that a group $(G, \circ)$ in this sense contains exactly one *unit element*, that is, an element $e$ such that $x \circ e = e \circ x = x$ for all $x \in G$, also called a *neutral element*. A well-known example is the group of bijections of a set $M$. If the group operation $\circ$ is commutative, we speak of a *commutative* or *abelian* group.

Here are some examples of semigroups that are not groups: (a) the set of strings on some alphabet $\mathsf{A}$ with respect to concatenation, the *word-semigroup* or *free semigroup generated from* $\mathsf{A}$. (b) the set $M^M$ of mappings from $M$ to itself with respect to composition. (c) $(\mathbb{N}, +)$ and $(\mathbb{N}, \cdot)$; these two are commutative semigroups. With the exception of $(M^M, \circ)$, all mentioned examples of semigroups are *regular*, which is to mean $x \circ y = x \circ z \Rightarrow y = z$, and $x \circ z = y \circ z \Rightarrow x = y$, for all $x, y, z$.

Substructures of semigroups are again semigroups. Substructures of groups are in general only semigroups, as seen from $(\mathbb{N}, +) \subseteq (\mathbb{Z}, +)$. Not so in the signature $\{\circ, e, ^{-1}\}$, where $e$ denotes the unit element and $x^{-1}$ the inverse of $x$. Here all substructures are indeed subgroups. The reason is that in $\{\circ, e, ^{-1}\}$, the group axioms can be written as universally quantified equations, where for brevity, we omit the writing of "for all $x, y, z$," namely as $x \circ (y \circ z) = (x \circ y) \circ z$, $x \circ e = x$, $x \circ x^{-1} = e$. These equations certainly retain their validity in the transition to substructures. We mention that from the last three equations, $e \circ x = x$ and $x^{-1} \circ x = e$ are derivable, although $\circ$ is not supposed to be commutative.

*Ordered* semigroups and groups possess along with $\circ$ some order, with respect to which $\circ$ is monotonic in both arguments, like $(\mathbb{N}, +, 0, \leqslant)$. A commutative ordered semigroup $(A, +, 0, \leqslant)$ with zero element $0$, which at the same time is the smallest element in $A$, and where $a \leqslant b$ iff there is some $c$ with $a + c = b$, is called a *domain of magnitude*. Everyday examples are the domains of *length, mass, money*, etc.

**3. Rings and fields.** These belong to the most commonly known structures. Below we list the axioms for the theory $T_F$ of fields in $+, \cdot, 0, 1$. A *field* is a model of $T_F$. A *ring* is a model of the axiom system $T_R$ for rings that derives from $T_F$ by dropping the constant $1$ from the signature and the axioms $\mathsf{N}^\times$, $\mathsf{C}^\times$, and $\mathsf{I}^\times$ from $T_F$. Here are the axioms of $T_F$:

$$\mathrm{N^+}: \quad x+0=x \qquad\qquad\qquad \mathrm{N^\times}: \quad x\cdot 1=x$$
$$\mathrm{C^+}: \quad x+y=y+x \qquad\qquad \mathrm{C^\times}: \quad x\cdot y=y\cdot x$$
$$\mathrm{A^+}: \quad (x+y)+z=x+(y+z) \quad \mathrm{A^\times}: \quad (x\cdot y)\cdot z=x\cdot(y\cdot z)$$
$$\mathrm{D}\ \ : \quad x\cdot(y+z)=x\cdot y+x\cdot z \quad \mathrm{D'}\ : \quad (y+z)\cdot x=y\cdot x+z\cdot x$$
$$\mathrm{I^+}: \quad \forall x\exists y\, x+y=0 \qquad\qquad \mathrm{I^\times}: \quad 0\neq 1 \wedge (\forall x\neq 0)\exists y\, x\cdot y=1$$

In view of $\mathrm{C^\times}$, axiom $\mathrm{D'}$ is dispensable for $T_F$ but not for $T_R$. When removing $\mathrm{I^+}$ from $T_R$, we obtain the theory of *semirings*. A well-known example is $(\mathbb{N},+,\cdot,0)$. A commutative ring that has a unit element 1 but no *zero-divisor* (i.e., $\neg\exists x\exists y(x,y\neq 0 \wedge x\cdot y=0)$ is called an *integral domain*. A typical example is $(\mathbb{Z},+,\cdot,0,1)$.

Let $\mathcal{K},\mathcal{K}'$ be any fields with $\mathcal{K}\subset\mathcal{K}'$. We call $a\in K'\backslash K$ *algebraic* or *transcendental* on $\mathcal{K}$, depending on whether $a$ is a zero of a polynomial with coefficients in $K$ or not. If every polynomial of degree $\geqslant 1$ with coefficients in $K$ breaks down into linear factors, as is the case for the field of complex numbers, then $\mathcal{K}$ is called *algebraically closed*, in short, $\mathcal{K}$ is a.c. These fields will be more closely inspected in **3.3** and Chapter **5**. Each field $\mathcal{K}$ has a smallest subfield $\mathcal{P}$, called a *prime field*. One says that $\mathcal{K}$ *has characteristic* 0 or $p$ (a prime number), depending on whether $\mathcal{P}$ is isomorphic to the field $\mathbb{Q}$ or the finite field of $p$ elements. No other prime fields exist. It is not hard to show that $\mathcal{K}$ has the characteristic $p$ iff the sentence $\mathrm{char}_p: \underbrace{1+\cdots+1}_{p}=0$ holds in $\mathcal{K}$.

Rings, fields, etc. may also be *ordered*, whereby the usual monotonicity laws are required. For example, $(\mathbb{Z},<,+,\cdot,0,1)$ is the *ordered ring* of integers and $(\mathbb{N},<,+,\cdot,0,1)$ the *ordered semiring* of natural numbers.

**4. Semilattices and lattices.** $\mathcal{A}=(A,\circ)$ is called a *semilattice* if $\circ$ is associative, commutative, and idempotent. An example is $(\{0,1\},\circ)$ with $\circ = \wedge$. If we define $a\leqslant b :\Leftrightarrow a\circ b=a$ then $\leqslant$ is a reflexive partial order on $A$. Reflexivity holds, since $a\circ a=a$. As can be easily verified, $a\circ b$ is in fact the *infimum* of $a,b$ with respect to $\leqslant$, $a\circ b=\inf\{a,b\}$, that is, $a\circ b\leqslant a,b$, and $c\leqslant a,b\Rightarrow c\leqslant a\circ b$, for all $a,b,c\in A$.

$\mathcal{A}=(A,\cap,\cup)$ is called a *lattice* if $(A,\cap)$ and $(A,\cup)$ are both semilattices and the following so-called *absorption laws* hold: $a\cap(a\cup b)=a$ and $a\cup(a\cap b)=a$. These imply $a\cap b=a\Leftrightarrow a\cup b=b$. As above, $a\leqslant b :\Leftrightarrow a\cap b=a$ defines a partial order such that $a\cap b=\inf\{a,b\}$.

In addition, one has $a \cup b = \sup\{a, b\}$ (the *supremum* of $a, b$), which is to mean $a, b \leqslant a \cup b$, and $a, b \leqslant c \Rightarrow a \cup b \leqslant c$, for all $c \in A$. If $\mathcal{A}$ satisfies, moreover, the *distributive laws* $x \cap (y \cup c) = (x \cap y) \cup (x \cap c)$ and $x \cup (y \cap c) = (x \cup y) \cap (x \cup c)$, then $\mathcal{A}$ is termed a *distributive* lattice. For instance, the power set $\mathfrak{P}M$ with the operations $\cap$ and $\cup$ for $\cap$ and $\cup$ respectively is a distributive lattice, as is every nonempty family of subsets of $M$ closed under $\cap$ and $\cup$, a so-called *lattice of sets*. Another important example is $(\mathbb{N}, \gcd, \text{lcm})$. Here $\gcd(a, b)$ and $\text{lcm}(a, b)$ denote the greatest common divisor and the least common multiple of $a, b \in \mathbb{N}$.

**5. Boolean algebras.** An algebra $\mathcal{A} = (A, \cap, \cup, \neg)$ where $(A, \cap, \cup)$ is a distributive lattice and in which at least the equations

$$\neg\neg x = x, \quad \neg(x \cap y) = \neg x \cup \neg y, \quad x \cap \neg x = y \cap \neg y$$

are valid is called a *Boolean algebra*. The paradigm structure is the two-element Boolean algebra $\mathbf{2} := (\{0, 1\}, \wedge, \vee, \neg)$, with $\cap, \cup$ interpreted as $\wedge, \vee$, respectively. One defines the constants 0 and 1 by $0 := a \cap \neg a$ for any $a \in A$ and $1 := \neg 0$. There are many ways to characterize Boolean algebras $\mathcal{A}$, for instance, by saying that $\mathcal{A}$ satisfies all equations valid in $\mathbf{2}$. The signature can also be variously selected. For example, the signature $\wedge, \vee, \neg$ is well suited to deal algebraically with two-valued propositional logic. Terms of this signature are, up to the denotation of variables, precisely the Boolean formulas from **1.1**, and a valid logical equivalence $\alpha \equiv \beta$ corresponds to the equation $\alpha = \beta$, valid in $\mathbf{2}$. Further examples of Boolean algebras are the *algebras of sets* $\mathcal{A} = (A, \cap, \cup, \neg)$. Here $A$ consists of a nonempty system of subsets of a set $I$, closed under $\cap$, $\cup$, and $\neg$ (complementation in $I$). These are the most general examples; a famous theorem, *Stone's representation theorem*, says that each Boolean algebra is isomorphic to an algebra of sets.

**6. Logical $L$-matrices.** These are structures $\mathcal{A} = (A, L^{\mathcal{A}}, D^{\mathcal{A}})$, where $L$ contains only operation symbols (the "logical" symbols) and $D$ denotes a unary predicate, *the set of distinguished values* of $\mathcal{A}$. Best known is the two-valued *Boolean matrix* $\mathcal{B} = (\mathbf{2}, D^{\mathcal{B}})$ with $D^{\mathcal{B}} = \{1\}$. The consequence relation $\vDash_{\mathcal{A}}$ in the propositional language $\mathcal{F}$ of signature $L$ is defined as in the two-valued case: Let $X \subseteq \mathcal{F}$ and $\varphi \in \mathcal{F}$. Then $X \vDash_{\mathcal{A}} \varphi$ if $w\varphi \in D^{\mathcal{A}}$ for every $w : PV \to A$ with $wX \subseteq D^{\mathcal{A}}$ ($wX := \{w\alpha \mid \alpha \in X\}$). In words, if the values of all $\alpha \in X$ are distinguished, then so too is the value of $\varphi$.

**Homomorphisms and isomorphisms.** The following notions are important for both mathematical and logical investigations. Much of the material presented here will be needed in Chapter **5**. In the following definition, $n$ $(>0)$ denotes as always the arity of $f$ or $r$.

**Definition.** Let $\mathcal{A}, \mathcal{B}$ be $L$-structures and $h: \mathcal{A} \to \mathcal{B}$ (strictly speaking $h: A \to B$) a mapping such that for all $f, c, r \in L$ and $\vec{a} \in A^n$,

(H): $hf^{\mathcal{A}}\vec{a} = f^{\mathcal{B}}h\vec{a}, \ hc^{\mathcal{A}} = c^{\mathcal{B}}, \ r^{\mathcal{A}}\vec{a} \Rightarrow r^{\mathcal{B}}h\vec{a} \quad \left(h\vec{a} = (ha_1, \ldots, ha_n)\right).$

Then $h$ is called a *homomorphism*. If the third condition in (H) is replaced by the stronger condition (S): $(\exists \vec{b} \in A^n)(h\vec{a}=h\vec{b} \ \& \ r^{\mathcal{A}}\vec{b}) \Leftrightarrow r^{\mathcal{B}}h\vec{a}$ [2] then $h$ is said to be a *strong homomorphism* (for algebras, the word "strong" is dispensable). An injective strong homomorphism $h: \mathcal{A} \to \mathcal{B}$ is called an *embedding* of $\mathcal{A}$ into $\mathcal{B}$. If, in addition, $h$ is bijective then $h$ is called an *isomorphism*, and in case $\mathcal{A} = \mathcal{B}$, an *automorphism*.

An embedding or isomorphism $h: \mathcal{A} \to \mathcal{B}$ satisfies $r^{\mathcal{A}}\vec{a} \Leftrightarrow r^{\mathcal{B}}h\vec{a}$. Indeed, since $h\vec{a}=h\vec{b} \Leftrightarrow \vec{a}=\vec{b}$, (S) yields $r^{\mathcal{B}}h\vec{a} \Rightarrow (\exists \vec{b} \in A^n)(\vec{a}=\vec{b} \ \& \ r^{\mathcal{A}}\vec{b}) \Rightarrow r^{\mathcal{A}}\vec{a}$. $\mathcal{A}, \mathcal{B}$ are said to be *isomorphic*, in symbols $\mathcal{A} \simeq \mathcal{B}$, if there is an isomorphism from $\mathcal{A}$ to $\mathcal{B}$. It is readily verified that $\simeq$ is reflexive, symmetric, and transitive, hence an equivalence relation on the class of all $\mathcal{L}$-structures.

**Examples 1.** (a) A valuation $w$ considered in **1.1** can be regarded as a homomorphism of the propositional formula algebra $\mathcal{F}$ into the two-element Boolean algebra $\mathbf{2}$. Such a $w: \mathcal{F} \to \mathbf{2}$ is necessarily onto.

(b) Let $\mathcal{A} = (A, *)$ be a word semigroup with the concatenation operation $*$ and $\mathcal{B}$ the additive semigroup of natural numbers, considered as $L$-structures for $L = \{\circ\}$ with $\circ^{\mathcal{A}} = *$ and $\circ^{\mathcal{B}} = +$. Let $\mathrm{lh}(\xi)$ denote the length of a word or string $\xi \in A$. Then $\xi \mapsto \mathrm{lh}(\xi)$ is a homomorphism since $\mathrm{lh}(\xi * \eta) = \mathrm{lh}(\xi) + \mathrm{lh}(\eta)$, for all $\xi, \eta \in A$. If $\mathcal{A}$ is generated from a single letter, $\mathrm{lh}$ is evidently bijective, hence an isomorphism.

(c) The mapping $a \mapsto (a, 0)$ from $\mathbb{R}$ to $\mathbb{C}$ (= set of complex numbers, understood as ordered pairs of real numbers) is a good example of an embedding of the field $\mathbb{R}$ into the field $\mathbb{C}$. Nonetheless, in this case, we are used to saying that $\mathbb{R}$ is a subfield of $\mathbb{C}$, and that $\mathbb{R}$ is a subset of $\mathbb{C}$.

---

[2] $(\exists \vec{b} \in A^n)(h\vec{a}=h\vec{b} \ \& \ r^{\mathcal{A}}\vec{b})$ abbreviates 'there is some $\vec{b} \in A^n$ with $h\vec{a} = h\vec{b}$ and $r^{\mathcal{A}}\vec{b}$'. If $h: \mathcal{A} \to \mathcal{B}$ is onto (and only this case will occur in our applications) then (S) is equivalent to the more suggestive condition $r^{\mathcal{B}} = \{h\vec{a} \mid r^{\mathcal{A}}\vec{a}\}$.

(d) Let $\mathcal{A} = (\mathbb{R}, +, <)$ be the ordered additive group of real numbers and $\mathcal{B} = (\mathbb{R}_+, \cdot, <)$ the multiplicative group of positive reals. Then for any $b \in \mathbb{R}_+ \backslash \{1\}$ there is precisely one isomorphism $\eta : \mathcal{A} \to \mathcal{B}$ such that $\eta 1 = b$, namely $\eta : x \mapsto b^x$, the exponential function $\exp_b$ to the base $b$. It is even possible to *define* $\exp_b$ as this isomorphism, by first proving that—up to isomorphism—there is only one continuously ordered abelian group (first noticed in [Ta2] though not explicitly put into words).

(e) The algebras $\mathcal{A} = (\{0, 1\}, +)$ and $\mathcal{B} = (\{0, 1\}, \leftrightarrow)$ are only apparently different, but are in fact isomorphic, with the isomorphism $\delta$ where $\delta 0 = 1$, $\delta 1 = 0$. Thus, since $\mathcal{A}$ is a group, $\mathcal{B}$ is a group as well, which is not obvious at first glance. By adjoining the unary predicate $D = \{1\}$, $\mathcal{A}$ and $\mathcal{B}$ become (nonisomorphic) logical matrices. These actually define the two "dual" fragmentary two-valued logics for the connectives *either ... or ...*, and *... if and only if ...*, which have many properties in common.

**Congruences.** A *congruence relation* (or simply a *congruence*) in a structure $\mathcal{A}$ of signature $L$ is an equivalence relation $\approx$ in $A$ such that for all $n > 0$, all $f \in L$ of arity $n$, and all $\vec{a}, \vec{b} \in A^n$,

$$\vec{a} \approx \vec{b} \ \Rightarrow \ f^A \vec{a} \approx f^A \vec{b}.$$

Here $\vec{a} \approx \vec{b}$ means $a_i \approx b_i$ for $i = 1, \ldots, n$. A trivial example is the identity in $\mathcal{A}$. If $h : \mathcal{A} \to \mathcal{B}$ is a homomorphism then $\approx_h \subseteq A^2$, defined by $a \approx_h b \Leftrightarrow ha = hb$, is a congruence in $\mathcal{A}$, called the *kernel* of $h$. Let $A'$ be the set of equivalence classes $a/\approx := \{x \in A \mid a \approx x\}$ for $a \in A$, also called the *congruence classes* of $\approx$, and set $\vec{a}/\approx := (a_1/\approx, \ldots, a_n/\approx)$ for $\vec{a} \in A^n$. Define $f^{A'}(\vec{a}/\approx) := (f^A \vec{a})/\approx$ and let $r^{A'} \vec{a}/\approx :\Leftrightarrow (\exists \vec{b} \approx \vec{a}) r^A \vec{b}$. These definitions are *sound*, that is, independent of the choice of the $n$-tuple $\vec{a}$ of representatives. Then $A'$ becomes an $L$-structure $\mathcal{A}'$, the *factor structure of $\mathcal{A}$ modulo $\approx$*, denoted by $\mathcal{A}/\approx$. Interesting, in particular for Chapter **5**, is the following very general and easily provable

**Homomorphism theorem.** *Let $\mathcal{A}$ be $L$-structure and $\approx$ a congruence in $\mathcal{A}$. Then $k : a \mapsto a/\approx$ is a strong homomorphism from $\mathcal{A}$ onto $\mathcal{A}/\approx$, the canonical homomorphism. Conversely, if $h : \mathcal{A} \to \mathcal{B}$ is a strong homomorphism from $\mathcal{A}$ onto an $\mathcal{L}$-structure $\mathcal{B}$ with kernel $\approx$ then $\imath : a/\approx \mapsto ha$ is an isomorphism from $\mathcal{A}/\approx$ to $\mathcal{B}$, and $h = \imath \circ k$.*

**Proof.** We omit here the superscripts for $f$ and $r$ just for the sake of legibility. Clearly, $kf\vec{a} = (f\vec{a})/\approx = f(\vec{a}/\approx) = fk\vec{a} \ (=f(ka_1, \ldots, ka_n))$, and $(\exists \vec{b} \in A^n)(k\vec{a} = k\vec{b}\ \&\ r\vec{b}) \Leftrightarrow (\exists \vec{b} \approx \vec{a})r\vec{b} \Leftrightarrow r\,\vec{a}/\approx\ \Leftrightarrow r\,k\vec{a}$ by definition. Hence $k$ is what we claimed. The definition of $\imath$ is sound, and $\imath$ is bijective since $ha = hb \Rightarrow a/\approx\ =\ b/\approx$. Furthermore, $\imath$ is an isomorphism because

$$\imath f(\vec{a}/\approx) = hf\vec{a} = f h\vec{a} = f\imath(\vec{a}/\approx)\ \text{and}\ r\,\vec{a}/\approx\ \Leftrightarrow r\,h\vec{a} \Leftrightarrow r\,\imath(\vec{a}/\approx).$$

Finally, $h$ is the composition $\imath \circ k$ by the definitions of $\imath$ and $k$. ◻

**Remark.** For algebras $\mathcal{A}$, this theorem is the usual homomorphism theorem of universal algebra. $\mathcal{A}/\approx$ is then named the *factor algebra*. The theorem covers groups, rings, etc. In groups, the kernel of a homomorphism is already determined by the congruence class of the unit element, called a *normal subgroup*, in rings by the congruence class of 0, called an *ideal*. Hence, in textbooks on basic algebra the homomorphism theorem is separately formulated for groups and rings, but is easily derivable from the general theorem present here.

**Direct products.** These provide the basis for many constructions of new structures, especially in **5.7**. A well-known example is the $n$-dimensional vector group $(\mathbb{R}^n, 0, +)$. This is the $n$-fold direct product of the group $(\mathbb{R}, 0, +)$ with itself. The addition in $\mathbb{R}^n$ is defined componentwise, as is also the case in the following

**Definition.** Let $(\mathcal{A}_i)_{i \in I}$ be a nonempty family of $L$-structures. The *direct product* $\mathcal{B} = \prod_{i \in I} \mathcal{A}_i$ is the structure defined as follows: Its domain is $B = \prod_{i \in I} A_i$, called the *direct product* of the sets $A_i$. The elements $a = (a_i)_{i \in I}$ of $B$ are functions defined on $I$ with $a_i \in A_i$ for each $i \in I$. Relations and operations in $\mathcal{B}$ are defined componentwise, that is,

$$r^{\mathcal{B}}\vec{a} \Leftrightarrow r^{\mathcal{A}_i}\vec{a}_i\ \text{for all}\ i \in I, \quad f^{\mathcal{B}}\vec{a} = (f^{\mathcal{A}_i}\vec{a}_i)_{i \in I}, \quad c^{\mathcal{B}} = (c^{\mathcal{A}_i})_{i \in I},$$

where $\vec{a} = (a^1, \ldots, a^n) \in B^n$ (here the superscripts count the components) with $a^\nu := (a_i^\nu)_{i \in I}$ for $\nu = 1, \ldots, n$, and $\vec{a}_i := (a_i^1, \ldots, a_i^n) \in A_i^n$.

Whenever $\mathcal{A}_i = \mathcal{A}$ for all $i \in I$, then $\prod_{i \in I} \mathcal{A}_i$ is denoted by $\mathcal{A}^I$ and called a *direct power* of the structure $\mathcal{A}$. Note that $\mathcal{A}$ is embedded in $\mathcal{A}^I$ by the mapping $a \mapsto (a)_{i \in I}$, where $(a)_{i \in I}$ denotes the $I$-tuple with the constant value $a$, that is, $(a)_{i \in I} = (a, a, \ldots)$. For $I = \{1, \ldots, m\}$, the product $\prod_{i \in I} \mathcal{A}_i$ is also written as $\mathcal{A}_1 \times \cdots \times \mathcal{A}_m$. If $I = \{0, \ldots, n-1\}$ one mostly writes $\mathcal{A}^n$ for $\mathcal{A}^I$.

**Examples 2.** (a) Let $I = \{1, 2\}$, $\mathcal{A}_i = (A_i, <^i)$, and $\mathcal{B} = \prod_{i \in I} \mathcal{A}_i$. Then $a <^{\mathcal{B}} b \Leftrightarrow a_1 <^1 b_1\ \&\ a_2 <^2 b_2$, for all $a, b \in B = A_1 \times A_2$. Note that if $\mathcal{A}_1, \mathcal{A}_2$ are ordered sets then $\mathcal{B}$ is only a partial order. The deeper reason for this observation will become clear in Chapter **5**.

(b) Let $\mathcal{B} = 2^I$ be a direct power of the two-element Boolean algebra $2$. The elements $a \in B$ are $I$-tuples of 0 and 1. These uniquely correspond to the subsets of $I$ via the mapping $\imath: a \mapsto I_a := \{i \in I \mid a_i = 1\}$. As a matter of fact, $\imath$ is an isomorphism from $\mathcal{B}$ to $(\mathfrak{P}I, \cap, \cup, \neg)$, as can readily be verified; Exercise 4.

### Exercises

1. Show that there are (up to isomorphism) exactly five two-element proper groupoids. Here a groupoid $(H, \cdot)$ is termed *proper* if the operation $\cdot$ is essentially binary.

2. $\approx$ $(\subseteq A^2)$ is termed *Euclidean* if $a \approx b$ & $a \approx c \Rightarrow b \approx c$, for all $a, b, c \in A$. Show that $\approx$ is an equivalence relation in $A$ if and only if $\approx$ is reflexive and Euclidean.

3. Prove that an equivalence relation $\approx$ on an algebraic $L$-structure $\mathcal{A}$ is a congruence iff for all $f \in L$ of arity $n$, all $i = 1, \ldots, n$, and all $a_1, \ldots, a_{i-1}, a, a', a_{i+1}, \ldots, a_n \in A$ with $a \approx a'$,
$$f(a_1, \ldots, a_{i-1}, a, a_{i+1}, \ldots, a_n) \approx f(a_1, \ldots, a_{i-1}, a', a_{i+1}, \ldots, a_n).$$

4. Prove in detail that $2^I \simeq (\mathfrak{P}I, \cap, \cup, \neg)$ for a nonempty index set $I$. Prove the corresponding statement for any subalgebra of $2^I$.

5. Show that $h: \prod_{i \in I} \mathcal{A}_i \to \mathcal{A}_j$ with $ha = a_j$ is a homomorphism for each $j \in I$.

## 2.2  Syntax of First-Order Languages

Standard mathematical language enables us to talk precisely about structures, such as the field of real numbers. However, for logical (and meta-mathematical) issues it is important to delimit the theoretical framework to be considered; this is achieved most simply by means of a formalization. In this way one obtains an *object language*; that is, the formalized elements of the language, such as the components of a structure, are *objects* of our consideration. To formalize interesting properties of a structure in this language, one requires at least variables for the elements of its domain, called *individual variables*. Further are required sufficiently many logical symbols, along with symbols for the distinguished relations, functions, and constants of the structure. These *extralogical* symbols constitute the signature $L$ of the formal language that we are going to define.

In this manner one arrives at the *first-order languages*, also termed *elementary* languages. Nothing is lost in terms of generality if the set of variables is the same for all elementary languages; we denote this set by *Var* and take it to consist of the countably many symbols $v_0, v_1, \ldots$ Two such languages therefore differ only in the choice of their extralogical symbols. Variables for subsets of the domain are consciously excluded, since languages containing variables both for individuals and sets of these individuals—second-order languages, discussed in **3.8**—have different semantic properties from those investigated here.

We first determine the *alphabet*, the set of *basic symbols* of a first-order language determined by a signature $L$. It includes, of course, the already specified variables $v_0, v_1, \ldots$ In what follows, these will mostly be denoted by $x, y, z, u, v$, though sometimes other letters with or without indices may serve the same purpose. The boldface printed original variables are useful in writing down a formula in the variables $v_{i_1}, \ldots, v_{i_n}$, for these can then be denoted, for instance, by $v_1, \ldots, v_n$, or by $x_1, \ldots, x_n$.

Further, the *logical* symbols $\wedge$ (and), $\neg$ (not), $\forall$ (*for all*), the equality sign $=$, and, of course, all extralogical symbols from $L$ should belong to the alphabet. Note that the boldface symbol $=$ is taken as a basic symbol; simply taking $=$ could lead to unintended mix-ups with the metamathematical use of the equality symbol $=$ (in Chapter **4** also *identity-free* languages without $=$ will be considered). Finally, the parentheses $( , )$ are included in the alphabet. Other symbols are introduced by definition, e.g., $\vee$, $\rightarrow$, $\leftrightarrow$ are defined as in **1.4** and the symbols $\exists$ (*there exists*) and $\exists!$ (*there exists exactly one*) will be defined later. Let $\mathcal{S}_\mathcal{L}$ denote the set of all strings made up of symbols that belong to the alphabet of $\mathcal{L}$.

From the set $\mathcal{S}_\mathcal{L}$ of all strings we pick out the meaningful ones, namely terms and formulas, according to certain rules. A term, under an interpretation of the language, will always denote an element of a domain, provided an assignment of the occurring variables to elements of that domain has been given. In order to keep the syntax as simple as possible, terms will be understood as certain parenthesis-free strings, although this kind of writing may look rather unusual at the first glance.

**Terms in $L$:**

(T1) Variables and constants, considered as atomic strings, are terms, also called *prime terms*.

(T2) If $f \in L$ is $n$-ary and $t_1, \ldots, t_n$ are terms, then $ft_1 \cdots t_n$ is a term.

This is a recursive definition of the set of terms as a subset of $\mathcal{S}_{\mathcal{L}}$. Any string that is not generated by (T1) and (T2) is not a term in this context (cf. the related definition of $\mathcal{F}$ in **1.1**). Parenthesis-free term notation simplifies the syntax, but for binary operations we proceed differently in practice and write, for example, the term $\cdot + xyz$ as $(x + y) \cdot z$. The reason is that a high density of information in the notation complicates reading. Our brain does not process information sequentially like a computer. Officially, terms are parenthesis-free, and the parenthesized notation is just an alternative way of rewriting terms. Similarly to the unique reconstruction property of propositional formulas in **1.1**, here the *unique term reconstruction* property holds, that is,

$$ft_1 \cdots t_n = fs_1 \cdots s_n \text{ implies } s_i = t_i \text{ for } i = 1, \ldots, n \quad (t_i, s_i \text{ terms}),$$

which immediately follows from the *unique term concatenation* property

$$t_1 \cdots t_n = s_1 \cdots s_m \text{ implies } n = m \text{ and } t_i = s_i \text{ for } i = 1, \ldots, n.$$

The latter is shown in Exercise 2. $\mathcal{T}\ (= \mathcal{T}_L)$ denotes the set of all terms of a given signature $L$. Variable-free terms, which can exist only with the availability of constant symbols, are called *constant terms* or *ground terms*, mainly in logic programming. With the operations given in $\mathcal{T}$ by setting $f^{\mathcal{T}}(t_1, \ldots, t_n) = ft_1 \cdots t_n$, $\mathcal{T}$ forms an algebra, the *term algebra*. From the definition of terms immediately follows the useful

**Principle of proof by term induction.** *Let $\mathcal{E}$ be a property of strings such that $\mathcal{E}$ holds for all prime terms, and for each $n > 0$ and each $n$-ary function symbol $f$, the assumptions $\mathcal{E}t_1, \ldots, \mathcal{E}t_n$ imply $\mathcal{E}ft_1 \cdots t_n$. Then all terms have the property $\mathcal{E}$.*

Indeed, $\mathcal{T}$ is by definition the smallest set of strings satisfying the conditions of this principle, and hence a subset of the set of all strings with the property $\mathcal{E}$. A simple application of term induction is the proof that each compound term $t$ is a *function term* in the sense that $t = ft_1 \cdots t_n$ for some $n$-ary function symbol $f$ and some terms $t_1, \ldots, t_n$. Simply consider the property '$t$ is either prime or a function term'. Term induction can also be executed on certain subsets of $\mathcal{T}$, for instance on ground terms.

We also have at our disposal a *definition principle* by term recursion which, rather than defining it generally, we present through examples. The set $\text{var}\, t$ of variables occurring in a term $t$ is recursively defined by

$$\text{var}\, c = \emptyset \ ; \quad \text{var}\, x = \{x\} \ ; \quad \text{var}\, ft_1 \cdots t_n = \text{var}\, t_1 \cup \cdots \cup \text{var}\, t_n.$$

$\text{var}\, t$, and even $\text{var}\, \xi$ for any $\xi \in \mathcal{S}_{\mathcal{L}}$, can also be defined explicitly using concatenation. $\text{var}\, \xi$ is the set of all $x \in \text{Var}$ for which there are strings $\eta, \vartheta$ with $\xi = \eta x \vartheta$. The notion of a *subterm* of a term can also be defined recursively. Again, we can also do it more briefly using concatenation. Definition by term induction should more precisely be called *definition by term recursion*. But most authors are sloppy in this respect.

We now define recursively those strings of the alphabet of $L$ to be called *formulas*, also termed (first-order) *expressions* or *well-formed formulas*.

**Formulas in $L$:**

(F1) If $s, t$ are terms, then the string $s = t$ is a formula.

(F2) If $t_1, \ldots, t_n$ are terms and $r \in L$ is $n$-ary, then $rt_1 \cdots t_n$ is a formula.

(F3) If $\alpha, \beta$ are formulas and $x$ is a variable, then $(\alpha \wedge \beta)$, $\neg\alpha$, and $\forall x \alpha$ are formulas.

Any string not generated according to (F1), (F2), (F3) is in this context not a formula. Other logical symbols serve throughout merely as abbreviations, namely $\exists x \alpha := \neg \forall x \neg \alpha$, $(\alpha \vee \beta) := \neg(\neg\alpha \wedge \neg\beta)$, and as in **1.1**, $(\alpha \rightarrow \beta) := \neg(\alpha \wedge \neg\beta)$, and $(\alpha \leftrightarrow \beta) := ((\alpha \rightarrow \beta) \wedge (\beta \rightarrow \alpha))$. In addition, $s \neq t$ will throughout be written for $\neg\, s = t$. The formulas $\forall x \alpha$ and $\exists x \alpha$ are said to arise from $\alpha$ by *quantification*.

**Examples.** (a) $\forall x \exists y\, x + y = 0$ (more explicitly, $\forall x \neg \forall y \neg\, x + y = 0$) is a formula, expressing 'for all $x$ there exists a $y$ such that $x + y = 0$'. Here we assume tacitly that $x, y$ denote distinct variables. The same is assumed in all of the following whenever this can be made out from the context.

(b) $\forall x \forall x\, x = y$ is a formula, since repeated quantification of the same variable is not forbidden. $\forall z\, x = y$ is a formula also if $z \neq x, y$, although $z$ does then not appear in the formula $x = y$.

Example (b) indicates that the grammar of our formal language is more liberal than one might expect. This will spare us a lot of writing. The formulas $\forall x \forall x\, x = y$ and $\exists x \forall x\, x = y$ both have the same meaning as $\forall x\, x = y$.

These three formulas are logically equivalent (in a sense still to be defined), as are $\forall z\, x = y$ and $x = y$. It would be to our disadvantage to require any restriction here. In spite of this liberality, the formula syntax corresponds roughly to the syntax of natural language.

The formulas procured by (F1) and (F2) are said to be *prime* or *atomic* formulas, or simply called *prime*. As in propositional logic, prime formulas and their negations are called *literals*.

Prime formulas of the form $s = t$ are called *equations*. These are the only prime formulas if $L$ contains no relation symbols, in which case $L$ is called an *algebraic* signature. Prime formulas that are not equations begin with a relation symbol, although in practice a binary symbol tends to separate the two arguments as, for example, in $x \leqslant y$. The official notation is, however, that of clause (F2). The unique term concatenation property clearly implies the *unique prime formula reconstruction* property

$$rt_1 \cdots t_n = rs_1 \cdots s_n \text{ implies } t_i = s_i \text{ for } i = 1, \ldots, n.$$

The set of all formulas in $L$ is denoted by $\mathcal{L}$. If $L = \{\in\}$ or $L = \{\circ\}$ then $\mathcal{L}$ is also denoted by $\mathcal{L}_\in$ or $\mathcal{L}_\circ$, respectively. If $L$ is more complex, e.g. $L = \{\circ, e\}$, we write $\mathcal{L} = \mathcal{L}\{\circ, e\}$. The case $L = \emptyset$ is also permitted; it defines the *language of pure identity*, denoted by $\mathcal{L}_=$.

Instead of terms, formulas, and structures of signature $L$, we will talk of $\mathcal{L}$-terms (writing $\mathcal{T}_\mathcal{L}$ for $\mathcal{T}_L$), $\mathcal{L}$-formulas, and $\mathcal{L}$-structures respectively. We also omit the prefix if $\mathcal{L}$ has been given earlier and use the same conventions of parenthesis economy as in **1.1**. We will also allow ourselves other informal aids in order to increase readability. For instance, variously shaped brackets may be used as in $\forall x \exists y \forall z [z \in y \leftrightarrow \exists u (z \in u \wedge u \in x)]$. Even verbal descriptions (partial or complete) are permitted, as long as the intended formula is uniquely recognizable.

The strings $\forall x$ and $\exists x$ (read "for all $x$" respectively "there is an $x$") are called *prefixes*. Also concatenations of these such as $\forall x \exists y$ are prefixes. No other prefixes are considered here. Formulas in which $\forall, \exists$ do not occur are termed *quantifier-free* or *open*. These are the Boolean combinations of prime formulas. Generally, the *Boolean combinations* of formulas from a set $X \subseteq \mathcal{L}$ are the ones generated by $\neg$, $\wedge$ (and $\vee$) from those of $X$.

$X, Y, Z$ always denote sets of formulas, $\alpha, \beta, \gamma, \delta, \pi, \varphi, \ldots$ denote formulas, and $s, t$ terms, while $\Phi, \Psi$ are reserved to denote finite sequences of formulas and formal proofs. Substitutions (to be defined below) will be denoted by $\sigma, \tau, \omega, \rho$, and $\iota$.

Principles of *proof by formula induction* and of *definition by formula induction* (more precisely *formula recursion*) also exist for first-order and other formal languages. After the explanation of these principles for propositional languages in **1.1**, it suffices to present here some examples, adhering to the maxim **verba docent, exempla trahunt**. Formula recursion is based on the *unique formula reconstruction*, which is similar to the corresponding property in **1.1**: Each composed $\varphi \in \mathcal{L}$ can uniquely be written as $\varphi = \neg\alpha$, $\varphi = (\alpha \wedge \beta)$, or $\forall x \alpha$ for some $\alpha, \beta \in \mathcal{L}$ and $x \in \text{Var}$. A simple example of a recursive definition is $\text{rk}\,\varphi$, the *rank* of a formula $\varphi$. Starting with $\text{rk}\,\pi = 0$ for prime formulas $\pi$ it is defined as on page 8, with the additional clause $\text{rk}\,\forall x \alpha = \text{rk}\,\alpha + 1$. Functions on $\mathcal{L}$ are sometimes defined by recursion on $\text{rk}\,\varphi$, not on $\varphi$, as for instance on page 60.

Useful for some purposes is also the *quantifier rank*, $\text{qr}\,\varphi$. It represents a measure of nested quantifiers in $\varphi$. For prime $\pi$ let $\text{qr}\,\pi = 0$, and let $\text{qr}\,\neg\alpha = \text{qr}\,\alpha$, $\text{qr}(\alpha \wedge \beta) = \max\{\text{qr}\,\alpha, \text{qr}\,\beta\}$, $\text{qr}\,\forall x \alpha = \text{qr}\,\alpha + 1$.

Note that $\text{qr}\,\exists x \varphi = \text{qr}\,\neg\forall x \neg\varphi = \text{qr}\,\forall x \varphi$. A *subformula* of a formula is defined analogously to the definition in **1.1**. Hence, we need say no more on this. We write $x \in \text{bnd}\,\varphi$ (or $x$ occurs bound in $\varphi$) if $\varphi$ contains the prefix $\forall x$. In subformulas of $\varphi$ of the form $\forall x \alpha$, the formula $\alpha$ is called the *scope* of $\forall x$. The same prefix can occur repeatedly and with nested scopes in $\varphi$, as for instance in $\forall x(\forall x\, x = 0 \wedge x < y)$. In practice we avoid this way of writing, though for a computer this would pose no problem.

Intuitively, the formulas (a) $\forall x \exists y\, x + y = 0$ and (b) $\exists y\, x + y = 0$ are different in that in every context with a given meaning for $+$ and $0$, the former is either true or false, whereas in (b) the variable $x$ is waiting to be assigned a value. One also says that all variables in (a) are bound, while (b) contains the "free" variable $x$. The syntactic predicate '$x$ occurs free in $\varphi$', or '$x \in \text{free}\,\varphi$' is defined inductively: Let $\text{free}\,\alpha = \text{var}\,\alpha$ for prime formulas $\alpha$ ($\text{var}\,\alpha$ was defined on page 56), and

$$\text{free}\,(\alpha \wedge \beta) = \text{free}\,\alpha \cup \text{free}\,\beta, \quad \text{free}\,\neg\alpha = \text{free}\,\alpha, \quad \text{free}\,\forall x \alpha = \text{free}\,\alpha \setminus \{x\}.$$

For instance, $\text{free}\,(\forall x \exists y\, x + y = 0) = \emptyset$, while $\text{free}\,(x \leqslant y \wedge \forall x \exists y\, x + y = 0)$ equals $\{x, y\}$. As the last formula shows, $x$ can occur both free and bound in a formula. This too will be avoided in practice whenever possible. In some proof-theoretically oriented presentations, even different symbols are chosen for free and bound variables. Each of these approaches has its advantages and its disadvantages.

Formulas without free variables are called *sentences*, or *closed formulas*. $1+1=0$ and $\forall x \exists y \, x + y = 0$ $(= \forall x \neg \forall y \neg x + y = 0)$ are examples. Throughout take $\mathcal{L}^0$ to denote the set of all sentences of $\mathcal{L}$. More generally, let $\mathcal{L}^k$ be the set of all formulas $\varphi$ such that $free\,\varphi \subseteq Var_k := \{v_0, \ldots, v_{k-1}\}$. Clearly, $\mathcal{L}^0 \subseteq \mathcal{L}^1 \subseteq \cdots$ and $\mathcal{L} = \bigcup_{k \in \mathbb{N}} \mathcal{L}^k$.

At this point we meet a for the remainder of the book valid

**Convention.** As long as not otherwise stated, the notation $\varphi = \varphi(x)$ means that the formula $\varphi$ contains at most $x$ as a free variable; more generally, $\varphi = \varphi(x_1, \ldots, x_n)$ or $\varphi = \varphi(\vec{x})$ is to mean $free\,\varphi \subseteq \{x_1, \ldots, x_n\}$, where $x_1, \ldots, x_n$ stand for arbitrary but distinct variables. Not all of these variables need actually occur in $\varphi$. Further, $t = t(\vec{x})$ for terms $t$ is to be read completely analogously.

The term $ft_1 \cdots t_n$ is often denoted by $f\vec{t}$, the prime formula $rt_1 \cdots t_n$ by $r\vec{t}$. Here $\vec{t}$ denotes the string concatenation $t_1 \cdots t_n$. Fortunately, $\vec{t}$ behaves exactly like the sequence $(t_1, \ldots, t_n)$ as was pointed out already; it has the unique term concatenation property, see page 55.

**Substitutions.** We begin with the substitution $\frac{t}{x}$ of some term $t$ for a single variable $x$, called a *simple substitution*. Put intuitively, $\varphi \frac{t}{x}$ (also denoted by $\varphi_x(t)$ and read "$\varphi \, t$ for $x$") is the formula that results from replacing all free occurrences of $x$ in $\varphi$ by the term $t$. This intuitive characterization is made precise recursively, first for terms by

$$x \tfrac{t}{x} = t, \quad y \tfrac{t}{x} = y \ (x \neq y), \quad c \tfrac{t}{x} = c, \quad (ft_1 \cdots t_n) \tfrac{t}{x} = ft'_1 \cdots t'_n,$$

where, for brevity, $t'_i$ stands for $t_i \frac{t}{x}$, and next for formulas as follows:

$$(t_1 = t_2)\tfrac{t}{x} = t'_1 = t'_2, \quad (r\vec{t})\tfrac{t}{x} = rt'_1 \cdots t'_n, \qquad (\forall y \alpha)\tfrac{t}{x} = \begin{cases} \forall y \alpha \text{ if } x = y, \\ \forall y (\alpha \tfrac{t}{x}) \text{ otherwise.} \end{cases}$$
$$(\alpha \wedge \beta)\tfrac{t}{x} = \alpha \tfrac{t}{x} \wedge \beta \tfrac{t}{x}, \quad (\neg \alpha)\tfrac{t}{x} = \neg(\alpha \tfrac{t}{x}),$$

Then also $(\alpha \to \beta)\frac{t}{x} = \alpha \frac{t}{x} \to \beta \frac{t}{x}$, and the corresponding holds for $\vee$, while $(\exists y \alpha)\frac{t}{x} = \exists y \alpha$ for $y = x$, and $\exists y(\alpha \frac{t}{x})$ otherwise. Simple substitutions are special cases of so-called *simultaneous substitutions*

$$\varphi \tfrac{t_1 \cdots t_n}{x_1 \cdots x_n} \qquad (x_1, \ldots, x_n \text{ distinct}).$$

For brevity, this will be written $\varphi \frac{\vec{t}}{\vec{x}}$ or $\varphi_{\vec{x}}(\vec{t})$ or just $\varphi(\vec{t})$, provided there is no danger of misunderstanding. Here the variables $x_i$ are simultaneously replaced by the terms $t_i$ at free occurrences. Simultaneous substitutions easily generalize to *global* substitutions $\sigma$. Such a $\sigma$ assigns to *every* variable $x$ a term $x^\sigma \in \mathcal{T}$. It extends to the whole of $\mathcal{T}$ by the clauses $c^\sigma = c$ and $(f\vec{t})^\sigma = ft_1^\sigma \cdots t_n^\sigma$, and subsequently to $\mathcal{L}$ by recursion on $\mathrm{rk}\,\varphi$, so that $\sigma$ is defined for the whole of $\mathcal{T} \cup \mathcal{L}$: $(t_1 = t_2)^\sigma = t_1^\sigma = t_2^\sigma$, $(r\vec{t})^\sigma = rt_1^\sigma \cdots t_n^\sigma$, $(\alpha \wedge \beta)^\sigma = \alpha^\sigma \wedge \beta^\sigma$, $(\neg\alpha)^\sigma = \neg\alpha^\sigma$, and $(\forall x\varphi)^\sigma = \forall x\varphi^\tau$, where $\tau$ is defined by $x^\tau = x$ and $y^\tau = y^\sigma$ for $y \neq x$.[3]

These clauses cover also the case of a simultaneous substitution, because $\frac{\vec{t}}{\vec{x}}$ can be identified with the global substitution $\sigma$ such that $x_i^\sigma = t_i$ for $i = 1, \ldots, n$ and $x^\sigma = x$ otherwise. In other words, a simultaneous substitution can be understood as a global substitution $\sigma$ such that $x^\sigma = x$ for *almost all* variables $x$, i.e., with the exception of finitely many. The *identical substitution*, always denoted by $\iota$, is defined by $x^\iota = x$ for all $x$; hence $t^\iota = t$ and $\varphi^\iota = \varphi$ for all terms $t$ and formulas $\varphi$.

Clearly, a global substitution yields *locally*, i.e. with respect to individual formulas, the same as a suitable simultaneous substitution. Moreover, it will turn out below that simultaneous substitutions are products of simple ones. Nonetheless, a separate study of simultaneous substitutions is useful mainly for Chapter **4**.

It always holds that $\frac{t_1 t_2}{x_1 x_2} = \frac{t_2 t_1}{x_2 x_1}$, whereas the compositions $\frac{t_1}{x_1}\frac{t_2}{x_2}$ and $\frac{t_2}{x_2}\frac{t_1}{x_1}$ are distinct, in general. Let us elaborate by explaining the difference between $\varphi\frac{t_1 t_2}{x_1 x_2}$ and $\varphi\frac{t_1}{x_1}\frac{t_2}{x_2}$ $\left(= (\varphi\frac{t_1}{x_1})\frac{t_2}{x_2}\right)$. For example, if one wants to swap $x_1, x_2$ at their free occurrences in $\varphi$ then the desired formula is $\varphi\frac{x_2 x_1}{x_1 x_2}$, but not, in general, $\varphi\frac{x_2}{x_1}\frac{x_1}{x_2}$ (choose for instance $\varphi = x_1 < x_2$). Rather $\varphi\frac{x_2 x_1}{x_1 x_2} = \varphi\frac{y}{x_2}\frac{x_2}{x_1}\frac{x_1}{y}$ for any $y \notin \mathrm{var}\,\varphi \cup \{x_1, x_2\}$, as is readily shown by induction on $\varphi$ after first treating terms. We recommend to carry out this induction in detail. In the same way we obtain

(1)    $\varphi\frac{\vec{t}}{\vec{x}} = \varphi\frac{y}{x_n}\frac{t_1 \cdots t_{n-1}}{x_1 \cdots x_{n-1}}\frac{t_n}{y}$     ($y \notin \mathrm{var}\,\varphi \cup \mathrm{var}\,\vec{x} \cup \mathrm{var}\,\vec{t}, \ n \geqslant 2$).

This formula shows that a simultaneous substitution is a suitable product (composition) of simple substitutions. Conversely, it can be shown that each such product can be written as a single simultaneous substitution. In some cases (1) can be simplified. Useful, for example, is the following equation which holds in particular when all terms $t_i$ are variable-free:

(2)    $\varphi\frac{\vec{t}}{\vec{x}} = \varphi\frac{t_1}{x_1} \cdots \frac{t_n}{x_n}$     ($x_i \notin \mathrm{var}\,t_j$ for $i \neq j$).

---

[3] Since $\mathrm{rk}\,\varphi < \mathrm{rk}\,\forall x\varphi$, we may assume according to the recursive construction of $\sigma$ that $\varphi^\tau$ is already defined for all global substitutions $\tau$.

Getting on correctly with substitutions is not altogether simple; it requires practice, because our ability to regard complex strings is not especially trustworthy. A computer is not only much faster but also more reliable in this respect.

### Exercises

1. Show by term induction that a terminal segment of a term $t$ is a concatenation $s_1 \cdots s_m$ of terms $s_i$ for some $m \geqslant 1$. Thus, a symbol in $t$ is at each position in $t$ the initial symbol of a unique subterm $s$ of $t$. The uniqueness of $s$ is an easy consequence of Exercise 2(a).

2. Let $\mathcal{L}$ be a first-order language, $\mathcal{T} = \mathcal{T}_\mathcal{L}$, and $\mathcal{E}t$ the property 'No proper initial segment of $t$ ($\in \mathcal{T}$) is a term, nor is $t$ a proper initial segment of a term from $\mathcal{T}$'. Prove (a) $\mathcal{E}t$ for all $t \in \mathcal{T}$, hence $t\xi = t'\xi' \Rightarrow t = t'$ for all $t, t' \in \mathcal{T}$ and arbitrary $\xi, \xi' \in \mathcal{S}_\mathcal{L}$, and (b) the unique term concatenation property (page 55).

3. Prove (a) No proper initial segment of a formula $\varphi$ is a formula. (b) The unique formula reconstruction property stated on page 58. (c) $\neg\xi \in \mathcal{L} \Rightarrow \xi \in \mathcal{L}$ and $\alpha, (\alpha \wedge \xi) \in \mathcal{L} \Rightarrow \xi \in \mathcal{L}$. (c) easily yields (d) $\alpha, (\alpha \rightarrow \xi) \in \mathcal{L} \Rightarrow \xi \in \mathcal{L}$, for all $\xi \in \mathcal{S}_\mathcal{L}$.

4. Prove $\varphi \frac{t}{x} = \varphi$ for $x \notin \operatorname{free} \varphi$, and $\varphi \frac{y}{x} \frac{t}{y} = \varphi \frac{t}{x}$ for $y \notin \operatorname{var} \varphi$. It can be shown that these restrictions are indispensable, provided $t \neq x$.

5. Let $X \subseteq \mathcal{L}$ be a nonempty formula set and $X^* = X \cup \{\neg\varphi \mid \varphi \in X\}$. Show that a Boolean combination of formulas from $X$ is equivalent to a disjunction of conjunctions of formulas from $X^*$.

## 2.3   Semantics of First-Order Languages

Intuitively it is clear that the formula $\exists y \, y + y = x$ can be allocated a truth value in the domain $(\mathbb{N}, +)$ only if to the free variable $x$ there corresponds a value in $\mathbb{N}$. Thus, along with an interpretation of the extralogical symbols, a truth value allocation for a formula $\varphi$ requires a valuation of at least the variables occurring free in $\varphi$. However, it is technically more convenient to work with a global assignment of values to all variables, even if in a concrete case only the values of finitely many variables are needed. We therefore begin with the following

**Definition.** A *model* $\mathcal{M}$ is a pair $(\mathcal{A}, w)$ consisting of an $\mathcal{L}$-structure $\mathcal{A}$ and a *valuation* $w \colon \text{Var} \to A$, $w \colon x \mapsto x^w$. We denote $r^{\mathcal{A}}, f^{\mathcal{A}}, c^{\mathcal{A}}$, and $x^w$ also by $r^{\mathcal{M}}, f^{\mathcal{M}}, c^{\mathcal{M}}$, and $x^{\mathcal{M}}$, respectively. The domain of $\mathcal{A}$ will also called the *domain of* $\mathcal{M}$.

Models are sometimes called *interpretations*, occasionally also $\mathcal{L}$-*models* if the connection to $\mathcal{L}$ is to be highlighted. Some authors identify models with structures from the outset. This also happens in **2.5**, where we are talking about models of theories. The notion of a model is to be maintained sufficiently flexible in logic and mathematics.

A model $\mathcal{M}$ allocates in a natural way to every term $t$ a value in $A$, denoted by $t^{\mathcal{M}}$ or $t^{\mathcal{A},w}$ or just by $t^w$. Clearly, for prime terms the value is already given by $\mathcal{M}$. This evaluation extends to compound terms by term induction as follows: $(f\vec{t})^{\mathcal{M}} = f^{\mathcal{M}}\vec{t}^{\mathcal{M}}$, where $\vec{t}^{\mathcal{M}}$ abbreviates here the sequence $(t_1^{\mathcal{M}}, \ldots, t_n^{\mathcal{M}})$. If the context allows we neglect the superscripts and retain just an imaginary distinction between symbols and their interpretation. For instance, if $\mathcal{A} = (\mathbb{N}, +, \cdot, 0, 1)$ and $x^w = 2$, say, we write somewhat sloppily $(0 \cdot x + 1)^{\mathcal{A},w} = 0 \cdot 2 + 1 = 1$.

The value of $t$ under $\mathcal{M}$ depends only on the meaning of the symbols that effectively occur in $t$; using induction on $t$, the following slightly more general claim is obtained: if $\text{var}\, t \subseteq V \subseteq \text{Var}$ and $\mathcal{M}, \mathcal{M}'$ are models with the same domain such that $x^{\mathcal{M}} = x^{\mathcal{M}'}$ for all $x \in V$ and $\mathsf{s}^{\mathcal{M}} = \mathsf{s}^{\mathcal{M}'}$ for all remaining symbols $\mathsf{s}$ occurring in $t$, then $t^{\mathcal{M}} = t^{\mathcal{M}'}$. Clearly, $t^{\mathcal{A},w}$ may simply be denoted by $t^{\mathcal{A}}$, provided the term $t$ contains no variables.

We now are going to define a satisfiability relation $\vDash$ between models $\mathcal{M} = (\mathcal{A}, w)$ and formulas $\varphi$, using induction on $\varphi$ as in **1.3**. We read $\mathcal{M} \vDash \varphi$ as $\mathcal{M}$ *satisfies* $\varphi$, or $\mathcal{M}$ *is a model for* $\varphi$.

Sometimes $\mathcal{A} \vDash \varphi\,[w]$ is written instead of $\mathcal{M} \vDash \varphi$. A similar notation, just as frequently encountered, is introduced later. Each of these notations has its advantages, depending on the context. If $\mathcal{M} \vDash \varphi$ for all $\varphi \in X$ we write $\mathcal{M} \vDash X$ and call $\mathcal{M}$ a *model for* $X$. For the formulation of the satisfaction clauses below (taken from [Ta1]) we consider for given $\mathcal{M} = (\mathcal{A}, w)$, $x \in \text{Var}$, and $a \in A$ also the model $\mathcal{M}_x^a$ (generalized to $\mathcal{M}_{\vec{x}}^{\vec{a}}$

below). $\mathcal{M}_x^a$ differs from $\mathcal{M}$ only in that the variable $x$ receives the value $a \in A$ instead of $x^{\mathcal{M}}$. Thus, $\mathcal{M}_x^a = (\mathcal{A}, w')$ with $x^{w'} = a$ and $y^{w'} = y^w$ otherwise. The satisfaction clauses then look as follows:

$$\mathcal{M} \vDash s = t \quad \Leftrightarrow \quad s^{\mathcal{M}} = t^{\mathcal{M}},$$
$$\mathcal{M} \vDash r\vec{t} \quad \Leftrightarrow \quad r^{\mathcal{M}}\vec{t}^{\mathcal{M}},$$
$$\mathcal{M} \vDash (\alpha \wedge \beta) \quad \Leftrightarrow \quad \mathcal{M} \vDash \alpha \text{ and } \mathcal{M} \vDash \beta,$$
$$\mathcal{M} \vDash \neg\alpha \quad \Leftrightarrow \quad \mathcal{M} \nvDash \alpha,$$
$$\mathcal{M} \vDash \forall x \alpha \quad \Leftrightarrow \quad \mathcal{M}_x^a \vDash \alpha \text{ for all } a \in A.$$

**Remark 1.** The last satisfaction clause can be stated differently if a name for each $a \in A$, say $\boldsymbol{a}$, is available in the signature: $\mathcal{M} \vDash \forall x \alpha \Leftrightarrow \mathcal{M} \vDash \alpha \frac{\boldsymbol{a}}{x}$ for all $a \in A$. This assumption permits the definition of the satisfaction relation for sentences using induction on sentences while bypassing arbitrary formulas. If not every $a \in A$ has a name in $L$, one could "fill up" $L$ in advance by adjoining to $L$ a name $\boldsymbol{a}$ for each $a$. But expanding the language is not always wanted and does not really simplify the matter.

$\mathcal{M}_x^a$ is slightly generalized to $\mathcal{M}_{\vec{x}}^{\vec{a}} := \mathcal{M}_{x_1 \cdots x_n}^{a_1 \cdots a_n}$ $(= (\mathcal{M}_{x_1}^{a_1})_{x_2}^{a_2} \cdots)$, which differs from $\mathcal{M}$ in the values of a sequence $x_1, \ldots, x_n$ of distinct variables. This and writing $\forall \vec{x} \varphi$ for $\forall x_1 \cdots \forall x_n \varphi$ permits a short notation of a useful generalization of the last clause above, namely

$$\mathcal{M} \vDash \forall \vec{x} \varphi \quad \Leftrightarrow \quad \mathcal{M}_{\vec{x}}^{\vec{a}} \vDash \varphi \text{ for all } \vec{a} \in A^n.$$

The definitions of $\alpha \vee \beta$, $\alpha \to \beta$, and $\alpha \leftrightarrow \beta$ from page 56 readily imply the additional clauses $\mathcal{M} \vDash \alpha \vee \beta$ *iff* $\mathcal{M} \vDash \alpha$ or $\mathcal{M} \vDash \beta$, $\mathcal{M} \vDash \alpha \to \beta$ *iff* $\mathcal{M} \vDash \alpha \Rightarrow \mathcal{M} \vDash \beta$, and analogously for $\leftrightarrow$. Clearly, if $\vee, \to, \leftrightarrow$ were treated as independent connectives, these equivalences would have to be added to the above ones. Further, the definition of $\exists x \varphi$ in **2.2** corresponds to its intended meaning, because $\mathcal{M} \vDash \exists x \varphi \Leftrightarrow \mathcal{M}_x^a \vDash \varphi$ for some $a \in A$. Indeed, whenever $\mathcal{M} \vDash \neg\forall x \neg\varphi$ $(= \exists x \varphi)$ then $\mathcal{M}_x^a \vDash \neg\varphi$ does not hold for all $a$; hence there is some $a \in A$ such that $\mathcal{M}_x^a \nvDash \neg\varphi$, or equivalently, $\mathcal{M}_x^a \vDash \varphi$. And this chain of reasoning is obviously reversible.

**Example 1.** $\mathcal{M} \vDash \exists x\, x = t$ for arbitrary $\mathcal{M}$, provided $x \notin \mathrm{var}\, t$. Indeed, $\mathcal{M}_x^a \vDash x = t$ with $a := t^{\mathcal{M}}$, since $x^{\mathcal{M}_x^a} = a = t^{\mathcal{M}} = t^{\mathcal{M}_x^a}$ in view of $x \notin \mathrm{var}\, t$. The assumption $x \notin \mathrm{var}\, t$ is essential. For instance, $\mathcal{M} \vDash \exists x\, x = fx$ holds only if the function $f^{\mathcal{M}}$ has a fixed point.

We now introduce several fundamental notions that will be treated more systematically in **2.4** and **2.5**, once certain necessary preparations have been completed.

**Definition.** A formula or set of formulas in $\mathcal{L}$ is termed *satisfiable* if it has a model. $\varphi \in \mathcal{L}$ is called *generally valid, logically valid*, or a *tautology*, in short, $\vDash \varphi$, if $\mathcal{M} \vDash \varphi$ for every model $\mathcal{M}$. Formulas $\alpha, \beta$ are called (logically or semantically) *equivalent*, in symbols, $\alpha \equiv \beta$, if

$$\mathcal{M} \vDash \alpha \;\Leftrightarrow\; \mathcal{M} \vDash \beta, \text{ for each } \mathcal{L}\text{-model } \mathcal{M}.$$

Further, let $\mathcal{A} \vDash \varphi$ (read $\varphi$ *holds in* $\mathcal{A}$ or $\mathcal{A}$ *satisfies* $\varphi$) if $(\mathcal{A}, w) \vDash \varphi$ for all $w \colon \mathrm{Var} \to A$. One writes $\mathcal{A} \vDash X$ in case $\mathcal{A} \vDash \varphi$ for all $\varphi \in X$. Finally, let $X \vDash \varphi$ (read *from $X$ follows $\varphi$*, or $\varphi$ *is a consequence of $X$*) if every model $\mathcal{M}$ of $X$ also satisfies the formula $\varphi$, i.e., $\mathcal{M} \vDash X \Rightarrow \mathcal{M} \vDash \varphi$.

As in Chapter **1**, $\vDash$ denotes both the satisfaction and the consequence relation. Here, as there, we write $\varphi_1, \ldots, \varphi_n \vDash \varphi$ for $\{\varphi_1, \ldots, \varphi_n\} \vDash \varphi$. Note that in addition, $\vDash$ denotes the validity relation in structures, which is illustrated by the following

**Example 2.** We show that $\mathcal{A} \vDash \forall x \exists y\, x \neq y$, where the domain of $\mathcal{A}$ contains at least two elements. Indeed, let $\mathcal{M} = (\mathcal{A}, w)$ and let $a \in A$ be given arbitrarily. Then there exists some $b \in A$ with $a \neq b$. Hence, $(\mathcal{M}_x^a)_y^b = \mathcal{M}_{xy}^{ab} \vDash x \neq y$, and so $\mathcal{M}_x^a \vDash \exists y\, x \neq y$. Since $a$ was arbitrary, $\mathcal{M} \vDash \forall x \exists y\, x \neq y$. Clearly the actual values of $w$ are irrelevant in this argument. Hence $(\mathcal{A}, w) \vDash \forall x \exists y\, x \neq y$ for all $w$, that is, $\mathcal{A} \vDash \forall x \exists y\, x \neq y$.

Here some care is needed. While $\mathcal{M} \vDash \varphi$ or $\mathcal{M} \vDash \neg\varphi$ for all formulas, $\mathcal{A} \vDash \varphi$ or $\mathcal{A} \vDash \neg\varphi$ (the law of the excluded middle for validity in structures) is in general correct only for sentences $\varphi$, as Theorem 3.1 will show. If $\mathcal{A}$ contains more than one element, then, for example, neither $\mathcal{A} \vDash x = y$ nor $\mathcal{A} \vDash x \neq y$. Indeed, $x = y$ is falsified by any $w$ such that $x^w \neq y^w$, and $x \neq y$ by any $w$ with $x^w = y^w$. This is one of the reasons why models were not simply identified with structures.

For $\varphi \in \mathcal{L}$ let $\varphi^g$ be the sentence $\forall x_1 \cdots \forall x_m \varphi$, where $x_1, \ldots, x_m$ is an enumeration of *free* $\varphi$ according to index size, say. $\varphi^g$ is called the *generalized of* $\varphi$, also called its *universal closure*. For $\varphi \in \mathcal{L}^0$ clearly $\varphi^g = \varphi$. From the definitions immediately results

(1)  $\mathcal{A} \vDash \varphi \;\Leftrightarrow\; \mathcal{A} \vDash \varphi^g$,

and more generally, $\mathcal{A} \vDash X \Leftrightarrow \mathcal{A} \vDash X^g$ ($:= \{\varphi^g \mid \varphi \in X\}$). (1) explains why $\varphi$ and $\varphi^g$ are often notionally identified, and the information that formally runs $\varphi^g$ is often shortened to $\varphi$. It must always be clear from the context whether our eye is on validity in a structure, or on validity in a model with its fixed valuation. Only in the first case can a generalization (or globalization) of the free variables be thought of as carried out. However, independent of this discussion, $\vDash \varphi \Leftrightarrow \vDash \varphi^g$ always holds.

Even after just these incomplete considerations it is already clear that numerous properties of structures and whole systems of axioms can adequately be described by first-order formulas and sentences. Thus, for example, an axiom system for groups in $\circ, e, ^{-1}$, mentioned already in **2.1**, can be formulated as follows:

$$\forall x \forall y \forall z \ x \circ (y \circ z) = (x \circ y) \circ z; \quad \forall x \ x \circ e = x; \quad \forall x \ x \circ x^{-1} = e.$$

Precisely, the sentences that follow from these axioms form the *elementary group theory in* $\circ, e, ^{-1}$. It will be denoted by $T_G^=$. In the sense elaborated in Exercise 3 in **2.6** an equivalent formulation of the theory of groups in $\circ, e$, denoted by $T_G$, is obtained if the third $T_G^=$-axiom is replaced by $\forall x \exists y \ x \circ y = e$. Let us mention that $\forall x \ e \circ x = x$ and $\forall x \exists y \ y \circ x = e$ are provable in $T_G$ and also in $T_G^=$.

An axiom system for ordered sets can also easily be provided, in that one formalizes the properties of being irreflexive, transitive, and connex. Here and elsewhere, $\forall x_1 \cdots x_n \varphi$ stands for $\forall x_1 \cdots \forall x_n \varphi$:

$$\forall x \ x \not< x; \ \forall xyz(x < y \wedge y < z \rightarrow x < z); \ \forall xy(x \neq y \rightarrow x < y \vee y < x).$$

In writing down these and other axioms the outer $\forall$-prefixes are very often omitted so as to save on writing, and we think implicitly of the generalization of variables as having been carried out. This kind of economical writing is employed also in the formulation of (1) above, which strictly speaking runs 'for all $\mathcal{A}, \varphi : \mathcal{A} \vDash \varphi \Leftrightarrow \mathcal{A} \vDash \varphi^g$'.

For sentences $\alpha$ of a given language it is intuitively clear that the values of the variables of $w$ for the relation $(\mathcal{A}, w) \vDash \alpha$ are irrelevant. The precise proof is extracted from the following theorem for $V = \emptyset$. Thus, either $(\mathcal{A}, w) \vDash \alpha$ for all $w$ and hence $\mathcal{A} \vDash \alpha$, or else $(\mathcal{A}, w) \vDash \alpha$ for no $w$, i.e., $(\mathcal{A}, w) \vDash \neg\alpha$ for all $w$, and hence $\mathcal{A} \vDash \neg\alpha$. Sentences therefore obey the already-cited tertium non datur.

**Theorem 3.1 (Coincidence theorem).** *Let* $V \subseteq \mathrm{Var}$, *free* $\varphi \subseteq V$, *and* $\mathcal{M}, \mathcal{M}'$ *be models on the same domain* $A$ *such that* $x^{\mathcal{M}} = x^{\mathcal{M}'}$ *for all* $x \in V$, *and* $\mathsf{s}^{\mathcal{M}} = \mathsf{s}^{\mathcal{M}'}$ *for all extralogical symbols* $\mathsf{s}$ *occurring in* $\varphi$. *Then* $\mathcal{M} \vDash \varphi \Leftrightarrow \mathcal{M}' \vDash \varphi$.

**Proof** by induction on $\varphi$. Let $\varphi = r\vec{t}$ be prime, so that $\operatorname{var}\vec{t} \subseteq V$. As was mentioned earlier, the value of a term $t$ depends only on the meaning of the symbols occurring in $t$. But in view of the suppositions, these meanings are the same in $\mathcal{M}$ and $\mathcal{M}'$. Therefore, $\vec{t}^{\mathcal{M}} = \vec{t}^{\mathcal{M}'}$ (i.e., $t_i^{\mathcal{M}} = t_i^{\mathcal{M}'}$ for $i = 1, \ldots, n$), and so $\mathcal{M} \vDash r\vec{t} \Leftrightarrow r^{\mathcal{M}}\vec{t}^{\mathcal{M}} \Leftrightarrow r^{\mathcal{M}'}\vec{t}^{\mathcal{M}'} \Leftrightarrow \mathcal{M}' \vDash r\vec{t}$. For equations $t_1 = t_2$ one reasons analogously. Further, the induction hypothesis for $\alpha, \beta$ yields $\mathcal{M} \vDash \alpha \wedge \beta \Leftrightarrow \mathcal{M} \vDash \alpha, \beta \Leftrightarrow \mathcal{M}' \vDash \alpha, \beta \Leftrightarrow \mathcal{M}' \vDash \alpha \wedge \beta$. In the same way one obtains $\mathcal{M} \vDash \neg\alpha \Leftrightarrow \mathcal{M}' \vDash \neg\alpha$. By the induction step on $\forall$ it becomes clear that the induction hypothesis needs to be skillfully formulated. It must be given with respect to any pair $\mathcal{M}, \mathcal{M}'$ of models and any subset $V$ of Var.

Therefore let $a \in A$ and $\mathcal{M}_x^a \vDash \varphi$. Since for $V' := V \cup \{x\}$ certainly $\operatorname{free}\varphi \subseteq V'$ and the models $\mathcal{M}_x^a$, $\mathcal{M}'^a_x$ coincide for all $y \in V'$ (although in general $x^{\mathcal{M}} \neq x^{\mathcal{M}'}$), by the induction hypothesis $\mathcal{M}_x^a \vDash \varphi \Leftrightarrow \mathcal{M}'^a_x \vDash \varphi$, for each $a \in A$. This clearly implies

$$\mathcal{M} \vDash \forall x\varphi \Leftrightarrow \mathcal{M}_x^a \vDash \varphi \text{ for all } a \Leftrightarrow \mathcal{M}'^a_x \vDash \varphi \text{ for all } a \Leftrightarrow \mathcal{M}' \vDash \forall x\varphi. \qquad \square$$

It follows from this theorem that an $\mathcal{L}$-model $\mathcal{M} = (\mathcal{A}, w)$ of $\varphi$ for the case that $\varphi \in \mathcal{L} \subseteq \mathcal{L}'$ can be completely arbitrarily expanded to an $\mathcal{L}'$-model $\mathcal{M}' = (\mathcal{A}', w)$ of $\varphi$, i.e., arbitrarily fixing $\mathsf{s}^{\mathcal{M}'}$ for $\mathsf{s} \in L'\backslash L$ gives $\mathcal{M} \vDash \varphi \Leftrightarrow \mathcal{M}' \vDash \varphi$ by the above theorem with $V = \mathrm{Var}$. This readily implies that the consequence relation $\vDash_{\mathcal{L}'}$ with respect to $\mathcal{L}'$ is a *conservative* extension of $\vDash_{\mathcal{L}}$ in that $X \vDash_{\mathcal{L}} \varphi \Leftrightarrow X \vDash_{\mathcal{L}'} \varphi$, for all sets $X \subseteq \mathcal{L}$ and all $\varphi \in \mathcal{L}$. Hence, there is no need here for using indices. In particular, the satisfiability or general validity of $\varphi$ depends only on the symbols effectively occurring in $\varphi$.

Another application of Theorem 3.1 is the following fact, which justifies the already mentioned "omission of superfluous quantifiers."

(2) $\quad \forall x\varphi \equiv \varphi \equiv \exists x\varphi$ whenever $x \notin \operatorname{free}\varphi$.

Indeed, $x \notin \operatorname{free}\varphi$ implies $\mathcal{M} \vDash \varphi \Leftrightarrow \mathcal{M}_x^a \vDash \varphi$ (here $a \in A$ is arbitrary) according to Theorem 3.1; choose $\mathcal{M}' = \mathcal{M}_x^a$ and $V = \operatorname{free}\varphi$. Therefore,

$$\mathcal{M} \vDash \forall x\varphi \Leftrightarrow \mathcal{M}_x^a \vDash \varphi \text{ for all } a \Leftrightarrow \mathcal{M} \vDash \varphi$$
$$\Leftrightarrow \mathcal{M}_x^a \vDash \varphi \text{ for some } a \Leftrightarrow \mathcal{M} \vDash \exists x\varphi.$$

Very important for the next theorem and elsewhere is

(3) $\quad$ If $\mathcal{A} \subseteq \mathcal{B}$, $\mathcal{M} = (\mathcal{A}, w)$, $\mathcal{M}' = (\mathcal{B}, w)$ and $w \colon \mathrm{Var} \to A$ then $t^{\mathcal{M}} = t^{\mathcal{M}'}$.

This is clear for prime terms, and the induction hypothesis $t_i^{\mathcal{M}} = t_i^{\mathcal{M}'}$ for $i = 1, \ldots, n$ together with $f^{\mathcal{M}} = f^{\mathcal{M}'}$ imply

$$(f\vec{t})^{\mathcal{M}} = f^{\mathcal{M}}(t_1^{\mathcal{M}}, \ldots, t_n^{\mathcal{M}}) = f^{\mathcal{M}'}(t_1^{\mathcal{M}'}, \ldots, t_n^{\mathcal{M}'}) = (f\vec{t})^{\mathcal{M}'}.$$

For $\mathcal{M} = (\mathcal{A}, w)$ and $x_i^w = a_i$ let $t^{\mathcal{A}, \vec{a}}$, or more suggestively $t^{\mathcal{A}}(\vec{a})$ denote the value of $t = t(\vec{x})$. Then (3) can somewhat more simply be written as

(4)  $\mathcal{A} \subseteq \mathcal{B}$ and $t = t(\vec{x})$ imply $t^{\mathcal{A}}(\vec{a}) = t^{\mathcal{B}}(\vec{a})$ for all $\vec{a} \in A^n$.

Thus, along with the basic functions, also the so-called *term functions* $\vec{a} \mapsto t^{\mathcal{A}}(\vec{a})$ are the restrictions to their counterparts in $\mathcal{B}$. Clearly, if $n = 0$ or $t$ is variable-free, one may write $t^{\mathcal{A}}$ for $t^{\mathcal{A}}(\vec{a})$. Note that in these cases $t^{\mathcal{A}} = t^{\mathcal{B}}$ whenever $\mathcal{A} \subseteq \mathcal{B}$, according to (4).

By Theorem 3.1 the satisfaction of $\varphi$ in $(\mathcal{A}, w)$ depends only on the values of the $x \in$ free $\varphi$. Let $\varphi = \varphi(\vec{x})^4$ and $\vec{a} = (a_1, \ldots, a_n) \in A^n$. Then the statement

$$(\mathcal{A}, w) \vDash \varphi \text{ for a valuation } w \text{ with } x_1^w = a_1, \ldots, x_n^w = a_n$$

can more suggestively be expressed by writing

$$(\mathcal{A}, \vec{a}) \vDash \varphi \quad \text{or} \quad \mathcal{A} \vDash \varphi[a_1, \ldots, a_n] \quad \text{or} \quad \mathcal{A} \vDash \varphi[\vec{a}]$$

without mentioning $w$ as a global valuation. Such notation also makes sense if $w$ is restricted to a valuation on $\{x_1, \ldots, x_n\}$. One may accordingly extend the concept of a model and call a pair $(\mathcal{A}, \vec{a})$ a model for a formula $\varphi(\vec{x})$ whenever $(\mathcal{A}, \vec{a}) \vDash \varphi(\vec{x})$, in particular if $\varphi \in \mathcal{L}^n$. We return to this extended concept in **4.1**. Until then we use it only for $n = 0$. That is, besides $\mathcal{M} = (\mathcal{A}, w)$ also the structure $\mathcal{A}$ itself is occasionally called a model for a set $S \subseteq \mathcal{L}^0$ of sentences, provided $\mathcal{A} \vDash S$.

As above let $\varphi = \varphi(\vec{x})$. Then $\varphi^{\mathcal{A}} := \{\vec{a} \in A^n \mid \mathcal{A} \vDash \varphi[\vec{a}]\}$ is called *the predicate defined by the formula* $\varphi$ *in the structure* $\mathcal{A}$. For instance, the $\leqslant$-predicate in $(\mathbb{N}, +)$ is defined by $\varphi(x, y) = \exists z\, z + x = y$, but also by several other formulas.

More generally, a predicate $P \subseteq A^n$ is termed (explicitly or elementarily or first-order) *definable in* $\mathcal{A}$ if there is some $\varphi = \varphi(\vec{x})$ with $P = \varphi^{\mathcal{A}}$, and $\varphi$ is called a *defining formula* for $P$. Analogously, $f : A^n \to A$ is called definable in $\mathcal{A}$ if $\varphi^{\mathcal{A}} = $ graph $f$ for some $\varphi(\vec{x}, y)$. One often talks in this

---

[4] Since this equation is to mean *free* $\varphi \subseteq \{x_1, \ldots, x_n\}$, $\vec{x}$ is not uniquely determined by $\varphi$. Hence, the phrase "Let $\varphi = \varphi(\vec{x})\ldots$" implicitly includes along with a given $\varphi$ also a tuple $\vec{x}$ given in advance. The notation $\varphi = \varphi(\vec{x})$ does not even state that $\varphi$ contains free variables at all.

case of *explicit* definability of $f$ in $\mathcal{A}$, to distinguish it from other kinds of definability. Much information is gained from the knowledge of which sets, predicates, or functions are definable in a structure. For instance, the sets definable in $(\mathbb{N}, 0, 1, +)$ are the eventually periodic ones (periodic from some number on). Thus, $\cdot$ cannot explicitly be defined by $+, 0, 1$ because the set of square numbers is not eventually periodic.

$\mathcal{A} \subseteq \mathcal{B}$ and $\varphi = \varphi(\vec{x})$ do not imply $\varphi^{\mathcal{A}} = \varphi^{\mathcal{B}} \cap A^n$, in general. For instance, let $\mathcal{A} = (\mathbb{N}, +)$, $\mathcal{B} = (\mathbb{Z}, +)$, and $\varphi = \exists z\, z + x = y$. Then $\varphi^{\mathcal{A}} = \leq^{\mathcal{A}}$, while $\varphi^{\mathcal{B}}$ contains all pairs $(a, b) \in \mathbb{Z}^2$. As the next theorem will show, $\varphi^{\mathcal{A}} = \varphi^{\mathcal{B}} \cap A^n$ holds in general only for open formulas $\varphi$, and is even characteristic for $\mathcal{A} \subseteq \mathcal{B}$ provided $A \subseteq B$. Clearly, $A \subseteq B$ is much weaker a condition than $\mathcal{A} \subseteq \mathcal{B}$:

**Theorem 3.2 (Substructure theorem).** *For structures $\mathcal{A}, \mathcal{B}$ such that $A \subseteq B$ the following conditions are equivalent:*

(i)   $\mathcal{A} \subseteq \mathcal{B}$,

(ii)  $\mathcal{A} \vDash \varphi\,[\vec{a}] \Leftrightarrow \mathcal{B} \vDash \varphi\,[\vec{a}]$, *for all open $\varphi = \varphi(\vec{x})$ and all $\vec{a} \in A^n$,*

(iii) $\mathcal{A} \vDash \varphi\,[\vec{a}] \Leftrightarrow \mathcal{B} \vDash \varphi\,[\vec{a}]$, *for all prime formulas $\varphi(\vec{x})$ and $\vec{a} \in A^n$.*

**Proof.** (i)$\Rightarrow$(ii): It suffices to prove that $\mathcal{M} \vDash \varphi \Leftrightarrow \mathcal{M}' \vDash \varphi$, with $\mathcal{M} = (\mathcal{A}, w)$ and $\mathcal{M}' = (\mathcal{B}, w)$, where $w \colon \mathrm{Var} \to A$. In view of (3) the claim is obvious for prime formulas, and the induction steps for $\wedge, \neg$ are carried out just as in Theorem 3.1. (ii)$\Rightarrow$(iii): Trivial. (iii)$\Rightarrow$(i): By (iii), $r^{\mathcal{A}}\vec{a} \Leftrightarrow \mathcal{A} \vDash r\vec{x}\,[\vec{a}] \Leftrightarrow \mathcal{B} \vDash r\vec{x}\,[\vec{a}] \Leftrightarrow r^{\mathcal{B}}\vec{a}$. Analogously,

$$f^{\mathcal{A}}\vec{a} = b \;\Leftrightarrow\; \mathcal{A} \vDash f\vec{x} = y\,[\vec{a}, b] \;\Leftrightarrow\; \mathcal{B} \vDash f\vec{x} = y\,[\vec{a}, b] \;\Leftrightarrow\; f^{\mathcal{B}}\vec{a} = b,$$

for all $\vec{a} \in A^n$, $b \in A$. These conclusions state precisely that $\mathcal{A} \subseteq \mathcal{B}$.  ∎

Let $\alpha$ be of the form $\forall \vec{x}\beta$ with open $\beta$, where $\forall \vec{x}$ may also be the empty prefix. Then $\alpha$ is a *universal* or $\forall$-*formula* (spoken "A-formula"), and for $\alpha \in \mathcal{L}^0$ also a *universal* or $\forall$-*sentence*. A simple example is $\forall x \forall y\, x = y$, which holds in $\mathcal{A}$ iff $A$ contains precisely one element. Dually, $\exists \vec{x}\beta$ with $\beta$ open is termed an $\exists$-*formula*, and an $\exists$-*sentence* whenever $\exists \vec{x}\beta \in \mathcal{L}^0$. Examples are the "how-many sentences"

$$\exists_1 := \exists v_0\, v_0 = v_0; \quad \exists_n := \exists v_0 \cdots \exists v_{n-1} \bigwedge_{i<j<n} v_i \neq v_j \quad (n > 1).$$

$\exists_n$ states 'there exist at least $n$ elements', $\neg\exists_{n+1}$ thus that 'there exist at most $n$ elements', and $\exists_{=n} := \exists_n \wedge \neg\exists_{n+1}$ says 'there exist exactly

$n$ elements'. Since $\exists_1$ is a tautology, it is convenient to set $\top := \exists_1$, and $\exists_0 := \bot := \neg \top$ in all first-order languages with equality. Clearly, equivalent definitions of $\top$, $\bot$ may be used as well.

**Corollary 3.3.** *Let $\mathcal{A} \subseteq \mathcal{B}$. Then every $\forall$-sentence $\forall \vec{x} \alpha$ valid in $\mathcal{B}$ is also satisfied in $\mathcal{A}$. Dually, every $\exists$-sentence $\exists \vec{x} \beta$ valid in $\mathcal{A}$ is also valid in $\mathcal{B}$.*

**Proof.** Let $\mathcal{B} \vDash \forall \vec{x} \beta$ and $\vec{a} \in A^n$. Then $\mathcal{B} \vDash \beta [\vec{a}]$, hence $\mathcal{A} \vDash \beta [\vec{a}]$ by Theorem 3.2. $\vec{a}$ was arbitrary and therefore $\mathcal{A} \vDash \forall \vec{x} \beta$. Now let $\mathcal{A} \vDash \exists \vec{x} \beta$. Then $\mathcal{A} \vDash \beta [\vec{a}]$ for some $\vec{a} \in A^n$, hence $\mathcal{B} \vDash \beta [\vec{a}]$ by Theorem 3.2, and consequently $\mathcal{B} \vDash \exists \vec{x} \beta$. ◻

We now formulate a generalization of certain individual often-used arguments about the invariance of properties under isomorphisms:

**Theorem 3.4 (Invariance theorem).** *Let $\mathcal{A}, \mathcal{B}$ be isomorphic structures of signature $L$ and let $\imath \colon \mathcal{A} \to \mathcal{B}$ be an isomorphism. Then for all $\varphi = \varphi(\vec{x})$*

$$\mathcal{A} \vDash \varphi [\vec{a}] \;\Leftrightarrow\; \mathcal{B} \vDash \varphi [\imath \vec{a}] \quad (\vec{a} \in A^n, \; \imath \vec{a} = (\imath a_1, \dots, \imath a_n)).$$

*In particular $\mathcal{A} \vDash \varphi \Leftrightarrow \mathcal{B} \vDash \varphi$, for all sentences $\varphi$ of $\mathcal{L}$.*

**Proof.** It is convenient to reformulate the claim as

$$\mathcal{M} \vDash \varphi \;\Leftrightarrow\; \mathcal{M}' \vDash \varphi \quad (\mathcal{M} = (\mathcal{A}, w), \; \mathcal{M}' = (\mathcal{B}, w'), \; w' : x \mapsto \imath x^w).$$

This is easily confirmed by induction on $\varphi$ after first proving $\imath(t^{\mathcal{M}}) = t^{\mathcal{M}'}$ inductively on $t$. This proof clearly includes the case $\varphi \in \mathcal{L}^0$. ◻

Thus, for example, it is once and for all clear that the isomorphic image of a group is a group even if we know at first only that it is a groupoid. Simply let $\alpha$ in the theorem run through all axioms of group theory. Another application: Let $\imath$ be an isomorphism of the group $\mathcal{A} = (A, \circ)$ onto the group $\mathcal{A}' = (A', \circ)$ and let $e$ and $e'$ denote their unit elements, not named in the signature. We claim that nonetheless $\imath e = e'$, using the fact that the unit element of a group is the only solution of $x \circ x = x$ (Example 2, page 83). Thus, since $\mathcal{A} \vDash e \circ e = e$, we get $\mathcal{A}' \vDash \imath e \circ \imath e = \imath e$ by Theorem 3.4, hence $\imath e = e'$. Theorem 3.4, incidentally, holds for formulas of higher order as well. For instance, the property of being a continuously ordered set (formalizable in a second-order language, see **3.8**) is likewise invariant under isomorphism.

$\mathcal{L}$-structures $\mathcal{A}, \mathcal{B}$ are termed *elementarily equivalent* if $\mathcal{A} \vDash \alpha \Leftrightarrow \mathcal{B} \vDash \alpha$, for all $\alpha \in \mathcal{L}^0$. One then writes $\mathcal{A} \equiv \mathcal{B}$. We consider this important notion in **3.3** and more closely in **5.1**. Theorem 3.4 states in particular that $\mathcal{A} \simeq \mathcal{B} \Rightarrow \mathcal{A} \equiv \mathcal{B}$. The question immediately arises whether the converse of this also holds. For infinite structures the answer is negative (see **3.3**), for finite structures affirmative; a finite structure of a finite signature can, up to isomorphism, even be described by a single sentence. For example, the 2-element group $(\{0, 1\}, +)$ is up to isomorphism well determined by the following sentence, which tells us precisely how $+$ operates:

$$\exists v_0 \exists v_1 [v_0 \neq v_1 \wedge \forall x (x = v_0 \vee x = v_1)$$
$$\wedge\ v_0 + v_0 = v_1 + v_1 = v_0 \wedge v_0 + v_1 = v_1 + v_0 = v_1].$$

We now investigate the behavior of the satisfaction relation under substitution. The definition of $\varphi \frac{t}{x}$ in **2.2** pays no attention to *collision of variables*, which is taken to mean that some variables of the substitution term $t$ fall into the scope of quantifiers after the substitution has been performed. In this case $\mathcal{M} \vDash \forall x \varphi$ does not necessarily imply $\mathcal{M} \vDash \varphi \frac{t}{x}$, although this might have been expected. In other words, $\forall x \varphi \vDash \varphi \frac{t}{x}$ is not unrestrictedly correct. For instance, if $\varphi = \exists y\, x \neq y$ then certainly $\mathcal{M} \vDash \forall x \varphi\ (= \forall x \exists y\, x \neq y)$ whenever $\mathcal{M}$ has at least two elements, but $\mathcal{M} \vDash \varphi \frac{y}{x}\ (= \exists y\, y \neq y)$ is certainly false. Analogously $\varphi \frac{t}{x} \vDash \exists x \varphi$ is not correct, in general. For example, choose $\forall y\, x = y$ for $\varphi$ and $y$ for $t$.

One could forcibly obtain $\forall x \varphi \vDash \varphi \frac{t}{x}$ without any limitation by renaming bound variables by a suitable modification of the inductive definition of $\varphi \frac{t}{x}$ in the quantifier step. However, such measures are rather unwieldy for the arithmetization of proof method in **6.2**. It is therefore preferable to put up with minor restrictions when we are formulating rules of deduction later. The restrictions we will use are somewhat stronger than they need to be but can be handled more easily; they look as follows:

Call $\varphi, \frac{t}{x}$ *collision-free* if $y \notin \operatorname{bnd} \varphi$ for all $y \in \operatorname{var} t$ distinct from $x$. We need not require $x \notin \operatorname{bnd} \varphi$ because $t$ is substituted only at free occurrences of $x$ in $\varphi$, that is, $x$ cannot fall after substitution within the scope of a prefix $\forall x$, even if $x \in \operatorname{var} t$. For collision-free $\varphi, \frac{t}{x}$ we always get $\forall x \varphi \vDash \varphi \frac{t}{x}$ by Corollary 3.6 below.

If $\sigma$ is a global substitution (see **2.2**) then $\varphi, \sigma$ are termed *collision-free* if $\varphi, \frac{x^\sigma}{x}$ are collision-free for every $x \in \operatorname{Var}$. If $\sigma = \frac{\vec{t}}{\vec{x}}$, this condition clearly need be checked only for the pairs $\varphi, \frac{x^\sigma}{x}$ with $x \in \operatorname{var} \vec{x}$ and $x \in \operatorname{free} \varphi$.

For $\mathcal{M} = (\mathcal{A}, w)$ put $\mathcal{M}^\sigma := (\mathcal{A}, w^\sigma)$ with $x^{w^\sigma} := (x^\sigma)^{\mathcal{M}}$ for $x \in \mathrm{Var}$, so that $x^{\mathcal{M}^\sigma} = x^\sigma{}^{\mathcal{M}}$ $(= (x^\sigma)^{\mathcal{M}})$. This equation reproduces itself to

(5) $t^{\mathcal{M}^\sigma} = t^\sigma{}^{\mathcal{M}}$ for all terms $t$.

Indeed, $t^{\mathcal{M}^\sigma} = f^{\mathcal{M}}(t_1^{\mathcal{M}^\sigma}, \ldots, t_n^{\mathcal{M}^\sigma}) = f^{\mathcal{M}}(t_1^{\sigma\mathcal{M}}, \ldots, t_n^{\sigma\mathcal{M}}) = t^{\sigma\mathcal{M}}$ for $t = f\vec{t}$ in view of the induction hypothesis $t_i^{\mathcal{M}^\sigma} = t_i^{\sigma\mathcal{M}}$ $(i = 1, \ldots, n)$. Notice that $\mathcal{M}^\sigma$ coincides with $\mathcal{M}_{\vec{x}}^{\vec{t}\mathcal{M}}$ for the case $\sigma = \frac{\vec{t}}{\vec{x}}$.

**Theorem 3.5 (Substitution theorem).** *Let $\mathcal{M}$ be a model and $\sigma$ a global substitution. Then holds for all $\varphi$ such that $\varphi, \sigma$ are collision-free,*

(6) $\mathcal{M} \vDash \varphi^\sigma \Leftrightarrow \mathcal{M}^\sigma \vDash \varphi$.

*In particular, $\mathcal{M} \vDash \varphi\frac{\vec{t}}{\vec{x}} \Leftrightarrow \mathcal{M}_{\vec{x}}^{\vec{t}\mathcal{M}} \vDash \varphi$, provided $\varphi, \frac{\vec{t}}{\vec{x}}$ are collision-free.*

**Proof** by induction on $\varphi$. In view of (5), we obtain

$$\mathcal{M} \vDash (t_1 = t_2)^\sigma \Leftrightarrow t_1^{\sigma\mathcal{M}} = t_2^{\sigma\mathcal{M}} \Leftrightarrow t_1^{\mathcal{M}^\sigma} = t_2^{\mathcal{M}^\sigma} \Leftrightarrow \mathcal{M}^\sigma \vDash t_1 = t_2.$$

Prime formulas $r\vec{t}$ are treated analogously. The induction steps for $\wedge, \neg$ in the proof of (6) are harmless. Only the $\forall$-step is interesting. The reader should recall the definition of $(\forall x\alpha)^\sigma$ page 60 and realize that the induction hypothesis refers to an arbitrary global substitution $\tau$.

$$
\begin{aligned}
\mathcal{M} \vDash (\forall x\alpha)^\sigma &\Leftrightarrow \mathcal{M} \vDash \forall x\,\alpha^\tau && (x^\tau = x \text{ and } y^\tau = y^\sigma \text{ else})\\
&\Leftrightarrow \mathcal{M}_x^a \vDash \alpha^\tau \text{ for all } a && (\text{definition})\\
&\Leftrightarrow (\mathcal{M}_x^a)^\tau \vDash \alpha \text{ for all } a && (\text{induction hypothesis})\\
&\Leftrightarrow (\mathcal{M}^\sigma)_x^a \vDash \alpha \text{ for all } a && ((\mathcal{M}_x^a)^\tau = (\mathcal{M}^\sigma)_x^a, \text{ see below})\\
&\Leftrightarrow \mathcal{M}^\sigma \vDash \forall x\alpha.
\end{aligned}
$$

We show that $(\mathcal{M}_x^a)^\tau = (\mathcal{M}^\sigma)_x^a$. Since $\forall x\alpha, \sigma$ (hence $\forall x\alpha, \frac{y^\sigma}{y}$ for every $y$) are collision-free, we have $x \notin \mathrm{var}\,y^\sigma$ if $y \neq x$, and since $y^\tau = y^\sigma$ we get in this case $y^{(\mathcal{M}_x^a)^\tau} = y^{\tau\mathcal{M}_x^a} = y^{\sigma\mathcal{M}_x^a} = y^{\sigma\mathcal{M}} = y^{\mathcal{M}^\sigma} = y^{(\mathcal{M}^\sigma)_x^a}$. But also in the case $y = x$ we have $x^{(\mathcal{M}_x^a)^\tau} = x^{\tau\mathcal{M}_x^a} = x^{\mathcal{M}_x^a} = a = x^{(\mathcal{M}^\sigma)_x^a}$. ∎

**Corollary 3.6.** *For all $\varphi$ and $\frac{\vec{t}}{\vec{x}}$ such that $\varphi, \frac{\vec{t}}{\vec{x}}$ are collision-free, the following properties hold:*

(a) *$\forall \vec{x}\varphi \vDash \varphi\frac{\vec{t}}{\vec{x}}$, in particular $\forall x\varphi \vDash \varphi\frac{t}{x}$*,     (b) *$\varphi\frac{\vec{t}}{\vec{x}} \vDash \exists \vec{x}\varphi$*,

(c) *$\varphi\frac{s}{x}, s = t \vDash \varphi\frac{t}{x}$, provided $\varphi, \frac{s}{x}, \frac{t}{x}$ are collision-free.*

**Proof.** Let $\mathcal{M} \vDash \forall \vec{x}\varphi$, so that $\mathcal{M}_{\vec{x}}^{\vec{a}} \vDash \varphi$ for all $\vec{a} \in A^n$. In particular, $\mathcal{M}_{\vec{x}}^{\vec{t}\mathcal{M}} \vDash \varphi$. Therefore, $\mathcal{M} \vDash \varphi\frac{\vec{t}}{\vec{x}}$ by Theorem 3.5. (b) follows easily from $\neg\exists \vec{x}\varphi \vDash \neg\varphi\frac{\vec{t}}{\vec{x}}$. This holds by (a), for $\neg\exists \vec{x}\varphi \equiv \forall \vec{x}\neg\varphi$ and $\neg(\varphi\frac{\vec{t}}{\vec{x}}) \equiv (\neg\varphi)\frac{\vec{t}}{\vec{x}}$.

(c): Let $\mathcal{M} \vDash \varphi \frac{s}{x}, s = t$, so that $s^{\mathcal{M}} = t^{\mathcal{M}}$ and $\mathcal{M}_x^{s^{\mathcal{M}}} \vDash \varphi$ by the theorem. Clearly, then also $\mathcal{M}_x^{t^{\mathcal{M}}} \vDash \varphi$. Hence $\mathcal{M} \vDash \varphi \frac{t}{x}$. $\quad\square$

**Remark 2.** The identical substitution $\iota$ is obviously collision-free with every formula. Thus, $\forall x \varphi \vDash \varphi \ (= \varphi^{\iota})$ is always the case, while $\forall x \varphi \vDash \varphi \frac{t}{x}$ is correct in general only if $t$ contains at most the variable $x$, since $\varphi, \frac{t}{x}$ are then collision-free. Theorem 3.5 and Corollary 3.6 are easily strengthened. Define inductively a ternary predicate '$t$ is free for $x$ in $\varphi$', which intuitively is to mean that no free occurrence in $\varphi$ of the variable $x$ lies within the scope of a prefix $\forall y$ whenever $y \in \mathrm{var}\, t$. In this case Theorem 3.5 holds for $\sigma = \frac{t}{x}$ as well, so that nothing needs to be changed in the proofs based on this theorem if one works with '$t$ is free for $x$ in $\varphi$', or simply reads "$\varphi, \frac{t}{x}$ are collision-free" as "$t$ is free for $x$ in $\varphi$." Though collision-freeness is somewhat cruder and slightly more restrictive, it is for all that more easily manageable, which will pay off, for example, in **6.2**, where proofs will be arithmetized. Once one has become accustomed to the required caution, it is allowable not always to state explicitly the restrictions caused by collisions of variables, but rather to assume them tacitly.

Theorem 3.5 also shows that the quantifier "there exists exactly one," denoted by $\exists!$, is correctly defined by $\exists! x \varphi := \exists x \varphi \wedge \forall x \forall y (\varphi \wedge \varphi \frac{y}{x} \rightarrow x = y)$ with $y \notin \mathrm{var}\, \varphi$. Indeed, it is easily seen that $\mathcal{M} \vDash \forall x \forall y (\varphi \wedge \varphi \frac{y}{x} \rightarrow x = y)$ means just $\mathcal{M}_x^a \vDash \varphi$ & $\mathcal{M}_y^b \vDash \varphi \frac{y}{x} \Rightarrow a = b$. In short, $\mathcal{M}_x^a \vDash \varphi$ for at most one $a$. Putting everything together, $\mathcal{M} \vDash \exists! x \varphi$ iff there is precisely one $a \in A$ with $\mathcal{M}_x^a \vDash \varphi$. An example is $\mathcal{M} \vDash \exists! x\, x = t$ for arbitrary $\mathcal{M}$ and $x \notin \mathrm{var}\, t$. In other words, $\exists! x\, x = t$ is a tautology. Half of this, namely $\vDash \exists x\, x = t$, was shown in Example 1, and $\vDash \forall x \forall y (x = t \wedge y = t \rightarrow x = y)$ is obvious. There are various equivalent definitions of $\exists! x \varphi$. For example, a short and catchy formula is $\exists x \forall y (\varphi \frac{y}{x} \leftrightarrow x = y)$, where $y \notin \mathrm{var}\, \varphi$. The equivalence proof is left to the reader.

### Exercises

1. Let $X \vDash \varphi$ and $x \notin \mathrm{free}\, X$. Show that $X \vDash \forall x \varphi$.

2. Prove that $\forall x (\alpha \rightarrow \beta) \vDash \forall x \alpha \rightarrow \forall x \beta$, which is obviously equivalent to $\vDash \forall x (\alpha \rightarrow \beta) \rightarrow \forall x \alpha \rightarrow \forall x \beta$.

3. Suppose $\mathcal{A}'$ results from $\mathcal{A}$ by adjoining a constant symbol $\boldsymbol{a}$ for some $a \in A$. Prove $\mathcal{A} \vDash \alpha\,[a] \Leftrightarrow \mathcal{A}' \vDash \alpha(\boldsymbol{a})\ (= \alpha \frac{\boldsymbol{a}}{x})$ for $\alpha = \alpha(x)$, by first verifying $t(x)^{\mathcal{A},a} = t(\boldsymbol{a})^{\mathcal{A}'}$. This is easily generalized to the case of more than one free variable in $\alpha$.

4. Show that (a) A conjunction of the $\exists_i$ and their negations is equivalent to $\exists_n \wedge \neg\exists_m$ for suitable $n, m$ ($\exists_n \wedge \neg\exists_0 \equiv \exists_n$, $\exists_1 \wedge \neg\exists_m \equiv \neg\exists_m$).
   (b) A Boolean combination of the $\exists_i$ is equivalent to $\bigvee_{\nu\leqslant n} \exists_{=k_\nu}$ or to $\exists_k \vee \bigvee_{\nu\leqslant n} \exists_{=k_\nu}$, with $k_0 < \cdots < k_n < k$. Note that $\bigvee_{\nu\leqslant n} \exists_{=k_\nu}$ equals $\exists_{=0}$ ($\equiv \bot$) for $n=k_0=0$ and $\neg\exists_n \equiv \bigvee_{\nu<n} \exists_{=\nu}$ for $n>0$.

## 2.4   General Validity and Logical Equivalence

From the perspective of predicate logic $\alpha \vee \neg\alpha$ ($\alpha \in \mathcal{L}$) is a trivial example of a tautology, because it results by inserting $\alpha$ for $p$ from the propositional tautology $p \vee \neg p$. Every propositional tautology provides generally valid $\mathcal{L}$-formulas by the insertion of $\mathcal{L}$-formulas for the propositional variables. But there are tautologies not arising in this way. $\forall x(x < x \vee x \not< x)$ is an example, though it has still a root in propositional logic. Tautologies without a such a root are $\exists x\, x = x$ and $\exists x\, x = t$ for $x \notin \mathrm{var}\, t$. The former arises from the convention that structures are always nonempty, the latter from the restriction to totally defined basic operations. A particularly interesting tautology is given by the following

**Example 1 (Russell's antinomy).** We will show that the "Russellian set" $u$, consisting of all sets not containing themselves as a member, does not exist which clearly follows from $\vDash \neg\exists u \forall x(x \in u \leftrightarrow x \notin x)$. We start with $\forall x(x \in u \leftrightarrow x \notin x) \vDash u \in u \leftrightarrow u \notin u$. This holds by Corollary 3.6(a). Clearly, $u \in u \leftrightarrow u \notin u$ is unsatisfiable. Hence, the same holds for $\forall x(x \in u \leftrightarrow x \notin x)$, and thus for $\exists u \forall x(x \in u \leftrightarrow x \notin x)$. Consequently, $\vDash \neg\exists u \forall x(x \in u \leftrightarrow x \notin x)$.

Note that we need not assume in the above argument that $\in$ means membership. The proof of $\vDash \neg\exists u \forall x(x \in u \leftrightarrow x \notin x)$ need not be related to set theory at all. Hence, our example represents rather a logical paradox than a set-theoretic antinomy. What looks like an antinomy here is the expectation that $\exists u \forall x(x \in u \leftrightarrow x \notin x)$ *should* hold in set theory if $\in$ is to mean membership and Cantor's definition of a set is taken literally.

The satisfaction clause for $\alpha \to \beta$ easily yields $\alpha \vDash \beta \Leftrightarrow \vDash \alpha \to \beta$, a special case of $X, \alpha \vDash \beta \Leftrightarrow X \vDash \alpha \to \beta$. This can be very useful in checking whether formulas given in implicative form are tautologies, as was mentioned already in **1.3**. For instance, from $\forall x\alpha \vDash \alpha \frac{t}{x}$ (which holds for collision-free $\alpha, \frac{t}{x}$) we immediately get $\vDash \forall x\alpha \to \alpha \frac{t}{x}$.

As in propositional logic, $\alpha \equiv \beta$ is again equivalent to $\vDash \alpha \leftrightarrow \beta$. By inserting $\mathcal{L}$-formulas for the variables of a propositional equivalence one automatically procures one of predicate logic. Thus, for instance, $\alpha \to \beta \equiv \neg\alpha \vee \beta$, because certainly $p \to q \equiv \neg p \vee q$. Since every $\mathcal{L}$-formula results from the insertion of propositionally irreducible $\mathcal{L}$-formulas in a formula of propositional logic, one also sees that every $\mathcal{L}$-formula can be converted into a conjunctive normal form. But there are also numerous other equivalences, for example $\neg\forall x\alpha \equiv \exists x\neg\alpha$ and $\neg\exists x\alpha \equiv \forall x\neg\alpha$. The first of these means just $\neg\forall x\alpha \equiv \neg\forall x\neg\neg\alpha \ (= \exists x\neg\alpha)$, obtained by replacing $\alpha$ by the equivalent formula $\neg\neg\alpha$ under the prefix $\forall x$. This is a simple application of Theorem 4.1 below with $\equiv$ for $\approx$.

As in propositional logic, semantic equivalence is an equivalence relation in $\mathcal{L}$ and, moreover, a *congruence in* $\mathcal{L}$. Speaking more generally, an equivalence relation $\approx$ in $\mathcal{L}$ satisfying the congruence property

CP: $\quad \alpha \approx \alpha', \ \beta \approx \beta' \ \Rightarrow \ \alpha \wedge \beta \approx \alpha' \wedge \beta', \ \neg\alpha \approx \neg\alpha', \ \forall x\alpha \approx \forall x\alpha'$

is termed a *congruence in* $\mathcal{L}$. Its most important property is expressed by

**Theorem 4.1 (Replacement theorem).** *Let $\approx$ be a congruence in $\mathcal{L}$ and $\alpha \approx \alpha'$. If $\varphi'$ results from $\varphi$ by replacing the formula $\alpha$ at one or more of its occurrences in $\varphi$ by the formula $\alpha'$, then $\varphi \approx \varphi'$.*

**Proof** by induction on $\varphi$. Suppose $\varphi$ is a prime formula. Both for $\varphi = \alpha$ and $\varphi \neq \alpha$, $\varphi \approx \varphi'$ clearly holds. Now let $\varphi = \varphi_1 \wedge \varphi_2$. In case $\varphi = \alpha$ holds trivially $\varphi \approx \varphi'$. Otherwise $\varphi' = \varphi_1' \wedge \varphi_2'$, where $\varphi_1', \varphi_2'$ result from $\varphi_1, \varphi_1$ by possible replacements. By the induction hypothesis $\varphi_1 \approx \varphi_1'$ and $\varphi_2 \approx \varphi_2'$. Hence, $\varphi = \varphi_1 \wedge \varphi_2 \approx \varphi_1' \wedge \varphi_2' = \varphi'$ according to CP above. The induction steps for $\neg, \forall$ follow analogously. ☐

This theorem will constantly be used, mainly with $\equiv$ for $\approx$, without actually specifically being cited, just as in the arithmetical rearrangement of terms, where the laws of arithmetic used are hardly ever named explicitly. The theorem readily implies that CP is provable for all defined connectives such as $\to$ and $\exists$. For example, $\alpha \approx \alpha' \Rightarrow \exists x\alpha \approx \exists x\alpha'$, because $\alpha \approx \alpha' \Rightarrow \exists x\alpha = \neg\forall x\neg\alpha \approx \neg\forall x\neg\alpha' = \exists x\alpha'$.

First-order languages have a finer structure than those of propositional logic. There are consequently further interesting congruences in $\mathcal{L}$. In particular, formulas $\alpha, \beta$ are *equivalent in an $\mathcal{L}$-structure $\mathcal{A}$*, in symbols

$\alpha \equiv_{\mathcal{A}} \beta$, if $\mathcal{A} \vDash \alpha\,[w] \Leftrightarrow \mathcal{A} \vDash \beta\,[w]$, for all $w$. Hence, in $\mathcal{A} = (\mathbb{N}, <, +, 0)$ the formulas $x < y$ and $\exists z\,(z \neq 0 \wedge x + z = y)$ are equivalent. The proof of CP for $\equiv_{\mathcal{A}}$ is very simple and is therefore left to the reader.

Clearly, $\alpha \equiv_{\mathcal{A}} \beta$ is equivalent to $\mathcal{A} \vDash \alpha \leftrightarrow \beta$. Because of $\equiv\,\subseteq\,\equiv_{\mathcal{A}}$, properties such as $\neg \forall x \alpha \equiv \exists x \neg \alpha$ carry over from $\equiv$ to $\equiv_{\mathcal{A}}$. But there are often new interesting equivalences in certain structures. For instance, there are structures in which every formula is equivalent to a formula without quantifiers, as we will see in **5.6**.

A very important fact with an almost trivial proof is that the intersection of a family of congruences is itself a congruence. Consequently, for any class $\boldsymbol{K} \neq \emptyset$ of $\mathcal{L}$-structures, $\equiv_{\boldsymbol{K}} := \bigcap \{\equiv_{\mathcal{A}} \mid \mathcal{A} \in \boldsymbol{K}\}$ is necessarily a congruence. For the class $\boldsymbol{K}$ of *all* $\mathcal{L}$-structures, $\equiv_{\boldsymbol{K}}$ equals the logical equivalence $\equiv$, which in this section we deal with exclusively. Below we list its most important features; these should be committed to memory, since they will continually be applied.

$$(1) \quad \forall x(\alpha \wedge \beta) \equiv \forall x \alpha \wedge \forall x \beta, \qquad (2) \quad \exists x(\alpha \vee \beta) \equiv \exists x \alpha \vee \exists x \beta,$$

$$(3) \quad \forall x \forall y \alpha \equiv \forall y \forall x \alpha, \qquad (4) \quad \exists x \exists y \alpha \equiv \exists y \exists x \alpha.$$

If $x$ does not occur free in the formula $\beta$, then also

$$(5) \quad \forall x(\alpha \vee \beta) \equiv \forall x \alpha \vee \beta, \qquad (6) \quad \exists x(\alpha \wedge \beta) \equiv \exists x \alpha \wedge \beta,$$

$$(7) \quad \forall x \beta \equiv \beta, \qquad (8) \quad \exists x \beta \equiv \beta,$$

$$(9) \quad \forall x(\alpha \to \beta) \equiv \exists x \alpha \to \beta, \qquad (10) \quad \exists x(\alpha \to \beta) \equiv \forall x \alpha \to \beta.$$

The simple proofs are left to the reader. (7) and (8) were stated in (2) in **2.3**. Only (9) and (10) look at first sight surprising. But in practice these equivalences are very frequently used. For instance, consider for a fixed set of formulas $X$ the evidently true metalogical assertion 'for all $\alpha$: if $X \vDash \alpha, \neg \alpha$ then $X \vDash \forall x\, x \neq x$'. This clearly states the same as 'If there is some $\alpha$ such that $X \vDash \alpha, \neg \alpha$ then $X \vDash \forall x\, x \neq x$'.

**Remark.** In everyday speech variables tend to remain unquantified, partly because in some cases the same meaning results from quantifying with "there exists a" as with "for all." For instance, consider the following three sentences, which obviously tell us the same thing, and of which the last two correspond to the logical equivalence (9):

- If a lawyer finds a loophole in the law it must be changed.
- If there is a lawyer who finds a loophole in the law it must be changed.
- For all lawyers: if one of them finds a loophole in the law then it must be changed.

Often, the type of quantification in linguistic bits of information can be made out only from the context, and this leads not all too seldom to unintentional (or intentional) misunderstandings. "Logical relations in language are almost always just alluded to, left to guesswork, and not actually expressed" (G. Frege).

Let $x, y$ be distinct variables and $\alpha \in \mathcal{L}$. One of the most important logical equivalences is *renaming of bound variables* (in short, *bound renaming*), stated in

(11)　(a) $\forall x \alpha \equiv \forall y (\alpha \frac{y}{x})$,　(b) $\exists x \alpha \equiv \exists y (\alpha \frac{y}{x})$　$(y \notin \mathrm{var}\,\alpha)$.

(b) follows from (a) by rearranging equivalently. Note that $y \notin \mathrm{var}\,\alpha$ is equivalent to $y \notin \mathrm{free}\,\alpha$ and $\alpha, \frac{y}{x}$ collision-free. Writing $\mathcal{M}_x^y$ for $\mathcal{M}_x^{y^{\mathcal{M}}}$, (a) derives as follows:

$$
\begin{aligned}
\mathcal{M} \vDash \forall x \alpha \;&\Leftrightarrow\; \mathcal{M}_x^a \vDash \alpha && \text{for all } a \quad \text{(definition)} \\
&\Leftrightarrow\; (\mathcal{M}_y^a)_x^a \vDash \alpha && \text{for all } a \quad \text{(Theorem 3.1)} \\
&\Leftrightarrow\; (\mathcal{M}_y^a)_x^y \vDash \alpha && \text{for all } a \quad ((\mathcal{M}_y^a)_x^y = (\mathcal{M}_y^a)_x^a) \\
&\Leftrightarrow\; \mathcal{M}_y^a \vDash \alpha \tfrac{y}{x} && \text{for all } a \quad \text{(Theorem 3.5)} \\
&\Leftrightarrow\; \mathcal{M} \vDash \forall y (\alpha \tfrac{y}{x}).
\end{aligned}
$$

(12) and (13) below are also noteworthy. According to (13), substitutions are completely described up to logical equivalence by so-called *free renamings* (substitutions of the form $\frac{y}{x}$). (13) also embraces the case $x \in \mathrm{var}\,t$. In (12) and (13) we tacitly assume that $\alpha, \frac{t}{x}$ are collision-free.

(12) $\forall x (x = t \to \alpha) \equiv \alpha \frac{t}{x} \equiv \exists x (x = t \wedge \alpha)$　$(x \notin \mathrm{var}\,t)$.

(13) $\forall y (y = t \to \alpha \frac{y}{x}) \equiv \alpha \frac{t}{x} \equiv \exists y (y = t \wedge \alpha \frac{y}{x})$　$(y \notin \mathrm{var}\,\alpha, t)$.

*Proof of* (12): $\forall x (x = t \to \alpha) \vDash (x = t \to \alpha) \frac{t}{x} = t = t \to \alpha \frac{t}{x} \vDash \alpha \frac{t}{x}$ by Corollary 3.6. Conversely, let $\mathcal{M} \vDash \alpha \frac{t}{x}$. If $\mathcal{M}_x^a \vDash x = t$ then clearly $a = t^{\mathcal{M}}$. Hence also $\mathcal{M}_x^a \vDash \alpha$, since $\mathcal{M}_x^{t^{\mathcal{M}}} \vDash \alpha$. Thus, $\mathcal{M}_x^a \vDash x = t \to \alpha$ for any $a \in A$, i.e., $\mathcal{M} \vDash \forall x (x = t \to \alpha)$. This proves the left equivalence in (12). The right equivalence reduces to the left one because

$$\exists x (x = t \wedge \alpha) = \neg \forall x \neg (x = t \wedge \alpha) \equiv \neg \forall x (x = t \to \neg \alpha) \equiv \neg \neg \alpha \tfrac{t}{x} \equiv \alpha \tfrac{t}{x}.$$

Item (13) is proved similarly. Note that $\forall y (y = t \to \alpha \frac{y}{x}) \vDash \alpha \frac{y}{x} \frac{t}{y} = \alpha \frac{t}{x}$ by Corollary 3.6 and Exercise 4 in **2.2**.

With the above equivalences we can now regain an equivalent formula starting with any formula in which all quantifiers are standing at the beginning. But this result requires both quantifiers $\forall$ and $\exists$, in the following denoted by $\mathsf{Q}, \mathsf{Q}_1, \mathsf{Q}_2, \ldots$

A formula of the form $\alpha = \mathsf{Q}_1 x_1 \cdots \mathsf{Q}_n x_n \beta$ with an open formula $\beta$ is termed a *prenex formula* or a *prenex normal form*, in short, a PNF. $\beta$ is called the *kernel* of $\alpha$. W.l.o.g. $x_1, \ldots, x_n$ are distinct and $x_i$ occurs free in $\beta$ since we may drop "superfluous quantifiers," see (2) page 66. Prenex normal forms are very important for classifying definable number-theoretic predicates in **6.3**, and for other purposes. The already mentioned $\forall$- and $\exists$-formulas are the simplest examples.

**Theorem 4.2 (on the prenex normal form).** *Every formula $\varphi$ is equivalent to a formula in prenex normal form that can effectively be constructed from $\varphi$.*

**Proof.** Without loss of generality let $\varphi$ contain only the logical symbols $\neg, \wedge, \forall, \exists$ (besides $=$). For each prefix $\mathsf{Q}x$ in $\varphi$ consider the number of symbols $\neg$ or $\wedge$ occurring to the left of $\mathsf{Q}x$. Let $s\varphi$ be the sum of these numbers, summed over all prefixes occurring in $\varphi$. Clearly, $\varphi$ is a PNF iff $s\varphi = 0$. Let $s\varphi \neq 0$. Then $\varphi$ contains some prefix $\mathsf{Q}x$ and $\neg$ or $\wedge$ stands immediately in front of $\mathsf{Q}x$. A successive application of either

$$\neg \forall x \alpha \equiv \exists x \neg \alpha, \quad \neg \exists x \alpha \equiv \forall x \neg \alpha, \quad \text{or} \quad \beta \wedge \mathsf{Q}x\alpha \equiv \mathsf{Q}y(\, b \wedge \alpha \tfrac{y}{x}) \ (y \notin \text{var}\,\alpha, \beta),$$

inside $\varphi$ obviously reduces $s\varphi$ stepwise. $\qquad\blacksquare$

**Example 2.** $\forall x \exists y (x \neq 0 \to x \cdot y = 1)$ is a PNF for $\forall x (x \neq 0 \to \exists y\, x \cdot y = 1)$. And $\exists x \forall y \forall z (\varphi \wedge (\varphi \tfrac{y}{x} \wedge \varphi \tfrac{z}{x} \to y = z))$ for $\exists x \varphi \wedge \forall y \forall z (\varphi \tfrac{y}{x} \wedge \varphi \tfrac{z}{x} \to y = z)$, provided $y, z \notin \text{free}\, \varphi$; if not, a bound renaming will help. An equivalent PNF for this formula with minimal quantifier rank is $\exists x \forall y (\varphi \tfrac{y}{x} \leftrightarrow x = y)$.

The formula $\forall x (x \neq 0 \to \exists y\, x \cdot y = 1)$ from Example 2 may be abbreviated by $(\forall x \neq 0) \exists y\, x \cdot y = 1$. More generally, we shall often write $(\forall x \neq t)\alpha$ for $\forall x (x \neq t \to \alpha)$ and $(\exists x \neq t)\alpha$ for $\exists x (x \neq t \wedge \alpha)$. A similar notation is used for $\leqslant, <, \in$ and their negations. For instance, $(\forall x \leqslant t)\alpha$ and $(\exists x \leqslant t)\alpha$ are to mean $\forall x (x \leqslant t \to \alpha)$ and $\exists x (x \leqslant t \wedge \alpha)$, respectively. For any binary relation symbol $\triangleleft$, the "prefixes" $(\forall y \triangleleft x)$ and $(\exists y \triangleleft x)$ are related to each other, as are $\forall$ and $\exists$, see Exercise 2.

### Exercises

1. Let $\alpha \equiv \beta$. Prove that $\alpha \tfrac{\vec{t}}{\vec{x}} \equiv \beta \tfrac{\vec{t}}{\vec{x}}$ $\quad (\alpha, \tfrac{\vec{t}}{\vec{x}}$ and $\beta, \tfrac{\vec{t}}{\vec{x}}$ collision-free$)$.

2. Prove that $\neg(\forall x \triangleleft y)\alpha \equiv (\exists x \triangleleft y)\neg \alpha$ and $\neg(\exists x \triangleleft y)\alpha \equiv (\forall x \triangleleft y)\neg \alpha$. Here $\triangleleft$ represents any binary relation symbol.

3. Show by means of bound renaming that both the conjunction and the disjunction of $\forall$-formulas $\alpha, \beta$ is equivalent to some $\forall$-formula. Prove the same for $\exists$-formulas.

4. Show that every formula $\varphi \in \mathcal{L}$ is equivalent to some $\varphi' \in \mathcal{L}$ built up from literals by means of $\wedge$, $\vee$, and $\exists$.

5. Let $P$ be a unary predicate symbol. Prove that $\exists x(Px \to \forall y Py)$ is a tautology.

6. Call $\alpha, \beta \in \mathcal{L}$ *tautologically equivalent* if $\vDash \alpha \Leftrightarrow \vDash \beta$. Confirm that the following (in general not logically equivalent) formulas are tautologically equivalent: $\alpha$, $\forall x \alpha$, and $\alpha \frac{c}{x}$, where the constant symbol $c$ does not occur in $\alpha$.

## 2.5   Logical Consequence and Theories

Whenever $\mathcal{L}' \supseteq \mathcal{L}$, the language $\mathcal{L}'$ is called an *expansion* or *extension* of $\mathcal{L}$ and $\mathcal{L}$ a *reduct* or *restriction* of $\mathcal{L}'$. Recall the insensitivity of the consequence relation to extensions of a first-order language, mentioned in **2.3**. Theorem 3.1 yields that establishing $X \vDash \alpha$ does not depend on the language to which the set of formulas $X$ and the formula $\alpha$ belong. For this reason, indices for $\vDash$, such as $\vDash_{\mathcal{L}}$, are dispensable.

Because of the unaltered satisfaction conditions for $\wedge$ and $\neg$, all properties of the propositional consequence gained in **1.3** carry over to the first-order logical consequence relation. These include general properties such as, for example, the reflexivity and transitivity of $\vDash$, and the semantic counterparts of the rules $(\wedge 1)$, $(\wedge 2)$, $(\neg 1)$, $(\neg 2)$ from **1.4**, for instance the counterpart of $(\wedge 1)$, $\dfrac{X \vDash \alpha, \beta}{X \vDash \alpha \wedge \beta}$.[5]

In addition, Gentzen-style properties such as the deduction theorem automatically carry over. But there are also completely new properties. Some of these will be elevated to basic rules of a logical calculus for first-order languages in **3.1**, to be found among the following ones:

---

[5] A suggestive way of writing "$X \vDash \alpha, \beta$ implies $X \vDash \alpha \wedge \beta$," a notation that was introduced already in Exercise 3 in **1.3**. A corresponding notation will also be used in stating the properties of $\vDash$ on the next page.

**Some properties of the predicate logical consequence relation.**

(a) $\dfrac{X \vDash \forall x \alpha}{X \vDash \alpha \frac{t}{x}}$  $(\alpha, \frac{t}{x}$ collision-free),

(b) $\dfrac{X \vDash \alpha \frac{s}{x}, s = t}{X \vDash \alpha \frac{t}{x}}$  $(\alpha, \frac{s}{x}$ and $\alpha, \frac{t}{x}$ collision-free),

(c) $\dfrac{X, \beta \vDash \alpha}{X, \forall x \beta \vDash \alpha}$  (anterior generalization),

(d) $\dfrac{X \vDash \alpha}{X \vDash \forall x \alpha}$  $(x \notin free\, X$, posterior generalization),

(e) $\dfrac{X, \beta \vDash \alpha}{X, \exists x \beta \vDash \alpha}$  $(x \notin free\, X, \alpha$, anterior particularization),

(f) $\dfrac{X \vDash \alpha \frac{t}{x}}{X \vDash \exists x \alpha}$  $(\alpha, \frac{t}{x}$ collision-free, posterior particularization)

(a) follows from $X \vDash \forall x \alpha \vDash \alpha \frac{t}{x}$, for $\vDash$ is transitive. Similarly, (b) follows from $\alpha \frac{s}{x}, s = t \vDash \alpha \frac{t}{x}$, stated in Corollary 3.6. Analogously (c) results from $\forall x \beta \vDash \beta$. To prove (d), suppose that $X \vDash \alpha$, $\mathcal{M} \vDash X$, and $x \notin free\, X$. Then $\mathcal{M}_x^a \vDash X$ for any $a \in A$ by Theorem 3.1, which just means $\mathcal{M} \vDash \forall x \alpha$. As regards (e), let $X, \beta \vDash \alpha$. Observe that by contraposition and by (d),

$$X, \beta \vDash \alpha \;\Rightarrow\; X, \neg \alpha \vDash \neg \beta \;\Rightarrow\; X, \neg \alpha \vDash \forall x \neg \beta,$$

whence $X, \neg \forall x \neg \beta \vDash \alpha$. (e) captures deduction *from* an existence claim, while (f) *confirms* an existence claim. (f) holds since $\alpha \frac{t}{x} \vDash \exists x \alpha$ according to Corollary 3.6. Both (e) and (f) are permanently applied in mathematical reasoning and will briefly be discussed in Example 1 on the next page. All above properties have certain variants; for example, a variant of (d) is

(g) $\dfrac{X \vDash \alpha \frac{y}{x}}{X \vDash \forall x \alpha}$  $(y \notin free\, X \cup var\, \alpha)$.

This results from (d) with $\alpha \frac{y}{x}$ for $\alpha$ and $y$ for $x$, since $\forall y \alpha \frac{y}{x} \equiv \forall x \alpha$.

From the above properties, complicated chains of deduction can, where necessary, be justified step by step. But in practice this makes sense only in particular circumstances, because formalized proofs are readable only at the expense of a lot of time, just as with lengthy computer programs, even with well-prepared documentation. What is most important is that a proof, when written down, can be understood and reproduced. This is why mathematical deduction tends to proceed *informally*, i.e., both claims and their proofs are formulated in a mathematical "everyday" language with the aid of fragmentary and flexible formalization. To what degree a proof is to be formalized depends on the situation and need not be determined in advance. In this way the strict syntactic structure of formal proofs is slackened, compensating for the imperfection of our brains in regard to processing syntactic information.

Further, certain informal proof methods will often be described by a more or less clear reference to so-called background knowledge, and not actually carried out. This method has proven itself to be sufficiently reliable. As a matter of fact, apart from specific cases it has not yet been bettered by any of the existing automatic proof machines. Let us present a very simple example of an informal proof in a language $\mathcal{L}$ for natural numbers that along with $0, 1, +, \cdot$ contains the symbol $\mid$ for divisibility, defined by $m \mid n \Leftrightarrow \exists k\, m \cdot k = n$. In addition, let $\mathcal{L}$ contain a symbol $\mathsf{f}$ for some given function from $\mathbb{N}$ to $\mathbb{N}$. We need no closer information on this function, but we shall write $\mathsf{f}_i$ for $\mathsf{f}(i)$ in Example 1.

**Example 1.** We want to prove $\forall n \exists x (\forall i {\leqslant} n) \mathsf{f}_i \mid x$. That is, for every $n$, $\mathsf{f}_0, \ldots, \mathsf{f}_n$ have a common multiple. A careful proof proceeds by induction on $n$. Here we focus solely on $X, \exists x (\forall i {\leqslant} n) \mathsf{f}_i \mid x \vDash \exists x (\forall i {\leqslant} n{+}1) \mathsf{f}_i \mid x$, the induction step. $X$ represents our prior knowledge about familiar properties of divisibility. Informally we reason as follows: Suppose $\exists x (\forall i {\leqslant} n) \mathsf{f}_i \mid x$ and let $x$ denote any common multiple of $\mathsf{f}_0, \ldots, \mathsf{f}_n$. Then $x \cdot \mathsf{f}_{n+1}$ is clearly a common multiple of $\mathsf{f}_0, \ldots, \mathsf{f}_{n+1}$, hence $\exists x (\forall i {\leqslant} n{+}1) \mathsf{f}_i \mid x$. That's all. To argue here formally like a proof machine, let us start from the obvious $(\forall i {\leqslant} n) \mathsf{f}_i \mid x \vDash (\forall i {\leqslant} n{+}1) \mathsf{f}_i \mid (x \cdot \mathsf{f}_{n+1})$. Posterior particularization of $x$ yields $X, (\forall i {\leqslant} n) \mathsf{f}_i \mid x \vDash \exists x (\forall i {\leqslant} n{+}1) \mathsf{f}_i \mid x$. From this follows the desired $X, \exists x (\forall i {\leqslant} n) \mathsf{f}_i \mid x \vDash \exists x (\forall i {\leqslant} n{+}1) \mathsf{f}_i \mid x$ by anterior particularization. Thus, formalizing a nearly trivial informal argument may need a lot of writing and turns out to be nontrivial in some sense.

Some textbooks deal with a somewhat stricter consequence relation, which we denote here by $\overset{g}{\vDash}$. The reason is that in mathematics one largely considers derivations in theories. For $X \subseteq \mathcal{L}$ and $\varphi \in \mathcal{L}$ define $X \overset{g}{\vDash} \varphi$ if $\mathcal{A} \vDash X \Rightarrow \mathcal{A} \vDash \varphi$, for all $\mathcal{L}$-structures $\mathcal{A}$. In contrast to $\vDash$, which may be called the *local* consequence relation, $\overset{g}{\vDash}$ can be considered as the *global* consequence relation since it cares only about $\mathcal{A}$, not about a concrete valuation $w$ in $\mathcal{A}$ as does $\vDash$.

Let us collect a few properties of $\overset{g}{\vDash}$. Obviously, $X \vDash \varphi$ implies $X \overset{g}{\vDash} \varphi$, but the converse does not hold in general. For example, $x = y \overset{g}{\vDash} \forall xy\, x = y$, but $x = y \nvDash \forall xy\, x = y$. By (d) from page 79, $X \vDash \varphi \Rightarrow X \vDash \varphi^g$ holds in general only if the free variables of $\varphi$ do not occur free in $X$, while $X \overset{g}{\vDash} \varphi \Rightarrow X \overset{g}{\vDash} \varphi^g$ (hence $\varphi \overset{g}{\vDash} \varphi^g$) holds unrestrictedly. A reduction of $\overset{g}{\vDash}$ to $\vDash$ is provided by the following equivalence, which easily follows from $\mathcal{M} \vDash X^g \Leftrightarrow \mathcal{A} \vDash X^g$, for each model $\mathcal{M} = (\mathcal{A}, w)$:

(1) $\quad X \overset{g}{\vDash} \varphi \Leftrightarrow X^g \vDash \varphi$.

Because of $S^g = S$ for sets of sentences $S$, we clearly obtain from (1)

(2) $\quad S \overset{g}{\vDash} \varphi \Leftrightarrow S \vDash \varphi \quad (S \subseteq \mathcal{L}^0)$.

In particular, $\overset{g}{\vDash} \varphi \Leftrightarrow \vDash \varphi$. Thus, a distinction between $\vDash$ and $\overset{g}{\vDash}$ is apparent only when premises are involved that are not sentences. In this case the relation $\overset{g}{\vDash}$ must be treated with the utmost care. Neither the rule of case distinction $\dfrac{X, \alpha \overset{g}{\vDash} \beta \mid X, \neg\alpha \overset{g}{\vDash} \beta}{X \overset{g}{\vDash} \beta}$ nor the deduction theorem $\dfrac{X, \alpha \overset{g}{\vDash} \beta}{X \overset{g}{\vDash} \alpha \to \beta}$ is unrestrictedly correct. For example $x = y \overset{g}{\vDash} \forall xy\, x = y$, but it is false that $\overset{g}{\vDash} x = y \to \forall xy\, x = y$. This means that the deduction theorem fails to hold for the relation $\overset{g}{\vDash}$. It holds only under certain restrictions.

One of the reasons for our preference of $\vDash$ over $\overset{g}{\vDash}$ is that $\vDash$ extends the propositional consequence relation conservatively, so that features such as the deduction theorem carry over unrestrictedly, while this is not the case for $\overset{g}{\vDash}$. It should also be said that $\overset{g}{\vDash}$ does not reflect the actual procedures of natural deduction in which formulas with free variables are frequently used also in deductions of sentences from sentences, for instance in Example 1.

We now make more precise the notion of a formalized theory in $\mathcal{L}$, where it is useful to think of the examples in **2.3**, such as group theory. Again, the definitions by different authors may look somewhat differently.

**Definition.** An *elementary theory* or *first-order theory* in $\mathcal{L}$, also termed an *$\mathcal{L}$-theory*, is a set of sentences $T \subseteq \mathcal{L}^0$ *deductively closed in* $\mathcal{L}^0$, i.e., $T \vDash \alpha \Leftrightarrow \alpha \in T$, for all $\alpha \in \mathcal{L}^0$. If $\alpha \in T$ then we say that $\alpha$ *is valid* or *true* or *holds in* $T$, or $\alpha$ *is a theorem* of $T$. The extralogical symbols of $\mathcal{L}$ are called the *symbols of* $T$. If $T \subseteq T'$ then $T$ is called a *subtheory* of $T'$, and $T'$ an *extension* of $T$. An $\mathcal{L}$-structure $\mathcal{A}$ such that $\mathcal{A} \vDash T$ is also termed a *model of* $T$, briefly a *$T$-model*. $\mathrm{Md}\, T$ denotes the class of all models of $T$ in this sense; $\mathrm{Md}\, T$ consist of $\mathcal{L}$-structures only.

For instance, $\{\alpha \in \mathcal{L}^0 \mid X \vDash \alpha\}$ is a theory for any set $X \subseteq \mathcal{L}$, since $\vDash$ is transitive. A theory $T$ in $\mathcal{L}$ satisfies $T \vDash \varphi \Leftrightarrow \mathcal{A} \vDash \varphi$ for all $\mathcal{A} \vDash T$, where $\varphi \in \mathcal{L}$ is any formula. Important is also $T \vDash \varphi \Leftrightarrow T \vDash \varphi^g$. These readily confirmed facts should be taken in and remembered, since they are constantly used. Different authors may use different definitions for a theory. For example, they may not demand that theories contain sentences only, as we do. Conventions of this type each have their advantages and disadvantages. Proofs regarding theories are always adaptable enough to accommodate small modifications of the definition. Using the definition given above we set the following

**Convention.** In talking of the *theory $S$*, where $S$ is a set of sentences, we always mean the theory determined by $S$, that is, $\{\alpha \in \mathcal{L}^0 \mid S \vDash \alpha\}$. A set $X \subseteq \mathcal{L}$ is called an *axiom system* for $T$ whenever $T = \{\alpha \in \mathcal{L}^0 \mid X^g \vDash \alpha\}$, i.e., we tacitly generalize all possibly open formulas in $X$. We have always to think of free variables occurring in axioms as being generalized.

Thus, axioms of a theory are always sentences. But we conform to standard practice of writing long axioms as formulas. We will later consider extensive axiom systems (in particular, for arithmetic and set theory) whose axioms are partly written as open formulas just for economy.

There exists a smallest theory in $\mathcal{L}$, namely the set **Taut** (= **Taut**$_\mathcal{L}$) of all generally valid sentences in $\mathcal{L}$, also called the "logical" theory. An axiom system for **Taut** is the empty set of axioms. There is also a largest theory: the set $\mathcal{L}^0$ *of all* sentences, the *inconsistent* theory, which possesses no models. All remaining theories are called *satisfiable* or *consistent*.[6] Moreover, the intersection $T = \bigcap_{i \in I} T_i$ of a nonempty family of theories $T_i$ is in turn a theory: if $T \vDash \alpha \in \mathcal{L}^0$ then clearly $T_i \vDash \alpha$ and so $\alpha \in T_i$ for each $i \in I$, hence $\alpha \in T$ as well. In this book $T$ and $T'$, with or without indices, exclusively denote theories.

For $T \subseteq \mathcal{L}^0$ and $\alpha \in \mathcal{L}^0$ let $T + \alpha$ denote the smallest theory that extends $T$ and contains $\alpha$. Similarly let $T + S$ for $S \subseteq \mathcal{L}^0$ be the smallest theory containing $T \cup S$. If $S$ is finite then $T' = T + S = T + \bigwedge S$ is called a *finite extension of* $T$. Here $\bigwedge S$ denotes the conjunction of all sentences in $S$. A sentence $\alpha$ is termed *compatible* or *consistent* with $T$ if $T + \alpha$ is

---

[6] *Consistent* mostly refers to a logic calculus, e.g., the calculus in **3.1**. However, it will be shown in **3.2** that consistency and satisfiability of a theory coincide, thus justifying the word's ambiguous use.

satisfiable, and *refutable in T* if $T + \neg\alpha$ is satisfiable. Thus, the theory $T_F$ of fields is *compatible* with the sentence $1 + 1 = 0$. Equivalently, $1 + 1 \neq 0$ is *refutable* in $T_F$, since the 2-element field satisfies $1 + 1 = 0$.

If both $\alpha$ and $\neg\alpha$ are compatible with $T$ then the sentence $\alpha$ is termed *independent* of $T$. The classic example is the independence of the parallel axiom from the remaining axioms of Euclidean plane geometry, which define *absolute* geometry. Much more difficult is the independence proof of the continuum hypothesis from the axioms for set theory. These axioms are presented and discussed in **3.4**.

At this point we introduce another important concept; $\alpha, \beta \in \mathcal{L}$ are said to be *equivalent in* or *modulo T*, $\alpha \equiv_T \beta$, if $\alpha \equiv_{\mathcal{A}} \beta$ for all $\mathcal{A} \vDash T$. Being an intersection of congruences, $\equiv_T$ is itself a congruence and hence satisfies the replacement theorem. This will henceforth be used without mention, as will the obvious equivalence of $\alpha \equiv_T \beta$, $T \vDash \alpha \leftrightarrow \beta$, and of $T \vDash (\alpha \leftrightarrow \beta)^g$. A suggestive writing of $\alpha \equiv_T \beta$ would also be $\alpha =_T \beta$.

**Example 2.** Let $T_G$ be as on p. 65. Claim: $x \circ x = x \equiv_{T_G} x = e$. The only tricky proof step is $T_G \vDash x \circ x = x \rightarrow x = e$. Let $x \circ x = x$ and choose some $y$ with $x \circ y = e$. The claim then follows from $x = x \circ e = x \circ x \circ y = x \circ y = e$. A strict formal proof of the latter uses anterior particularization.

Another important congruence is term equivalence. Call terms $s, t$ *equivalent* modulo (or in) $T$, in symbols $s \approx_T t$, if $T \vDash s = t$, that is, $\mathcal{A} \vDash s = t [w]$ for all $\mathcal{A} \vDash T$ and $w \colon Var \rightarrow A$. For instance, in $T = T_G^{=}$, $(x \circ y)^{-1} = y^{-1} \circ x^{-1}$ is easily provable, so that $(x \circ y)^{-1} \approx_T y^{-1} \circ x^{-1}$. Another example: in the theory of fields, each term is equivalent to a polynomial in several variables with integer coefficients.

If all axioms of a theory $T$ are $\forall$-sentences then $T$ is called a *universal* or $\forall$-*theory*. Examples are partial orders, orders, rings, lattices, and Boolean algebras. For such a theory, $\mathrm{Md}\, T$ is closed with respect to substructures, which means $\mathcal{A} \subseteq \mathcal{B} \vDash T \Rightarrow \mathcal{A} \vDash T$. This follows at once from Corollary 3.3. Conversely, a theory closed with respect to substructures is necessarily a universal one, as will turn out in **5.4**. $\forall$-theories are further classified. The most important subclasses are equational, quasi-equational, and universal Horn theories, all of which will be considered to some extent in later chapters. Besides $\forall$-theories, the $\forall\exists$-theories (those having $\forall\exists$-sentences as axioms) are of particular interest for mathematics. More about all these theories will be said in **5.4**.

Theories are frequently given by structures or classes of structures. The elementary theory $Th\,\mathcal{A}$ and the theory $Th\,\boldsymbol{K}$ of a nonempty class $\boldsymbol{K}$ of structures are defined respectively by

$$Th\,\mathcal{A} := \{\alpha \in \mathcal{L}^0 \mid \mathcal{A} \vDash \alpha\}, \quad Th\,\boldsymbol{K} := \bigcap\{Th\,\mathcal{A} \mid \mathcal{A} \in \boldsymbol{K}\}.$$

It is easily seen that $Th\,\mathcal{A}$ and $Th\,\boldsymbol{K}$ are theories in the precise sense defined above. Instead of $\alpha \in Th\,\boldsymbol{K}$ one often writes $\boldsymbol{K} \vDash \alpha$. In general, Md $Th\,\boldsymbol{K}$ is larger than $\boldsymbol{K}$, as we shall see.

One easily confirms that the set of formulas breaks up modulo $T$ (more precisely, modulo $\equiv_T$) into equivalence classes; their totality is denoted by $B_\omega T$. Based on these we can define in a natural manner operations $\wedge, \vee, \neg$. For instance, $\bar{\alpha} \wedge \bar{\beta} = \overline{\alpha \wedge \beta}$, where $\bar{\varphi}$ denotes the equivalence class to which $\varphi$ belongs. One shows easily that $B_\omega T$ forms a Boolean algebra with respect to $\wedge, \vee, \neg$. For every $n$, the set $B_n T$ of all $\bar{\varphi}$ in $B_\omega T$ such that the free variables of $\varphi$ belong to $\mathrm{Var}_n$ $(= \{\boldsymbol{v}_0, \ldots, \boldsymbol{v}_{n-1}\})$ is a subalgebra of $B_\omega T$. Note that $B_0 T$ is isomorphic to the Boolean algebra of all sentences modulo $\equiv_T$, also called the *Tarski–Lindenbaum algebra* of $T$. The significance of the Boolean algebras $B_n T$ is revealed only in the somewhat higher reaches of model theory, and they are therefore mentioned only incidentally.

## Exercises

1. Suppose $x \notin \mathrm{free}\,X$ and $c$ is not in $X, \alpha$. Prove the equivalence of

   (i) $X \vDash \alpha$,      (ii) $X \vDash \forall x \alpha$,      (iii) $X \vDash \alpha\frac{c}{x}$.

   This holds then in particular if $X$ is the axiom system of a theory or itself a theory. Then $x \notin \mathrm{free}\,X$ is trivially satisfied.

2. Let $S$ be a set of sentences, $\alpha$ and $\beta$ formulas, $x \notin \mathrm{free}\,\beta$, and let $c$ be a constant not occurring in $S$, $\alpha$, $\beta$. Show that

   $$S \vDash \alpha\tfrac{c}{x} \to \beta \iff S \vDash \exists x \alpha \to \beta.$$

3. Verify for all $\alpha, \beta \in \mathcal{L}^0$ that $\beta \in T + \alpha \iff \alpha \to \beta \in T$.

4. Let $T \subseteq \mathcal{L}$ be a theory, $\mathcal{L}_0 \subseteq \mathcal{L}$, and $T_0 := T \cap \mathcal{L}_0$. Prove that $T_0$ is also a theory (the so-called *reduct theory* in the language $\mathcal{L}_0$).

## 2.6   Explicit Definitions—Language Expansions

The deductive development of a theory, be it given by an axiom system or a single structure or classes of those, nearly always goes hand in hand with expansions of the language carried out step by step. For example, in developing elementary number theory in the language $\mathcal{L}(0, 1, +, \cdot)$, the introduction of the divisibility relation by means of the (explicit) definition $x\,|\,y \leftrightarrow \exists z\, x \cdot z = y$ has certainly advantages not only for purely technical reasons. This and similar examples motivate the following

**Definition I.** Let $r$ be an $n$-ary relation symbol not occurring in $\mathcal{L}$. An *explicit definition of $r$ in $\mathcal{L}$* is to mean a formula of the form

$$\eta_r : \quad r\vec{x} \leftrightarrow \delta(\vec{x})$$

with $\delta(\vec{x}) \in \mathcal{L}$ and distinct variables in $\vec{x}$, called the *defining formula*. For a theory $T$, the extension $T_r := T + \eta_r^g$ is then called a *definitorial extension* (or *expansion*) of $T$ by $r$, more precisely, by $\eta_r$.

$T_r$ is a theory in $\mathcal{L}[r]$, the language resulting from $\mathcal{L}$ by adjoining the symbol $r$. It will turn out that $T_r$ is a *conservative extension* of $T$, which, in the general case, means a theory $T' \supseteq T$ in $\mathcal{L}' \supseteq \mathcal{L}$ such that $T' \cap \mathcal{L} = T$. Thus, $T_r$ contains exactly the same $\mathcal{L}$-sentences as does $T$. In this sense, $T_r$ is a harmless extension of $T$. Our claim constitutes part of Theorem 6.1. For $\varphi \in \mathcal{L}[r]$ define the *reduced formula* $\varphi^{rd} \in \mathcal{L}$ as follows: Starting from the left, replace every prime formula $r\vec{t}$ occurring in $\varphi$ by $\delta_{\vec{x}}(\vec{t})$. Clearly, $\varphi^{rd} = \varphi$, provided $r$ does not appear in $\varphi$.

**Theorem 6.1 (Elimination theorem).** *Let $T_r \subseteq \mathcal{L}[r]$ be a definitorial extension of the theory $T \subseteq \mathcal{L}^0$ by the explicit definition $\eta_r$. Then for all formulas $\varphi \in \mathcal{L}[r]$ holds the equivalence*

$$(*) \quad T_r \vDash \varphi \Leftrightarrow T \vDash \varphi^{rd}.$$

*For $\varphi \in \mathcal{L}$ we get in particular $T_r \vDash \varphi \Leftrightarrow T \vDash \varphi$ (since $\varphi^{rd} = \varphi$). Hence, $T_r$ is a conservative extension of $T$, i.e., $\alpha \in T_r \Leftrightarrow \alpha \in T$, for all $\alpha \in \mathcal{L}^0$.*

**Proof.** Each $\mathcal{A} \vDash T$ is expandable to a model $\mathcal{A}' \vDash T_r$ with the same domain, setting $r^{\mathcal{A}'}\vec{a} :\Leftrightarrow \mathcal{A} \vDash \delta\,[\vec{a}]\ (\vec{a} \in A^n)$. Since $r\vec{t} \equiv_{T_r} \delta(\vec{t})$ for any $\vec{t}$, we obtain $\varphi \equiv_{T_r} \varphi^{rd}$ for all $\varphi \in \mathcal{L}[r]$ by the replacement theorem. Thus, $(*)$ follows from

$$T_r \vDash \varphi \Leftrightarrow \mathcal{A}' \vDash \varphi \text{ for all } \mathcal{A} \vDash T \quad (\text{Md}\, T_r = \{\mathcal{A}' \mid \mathcal{A} \vDash T\})$$
$$\Leftrightarrow \mathcal{A}' \vDash \varphi^{rd} \text{ for all } \mathcal{A} \vDash T \quad (\text{because } \varphi \equiv_{T_r} \varphi^{rd})$$
$$\Leftrightarrow \mathcal{A} \vDash \varphi^{rd} \text{ for all } \mathcal{A} \vDash T \quad (\text{Theorem 3.1})$$
$$\Leftrightarrow T \vDash \varphi^{rd}. \quad \square$$

Operation symbols and constants can be similarly introduced, though in this case there are certain conditions to observe. For instance, in $T_G$ (see page 65) the operation $^{-1}$ is defined by $\eta : \; y = x^{-1} \leftrightarrow x \circ y = e$. This definition is legitimate, since $T_G \vDash \forall x \exists! y\, x \circ y = e$; Exercise 3. Only this requirement (which by the way is a logical consequence of $\eta$) ensures that $T_G + \eta^g$ is a conservative extension of $T_G$. We therefore extend Definition I as follows, keeping in mind that to the end of this section constant symbols are to be counted among the operation symbols.

**Definition II.** An *explicit definition* of an $n$-ary operation symbol $f$ not occurring in $\mathcal{L}$ is a formula of the form

$$\eta_f : \quad y = f\vec{x} \leftrightarrow \delta(\vec{x}, y) \qquad (\delta \in \mathcal{L} \text{ and } y, x_1, \ldots, x_n \text{ distinct}).$$

$\eta_f$ is called *legitimate* in $T \subseteq \mathcal{L}$ if $T \vDash \forall \vec{x} \exists! y \delta$, and $T_f := T + \eta_f^g$ is then called a *definitorial extension by* $f$, more precisely by $\eta_f$. In the case $n = 0$ we write $c$ for $f$ and speak of an *explicit definition of the constant symbol* $c$. Written more suggestively $y = c \leftrightarrow \delta(y)$.

Some of the free variables of $\delta$ are often not explicitly named, and thus downgraded to parameter variables. More on this will be said in the discussion of the axioms for set theory in **3.4**. The elimination theorem is proved in almost exactly the same way as above, provided $\eta_f$ is legitimate in $T$. The reduced formula $\varphi^{rd}$ is defined correspondingly. For a constant $c$ ($n = 0$ in Definition II), let $\varphi^{rd} := \exists z(\varphi \frac{z}{c} \wedge \delta \frac{z}{y})$, where $\varphi \frac{z}{c}$ denotes the result of replacing $c$ in $\varphi$ by $z$ ($\notin \mathrm{var}\,\varphi$). Now let $n > 0$. If $f$ does not appear in $\varphi$, set $\varphi^{rd} = \varphi$. Otherwise, looking at the first occurrence of $f$ in $\varphi$ from the left, we certainly may write $\varphi = \varphi_0 \frac{f\vec{t}}{y}$ for appropriate $\varphi_0$, $\vec{t}$, and $y \notin \mathrm{var}\,\varphi$. Clearly, $\varphi \equiv_{T_f} \exists y(\varphi_0 \wedge y = f\vec{t}) \equiv_{T_f} \varphi_1$, with $\varphi_1 := \exists y(\varphi_0 \wedge \delta_f(\vec{t}, y))$. If $f$ still occurs in $\varphi_1$ then repeat this procedure, which ends in, say, $m$ steps in a formula $\varphi_m$ that no longer contains $f$. Then put $\varphi^{rd} := \varphi_m$.

Frequently, operation symbols $f$ are introduced in more or less strictly formalized theories by definitions of the form

$$(*) \quad f\vec{x} := t(\vec{x}),$$

where of course $f$ does not occur in the term $t(\vec{x})$. This procedure is in fact subsumed by Definition II, because the former is nothing more than a definitorial extension of $T$ with the explicit definition

$$\eta_f : \; y = f\vec{x} \leftrightarrow y = t(\vec{x}).$$

This definition is legitimate, since $\forall \vec{x}\, \exists! y\, y = t(\vec{x})$ is a tautology. It can readily be shown that $\eta_f^g$ is logically equivalent to $\forall \vec{x}\, f\vec{x} = t(\vec{x})$. Hence, (∗) can indeed be regarded as a kind of an informative abbreviation of a legitimate explicit definition with the defining formula $y = t(\vec{x})$.

**Remark 1.** Instead of introducing new operation symbols, so-called *iota-terms* from [HB] could be used. For any formula $\varphi = \varphi(\vec{x}, y)$ in a given language, let $\iota y \varphi$ be a term in which $y$ appears as a variable *bound by* $\iota$. Whenever $T \vDash \forall \vec{x}\, \exists! y \varphi$, then $T$ is extended by the axiom $\forall \vec{x}\, \forall y\, [y = \iota y \varphi(\vec{x}, y) \leftrightarrow \varphi(\vec{x}, y)]$, so that $\iota y \varphi(\vec{x}, y)$ so to speak stands for the function term $f\vec{x}$, which could have been introduced by an explicit definition. We mention that a definitorial language expansion is not a necessity. In principle, formulas of the expanded language can always be understood as abbreviations in the original language. This is in some presentations the actual procedure, though our imagination prefers additional notions over long sentences that would arise if we were to stick to a minimal set of basic notions.

Definitions I and II can be unified in a more general declaration. Let $T, T'$ be theories in the languages $\mathcal{L}, \mathcal{L}'$, respectively. Then $T'$ is called a *definitorial extension* (or *expansion*) of $T$ whenever $T' = T + \Delta$ for some list $\Delta$ of explicit definitions of new symbols legitimate in $T$, given in terms of those of $T$ (here *legitimate* refers to operation symbols and constants only). $\Delta$ need not be finite, but in most cases it is finite. A reduced formula $\varphi^{rd} \in \mathcal{L}$ is stepwise constructed as above, for every $\varphi \in \mathcal{L}'$. In this way the somewhat long-winded proof of the following theorem is reduced each time to the case of an extension by a single symbol:

**Theorem 6.2 (General elimination theorem).** *Let $T'$ be a definitorial extension of $T$. Then $\alpha \in T' \Leftrightarrow \alpha^{rd} \in T$. In particular, $\alpha \in T' \Leftrightarrow \alpha \in T$ whenever $\alpha \in \mathcal{L}$, i.e., $T'$ is a conservative extension of $T$.*

A relation or operation symbol s occurring in $T \subseteq \mathcal{L}$ is termed *explicitly definable in $T$* if $T$ contains an explicit definition of s whose defining formula belongs to $\mathcal{L}_0$, the language of symbols of $T$ without s. For example, in the theory $T_G$ of groups the constant $e$ is explicitly defined by $x = e \leftrightarrow x \circ x = x$; Example 2 page 83. Another example is presented in Exercise 3. In such a case each model of $T_0 := T \cap \mathcal{L}_0$ can be expanded in only one way to a $T$-model. If this special condition is fulfilled then $\mathsf{s}$ is said to be *implicitly definable* in $T$. This could also be stated as follows: if $T'$ is distinct from $T$ only in that the symbol $\mathsf{s}$ is everywhere replaced by a new symbol $\mathsf{s}'$, then either $T \cup T' \vDash \forall \vec{x}(\mathsf{s}\vec{x} \leftrightarrow \mathsf{s}'\vec{x})$ or $T \cup T' \vDash \forall \vec{x}(\mathsf{s}\vec{x} = \mathsf{s}'\vec{x})$, depending on whether $\mathsf{s}, \mathsf{s}'$ are relation or operation symbols. It is highly interesting that this kind of definability is already sufficient for the explicit definability of $\mathsf{s}$ in $T$. But we will go without the proof and only quote the following theorem.

**Beth's definability theorem.** *A relation or operation symbol implicitly definable in a theory $T$ is also explicitly definable in $T$.*

Definitorial expansions of a language should be conscientiously distinguished from expansions of languages that arise from the introduction of so-called *Skolem functions*. These are useful for many purposes and are therefore briefly described.

**Skolem normal forms.** According to Theorem 4.2, every formula $\alpha$ can be converted into an equivalent PNF, $\alpha \equiv \mathsf{Q}_1 x_1 \cdots \mathsf{Q}_k x_k \alpha'$, where $\alpha'$ is open. Obviously then $\neg\alpha \equiv \overline{\mathsf{Q}}_1 x_1 \cdots \overline{\mathsf{Q}}_k x_k \neg\alpha'$, where $\overline{\forall} = \exists$ and $\overline{\exists} = \forall$. Because $\vDash \alpha$ if and only if $\neg\alpha$ is unsatisfiable, the decision problem for general validity can first of all be reduced to the satisfiability problem for formulas in PNF. Using Theorem 6.3 below, the latter—at the cost of introducing new operation symbols—is then completely reduced to the satisfiability problem for $\forall$-formulas.

Call formulas $\alpha$ and $\beta$ *satisfiably equivalent* if both are satisfiable (not necessarily in the same model), or both are unsatisfiable. We construct for every formula, which w.l.o.g. is assumed to be given in prenex form $\alpha = \mathsf{Q}_1 x_1 \cdots \mathsf{Q}_k x_k \beta$, a satisfiably equivalent $\forall$-formula $\hat{\alpha}$ with additional operation symbols such that $free\,\hat{\alpha} = free\,\alpha$. The construction of $\hat{\alpha}$ will be completed after $m$ steps, where $m$ is the number of $\exists$-quantifiers among the $\mathsf{Q}_1, \ldots, \mathsf{Q}_k$. Take $\alpha = \alpha_0$ and $\alpha_i$ to be already constructed. If $\alpha_i$ is already an $\forall$-formula let $\hat{\alpha} = \alpha_i$. Otherwise $\alpha_i$ has the form $\forall x_1 \cdots \forall x_n \exists y \beta_i$ for some $n \geqslant 0$. With an $n$-ary operation symbol $f$ (which is a constant in case $n{=}0$) not yet used let $\alpha_{i+1} = \forall \vec{x} \beta_i \frac{f\vec{x}}{y}$. Thus, after $m$ steps an $\forall$-formula $\hat{\alpha}$ is obtained such that $free\,\hat{\alpha} = free\,\alpha$; this formula $\hat{\alpha}$ is called a *Skolem normal form* (SNF) of $\alpha$.

**Example 1.** If $\alpha$ is the formula $\forall x \exists y\, x < y$ then $\hat{\alpha}$ is just $\forall x\, x < fx$.
For $\alpha = \exists x \forall y\, x \cdot y = y$ we have $\hat{\alpha} = \forall y\, c \cdot y = y$.
If $\alpha = \forall x \forall y \exists z (x < z \wedge y < z)$ then $\hat{\alpha} = \forall x \forall y (x < fxy \wedge y < fxy)$.

**Theorem 6.3.** *Let $\hat{\alpha}$ be a Skolem normal form for the formula $\alpha$. Then*

(a) $\hat{\alpha} \vDash \alpha$,     (b) $\alpha$ *is satisfiably equivalent to $\hat{\alpha}$.*

**Proof.** (a): It suffices to show that $\alpha_{i+1} \vDash \alpha_i$ for each of the described construction steps. $\beta_i \frac{f\vec{x}}{y} \vDash \exists y \beta_i$ implies $\alpha_{i+1} = \forall \vec{x} \beta_i \frac{f\vec{x}}{y} \vDash \forall \vec{x} \exists y \beta_i = \alpha_i$, by (c) and (d) in **2.5**. (b): If $\hat{\alpha}$ is satisfiable then by (a) so too is $\alpha$. Conversely, suppose $\mathcal{A} \vDash \forall \vec{x} \exists y \beta_i (\vec{x}, y, \vec{z}) [\vec{c}]$. For each $\vec{a} \in A^n$ we choose some $b \in A$ such that $\mathcal{A} \vDash \beta [\vec{a}, b, \vec{c}]$ (which is possible in view of the axiom of choice AC) and expand $\mathcal{A}$ to $\mathcal{A}'$ by setting $f^{\mathcal{A}'} \vec{a} = b$ for the new operation symbol. Then evidently $\mathcal{A}' \vDash \alpha_{i+1} [\vec{c}]$. Thus, we finally obtain a model for $\hat{\alpha}$ that expands the initial model. $\square$

Now, for each $\alpha$, a tautologically equivalent $\exists$-formula $\check{\alpha}$ is gained as well (that is, $\vDash \alpha \Leftrightarrow \vDash \check{\alpha}$). By the above theorem, we first produce for $\beta = \neg\alpha$ a satisfiably equivalent SNF $\hat{\beta}$ and put $\check{\alpha} := \neg\hat{\beta}$. Then indeed $\vDash \alpha \Leftrightarrow \vDash \check{\alpha}$, because

$$\vDash \alpha \;\Leftrightarrow\; \beta \text{ unsatisfiable} \;\Leftrightarrow\; \hat{\beta} \text{ unsatisfiable} \;\Leftrightarrow\; \vDash \check{\alpha}.$$

**Example 2.** For $\alpha := \exists x \forall y (ry \to rx)$ we have $\neg\alpha \equiv \beta := \forall x \exists y (ry \wedge \neg rx)$ and $\hat{\beta} = \forall x (rfx \wedge \neg rx)$. Thus, $\check{\alpha} = \neg\hat{\beta} \equiv \exists x (rfx \to rx)$. The last formula is a tautology. Indeed, if $r^{\mathcal{A}} \neq \emptyset$ then clearly $\mathcal{A} \vDash \exists x (rfx \to rx)$. But the same holds if $r^{\mathcal{A}} = \emptyset$, for then never $\mathcal{A} \vDash rfx$. Thus, $\check{\alpha}$ and hence also $\alpha$ is a tautology, which is not at all obvious after a first glance at $\alpha$. This shows how useful Skolem normal forms can be for discovering tautologies.

**Remark 2.** There are many applications of Skolem normal forms, mainly in model theory and in logic programming. For instance, Exercise 5 permits one to reduce the satisfiability problem of an arbitrary first-order formula set to a set of $\forall$-formulas (at the cost of adjoining new function symbols). Moreover, a set $X$ of $\forall$-formulas is satisfiably equivalent to a set $X'$ of open formulas as will be shown in **4.1**, and this problem can be reduced completely to the satisfiability of a suitable set of propositional formulas, see also Remark 1 in **4.1**. The examples of applications of the propositional compactness theorem in **1.5** give a certain feeling for how to proceed in this way.

## Exercises

1. Suppose that $T_f$ results from $T$ by adjoining an explicit definition $\eta$ for $f$ and let $\alpha^{rd}$ be constructed as explained in the text. Show that $T_f$ is a conservative extension of $T$ if and only if $\eta$ is a legitimate explicit definition.

2. Let $\mathsf{S}: n \mapsto n+1$ denote the successor function in $\mathcal{N} = (\mathbb{N}, 0, \mathsf{S}, +, \cdot)$. Show that $Th\mathcal{N}$ is a definitorial extension of $Th(\mathbb{N}, \mathsf{S}, \cdot)$; in other words, $0$ and $+$ are explicitly definable by $\mathsf{S}$ and $\cdot$ in $\mathcal{N}$.

3. Prove that $\eta : y = x^{-1} \leftrightarrow x \circ y = e$ is a legitimate explicit definition in $T_G$ (it suffices to prove $T_G \vDash x \circ y = x \circ z \to y = z$). Show in addition that $T_G^{=} = T_G + \eta$. Thus, $T_G^{=}$ is a definitorial and hence a conservative extension of $T_G$. In this sense, the theories $T_G^{=}$ and $T_G$ are equivalent formulations of the theory of groups.

4. As is well known, the natural $<$-relation of $\mathbb{N}$ is explicitly definable in $(\mathbb{N}, 0, +)$, for instance, by $x < y \leftrightarrow (\exists z \neq 0)z + x = y$. Prove that the $<$-relation of $\mathbb{Z}$ is not explicitly definable in $(\mathbb{Z}, 0, +)$.

5. Construct to each $\alpha \in X$ ($\subseteq \mathcal{L}$) an SNF $\hat{\alpha}$ such that $X$ is satisfiably equivalent to $\hat{X} = \{\hat{\alpha} \mid \alpha \in X\}$ and $\hat{X} \vDash X$, called a *Skolemization* of $X$. Since we do not suppose that $X$ is countable, the function symbols introduced in $\hat{X}$ must properly be indexed.

# Chapter 3

# Complete logical Calculi

Our first goal is to characterize the consequence relation in a first-order language by means of a calculus similar to that of propositional logic. That this goal is attainable at all was shown for the first time by Gödel in [Gö1]. The original version of Gödel's theorem refers to the axiomatization of tautologies only and does not immediately imply the compactness theorem of first-order logic; but a more general formulation of completeness in **3.2** does. The importance of the compactness theorem for mathematical applications was first revealed in 1936 by A. Malcev, see [Ma].

The characterizability of logical consequence by means of a calculus (the content of the completeness theorem) is a crucial result in mathematical logic with far-reaching applications. In spite of its metalogical origin, the completeness theorem is essentially a mathematical theorem. It satisfactorily explains the phenomenon of the well-definedness of logical deductive methods in mathematics. To seek any additional, possibly unknown methods or rules of inference would be like looking for perpetual motion in physics. Of course, this insight does not affect the development of new ideas in solving open questions. We will say somewhat more regarding the metamathematical aspect of the theorem and its applications, as well as the use of the model construction connected with its proof in a partly descriptive manner, in **3.3**, **3.4**, and **3.5**.

Without beating around the bush, we deal from the outset with the case of an arbitrary, not necessarily countable first-order language. Nonetheless, the proof given, based on Henkin's idea of a constant expansion [He], is kept relatively short, mainly thanks to an astute choice of its logical basis. Although mathematical theories are countable as a rule, a successful

W. Rautenberg, *A Concise Introduction to Mathematical Logic*,
Universitext, DOI 10.1007/978-1-4419-1221-3_3,
© Springer Science+Business Media, LLC 2010

application of methods of mathematical logic in algebra and analysis relies essentially on the unrestricted version of the completeness theorem. Only with such generality does the proof display the inherent unity that tends to distinguish the proofs of magnificent mathematical theorems.

## 3.1   A Calculus of Natural Deduction

As in Chapter **2**, let $\mathcal{L}$ be an arbitrary but fixed first-order language in the logical signature $\neg, \wedge, \forall, =$. We define a calculus $\vdash$ by the system of deductive rules enclosed in the box below. The calculus operates with sequents as in **1.4**. It supplements the basic rules given there with three predicate-logical rules. We also use the same modes of speaking, for instance, '$X \vdash \alpha$' is read as '$X$ derivable $\alpha$'. Note that the initial rule (IR) is subject to a minor extension. Using (MR), it could be pared down to $\dfrac{}{\alpha \vdash \alpha}$ and $\dfrac{}{\vdash t = t}$, which are rules without premises like (IR).

$$
\begin{array}{ll}
(\text{IR}) \ \dfrac{}{X \vdash \alpha} \ (\alpha \in X \cup \{t = t\}) & (\text{MR}) \ \dfrac{X \vdash \alpha}{X' \vdash \alpha} \ (X \subseteq X') \\[3mm]
(\wedge 1) \ \dfrac{X \vdash \alpha, \beta}{X \vdash \alpha \wedge \beta} & (\wedge 2) \ \dfrac{X \vdash \alpha \wedge \beta}{X \vdash \alpha, \beta} \\[3mm]
(\neg 1) \ \dfrac{X \vdash \beta, \neg \beta}{X \vdash \alpha} & (\neg 2) \ \dfrac{X, \beta \vdash \alpha \ \mid \ X, \neg \beta \vdash \alpha}{X \vdash \alpha} \\[3mm]
(\forall 1) \ \dfrac{X \vdash \forall x \alpha}{X \vdash \alpha \frac{t}{x}} \ (\alpha, \tfrac{t}{x} \ \text{collision-free}) & \\[3mm]
(\forall 2) \ \dfrac{X \vdash \alpha \frac{y}{x}}{X \vdash \forall x \alpha} \ (y \notin \text{free } X \cup \text{var } \alpha) & \\[3mm]
(=) \ \dfrac{X \vdash s = t, \alpha \frac{s}{x}}{X \vdash \alpha \frac{t}{x}} \ (\alpha \ \text{prime}) &
\end{array}
$$

By (IR), $X \vdash t = t$ for arbitrary $X$ and $t$, in particular $\vdash t = t$. Here as everywhere, $\vdash \varphi$ stands for $\emptyset \vdash \varphi$ (read '$\varphi$ is derivable'). The remaining notation from **1.4** is also used here; thus, $\alpha \vdash \beta$ abbreviates $\{\alpha\} \vdash \beta$, etc. Note also that $\alpha \vdash \beta \vdash \gamma$ can have only the meaning $\alpha \vdash \beta$ and $\beta \vdash \gamma$.

⊢ is called a *calculus of natural deduction* because it models logical inference in mathematics and other deductive sciences sufficiently well.[1] Our aim is to show that ⊨ is completely characterized by ⊢. The calculus is developed in the sequel only insofar as the completeness proof requires. While undertaking further derivations can be instructive (see the examples and exercises), this is not the principal point of formalizing proofs unless one is after specific proof-theoretic goals. It should also be said that an acute study of formalized proofs does not really promote our ability to draw correct conclusions in everyday life.

All basic rules are sound in the sense of **1.4**. The restrictions in the rules (∀1), (∀2), and (=) ensure their soundness as shown by the properties (a), (g), and (b) on page 79. Rule (=) could have been strengthened from the outset to allow $\alpha$ to be any formula such that $\alpha, \frac{s}{x}, \frac{t}{x}$ are collision-free, but we get along with the weak version. Also (∀1) could be strengthened by weakening its restriction that $\alpha, \frac{t}{x}$ are collision-free in various ways. As already stated in **2.3**, we could in fact avoid any kind of restriction by means of a more involved definition for substitution. However, such measures would unnecessarily strengthen the calculus. Weakly formulated logical calculi like the one given here often alleviate certain induction procedures, for example in verifying these rules in other logical calculi, as will be done in **3.6** for a certain Hilbert calculus.

Because ⊢ can be understood as an extension of the corresponding calculus from **1.4**, all the examples of provable rules given there carry over automatically, the cut rule included. All further sound rules, such as the formal versions of generalization and particularization in **2.5**, are provable thanks to the completeness of the calculus. This is also true of the rule $\dfrac{X \vdash \alpha}{X \vdash \forall x \alpha}$ $(x \notin \text{free } X)$, which is sound by (d) in **2.5**, though it does not result directly from (∀2). However, we do not want to spend too much time on the proofs of other rules; they are irrelevant for the completeness proof, which can then be used to justify these rules retrospectively.

Just as in the propositional case the following proof method referring to the base rules above will often be applied; it is legitimate because the proof

---

[1] We deal here with a version of the calculus NK from [Ge] adapted to our purpose; more involved descriptions of this and related sequent calculi are given in various textbooks on proof theory; see e.g. [Po].

of the corresponding principle in **1.4** depends on neither the language nor the concrete rules.

**Principle of rule induction.** *Let $\mathcal{E}$ be a property of sequents $(X, \alpha)$ such that*

(o)  $\mathcal{E}(X, \alpha)$ *provided* $\alpha \in X$ *or* $\alpha$ *is of the form* $t = t$,
(s)  $\mathcal{E}(X, \alpha) \Rightarrow \mathcal{E}(X', \alpha)$ *for* (MR)*, and similarly for* $(\wedge 1)$–$(=)$.

*Then* $X \vdash \alpha$ *implies* $\mathcal{E}(X, \alpha)$, *for all* $X, \alpha$.

Since the basic rules are clearly sound, the *soundness of the calculus*, that is to say $\vdash \subseteq \vDash$, follows immediately by rule induction. Similarly one obtains the following monotonicity property:

$(mon)$  $\mathcal{L} \subseteq \mathcal{L}' \Rightarrow \vdash_{\mathcal{L}} \subseteq \vdash_{\mathcal{L}'}$.

Here the derivability relation is indexed; note that every elementary language defines its own derivability relation, and for the time being we are concerned with the comparison of these relations in various languages. Only with the completeness theorem will we see that the indices are superfluous, just as for the consequence relation $\vDash$. To prove $(mon)$ let $\mathcal{E}(X, \alpha)$ be the property '$X \vdash_{\mathcal{L}'} \alpha$', for which the conditions (o) and (s) of rule induction are easily verified. To confirm at least one of the induction steps (i.e. the verification of (s) for each single rule), let $X \vdash_{\mathcal{L}} \alpha, \beta$ and suppose $X \vdash_{\mathcal{L}'} \alpha, \beta$. Then $(\wedge 1)$, applied in $\mathcal{L}'$, yields $X \vdash_{\mathcal{L}'} \alpha \wedge \beta$ as well.

As in propositional logic we have here the easily provable

**Finiteness theorem.** *If* $X \vdash \alpha$ *then* $X_0 \vdash \alpha$ *for some finite* $X_0 \subseteq X$.

The only difference to the proof from **1.4** is that a few more rules have to be considered. Remember that $L$ denotes the signature of $\mathcal{L}$, $L_0$ that of $\mathcal{L}_0$, etc. For the moment we require a somewhat stronger version of the finiteness theorem, namely

$(fin)$ *If* $X \vdash_{\mathcal{L}} \alpha$ *then there exist a finite signature* $L_0 \subseteq L$ *and a finite subset* $X_0 \subseteq X$ *such that* $X_0 \vdash_{\mathcal{L}_0} \alpha$.

Herein the claim $X_0 \vdash_{\mathcal{L}_0} \alpha$, of course, includes $X_0 \cup \{\alpha\} \subseteq \mathcal{L}_0$. For the proof, consider the property 'there exist some finite $X_0 \subseteq X$ and $L_0 \subseteq L$ such that $X_0 \vdash_{\mathcal{L}_0} \alpha$'. It suffices to confirm the conditions (o) and (s) of the principle of rule induction. For $\alpha \in X \cup \{t = t\}$ we clearly have $X_0 \vdash_{\mathcal{L}_0} \alpha$, where $X_0 = \{\alpha\}$ or $X_0 = \emptyset$. Thus, $L_0$ may be chosen to contain all the extralogical symbols occurring in $\alpha$ or in $t = t$, and their number is surely finite. This confirms (o). The induction step on (MR) is trivial. For ($\wedge 1$) suppose $X_1 \vdash_{\mathcal{L}_1} \alpha_1$ and $X_2 \vdash_{\mathcal{L}_2} \alpha_2$ for some finite $X_i \subseteq X$ and $L_i \subseteq L$, $i = 1, 2$. Then ($mon$) gives $X_0 \vdash_{\mathcal{L}_0} \alpha_i$, where $X_0 = X_1 \cup X_2$ and $L_0 = L_1 \cup L_2$. Applying ($\wedge 1$) in $\mathcal{L}_0$, we obtain $X_0 \vdash_{\mathcal{L}_0} \alpha_1 \wedge \alpha_2$, which is what we want. The induction steps for all remaining rules proceed similarly and are even somewhat simpler. This confirms condition (s), which in turn proves ($fin$).

In the foregoing proof, $\mathcal{L}_0$ contains at least the extralogical symbols of $X_0$ and $\alpha$ but perhaps also some others. Only with the completeness theorem can we know that the symbols occurring in $X_0, \alpha$ in fact suffice. This insensitivity of derivation with respect to language extensions can be derived purely proof-theoretically, albeit with considerable effort, but purely combinatorially and without recourse to the infinitistic means of semantics. A modest demonstration of such methods is the constant elimination by Lemmas 2.1 and 2.2 from the next section.

Now for some more examples of provable rules required later.

**Example 1.** (a) $\dfrac{X \vdash s = t, s = t'}{X \vdash t = t'}$, (b) $\dfrac{X \vdash s = t}{X \vdash t = s}$, (c) $\dfrac{X \vdash t = s, s = t'}{X \vdash t = t'}$.

To show (a) let $x \notin var\, t'$ and let $\alpha$ be the formula $x = t'$. Then the premise of (a) is written $X \vdash s = t, \alpha \frac{s}{x}$. Rule ($=$) yields $X \vdash \alpha \frac{t}{x}$. Now, $\alpha \frac{t}{x}$ equals $t = t'$, since $x \notin var\, t'$; hence $X \vdash t = t'$. (b) is obtained immediately from (a), choosing there $t' = s$ because $X \vdash s = s$. And with this follows (c), for thanks to (b), the premise of (c) now yields $X \vdash s = t, s = t'$ and hence, by (a), the conclusion of (c).

**Example 2.** In (a)–(d), $n$ is as usual the arity of the symbols $f$ and $r$. (a) and (c) are provable for $i = 1, \ldots, n$. In order to ease the writing, $X \vdash \vec{t} = \vec{t'}$ abbreviates $X \vdash t_1 = t'_1, \ldots, t_n = t'_n$, so that, for instance, rule (b) below has actually $n$ premises:

(a) $\dfrac{X \vdash t_i = t}{X \vdash f\vec{t} = ft_1 \cdots t_{i-1}tt_{i+1} \cdots t_n}$,  (b) $\dfrac{X \vdash \vec{t} = \vec{t'}}{X \vdash f\vec{t} = f\vec{t'}}$,

(c) $\dfrac{X \vdash t_i = t, r\vec{t}}{X \vdash rt_1 \cdots t_{i-1}tt_{i+1} \cdots t_n}$,  (d) $\dfrac{X \vdash \vec{t} = \vec{t'}, r\vec{t}}{X \vdash r\vec{t'}}$.

Proof of (a): Let $X \vdash t_i = t$ and $\alpha := f\vec{t} = ft_1 \cdots t_{i-1}xt_{i+1} \cdots t_n$, where $x$ is not to occur in any of the $t_j$. Since $X \vdash \alpha \frac{t_i}{x}$ ($= f\vec{t} = f\vec{t}$), it follows that $X \vdash \alpha \frac{t}{x}$ using (=). This is the conclusion of (a). (b) is then obtained by considering Example 1(c) and the $n$-fold iteration of (a), as can best be seen by first working through the case $n = 2$. Rule (c) is just another application of (=) by taking the formula $rt_1 \cdots t_{i-1} x t_{i+1} \cdots t_n$ for $\alpha$, where again, $x$ is supposed not to occur in any of the $t_j$. Applying (c) $n$ times then yields (d).

The next example indicates that sometimes considerable effort is needed in formally deriving what is nearly obvious from the semantic point of view. Of course, difficulties in formal proofs strongly depend on the calculus in which these proofs are carried out.

**Example 3.** (a) $\vdash \exists x\, t = x$, for all $x, t$ with $x \notin \mathrm{var}\, t$, (b) $\vdash \exists x\, x = x$.
(a) holds because ($\forall 1$) gives $\forall x\, t \neq x \vdash t \neq t$, for $t \neq t$ equals $(t \neq x)\frac{t}{x}$; here $x \notin \mathrm{var}\, t$ is required. Clearly, $\forall x\, t \neq x \vdash t = t$ as well. Thus, by ($\neg 1$),

$$\forall x\, t \neq x \vdash \exists x\, t = x.$$

Trivially, also $\neg \forall x\, t \neq x \vdash \exists x\, t = x$ $(= \neg \forall x\, t \neq x)$. Therefore, by ($\neg 2$), $\vdash \exists x\, t = x$. The assumption $x \notin \mathrm{var}\, t$ is in fact essential in order to derive $\exists x\, t = x$; cf. Example 1 in **2.3** page 63. (b) is verified similarly, starting with $\forall x\, x \neq x \vdash x \neq x, x = x$.

A set $X$ ($\subseteq \mathcal{L}$) is called *inconsistent* if $X \vdash \alpha$ for all $\alpha \in \mathcal{L}$, and otherwise *consistent*, exactly as in propositional logic. A satisfiable set $X$ is evidently consistent. By ($\neg 1$), the inconsistency of $X$ is equivalent to $X \vdash \alpha, \neg \alpha$ for any $\alpha$, hence also to $X \vdash \bot$, since $\bot = \neg \top$ and certainly $X \vdash \top$ $(= \exists v_0\, v_0 = v_0)$ by Example 3.

The relation $\vdash$ is completely characterized by some inconsistency condition, as in **1.4**. Indeed, the proofs of the two properties

$$\mathsf{C}^+: \quad X \vdash \alpha \; \Leftrightarrow \; X, \neg \alpha \vdash \bot, \qquad \mathsf{C}^-: \quad X \vdash \neg \alpha \; \Leftrightarrow \; X, \alpha \vdash \bot$$

from Lemma 1.4.2 remain correct for any meaningful definition of $\bot$. The properties $\mathsf{C}^+$ and $\mathsf{C}^-$ will permanently be used in the sequel without our explicitly referring to them.

As in propositional logic, $X \subseteq \mathcal{L}$ is called *maximally consistent* if $X$ is consistent but each proper extension of $X$ in $\mathcal{L}$ is inconsistent. There are various characterizations of this property, e.g. the one in Exercise 4, known already from **1.4**. Examples of maximally consistent sets are the $\{\varphi \in \mathcal{L} \mid \mathcal{M} \vDash \varphi\}$ for $\mathcal{L}$-models $\mathcal{M}$, the typical ones as will turn out.

### Exercises

1. Derive the rule $\dfrac{X \vdash \alpha \frac{t}{x}}{X \vdash \exists x \alpha}$ ($\alpha, \frac{t}{x}$ collision-free).

2. Prove $\forall x \alpha \vdash \forall y(\alpha \frac{y}{x})$ and $\forall y(\alpha \frac{y}{x}) \vdash \forall x \alpha$ provided $y \notin \operatorname{var} \alpha$.

3. Using Exercise 2 and the cut rule prove $\dfrac{X \vdash \forall y(\alpha \frac{y}{x})}{X \vdash \forall z(\alpha \frac{z}{x})}$ ($y, z \notin \operatorname{var} \alpha$).

4. Show that $X \subseteq \mathcal{L}$ is maximal consistent iff either $\varphi \in X$ or $\neg \varphi \in X$ for each $\varphi \in \mathcal{L}$. This easily implies that a maximally consistent set $X$ is deductively closed, i.e. $X \vdash \varphi \Rightarrow \varphi \in X$, for each $\varphi \in \mathcal{L}$.

## 3.2 The Completeness Proof

Let $\mathcal{L}$ be a language and $c$ a constant symbol. $\mathcal{L}c$ is the result of adjoining $c$ to $\mathcal{L}$. We have $\mathcal{L}c = \mathcal{L}$ if $c$ occurs already in $\mathcal{L}$. Similarly $\mathcal{L}C$ denotes the language resulting from $\mathcal{L}$ by adjoining a set $C$ of constants, a *constant expansion* of $\mathcal{L}$. We shall also come across such expansions in Chapter **5**. Let $\alpha \frac{z}{c}$ (read "$\alpha z$ for $c$") denote the formula arising from $\alpha$ by replacing $c$ with the variable $z$, and put $X \frac{z}{c} := \{\alpha \frac{z}{c} \mid \alpha \in X\}$. $c$ then no longer occurs in $X \frac{z}{c}$. We actually require the following assertion only for a single variable $z$, but as is often the case, we are able too prove by induction only a stronger version unproblematically.

**Lemma 2.1 (on constant elimination).** *Suppose $X \vdash_{\mathcal{L}c} \alpha$. Then $X \frac{z}{c} \vdash_{\mathcal{L}} \alpha \frac{z}{c}$ for almost all variables $z$.*

**Proof** by rule induction in $\vdash_{\mathcal{L}c}$. If $\alpha \in X$ then $\alpha \frac{z}{c} \in X \frac{z}{c}$ is clear; if $\alpha$ is of the form $t = t$, so too is $\alpha \frac{z}{c}$. Thus, $X \frac{z}{c} \vdash_{\mathcal{L}} \alpha \frac{z}{c}$ in either case, even for all $z$. Only the induction steps on ($\forall 1$), ($\forall 2$), and ($=$) are not immediately apparent. We restrict ourselves to ($\forall 1$), because the induction steps for ($\forall 2$) and ($=$) proceed analogously. Let $\alpha, \frac{t}{x}$ be collision-free, $X \vdash_{\mathcal{L}c} \forall x \alpha$, and assume that $X \frac{z}{c} \vdash_{\mathcal{L}} (\forall x \alpha) \frac{z}{c}$ for almost all $z$. In addition, we may suppose that $z \notin \operatorname{var} \{\forall x \alpha, t\}$ for almost all $z$. A separate induction on $\alpha$ readily confirms $\alpha \frac{t}{x} \frac{z}{c} = \alpha' \frac{t'}{x}$ with $\alpha' := \alpha \frac{z}{c}$ and $t' := t \frac{z}{c}$. Clearly $\alpha', \frac{t'}{x}$ are collision-free as well. By our assumption, $X \frac{z}{c} \vdash_{\mathcal{L}} (\forall x \, \alpha) \frac{z}{c} = \forall x \alpha'$. Rule ($\forall 1$) then clearly yields $X \frac{z}{c} \vdash_{\mathcal{L}} \alpha' \frac{t'}{x} = \alpha \frac{t}{x} \frac{z}{c}$, and this holds still for almost all $z$ which completes the proof of the induction step on ($\forall 1$). $\blacksquare$

This lemma leads to the following rule of "constant quantification," the semantic counterpart of which plays an essential role in Chapter **5**:

$$(\forall 3) \quad \frac{X \vdash \alpha \frac{c}{x}}{X \vdash \forall x \alpha} \quad (c \text{ not in } X, \alpha).$$

Indeed, suppose that $X \vdash \alpha \frac{c}{x}$. Because of the finiteness theorem we may assume that $X$ is finite. By Lemma 2.1, where in the case at hand $\mathcal{L}c = \mathcal{L}$, some $y$ not occurring in $X \cup \{\alpha\}$ can be found such that $X \frac{y}{c} \vdash \alpha \frac{c}{x} \frac{y}{c} = \alpha \frac{y}{x}$ (the latter holds because $c$ does not occur in $\alpha$). Since $X \frac{y}{c} = X$, we thus obtain $X \vdash \alpha \frac{y}{x}$. Hence $X \vdash \forall x \alpha$ by ($\forall 2$). This confirms ($\forall 3$). A likewise useful consequence of constant elimination is

**Lemma 2.2.** *Let $C$ be any set of constant symbols and $\mathcal{L}' = \mathcal{L}C$. Then $X \vdash_{\mathcal{L}} \alpha \Leftrightarrow X \vdash_{\mathcal{L}'} \alpha$, for all $X \subseteq \mathcal{L}$ and $\alpha \in \mathcal{L}$. Thus, $\vdash_{\mathcal{L}'}$ is a conservative expansion of $\vdash_{\mathcal{L}}$.*

**Proof.** (*mon*) states that $X \vdash_{\mathcal{L}} \alpha \Rightarrow X \vdash_{\mathcal{L}'} \alpha$. Suppose conversely that $X \vdash_{\mathcal{L}'} \alpha$. To prove $X \vdash_{\mathcal{L}} \alpha$ we may assume, thanks to (*fin*) and (MR), that $C$ is finite. Since the adjunction of finitely many constants can be undertaken stepwise, we may suppose for the purpose of the proof that $\mathcal{L}' = \mathcal{L}c$ for a single constant $c$ not occurring in $\mathcal{L}$. Lemma 2.1 then yields $X \frac{z}{c} \vdash_{\mathcal{L}} \alpha \frac{z}{c}$ for at least one variable $z$. Now, $X \frac{z}{c} \vdash_{\mathcal{L}} \alpha \frac{z}{c}$ means the same as $X \vdash_{\mathcal{L}} \alpha$ because $c$ occurs neither in $X$ nor in $\alpha$. ◻

In the following, we represent the derivability relation in $\mathcal{L}$ and in every constant expansion $\mathcal{L}'$ of $\mathcal{L}$ with the same symbol $\vdash$. By Lemma 2.2 no misunderstandings can arise from this notation. Since the consistency of $X$ is equivalent to $X \nvdash \bot$, there is also no need to distinguish between the consistency of $X \subseteq \mathcal{L}$ with respect to $\mathcal{L}$ or $\mathcal{L}'$. This is highly significant for the proofs of the next two lemmas.

The proof of the completeness theorem essentially proceeds with a model construction from the syntactic material of a certain constant expansion of $\mathcal{L}$. We first choose for each variable $x$ and each $\alpha \in \mathcal{L}$ a constant $c_{x,\alpha}$ not occurring in $\mathcal{L}$; more precisely, we choose exactly one such constant for each pair $x, \alpha$. Define

$$(1) \quad \alpha^x := \neg \forall x \alpha \wedge \alpha \frac{c}{x} \quad (c := c_{x,\alpha}).$$

Here it is insignificant how many free variables $\alpha$ contains, and whether $x$ occurs at all in $\alpha$. Note that $\neg \alpha^x \equiv \exists x \neg \alpha \to \neg \alpha \frac{c}{x}$. The formula on the right side tells us that under the hypothesis $\exists x \neg \alpha$ the constant $c$

represents a *counterexample* to the validity of $\alpha$, that is, an example for the validity of $\neg\alpha$. Note also that $\neg\alpha^x \equiv \top$ whenever $x \notin \text{free }\alpha$.

**Lemma 2.3.** *Let* $\Gamma_{\mathcal{L}} := \{\neg\alpha^x \mid \alpha \in \mathcal{L}, x \in Var\}$, *where* $\alpha^x$ *is defined as in* (1), *and let* $X \subseteq \mathcal{L}$ *be consistent. Then* $X \cup \Gamma_{\mathcal{L}}$ *is consistent as well.*

**Proof.** Assume that $X \cup \Gamma_{\mathcal{L}} \vdash \bot$. There exist some $n \geqslant 0$ and formulas $\neg\alpha_0^{x_0}, \ldots, \neg\alpha_n^{x_n} \in \Gamma_{\mathcal{L}}$ such that (a) $X \cup \{\neg\alpha_i^{x_i} \mid i \leqslant n\} \vdash \bot$. Since $X \nvdash \bot$, there is some minimal $n$ with (b) $X' := X \cup \{\neg\alpha_i^{x_i} \mid i < n\} \nvdash \bot$. Let $x := x_n$, $\alpha := \alpha_n$, and $c := c_{x,\alpha}$. By (a), $X' \cup \{\neg\alpha^x\} \vdash \bot$. Hence, $X' \vdash \alpha^x$, and so $X' \vdash \neg\forall x\alpha, \alpha\,\frac{c}{x}$, by ($\wedge 2$). But $X' \vdash \alpha\,\frac{c}{x}$ yields $X' \vdash \forall x\alpha$ using ($\forall 3$), since $c$ does not occur in $X'$ and $\alpha$. Thus, $X' \vdash \forall x\alpha, \neg\forall x\alpha$, whence $X' \vdash \bot$, contradicting (b) and hence our assumption. $\quad\square$

Call $X \subseteq \mathcal{L}$ a *Henkin set* if $X$ satisfies the following two conditions:

(H1) $\quad X \vdash \neg\alpha \quad\Leftrightarrow\quad X \nvdash \alpha$, $\quad$ (equivalently, $X \vdash \alpha \Leftrightarrow X \nvdash \neg\alpha$),

(H2) $\quad X \vdash \forall x\alpha \quad\Leftrightarrow\quad X \vdash \alpha\,\frac{c}{x}$ for all constants $c$ in $\mathcal{L}$.

(H1) and (H2) imply another useful property of a Henkin set $X$, namely

(H3) $\quad$ For each term $t$ there is a constant $c$ such that $X \vdash t = c$.

Indeed, $X \vdash \neg\forall x\,t \neq x$ ($= \exists x\,t = x$) for $x \notin \text{var }t$ by Example 3 in **3.1**. Hence, $X \nvdash \forall x\,t \neq x$ in view of (H1). Thus, $X \nvdash t \neq c$ for some $c$ by (H2), and so $X \vdash t = c$ by (H1).

**Lemma 2.4.** *Let* $X \subseteq \mathcal{L}$ *be consistent. Then there exists a Henkin set* $Y \supseteq X$ *in a suitable constant expansion* $\mathcal{L}C$ *of* $\mathcal{L}$.

**Proof.** Put $\mathcal{L}_0 := \mathcal{L}$, $X_0 := X$ and assume that $\mathcal{L}_n$ and $X_n$ have been given. Let $\mathcal{L}_{n+1}$ result from $\mathcal{L}_n$ by adopting new constants $c_{x,\alpha,n}$ for all $x \in Var$, $\alpha \in \mathcal{L}_n$; more precisely, $\mathcal{L}_{n+1} = \mathcal{L}_n C_n$, with the set $C_n$ of constants $c_{x,\alpha,n}$. Further, let $X_{n+1} = X_n \cup \Gamma_{\mathcal{L}_n}$. Here $\Gamma_{\mathcal{L}_n}$ is defined as in Lemma 2.3 so that $X_{n+1} \subseteq \mathcal{L}_{n+1}$. Using Lemma 2.3 we have $X_n \nvdash \bot$ for each $n$. Let $X' := \bigcup_{n\in\mathbb{N}} X_n$; hence $X' \subseteq \mathcal{L}' := \bigcup_{n\in\mathbb{N}} \mathcal{L}_n = \mathcal{L}C$, where $C := \bigcup_{n\in\mathbb{N}} C_n$. Then $X' \nvdash \bot$, since $X'$, as the union of a chain of consistent sets, is surely consistent (in $\mathcal{L}'$). Let $\alpha \in \mathcal{L}'$, $x \in Var$, and, say, $\alpha \in \mathcal{L}^n$ with minimal $n$, and let $\alpha^x$ be the formula defined as in (1) but with respect to $\mathcal{L}^n$. Then $\neg\alpha^x$ belongs to $X_{n+1}$. Hence $\neg\alpha^x \in X'$. Now let $(H, \subseteq)$ be the partial order of all consistent extensions of $X'$ in $\mathcal{L}'$. Every chain $K \subseteq H$ has the upper bound $\bigcup K$ in $H$, because if all members of $K$ are consistent then so is $\bigcup K$. Also $H \neq \emptyset$, e.g. $X' \in H$. By Zorn's lemma, $H$ therefore contains a maximal element $Y$. In short, $Y$ is a maximal consistent extension of $X'$. Since $\neg \alpha^x \in X' \subseteq Y$ it holds

$$(2) \quad Y \vdash \neg \alpha^x \text{ for all } \alpha \in \mathcal{L}'.$$

Further, $Y$ is at the same time a Henkin set. Here is the proof:

(H1) $\Rightarrow$: $Y \vdash \neg \alpha$ implies $Y \nvdash \alpha$ due to the consistency of $Y$. $\Leftarrow$: If $Y \nvdash \alpha$ then surely $\alpha \notin Y$. As a result $Y, \alpha \vdash \bot$, for $Y$ is maximally consistent. Thus $Y \vdash \neg \alpha$ by C$^-$. You may also use Exercise 4 in **3.1**

(H2) $\Rightarrow$: Clear by ($\forall 1$). $\Leftarrow$: Let $Y \vdash \alpha \frac{c}{x}$ for all $c$ in $\mathcal{L}'$, so also $Y \vdash \alpha \frac{c}{x}$ for $c := c_{x,\alpha,n}$, where $n$ is minimal with $\alpha \in \mathcal{L}_n$. Assume that $Y \nvdash \forall x \alpha$. Then $Y \vdash \neg \forall x \alpha$ by (H1). But $Y \vdash \neg \forall x \alpha, \alpha \frac{c}{x}$ implies $Y \vdash \neg \forall x \alpha \wedge \alpha \frac{c}{x} = \alpha^x$ using ($\wedge 1$). Now, since $Y$ is consistent, $Y \vdash \alpha^x$ which contradicts (2). Thus, our assumption was wrong and indeed $Y \vdash \forall x \alpha$. $\blacksquare$

**Remark 1.** In the original language $\mathcal{L}$, consistent sets are not generally embeddable in Henkin sets. For instance, let the signature of $\mathcal{L}$ consist of the constants $c_i$, $i \in I$ with any infinite set $I$. Then the consistent set $X = \{v_0 \neq c_i \mid i \in I\}$ represents a counterexample in $\mathcal{L}_=$. In no consistent extension of $X$ can be derived $v_0 = c_i$ for some $i \in I$. In other words, (H3) is violated.

**Lemma 2.5.** *Every Henkin set $Y \subseteq \mathcal{L}$ possesses a model.*

**Proof.** The model constructed in the following is called a *term model*. Let $t \approx t'$ whenever $Y \vdash t = t'$. The relation $\approx$ is a congruence in the term algebra $\mathcal{T}$ of $\mathcal{L}$. This means (repeating the definition on page 51) that

(a)  $\approx$ is an equivalence relation,

(b)  $t_1 \approx t'_1, \ldots, t_n \approx t'_n \Rightarrow f\vec{t} \approx f\vec{t'}$, for operation symbols $f$.

The claim (a) follows immediately from $Y \vdash t = t$ and Example 1 in **3.1**; (b) is just another way of formulating 2(b). Let $A := \{\bar{t} \mid t \in \mathcal{T}\}$. Here $\bar{t}$ denotes the equivalence class of $\approx$ containing the term $t$, so that

(c)  $\bar{t} = \bar{s} \Leftrightarrow t \approx s \Leftrightarrow Y \vdash t = s$.

This set $A$ is the domain of the sought model $\mathcal{M} = (\mathcal{A}, w)$ for $Y$. The factorization of $\mathcal{T}$ will ensure that $=$ means identity in the model. Let $C$ be the set of constants in $\mathcal{L}$. By (H3) there is for each term $t$ in $\mathcal{T}$ some $c \in C$ such that $c \approx t$. Therefore even $A = \{\bar{c} \mid c \in C\}$. Now put $x^{\mathcal{M}} := \bar{x}$ and $c^{\mathcal{M}} := \bar{c}$ for variables and constants in $\mathcal{L}$. An operation symbol $f$ occurring in $\mathcal{L}$ of arity $n > 0$ is interpreted by $f^{\mathcal{M}}$, defined by

$$f^{\mathcal{M}}(\bar{t}_1, \ldots, \bar{t}_n) := \overline{ft_1 \cdots t_n}.$$

This definition is sound because $\approx$ is a congruence in the term algebra $\mathcal{T}$. Finally, define $r^{\mathcal{M}}$ for an $n$-ary relation symbol $r$ by

$$r^{\mathcal{M}}\bar{t}_1\cdots\bar{t}_n \Leftrightarrow Y \vdash r\vec{t}.$$

This definition is also sound, since $Y \vdash r\vec{t}$ implies $Y \vdash r\vec{t'}$ whenever $t_1 \approx t'_1, \ldots, t_n \approx t'_n$. Here we use Example 2(d) in **3.1**. We shall prove

$$\text{(d) } t^{\mathcal{M}} = \bar{t}; \quad \text{(e) } \mathcal{M} \vDash \alpha \Leftrightarrow Y \vdash \alpha,$$

of which (e) may be regarded as the goal of the constructions. (d) follows by term induction. It is evident for prime terms, and the induction hypothesis $t_i^{\mathcal{M}} = \bar{t}_i$ for $i = 1, \ldots, n$ leads with $t = f\vec{t}$ to

$$t^{\mathcal{M}} = f^{\mathcal{M}}(t_1^{\mathcal{M}}, \ldots, t_n^{\mathcal{M}}) = f^{\mathcal{M}}(\bar{t}_1, \ldots, \bar{t}_n) = \overline{ft_1 \cdots t_n} = \bar{t}.$$

(e) follows by induction on $\mathrm{rk}\,\alpha$. We begin with formulas of rank 0 (prime formulas). Induction proceeds under consideration of $\mathrm{rk}\,\alpha < \mathrm{rk}\,\neg\alpha$, $\mathrm{rk}\,\alpha, \mathrm{rk}\,\beta < \mathrm{rk}(\alpha \wedge \beta)$, and $\mathrm{rk}\,\alpha\,\frac{t}{x} < \mathrm{rk}\,\forall x\alpha$, as in formula induction:

$$\mathcal{M} \vDash t = s \quad \Leftrightarrow \quad t^{\mathcal{M}} = s^{\mathcal{M}} \qquad \Leftrightarrow \quad \bar{t} = \bar{s} \qquad \text{(by (d))}$$

$$\Leftrightarrow \quad Y \vdash t = s \quad \text{(by (c))}.$$

$$\mathcal{M} \vDash r\vec{t} \quad \Leftrightarrow \quad r^{\mathcal{M}} t_1^{\mathcal{M}} \cdots t_n^{\mathcal{M}} \Leftrightarrow \quad r^{\mathcal{M}}\bar{t}_1 \cdots \bar{t}_n \quad \Leftrightarrow \quad Y \vdash r\vec{t}.$$

$$\mathcal{M} \vDash \alpha \wedge \beta \Leftrightarrow \mathcal{M} \vDash \alpha, \beta \qquad \Leftrightarrow \quad Y \vdash \alpha, \beta \qquad \text{(induction hypothesis)}$$

$$\Leftrightarrow \quad Y \vdash \alpha \wedge \beta \qquad \text{(using } (\wedge 1), (\wedge 2)).$$

$$\mathcal{M} \vDash \neg\alpha \quad \Leftrightarrow \mathcal{M} \nvDash \alpha \qquad \Leftrightarrow \quad Y \nvdash \alpha \qquad \text{(induction hypothesis)}$$

$$\Leftrightarrow \quad Y \vdash \neg\alpha \qquad \text{(using (H1))}.$$

$$\mathcal{M} \vDash \forall x\alpha \quad \Leftrightarrow \quad \mathcal{M}_x^{\bar{c}} \vDash \alpha \text{ for all } c \in C \quad \text{(because } A = \{\bar{c} \mid c \in C\})$$

$$\Leftrightarrow \quad \mathcal{M}_x^{c^{\mathcal{M}}} \vDash \alpha \text{ for all } c \in C \quad \text{(because } c^{\mathcal{M}} = \bar{c})$$

$$\Leftrightarrow \quad \mathcal{M} \vDash \alpha\,\frac{c}{x} \text{ for all } c \in C \quad \text{(substitution theorem)}$$

$$\Leftrightarrow \quad Y \vdash \alpha\,\frac{c}{x} \text{ for all } c \in C \quad \text{(induction hypothesis)}$$

$$\Leftrightarrow \quad Y \vdash \forall x\alpha \qquad \text{(using (H2))}.$$

Because of $Y \vdash \alpha$ for all $\alpha \in Y$, (e) immediately implies $\mathcal{M} \vDash Y$. ∎

Just as for propositional logic, the equivalence of consistency and satisfiability, and the completeness of $\vdash$, result from the above. Information about the size of the model constructed in the next theorem will be given in Theorem 4.1.

**Theorem 2.6 (Model existence theorem).** *Each consistent $X \subseteq \mathcal{L}$ (in particular, each consistent theory $T$ in $\mathcal{L}$) has a model.*

**Proof.** Let $Y \supseteq X$ be a *Henkin expansion of $X$*, i.e., a Henkin set in a suitable constant expansion $\mathcal{L}C$ according to Lemma 2.4. By Lemma 2.5, $Y$ and hence also $X$ has a model $\mathcal{M}'$ in $\mathcal{L}C$. Let $\mathcal{M}$ denote the $\mathcal{L}$-reduct of $\mathcal{M}'$. In other words, "forget" the interpretation of the constants not in $\mathcal{L}$. Then, by Theorem 2.3.1, $\mathcal{M} \vDash X$ holds as well. ∎

**Theorem 2.7 (Completeness theorem).** *Let $\mathcal{L}$ denote any first-order language. Then $X \vdash \alpha \Leftrightarrow X \vDash \alpha$, for all $X \subseteq \mathcal{L}$ and $\alpha \in \mathcal{L}$.*

**Proof.** The soundness of $\vdash$ states that $X \vdash \alpha \Rightarrow X \vDash \alpha$. The converse follows indirectly. Let $X \nvdash \alpha$, so that $X, \neg \alpha$ is consistent. Theorem 2.6 then provides a model for $X \cup \{\neg \alpha\}$, whence $X \nvDash \alpha$. ∎

Thus, $\vDash$ and $\vdash$ can henceforth be freely interchanged. We will often confirm $X \vdash \alpha$ by proving $X \vDash \alpha$ in a semi-formal manner as is common in mathematics. In particular, for theories $T$, $T \vDash \alpha$ is equivalent to $T \vdash \alpha$, for which in the following we mostly write $\vdash_T \alpha$. Clearly, $\vdash_T \alpha$ means the same as $\alpha \in T$ for sentences $\alpha$. More generally, let $X \vdash_T \alpha$ stand for $X \cup T \vdash \alpha$ and $\alpha \vdash_T \beta$ for $\{\alpha\} \vdash_T \beta$. We will also occasionally abbreviate $\alpha \vdash_T \beta \ \& \ \beta \vdash_T \gamma$ to $\alpha \vdash_T \beta \vdash_T \gamma$. In subsequent chapters, equivalences such as $\alpha \vdash_T \beta \Leftrightarrow \vdash_T \alpha \rightarrow \beta \Leftrightarrow \vdash_{T+\alpha} \beta$ and $\vdash_T \alpha \Leftrightarrow \vdash_T \alpha^g$ will be used without further mentioning and should be committed to memory. Several other useful equivalences are listed in Exercise 4.

**Remark 2.** The methods in this section easily provide also completeness of a logical calculus for *identity-free* (or $=$-free) languages in which $=$ does not appear, considered in the exercises and Chapter **4**. Simply discard from the calculus in **3.1** everything that refers to $=$. Most things run as before. The domain of $\mathcal{M}$ is now the set $\mathcal{T}$ of all terms of $\mathcal{L}C$ *without* a factorization of $\mathcal{T}$, so that $t^{\mathcal{M}} = t$. Note that (H3) is not to our disposal anymore so that the proof of Lemma 2.5 must be modified. We will not go into details, since we need in **4.1** only a slight generalization of Exercise 1. In any case, consistency of a $=$-free set $X$ means the same, no matter whether $X$ is regarded as belonging to a language with or without $=$, because $X$ has a model in either case. Moreover, if $X$ consists of $\forall$-formulas only, we come along without a Henkin expansion in constructing a model as will be shown in Theorem 4.1.1. The set of terms of the original language is sufficient for model construction in this case.

### Exercises

1. Let $\mathcal{L}$ be $=$-free, $T_0 \neq \emptyset$ the set of its ground terms, and $U \subseteq \mathcal{L}$ a consistent set of $\forall$-sentences. Construct a model $\mathfrak{T} \models U$ on the domain $T_0$ by setting $c^{\mathfrak{T}} = c$, $f^{\mathfrak{T}}\vec{t} := f\vec{t}$ (hence $t^{\mathfrak{T}} = t$ for all $t \in T_0$, shown by induction on $t$), and $r^{\mathfrak{T}}\vec{t} :\Leftrightarrow X \vdash r\vec{t}$ ($\vec{t} \in T_0^n$; $X \supseteq U$ maximally consistent, so that $X \vdash \neg\varphi \Leftrightarrow X \nvdash \alpha$, for all $\alpha \in \mathcal{L}$, cf. e.g. Lemma 1.4.4). $\mathfrak{T}$ is called a *Herbrand model*; see also **4.1**.

2. Let $K \neq \emptyset$ be a chain of theories in $\mathcal{L}$, i.e., $T \subseteq T'$ or $T' \subseteq T$, for all $T, T' \in K$. Show that $\bigcup K$ is a theory that is consistent iff all $T \in K$ are consistent.

3. Suppose $T$ is consistent and $Y \subseteq \mathcal{L}$. Prove the equivalence of
   (i) $Y \vdash_T \bot$,    (ii) $\vdash_T \neg\alpha$ for some conjunction $\alpha$ of formulas in $Y$.

4. Let $x \notin \text{var}\, t$ and $\alpha, \frac{t}{x}$ collision-free. Verify the equivalence of
   (i) $\vdash_T \alpha \frac{t}{x}$,    (ii) $x = t \vdash_T \alpha$,    (iii) $\vdash_T x = t \to \alpha$,
   (iv) $\vdash_T \forall x(x = t \to \alpha)$,    (v) $\vdash_T \exists x(x = t \wedge \alpha)$.

## 3.3    First Applications: Nonstandard Models

In this section we draw important conclusions from Theorem 2.7 and the model-construction for proving it. Since the finiteness theorem holds for the provability relation $\vdash$, Theorem 2.7 immediately yields

**Theorem 3.1 (Finiteness theorem for the consequence relation).** $X \models \alpha$ *implies* $X_0 \models \alpha$ *for some finite subset* $X_0 \subseteq X$.

Let us consider a first application. The first-order theory of fields of characteristic 0 is axiomatized by the set $X$ containing the axioms for fields and the formulas $\neg\text{char}_p$ (page 48). We claim that

(1) *A sentence $\alpha$ valid in all fields of characteristic 0 is also valid in all fields of sufficiently high prime characteristic $p$ that depends on $\alpha$.*

Indeed, since $X \models \alpha$, for some finite subset $X_0 \subseteq X$ we have $X_0 \models \alpha$. If $p$ is a prime number larger than all prime numbers $q$ with $\neg\text{char}_q \in X_0$, then $\alpha$ holds in all fields of characteristic $p$, since these satisfy $X_0$. Thus

(1) holds. From (1) we obtain, for instance, the information that two given polynomials coprime over all fields of characteristic 0 are also coprime over fields of sufficiently high prime characteristic. The statement that given polynomials are coprime is readily formalized in $\mathcal{L}\{0, 1, +, \cdot\}$.

A noteworthy consequence of Theorem 3.1 is also the nonfinite axiomatizability of many elementary theories. Before presenting examples, we explain finite axiomatizability in a somewhat broader context.

A set $Z$ of strings of a given alphabet $\mathsf{A}$ is called *decidable* if there is an algorithm (a mechanical decision procedure) that after finitely many calculation steps provides us with an answer to the question whether a string $\xi$ of symbols of $\mathsf{A}$ belongs to $Z$; otherwise $Z$ is called *undecidable*. Thus it is certainly decidable whether $\xi$ is a formula. While this is all intuitively plausible, it nonetheless requires more precision (undertaken in **6.2**). A theory $T$ is called *recursively axiomatizable*, or just *axiomatizable*, if it possesses a decidable axiom system. This is the case, for instance, if $T$ is *finitely axiomatizable*, that is, if it has a finite axiom system.

From (1) it follows straight away that the theory of fields of characteristic 0 is not finitely axiomatizable. For were $F$ a finite set of axioms, their conjunction $\alpha = \bigwedge F$ would, by (1), also have a field of finite characteristic as a model. Here is another instructive example. An abelian group $\mathcal{G}$ is called *$n$-divisible* if $\mathcal{G} \vDash \vartheta_n$ with $\vartheta_n := \forall x \exists y \, x = ny$, where $ny$ is the $n$-fold sum $y + \cdots + y$, and $\mathcal{G}$ is called *divisible* if $\mathcal{G} \vDash \vartheta_n$ for all $n \geqslant 1$. Thus, the theory of divisible abelian groups, $\mathsf{DAG}$, is axiomatized by the set $X$ consisting of the axioms for abelian groups plus all sentences $\vartheta_n$. Also $\mathsf{DAG}$ is not finitely axiomatizable. This follows as above from

    (2) *A sentence $\alpha \in \mathcal{L}\{+, 0\}$ valid in all divisible abelian groups is also valid in at least one nondivisible abelian group.*

To prove (2), let $\alpha \in \mathsf{DAG}$, or equivalently $X \vDash \alpha$. According to Theorem 3.1, $X_0 \vDash \alpha$ for some finite $X_0 \subseteq X$. Let $\mathbb{Z}_p$ be the cyclic group of order $p$, where $p$ is a prime $> n$ for all $n$ with $\vartheta_n \in X_0$. The mapping $x \mapsto nx$ from $\mathbb{Z}_p$ to itself is surjective for $0 < n < p$; otherwise $\{na \mid a \in \mathbb{Z}_p\}$ would be a nontrivial subgroup of $\mathbb{Z}_p$ which cannot be. Hence, $\mathbb{Z}_p \vDash \vartheta_n$ for $0 < n < p$. Thus, $\mathbb{Z}_p \vDash X_0$ and so $\mathbb{Z}_p \vDash \alpha$. On the other hand, $\mathbb{Z}_p$ is not $p$-divisible because $px = 0$ for all $x \in \mathbb{Z}_p$. In exactly the same way, we can show that the theory of *torsion-free* abelian groups is not finitely axiomatizable. In these groups is $na \neq 0$ whenever $n, a \neq 0$.

In a similar manner, it is possible to prove for many theories that they are not finitely axiomatizable. However, this often demands more involved methods. For instance, consider the theory ACF of a.c. fields (see p. 48). It results from adjoining to the (finitely axiomatizable) theory of fields the schema of all sentences $\forall \vec{a} \, \exists x \, p(\vec{a}, x) = 0$, where $p(\vec{a}, x)$ denotes the term $x^{n+1} + a_n x^n + \cdots + a_1 x + a_0$ $(n = 0, 1, \ldots)$, called a *monic polynomial of degree* $n + 1$. Here let $a_0, \ldots, a_n, x$ denote distinct variables. Thus, in an a.c. field every monic polynomial has a zero, and so does *every* polynomial of positive degree. Nonfinite axiomatizability of ACF follows from the by no means trivial existence proof of fields in which all polynomials up to a certain degree do factorize but irreducible polynomials still exist.

As in propositional logic, the finiteness theorem for the consequence relation leads immediately to the corresponding compactness result:

**Theorem 3.2 (Compactness theorem).** *Any set $X$ of first-order formulas is satisfiable, provided every finite subset of $X$ is satisfiable.*

Because of the finer structure of first-order languages, this theorem is somewhat more amenable to certain applications than its propositional counterpart. It can be proved in various ways, even quite independent of a logical calculus; for instance, by means of ultraproducts, as will be carried out in **5.7**. It can also be reduced to the propositional compactness theorem, see Remark 1 in **4.1**. For applications of Theorem 3.2 we will concentrate on the construction of nonstandard models.

A theory $T$ $(\subseteq \mathcal{L}^0)$ is called *complete* if it is consistent and has no consistent proper extension in $\mathcal{L}^0$. It is easily seen that the completeness of $T$ is equivalent to either $\vdash_T \alpha$ or $\vdash_T \neg\alpha$ but not both, for each $\alpha \in \mathcal{L}^0$. Hence, $Th\,\mathcal{A}$ is complete for each $\mathcal{L}$-structure $\mathcal{A}$. Other equivalences of completeness are given by Theorem 5.2.1. Note that completeness of a theory is not related to the completeness theorem in **3.2**.

We will frequently come across the theory $Th\,\mathcal{N}$ with $\mathcal{N} = (\mathbb{N}, 0, \mathsf{S}, +, \cdot)$. Here $\mathsf{S} : n \mapsto n + 1$ is the *successor function*. $\mathcal{N}$ is the standard structure for the arithmetical language $\mathcal{L}_{ar} := \mathcal{L}\{0, \mathsf{S}, +, \cdot\}$. The choice of signature is a matter of convenience and has a long tradition. Of relations definable in $\mathcal{N}$, we name just $\leqslant$ and $<$, defined by $x \leqslant y \leftrightarrow \exists z \, z + x = y$, and $x < y \leftrightarrow x \leqslant y \wedge x \neq y$. This will be our standard definitions of the symbols $\leqslant$ and $<$ in $\mathcal{L}_{ar}$.

Certain axiomatic subtheories of the complete theory $Th\mathcal{N}$ are even more frequently dealt with, in particular the so-called *Peano arithmetic* PA, a first-order theory in $\mathcal{L}_{ar}$ that is important both for mathematical foundations as well as for investigations in computer science; see e.g. [Kra]. The axioms of PA are as follows:

$$\forall x\, Sx \neq 0, \qquad\qquad \forall x\, x+0 = x, \qquad\qquad \forall x\, x \cdot 0 = 0,$$
$$\forall xy(Sx = Sy \rightarrow x = y), \quad \forall xy\, x + Sy = S(x+y), \quad \forall xy\, x \cdot Sy = x \cdot y + x,$$
$$\text{IS:} \quad \varphi\,\tfrac{0}{x} \wedge \forall x(\varphi \rightarrow \varphi\,\tfrac{Sx}{x}) \rightarrow \forall x\varphi.$$

IS is called the *induction schema* and should not be mixed up with the induction axiom IA discussed on page 108. In IS, $\varphi$ runs over all formulas from $\mathcal{L}_{ar}$, i.e., IS reads more precisely $[\varphi\,\tfrac{0}{x} \wedge \forall x(\varphi \rightarrow \varphi\,\tfrac{Sx}{x}) \rightarrow \forall x\varphi]^g$; see our convention in **2.5**. With IS one can prove $\vdash_{PA} \forall x\varphi$ *by induction on* $x$: First confirm $\vdash_{PA} \varphi\,\tfrac{0}{x}$ (*induction initiation*), and then $\vdash_{PA} \forall x(\varphi \rightarrow \varphi\,\tfrac{Sx}{x})$, or equivalently, $\varphi \vdash_{PA} \varphi\,\tfrac{Sx}{x}$ (*induction step*). The latter means the derivation of the *induction claim* $\varphi\,\tfrac{Sx}{x}$ from the *induction hypothesis* $\varphi$.

**Example.** Let $\varphi$ be the formula $x = 0 \vee \exists v\, Sv = x$. We prove $\vdash_{PA} \forall x\varphi$. In other words, each $x \neq 0$ has a predecessor, not something seen at once from the axioms. Clearly, $\vdash_{PA} \varphi\,\tfrac{0}{x}$. Since $Sv = x \vdash_{PA} SSv = Sx$, we get $\exists v Sv = x \vdash_{PA} \exists v Sv = Sx$ (particularization). Since $x = 0 \vdash_{PA} \exists v Sv = Sx$ as well, we obtain $\varphi \vdash_{PA} \exists v Sv = Sx \vdash_{PA} Sx = 0 \vee \exists v Sv = Sx = \varphi\,\tfrac{Sx}{x}$. This confirms the induction step $\varphi \vdash_{PA} \varphi\,\tfrac{Sx}{x}$. Hence, $\vdash_{PA} \forall x\varphi$ by IS. The above is easily supplemented by an inductive proof of $\vdash_{PA} \forall x\, Sx \neq x$.

**Remark 1.** Only a few arithmetical facts (for instance, $\forall x\, 0 \leqslant x$) are derivable in PA without IS. Already the derivation of such simple statements as $\forall x\, x \leqslant Sx$ and $\forall x\, Sx \neq x$ needs IS. The schema IS is extremely strong. In **7.2** it will then become clear that PA fully embraces elementary number theory and practically the whole of discrete mathematics. More about PA in the exercises; these are exclusively devoted to PA, in order to give the reader familiarity with this important theory as early as possible. Despite its strength, PA is incomplete, as will be shown in **6.5**. It is not of any import that subtraction is only partially defined in PA. A theory of integers, formulated similarly to PA, may be more convenient for number theory, but is actually not stronger than PA; it is interpretable in PA in the sense of **6.6**. We mention that PA is not finitely axiomatizable, shown for the first time in [Ry]. For this and other historical remarks see, e.g., [HP].

We will now prove that not only PA but also the complete theory $Th\mathcal{N}$ has alongside the standard model $\mathcal{N}$ other models not isomorphic to $\mathcal{N}$, called *nonstandard models*. In these models, exactly the same theorems hold as in $\mathcal{N}$. The existence proof of a nonstandard model $\mathcal{N}'$ of $Th\mathcal{N}$ is strikingly simple. Let $x \in Var$ and $X := Th\mathcal{N} \cup \{\underline{n} < x \mid n \in \mathbb{N}\}$. Here and throughout the text we use $\underline{n}$ to denote the term $\mathsf{S}^n 0 := \underbrace{\mathsf{S} \cdots \mathsf{S}}_{n} 0$.

Therefore, $\underline{1} = \mathsf{S}0$, $\underline{2} = \mathsf{S}\underline{1}, \ldots$, and generally $\underline{\mathsf{S}n} = \mathsf{S}\underline{n}$. The term $\underline{0}$ (that is, $\mathsf{S}^0 0$) is mostly denoted by 0. Note that $\underline{n} < x$ is the formula $\underline{n} \leqslant x \wedge \underline{n} \neq x$. One may replace $x$ here by a constant symbol $c$, thus expanding the language. But both approaches lead to the same result.

Every finite subset $X_0 \subseteq X$ possesses a model. Indeed, there is evidently some $m$ such that $X_0 \subseteq X_1 := Th\mathcal{N} \cup \{\underline{n} < x \mid n < m\}$, and $X_1$ certainly has a model: one need only assign to $x$ in $\mathcal{N}$ the number $m$. Thus, by Theorem 3.2, $X$ has a model $(\mathcal{N}', c)$ with the domain $\mathbb{N}'$, where $c \in \mathbb{N}'$ denotes the interpretation of $x$. We know that $\mathcal{N}'$ satisfies all sentences valid in $\mathcal{N}$, including in particular the sentences $\underline{\mathsf{S}n} = \mathsf{S}\underline{n}$, $\underline{n+m} = \underline{n}+\underline{m}$ and $\underline{n \cdot m} = \underline{n} \cdot \underline{m}$. Therefore, $n \mapsto \underline{n}^{\mathcal{N}'}$ constitutes an embedding from $\mathcal{N}$ into $\mathcal{N}'$ whose image can be thought of as coinciding with $\mathcal{N}$.[2] In other words, it is legitimate to assume that $\underline{n}^{\mathcal{N}'} = n$ so that $\mathcal{N} \subseteq \mathcal{N}'$.

Because $\mathcal{N}' \vDash X$, on the one hand $\mathcal{N}'$ is elementarily equivalent to $\mathcal{N}$, and on the other $n < a$ for all $n$ and any $a \in \mathbb{N}'\backslash\mathbb{N}$, since in $\mathcal{N}$ and hence in $\mathcal{N}'$ we have $(\forall x \leqslant \underline{n}) \bigvee_{i \leqslant n} x = \underline{i}$. In short, $\mathbb{N}$ is a (proper) initial segment of $\mathbb{N}'$, or $\mathcal{N}'$ is an *end extension* of $\mathcal{N}$. The elements of $\mathbb{N}'\backslash\mathbb{N}$ are called *nonstandard numbers*. Alongside $c$, other examples are $c + c$ and $c + \underline{n}$ for $n \in \mathbb{N}$. Clearly, $c$ has both an immediate successor and an immediate predecessor in the order, because $\mathcal{N}' \vDash (\forall x \neq 0)\exists! y\, x = \mathsf{S}y$. The figure gives a rough picture of a nonstandard model $\mathcal{N}'$:

$$\mathbb{N}' : \quad \overset{\overbrace{\qquad\qquad}^{\mathbb{N}}}{\underset{0\ \ 1}{\bullet\ \bullet\ \bullet\bullet\bullet\cdots}\ \cdots\bullet\bullet\bullet\bullet\bullet\cdots\ \underset{c}{\bullet}\ \bullet\ \bullet\ \bullet\bullet\bullet\cdots\ \cdots\bullet\bullet\bullet\ \underset{c+c}{\bullet}\ \bullet\ \bullet\ \bullet\bullet\bullet\cdots\ \cdots}$$

$\mathcal{N}'$ has the same number-theoretic features as $\mathcal{N}$, at least all those that can be formulated in $\mathcal{L}_{ar}$. These include nearly all the interesting ones, as will turn out to be the case in **7.1**. For example, $\forall x \exists y (x = \underline{2}y \vee x = \underline{2}y+\underline{1})$ holds in every model of $Th\mathcal{N}$, that is, every nonstandard number is either

---

[2] Whenever $\mathcal{A}$ is embeddable into $\mathcal{B}$ there is a structure $\mathcal{B}'$ isomorphic to $\mathcal{B}$ such that $\mathcal{A} \subseteq \mathcal{B}'$. The domain $B'$ arises from $B$ by interchanging the images of the elements of $\mathcal{A}$ with their originals.

even or odd. Clearly, the model $\mathbb{N}'$ contains gaps in the sense of **2.1**. The most obvious example is $(\mathbb{N}, \mathbb{N}' \backslash \mathbb{N})$.

**Remark 2.** Theorem 4.1 will show that $Th\mathcal{N}$ has also countable nonstandard models. The order of such a model $\mathcal{N}'$ is easy to make intuitive: it arises from the half-open interval $[0, 1)$ of rational numbers by replacing 0 with $\mathbb{N}$ and every other $r \in [0, 1)$ by a specimen from $\mathbb{Z}$. On the other hand, neither $+^{\mathcal{N}'}$ nor $\cdot^{\mathcal{N}'}$ is effectively describable; see for instance [HP].

Replacing the induction schema IS in the axiom system for PA by the so-called *induction axiom*

$$\text{IA:} \quad \forall P(P0 \wedge \forall x(Px \to PSx) \to \forall x Px) \qquad (P \text{ a predicate variable})$$

results in a *categorical* axiom system that, up to isomorphism, has just a single model (see e.g. [Ra5]). How is it possible that $\mathcal{N}$ is uniquely determined up to isomorphism by a few axioms, but at the same time nonstandard models exist for $Th\mathcal{N}$? The answer is simple: IA cannot be adequately formulated in $\mathcal{L}_{ar}$. That is, IA is not an axiom or perhaps an axiom schema of the first-order language of $\mathcal{N}$. It is a sentence of a second-order language, about which we shall say more in **3.8**. However, this intimated limitation regarding the possibilities of formulation in first-order languages is merely an apparent one, as the undertakings of the rest of the book will show, especially the considerations about axiomatic set theory in **3.4**.

In no nonstandard model $\mathcal{N}'$ is the initial segment $\mathbb{N}$ definable, indeed not even *parameter definable*, which means that there is no $\alpha = \alpha(x, \vec{y})$ and no $b_1, \ldots, b_n \in \mathbb{N}'$ such that $\mathbb{N} = \{a \in \mathbb{N}' \mid \mathcal{N}' \vDash \alpha [a, \vec{b}]\}$. Otherwise we would have $\mathcal{N}' \vDash \alpha \frac{0}{x} \wedge \forall x(\alpha \to \alpha \frac{Sx}{x}) [\vec{b}]$. This yields $\mathcal{N}' \vDash \forall x \alpha [\vec{b}]$ by IS, in contradiction to $\mathbb{N}' \backslash \mathbb{N} \neq \emptyset$. The same reasoning shows that no proper initial segment $A \subset \mathbb{N}'$ without a largest element is definable in $\mathbb{N}'$; such an $A$ would clearly define a gap in the order of $\mathbb{N}'$. The situation can also be described as *gaps in $\mathbb{N}$ are not recognizable from within*.

Introductory courses in real analysis tend to give the impression that a meaningful study of the subject requires the axiom of continuity: *Every nonempty bounded set of real numbers has a supremum*. On this basis, Cauchy and Weierstrass reformed analysis, thus banishing from mathematics the somewhat mysterious infinitesimal arguments of Leibniz, Newton, and Euler. But mathematical logic has developed methods that, to a large extent, justify the original arguments. This is undertaken in the framework of *nonstandard analysis*, developed above all by A. Robinson around 1950. In the sequel, we provide an indication of its basic idea.

The same construction as for $\mathcal{N}$ also provides a nonstandard model for the theory of $\mathcal{R} = (\mathbb{R}, +, \cdot, <, \{\boldsymbol{a} \mid a \in \mathbb{R}\})$, where for each real number $a$, a name $\boldsymbol{a}$ was added to the signature. Consider $X = Th\,\mathcal{R} \cup \{\boldsymbol{a} < x \mid a \in \mathbb{R}\}$. Every finite subset of $X$ has a model on the domain $\mathbb{R}$. Thus, $X$ is consistent, and as above, a model of $X$ represents a proper extension $\mathcal{R}^*$ of $\mathcal{R}$, a *nonstandard model of analysis*. In each such model the same theorems hold as in $\mathcal{R}$. For instance, in $\mathcal{R}^*$ every polynomial of positive degree can be decomposed into linear and quadratic factors. In Chapter **5** it will be shown that the nonstandard models of $Th\,\mathcal{R}$ are precisely the real closed extensions of $\mathcal{R}$. All these are elementarily equivalent to $\mathcal{R}$.

For analysis, it is now decisive that the language can be enriched from the very beginning, say by the adoption of the symbols $\exp, \ln, \sin, \cos$ for the exponential, logarithmic, and trigonometric functions, and further symbols for further functions. We denote a thus expanded standard model once again by $\mathcal{R}$ and a corresponding nonstandard model by $\mathcal{R}^*$. The mentioned real functions available in $\mathcal{R}$ carry over to $\mathcal{R}^*$ and maintain all properties that can be elementarily formulated. That means in fact almost all properties with interesting applications, for example

$$\forall xy\, \exp(x+y) = \exp x \cdot \exp y, \quad (\forall x>0)\, \exp \ln x = x, \quad \forall x\, \sin^2 x + \cos^2 x = 1,$$

as well as the addition theorems for the trigonometric functions and so on. All these functions remain continuous and repeatedly differentiable. However, the Bolzano–Weierstrass theorem and other topological properties cannot be salvaged in full generality. They are replaced by the aforementioned infinitesimal arguments.

In a nonstandard model $\mathcal{R}^*$ of $Th\,\mathcal{R}$ with $\mathcal{R} \subseteq \mathcal{R}^*$ there exist not only infinitely large numbers $c$ (i.e., $r < c$ for all $r \in \mathbb{R}$), but also infinitely many small positive numbers. Let $c$ be infinite. Since $\frac{1}{r} < c \Leftrightarrow \frac{1}{c} < r$ for all $r > 0$, $\frac{1}{c}$ is smaller than each positive real $r$, and yet positive. That is, $\frac{1}{c}$ is fairly precisely what Leibniz once named an *infinitesimal*. Taking a somewhat closer look reveals the following picture: Every real number $a$ is sitting in a nest of nonstandard numbers $a^* \in \mathcal{R}^*$ that are only infinitesimally distinct from $a$. In other words, $|a^* - a|$ is an infinitesimal. Hence, quantities such as $dx, dy$ *exist* in mathematical reality, and may once again be considered as infinitesimals in the sense of their inventor Leibniz. These quantities are exactly the elements of $\mathcal{R}^*$ infinitesimally distinct from 0.

From the existence of nonstandard models for $Th\,\mathcal{R}$, it can be concluded that the continuity axiom, just like IA, cannot be elementarily formulated. For by adjoining this axiom to those for ordered fields, $\mathcal{R}$ is characterized, up to isomorphism, as the only continuously ordered field; see e.g. [Ta4]. Hence, the order of a nonstandard model $\mathcal{R}^*$ of $Th\,\mathcal{R}$ possesses gaps. Here, too, the gaps are "not recognizable from within," since every nonempty, bounded parameter-definable subset of $\mathbb{R}^*$ has a supremum in $\mathbb{R}^*$. That is the case because in $\mathcal{R}$ and thus also in $\mathcal{R}^*$, the following *continuity schema* holds, which ensures the existence of a supremum for those sets; here $\varphi = \varphi(x, \vec{y})$ runs over all formulas such that $y, z \notin free\,\varphi$:

$$\exists x\varphi \wedge \exists y \forall x(\varphi \to x \leqslant y) \to \exists z \forall x[(\varphi \to x \leqslant z) \wedge \forall y((\varphi \to x \leqslant y) \to z \leqslant y)].$$

Analogous remarks can be made on complex numbers. There is an algebraically closed field $\mathcal{R}^*[i] \supseteq \mathcal{R}^*$ in which familiar facts such as Euler's formula $e^{ix} = \cos x + i \cdot \sin x$ continue to hold, in particular $e^{i\pi} = -1$.

## Exercises

1. Prove in PA the associativity, commutativity, and distributivity of $+, \cdot$. Before proving $x+y = y+x$ derive $Sx+y = x+Sy$ and $0+y = y$ by induction on $y$. The basic arithmetical laws provable in PA are collected in the axiom system N on page 235.

2. $\leqslant$ was defined in $\mathcal{L}_{ar}$ on page 105. Reflexivity and transitivity of $\leqslant$ easily derive in PA. Prove in PA the antisymmetry of $\leqslant$.

3. Prove $x < y \equiv_{PA} Sx \leqslant y$ (or equivalently, $x < Sy \equiv_{PA} x \leqslant y$). Use this to prove $\vdash_{PA} x \leqslant y \vee y \leqslant x$ by induction on $x$.

4. Verify (a),(b), and (c) for arbitrary formulas $\alpha, \beta, \gamma \in \mathcal{L}_{ar}$ such that $y \notin var\{\alpha, \beta\}$ and $z \notin var\,\gamma$.
   (a) $\vdash_{PA} \forall x((\forall y{<}x)\alpha\,\frac{y}{x} \to \alpha) \to \forall x\alpha$, the *schema of $<$-induction*,
   (b) $\vdash_{PA} \exists x\beta \to \exists x(\beta \wedge (\forall y{<}x)\neg\beta\,\frac{y}{x})$, the *minimum schema*,
   (c) $\vdash_{PA} (\forall x{<}v)\exists y\gamma \to \exists z(\forall x{<}v)(\exists y{<}z)\gamma$, the *schema of collection*.

## 3.4   ZFC and Skolem's Paradox

Before turning to further consequences of the results from **3.2**, we collect a few basic facts about countable sets. The proofs are simple and can be found in any textbook on basic set theory. A set $M$ is called *countable* if there is a surjection $f \colon \mathbb{N} \to M$ (i.e. $M = \{a_n \mid n \in \mathbb{N}\}$ provided $fn = a_n$) or $M = \emptyset$, and otherwise *uncountable*. Every subset of a countable set is itself countable. If $f \colon M \to N$ is surjective and $M$ is countable then clearly so too is $N$. Sets $M, N$ are termed *equipotent*, briefly $M \sim N$, if a bijection from $M$ to $N$ exists. If $M \sim \mathbb{N}$, then $M$ is said to be *countably infinite*. A countable set can only be countably infinite or *finite*, which is to mean equipotent to $\{1, \dots, n\}$ for some $n \in \mathbb{N}$.

The best-known uncountable set is $\mathbb{R}$. It is equipotent to $\mathfrak{P}\mathbb{N}$. The uncountability of $\mathfrak{P}\mathbb{N}$ is a particular case of an important theorem from Cantor: *The power set $\mathfrak{P}M$ of a set $M$ has a higher cardinality than $M$,* i.e., no injection from $M$ to $\mathfrak{P}M$ is surjective. The cardinality of sets will be explained to some extent in **5.1**. Here it suffices to know that two sets $M, N$ are of the same cardinality iff $M \sim N$, and that there are countable and uncountable infinite sets.

If $M, N$ are countable so too are $M \cup N$ and $M \times N$, as is easy to see. Moreover, as was shown already by Cantor, a countable union $U = \bigcup_{i \in \mathbb{N}} M_i$ of countable sets $M_i$ is again countable. Cantor's proof consists in writing down $U$ as an infinite matrix where the $n$th line enumerates of $M_n = \{a_{nm} \mid m \in \mathbb{N}\}$. Then enumerate the matrix in the zigzag manner indicated by the figure on the right, beginning with $a_{00}$. Accordingly, for countable $M$, in particular $U = \bigcup_{n \in \mathbb{N}} M^n$, the set of all

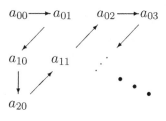

finite sequences of elements in $M$ is again countable, because every $M^n$ is countable. Hence, every first-order language with a countable signature is itself countable, more precisely countably infinite.[3]

By a *countable theory* we always mean a theory formalized in a countable language. We now formulate a theorem significant for many reasons.

---

[3] Here we use the axiom of choice, since for every $M_i$ some enumeration is chosen. It can be shown that without the axiom of choice the proof cannot be carried out.

**Theorem 4.1 (Löwenheim–Skolem).** *A countable consistent theory $T$ always has a countable model.*

**Proof.** By Theorem 2.6, $T$ $(\subseteq \mathcal{L})$ has a model $\mathcal{M}$ with domain $A$, consisting of the equivalence classes $\bar{c}$ for $c \in C$ in the set of all terms of $\mathcal{L}' = \mathcal{L}C$, where $C = \bigcup_{n \in \mathbb{N}} C_n$ is a set of new constants. By construction, $C_0$ is equipotent to $Var \times \mathcal{L}$ and thus countable. The same holds for every $C_n$, and so $C$ is also countable. The map $c \mapsto \bar{c}$ from $C$ to $A$ is trivially surjective, so that $\mathcal{M}$ has a countable (possibly finite) domain, and this was the very claim.   ∎

In **5.1** we will significantly generalize the theorem, but even in the above formulation it leads to noteworthy consequences. For example, there are also countable ordered fields as nonstandard models of the first-order theory $Th\,(\mathbb{R}, 0, 1, +, <, \cdot, \exp, \sin, \dots)$ in which the usual theorems about real functions retain their validity. Thus, one need not really overstep the countable to obtain a rich theory of analysis.

Especially surprising is the existence, ensured by Theorem 4.1, of countable models of formalized set theory. Although set theory can be regarded as the basis for the whole of presently existing mathematics, it embraces only a few set-building principles. The most important system of formalized set theory is **ZFC**, created at the beginning of the twentieth century.

**Remark 1.** Z stands for E. Zermelo, F for A. Fraenkel, and C for AC, the axiom of choice. ZF denotes the theory resulting from the removal of AC. ZFC sets out from the principle that every element of a set is again a set, so that a distinction between sets and systems of sets vanishes. Thus, ZFC speaks exclusively about sets, unlike Russell's type-theoretic system, in which, along with sets, so-called *urelements* (objects that are members of sets but aren't sets themselves) are considered. Set theory without urelements is fully sufficient as a foundation of mathematics and for nearly all practical purposes. Even from the epistemological point of view there is no evidence that urelements occur in reality: each object can be identified with the set of all properties that distinguish it from other objects. Nonetheless, urelements are still in use as a technical tool in certain set-theoretic investigations. We mention in passing that neither ZF nor ZFC is finitely axiomatizable. This seems plausible if we look at the axioms given below, but the proof is not quite easy.

To make clear that **ZFC** is a countable first-order theory and hence belongs to the scope of applications of Theorem 4.1, we present in the following its axioms. Each of the axioms will be briefly discussed. This will be at the same time an excellent exercise in advanced formalization technique. The set-theoretic language already denoted in **2.2** by $\mathcal{L}_\in$ is one of the most conceivably simple languages and is certainly countable. Alongside $=$ it contains only the membership symbol $\in$. This symbol should be distinguished from the somewhat larger $\in$ that is used throughout in our metatheory. The variables are now called *set variables*. These will as a rule be denoted by lowercase letters as in other first-order languages. In order to make the axioms and its consequences easily legible, we employ the widely used abbreviations

$$(\forall y \in x)\varphi := \forall y(y \in x \to \varphi), \quad (\exists y \in x)\varphi := \exists y(y \in x \wedge \varphi).$$

Besides, we define the relation of inclusion by $x \subseteq y \leftrightarrow \forall z(z \in x \to z \in y)$. Note also that all free variables occurring in the axioms below (e.g., $x, y$ in AE) have to be thought of as being generalized according to our convention in **2.5**. The ZFC axioms are then the following:

AE : $\forall z(z \in x \leftrightarrow z \in y) \to x = y$ (axiom of extensionality).

AS : $\exists y \forall z(z \in y \leftrightarrow \varphi \wedge z \in x)$ (axiom of separation).

Here $\varphi$ runs over all $\mathcal{L}_\in$-formulas with $y \notin$ free $\varphi$. AS is in fact a schema of axioms. Let $\varphi = \varphi(x, z, \vec{a})$. From AS and AE , $\forall x \exists! y \forall z(z \in y \leftrightarrow \varphi \wedge z \in x)$ is derivable. Indeed, observe for $y, y' \notin$ free $\varphi$ the obvious derivability of

$$(z \in y \leftrightarrow \varphi \wedge z \in x) \wedge (z \in y' \leftrightarrow \varphi \wedge z \in x) \to (z \in y \leftrightarrow z \in y').$$

This implies $\forall z(z \in y \leftrightarrow \varphi \wedge z \in x) \wedge \forall z(z \in y' \leftrightarrow \varphi \wedge z \in x) \to y = y'$ and hence the claim. Thus, $y = \{z \in x \mid \varphi\} \leftrightarrow \forall z(z \in y \leftrightarrow \varphi \wedge z \in x)$ is a legitimate definition in the sense of **2.6**. $\{z \in x \mid \varphi\}$ is called a *set term* and is just a suggestive writing of a function term $f_{\vec{a}}x$. This term still depends on the "parameter" vector $\vec{a}$. It collects the free variables of $\varphi$ distinct from $x, z$. Thus, instead of introducing each time a new operation symbol, one uses the more economical "curled bracket notation."

The empty set can explicitly be defined by $y = \emptyset \leftrightarrow \forall z\, z \notin y$. Indeed, thanks to AS, $\exists y \forall z(z \in y \leftrightarrow z \notin x \wedge z \in x)$ is provable. This formula is equivalent to $\exists y \forall z\, z \notin y$, since $z \in y \leftrightarrow z \notin x \wedge z \in x \equiv z \notin y$. Clearly, using AE, $\forall z\, z \notin y \wedge \forall z\, z \notin y' \to y = y'$ is provable. This, together with $\exists y \forall z\, z \notin y$, yields $\exists! y \forall z\, z \notin y$, which legitimates the explicit definition $y = \emptyset \leftrightarrow \forall z\, z \notin y$, as was explained in detail in **2.6**. The next axiom is

AU : $\forall x \exists y \forall z(z \in y \leftrightarrow (\exists u \in x)\, z \in u)$ (axiom of union).

Here again, because of AE, $\exists y$ can be replaced by $\exists! y$. As in **2.6**, we may therefore define an operator on the universe,[4] denoted by $x \mapsto \bigcup x$. We avoid the word "function," since functions are understood as special objects of the universe. AU is equivalent to $\forall x \exists y \forall z((\exists u \in x) z \in u \to z \in y)$, because $\bigcup x$ can be separated from such a set $y$ by means of AS. The following axiom could analogously be weakened.

$\quad$ AP : $\quad \forall x \exists y \forall z (z \in y \leftrightarrow z \subseteq x)$ $\quad$ (power set axiom).

Let $\mathfrak{P} x$ denote the $y$ that in view of AE is uniquely determined by $x$ in AP. What first can easily be proved is $\forall x (x \in \mathfrak{P} \emptyset \leftrightarrow x = \emptyset)$ as well as $\forall x (x \in \mathfrak{P} \mathfrak{P} \emptyset \leftrightarrow x = \emptyset \lor x = \mathfrak{P} \emptyset)$. Since $\emptyset \neq \mathfrak{P} \emptyset$, the set $\mathfrak{P} \mathfrak{P} \emptyset$ contains precisely two elements. This is decisive for defining $\{a, b\}$ below.

$\quad$ The following axiom was added to those of Zermelo by Fraenkel.

$\quad$ AR : $\forall x \exists! y \varphi \to \forall u \exists v \forall y (y \in v \leftrightarrow (\exists x \in u)\, \varphi)$ $\quad$ (axiom of replacement).

Here $\varphi = \varphi(x, y, \vec{a})$ and $u, v \notin \text{free}\, \varphi$. If $\forall x \exists! y \varphi$ is provable, then we know from **2.6** that an operator $x \mapsto Fx$ can be introduced. By AR, the image of a set $u$ under $F$ is again a set which, as a rule, is denoted by $\{Fx \mid x \in u\}$. $F$ may depend on further parameters $a_1, \ldots, a_n$, so we had better write $F_{\vec{a}}$ for $F$. AR is very strong; even AS is derivable from it, Exercise 4. An instructive example of an application of AR, for which $\forall x \exists! y \varphi$ is certainly provable, is provided by $\varphi = \varphi(x, y, a, b) := x = \emptyset \land y = a \lor x \neq \emptyset \land y = b$. The operator $F = F_{a,b}$ defined by $\varphi$ clearly satisfies $F \emptyset = a$ and $Fx = b$ if $x \neq \emptyset$. Accordingly, the image of the two-element set $\mathfrak{P} \mathfrak{P} \emptyset$ under $F_{a,b}$ contains the (not necessarily distinct) members $a, b$. We then define

$$\{a, b\} := \{F_{a,b}(x) \mid x \in \mathfrak{P} \mathfrak{P} \emptyset\}$$

and call this set the *pair set* of $a, b$. We next put $a \cup b := \bigcup \{a, b\}$ (while $a \cap b := \{z \in a \mid z \in b\}$ already exists from AS). Further, let $\{a\} := \{a, a\}$ and $\{a_1, \ldots, a_{n+1}\} = \{a_1, \ldots, a_n\} \cup \{a_{n+1}\}$ for $n \geqslant 2$. Now we can prove that $\mathfrak{P} \emptyset = \{\emptyset\}$, $\mathfrak{P} \mathfrak{P} \emptyset = \{\emptyset, \{\emptyset\}\}, \ldots$ The *ordered pair* of $a, b$ is defined as $(a, b) := \{\{a\}, \{a, b\}\}$. This definition may look artificial but it implies the basic property $(a, b) = (c, d) \leftrightarrow a = c \land b = c$. Only this is needed.

$\quad$ We now have at our disposal the tools necessary to develop elementary set theory. Beginning with sets of ordered pairs it is possible to model relations and functions and all concepts building upon them, even though

---

[4] A frequently used synonym for the domain of a ZFC-model, mostly denoted by $V$, and 'for all sets $a$' is then often expressed as 'for all $a \in V$'.

the existence of infinite sets is still unprovable. Mathematical requirements demand their existence, though then the borders of our experience with finite sets are transgressed. The easiest way to get infinite sets is using the set operator $x \mapsto \mathsf{S}x$, with $\mathsf{S}x := x \cup \{x\}$.

$\quad$ AI : $\quad \exists u[\emptyset \in u \wedge \forall x(x \in u \to \mathsf{S}x \in u)]$ $\quad$ (axiom of infinity).

Such a set $u$ contains $\emptyset$, $\mathsf{S}\emptyset = \emptyset \cup \{\emptyset\} = \{\emptyset\}$, $\mathsf{S}\mathsf{S}\emptyset = \{\emptyset, \{\emptyset\}\}, \ldots$ and is therefore infinite in the naive sense. This holds in particular for the smallest set $u$ of this type, denoted by $\omega$. In formalized set theory $\omega$ plays the role of the set of natural numbers. $\omega$ contains $0 := \emptyset$, $1 := \mathsf{S}0 = \{0\}$, $2 := \mathsf{S}1 = \{\emptyset, \{\emptyset\}\} = \{0,1\}, \ldots$ Generally, $\mathsf{S}n = \{0, 1, \ldots, n\}$. Thus, the $n \in \mathbb{N}$ are represented in **ZF** by certain variable-free terms, called $\omega$-*terms*.

$\quad$ In everyday mathematics the following axiom is basically dispensable:

$\quad$ AF : $\quad (\forall x \neq \emptyset)(\exists y \in x)\, x \cap y = \emptyset$ $\quad$ (axiom of foundation).

Put intuitively: Every set $x \neq \emptyset$ contains an $\in$-minimal element $y$. AF precludes the possibility of "$\in$-circularity" $x_0 \in \cdots \in x_n \in x_0$. In particular, there is no set $x$ with $x \in x$.

**Remark 2.** In axiomatic set theory, AF plays a highly important role. The most important consequence of AF is the existence of the *von Neumann hierarchy* $V = \bigcup_{\alpha \in On} V_\alpha$. Here $On$ denotes the class of all ordinal numbers. These are generalizations of natural numbers, defined in each textbook on set theory. $V_\alpha$ is a set for each $\alpha \in On$ and defined by recursion: $V_0 = \emptyset$, $V_{\alpha+1} = \mathfrak{P}V_\alpha$, and $V_\lambda = \bigcup_{\alpha < \lambda} V_\alpha$ for limit ordinals $\lambda$. All this is more important for the foundations of mathematics than for applications of set theory.

$\quad$ **ZF** is the theory with the above axioms. **ZFC** results from **ZF** by adjoining the *axiom of choice* AC:

$$\forall u[\emptyset \notin u \wedge (\forall x \in u)(\forall y \in u)(x \neq y \to x \cap y = \emptyset) \to \exists z(\forall x \in u)\exists! y(y \in x \cap z)].$$

AC states that for every set $u$ of disjoint nonempty sets $x$ there is a set $z$, a *choice set*, that picks up precisely one element from each $x$ in $u$. One of the many equivalences to AC is $\prod_{i \in I} A_i \neq \emptyset$, for *any* index set $I$.

$\quad$ The above expositions clearly show that **ZFC** can be understood as a first-order theory. In some sense, **ZFC** is even the purest such theory, because all sophisticated proof methods that occur in mathematics, for instance transfinite induction and recursion and every other type of induction and recursion, can be made explicit and derived in the first-order language $\mathcal{L}_\in$ of **ZFC** without particular difficulty.

Whereas mathematicians regularly transgress the framework of a theory, even one that is unambiguously defined by first-order axioms, in that they make use of combinatorial, number- or set-theoretic tools wherever it suits them, set theory, as it stands now, imposes upon itself an upper limit. Within ZFC, all sophisticated proof and definition techniques gain an elementary character, so to speak. As a matter of fact, there are no pertinent arguments against the claim that the whole of mathematics can be treated within the frame of ZFC as a single first-order theory, a claim based on general mathematical experience that is highly interesting for the philosophy of mathematics. However, one should not make a religion out of this insight, because for mathematical practice it is of limited significance only.

If ZFC is consistent—and no one really doubts this assumption although there is no way of proving it—then by Theorem 4.1, ZFC must have a countable model $\mathcal{V} = (V, \in^{\mathcal{V}})$. The existence of such a model $\mathcal{V}$ is at first glance paradoxical because the existence of uncountable sets is easily provable *within* ZFC. An example is $\mathfrak{P}\omega$. On the other hand, because of $(\mathfrak{P}\omega)^{\mathcal{V}} \subseteq V$, it must be true (judged from the outside) that also $(\mathfrak{P}\omega)^{\mathcal{V}}$ is countable. Thus, the notion *countable* has a different meaning "inside and outside the world $\mathcal{V}$," which comes rather unexpectedly. This is the so-called *Skolem's paradox*.

The explanation of Skolem's paradox is that the countable model $\mathcal{V}$, to put it figuratively, is "thinned out" and contains fewer sets and functions than expected. Indeed, roughly put, it contains just enough to satisfy the axioms, yet not, for instance, some bijection from $\omega^{\mathcal{V}}$ to $(\mathfrak{P}\omega)^{\mathcal{V}}$, which, seen from the outside, certainly exists. Therefore, the countable set $(\mathfrak{P}\omega)^{\mathcal{V}}$ is uncountable from the perspective of the world $\mathcal{V}$. In other words, countability is not an absolute concept.

Moreover, the universe $V$ of a ZFC-model is by definition a set, whereas $\vdash_{\mathsf{ZFC}} \neg \exists v \forall z\, z \in v$, i.e., there is no "universal set." Thus, seen from within, $V$ is too big to be a set. $\neg \exists v \forall z\, z \in v$ is derived as follows: the hypothesis $\exists v \forall z\, z \in v$ entails with AE and AS the existence of the "Russellian set" $u = \{x \in v \mid x \notin x\}$. That is, $\exists v \forall z\, z \in v \vdash_{\mathsf{ZFC}} \exists u \forall x (x \in u \leftrightarrow x \notin x)$.

On the other hand, by Example 1 page 73, $\vdash_{\mathsf{ZFC}} \neg \exists u \forall x (x \in u \leftrightarrow x \notin x)$. Thus, indeed $\vdash_{\mathsf{ZFC}} \neg \exists v \forall z\, z \in v$. Accordingly, even the notion of a set depends on the model. There is no absolute definition of a set.

None of the above has anything to do with ZFC's incompleteness.[5] Mathematics has no problem with the fact that its basic theory is incomplete and cannot be rendered complete, at least not in an axiomatic manner. More of a problem is the lack of undisputed criteria for extending ZFC in a way coinciding with truth or at least with our intuition.

### Exercises

1. Let $T$ be an elementary theory with arbitrarily large finite models. Prove that $T$ also has an infinite model.

2. Suppose $\mathcal{A} = (A, <)$ is an infinite well-ordered set (see **2.1**). Show that there is a not well-ordered set elementarily equivalent to $\mathcal{A}$. Thus, being well-ordered is not a first-order property.

3. Prove that a consistent theory $T$ coincides with the intersection of all its complete extensions, i.e., $T = \bigcap \{ T' \supseteq T \mid T' \text{ complete} \}$.

4. Derive the axiom AS of separation from the replacement axiom AR.

5. $\mathsf{fin}(a) := \forall s [\emptyset \in s \wedge (\forall u \in s)(a \setminus u \neq \emptyset \to (\exists e \in a \setminus u) u \cup \{e\} \in s) \to a \in s]$ is one of several definitions of '$a$ is finite'. Prove for each $\varphi \in \mathcal{L}_\in$: $\mathsf{fin}(a) \vdash_{\mathsf{ZF}} \varphi_x(\emptyset) \wedge \forall u \forall e (\varphi_x(u) \to \varphi_x(u \cup \{e\})) \to \varphi_x(a)$.

## 3.5   Enumerability and Decidability

Of all the far-reaching consequences of the completeness theorem, perhaps the most significant is the effective enumerability of all tautologies of a countable first-order language. Once Gödel had proved this, the hope grew that the decidability problem for tautologies might soon be resolved. Indeed, the wait was not long, and a few years after Gödel's result Church proved the problem to be unsolvable for sufficiently expressive languages. This section is intended to provide only a brief glimpse of enumeration and decision problems as they appear in logic, computer science, and elsewhere. We consider them more rigorously in Chapters **5** and **6**.

The term *effectively enumerable* will be made more precise in **6.1** by the notion of *recursive enumerability*. At this stage, our explanation of this

---

[5] In **6.6** the incompleteness of ZFC and all its axiomatic extensions will be proved.

notion must be somewhat superficial, though like that for a decidable set it is highly visualizable. Put roughly, a set $M$ of natural numbers, say, or syntactic objects, finite structures, or similar objects is called *effectively* (or *recursively*) *enumerable* if there exists an algorithm that delivers the elements of $M$ stepwise. Thus, in the case of an infinite set $M$, the algorithm does not stop its execution by itself.

The calculus of natural deduction enables first of all an effective enumeration of all provable finite sequences of a first-order language with at most countably many logical symbols, i.e., all pairs $(X, \alpha)$ such that $X \vdash \alpha$ and $X$ is finite, at least in principle. First of all, we imagine all initial sequents as enumerated in an ongoing, explicitly producible sequence $S_0, S_1, \ldots$ Then it is systematically checked whether one of the sequent rules is applicable; the resulting sequents are then enumerated in a second sequence and so on. Leaving aside problems concerning the storage capacity of such a deduction machine, as well as the difficulties involved in evaluating the flood of information that would pour from such a device, it is simply a question of organization to create a program that enumerates all provable finite sequents.

Moreover, it can be seen without difficulty that the tautologies of a countable language $\mathcal{L}$ are effectively enumerable; one need only pick out from an enumeration procedure of provable sequents $(X, \alpha)$ those such that $X = \emptyset$. In short, the aforementioned deduction machine delivers stepwise a sequence $\alpha_0, \alpha_1, \ldots$ (without repetitions if so desired) that consists of exactly the tautologies of $\mathcal{L}$. This would be somewhat easier with the calculus in **3.6**. However, we cannot in this way obtain a decision procedure as to whether any given formula $\alpha \in \mathcal{L}$ is a tautology, for we do not know whether $\alpha$ ever appears in the produced sequence. We will prove rigorously in **6.5** that in fact such an algorithm does not exist, provided $\mathcal{L}$ contains at least a binary predicate or operation symbol. Decision procedures exist only for $\mathcal{L}_=$ as will be shown in **5.2**, and expansions of $\mathcal{L}_=$ containing only unary predicate and constant symbols, and at most one unary operation symbol; see also [BGG].

The deduction machine can also be applied to enumerate the theorems of a given axiomatizable theory $T$, in that parallel to the enumeration process for all provable sequents of the language, a process is also set going that enumerates all axioms of $T$. It must then continually be checked for the enumerated sequents whether all their premises occur as already-enumerated assertions; if so, then the conclusion of the sequent in question is provable in $T$. The preceding considerations constitute an informal proof of the following theorem. A rigorous proof free of merely intuitive arguments is provided by Theorem 6.2.4.

**Theorem 5.1.** *The theorems of an axiomatizable theory are effectively enumerable.*

Almost all theories considered in mathematics are axiomatizable, including formalized set theory **ZFC** and Peano arithmetic **PA**. While the axiom systems of these two theories are infinite and cannot be replaced by finite ones, these sets of axioms are evidently decidable.

Our experience hitherto shows us that all theorems of mathematics held to be proved are also provable in **ZFC**. Hence, according to Theorem 5.1, all mathematical theorems can in principle be stepwise generated by a computer. This fact is theoretically highly important, even if it has little far-reaching practical significance at present.

Recall the notion of a complete theory. Among the most important examples is the theory of the real closed fields (Theorem 5.5.5). A noteworthy feature of complete and axiomatizable theories is their *decidability*. We call a theory *decidable* if the set of its theorems is a decidable set of formulas, and otherwise *undecidable*. We shall prove the next theorem in an intuitive manner. A strict proof, based on the rigorous definition of decidability based on the theory of recursive functions in **6.1**, will later be provided by Theorem 6.4.4 on page 247.

**Theorem 5.2.** *A complete axiomatizable theory $T$ is decidable.*

**Proof.** By Theorem 5.1 let $\alpha_0, \alpha_1, \ldots$ be an effective enumeration of all sentences provable in $T$. A decision procedure consists simply in comparing for given $\alpha \in \mathcal{L}^0$ the sentences $\alpha$ and $\neg\alpha$ in the $n$th construction step of $\alpha_0, \alpha_1, \ldots$ with $\alpha_n$. If $\alpha = \alpha_n$ then $\vdash_T \alpha$; if $\alpha = \neg\alpha_n$ then $\nvdash_T \alpha$. This process certainly terminates, because due to the completeness of $T$, either $\alpha$ or $\neg\alpha$ will appear in the enumeration sequence $\alpha_0, \alpha_1, \ldots$ of the theorems of $T$. ◻

Conversely, a complete decidable theory is trivially axiomatizable (by $T$ itself). Thus, for complete theories, "decidable" and "axiomatizable" mean one and the same thing. A consistent theory has a model and hence at least one *completion*, i.e., a complete extension in the same language. The only completion of a complete theory $T$ is $T$ itself. A remarkable generalization of Theorem 5.2 is Exercise 3.

A (countable) decidable theory has always a decidable completion, see Exercise 4. Hence, a theory all completions of which are undecidable is itself undecidable. We will meet such theories in **6.5**. On the other hand, if $T$ has only finitely many completions, $T_0, \ldots, T_n$ say, all of which are decidable, then so is $T$. Indeed, according to Exercise 3 in **3.4**, $\alpha \in T$ iff $\alpha \in T_i$ for all $i \leqslant n$.[6] See also Exercise 3 below.

In the early stages in the development of fast computing machines, high hopes were held concerning the practical carrying out of mechanized decision procedures. For various reasons, this optimism has since been muted, though skillfully employed computers can be helpful not only in verifying proofs but also in finding them. This area of applied logic is called *automated theorem proving* (ATP). Convincing examples include computer-supported proofs of the four-color theorem, the Robbins problem about a particular axiomatization of Boolean algebras, Bieberbach's conjecture in function theory, and the nonexistence of a projective plane of order 10. ATP is used today both in hardware and software verification, for instance in integrated circuit (chip) design and verification. A quick source of information about ATP is the Internet.

Despite these applications, even a developed artificial-intelligence system has presently no chance of simulating the heuristic approach in mathematics, where a precise proof from certain hypotheses is frequently only the culmination of a series of considerations flowing from the imagination. Creativity in mathematics of today is still a domain of human beings, not of automata. However, that is not to say that an automatic system may not be creative in a new way, for it is not necessarily the case that the human procedural method, influenced by all kinds of pictorial thoughts, is the sole means of gaining mathematical knowledge.

---

[6] The elementary absolute (plane) geometry $T$ has precisely two completions, Euclidean and non-Euclidean (or hyperbolic) geometry. Both are axiomatizable, hence decidable. Completeness follows in either case from the completeness of the elementary theory of real numbers, Theorem 5.5.5. Thus, absolute geometry is decidable as well. Further applications can be found in **5.2**.

**Exercises**

1. Let $T' = T + \alpha$ $(\alpha \in \mathcal{L}^0)$ be a finite extension of $T$. Show that if $T$ is decidable so too is $T'$ (cf. Lemma 6.5.3).

2. Assume that $T$ is consistent and has finitely many completions only. Prove that each completion of $T$ is a finite extension of $T$.

3. Show that an axiomatizable theory with finitely many completions is decidable (observe Exercise 2, Exercise 3 in **3.4**, and Theorem 5.2).

4. Using the Lindenbaum construction in **1.4**, show that a decidable countable theory $T$ has a decidable completion, [TMR, p. 15].

5. Show that a consistent theory $T$ that has finitely many completions has also only finitely many extensions. More precisely, if $T$ has $n$ completions then $T$ has $2^n - 1$ consistent extensions. Clearly, $n = 1$ if $T$ itself is complete.

# 3.6 Complete Hilbert Calculi

The sequent calculus of **3.1** models natural deduction sufficiently well. But it is nonetheless advantageous to use a Hilbert calculus for some purposes, for instance the arithmetization of formal proofs. Such calculi are based on logical axioms and rules of inference such as modus ponens MP: $\alpha, \alpha \to \beta / \beta$, also called *Hilbert-style rules*. These rules can be understood as sequent rules without premises. In a Hilbert calculus, deductions are drawn from a fixed set of formulas $X$, e.g., the axioms of a theory, with the inclusion of the logical axioms. The situation is basically the same as in **1.6**. In the case $X = \emptyset$ one deduces from the logical axioms alone, and only tautologies are derivable.

In the following we prove the completeness of a Hilbert calculus in the logical symbols $\neg, \wedge, \forall, =$. It will be denoted here by $\vdash$. MP is its only rule of inference. $\vdash$ refers to any first-order language $\mathcal{L}$ and is essentially an extension of the corresponding propositional Hilbert calculus treated in **1.6**. Once again, implication, defined by $\alpha \to \beta := \neg(\alpha \wedge \neg\beta)$, will play a useful part in presenting the calculus.

The *logical axiom system* $\Lambda$ of our calculus is taken to consist of all formulas $\forall x_1 \cdots \forall x_n \varphi$, where $\varphi$ is a formula of the form $\Lambda 1$–$\Lambda 10$ below, and $n \geqslant 0$. For example, due to $\Lambda 9$, $x = x$, $\forall x\, x = x$, $\forall y\, x = x$, $\forall x \forall y\, x = x$ are logical axioms, even though $\forall y$ is meaningless in the last two formulas. One may also say that $\Lambda$ is the set of all formulas that can be derived from $\Lambda 1$–$\Lambda 10$ by means of the rule MQ: $\alpha / \forall x \alpha$. **Attention:** MQ is not a rule of inference of the calculus $\vdash$, nor is it provable, although the set of tautologies is closed under MQ. We will later take a closer look at MQ.

$\Lambda 1$: $(\alpha \to \beta \to \gamma) \to (\alpha \to \beta) \to \alpha \to \gamma$, $\Lambda 2$: $\alpha \to \beta \to \alpha \wedge \beta$,

$\Lambda 3$: $\alpha \wedge \beta \to \alpha$, $\quad \alpha \wedge \beta \to \beta$, $\qquad \Lambda 4$: $(\alpha \to \neg \beta) \to \beta \to \neg \alpha$,

$\Lambda 5$: $\forall x \alpha \to \alpha \frac{t}{x}$ $(\alpha, \frac{t}{x}$ collision-free$)$, $\Lambda 6$: $\alpha \to \forall x \alpha$ $(x \notin \text{free } \alpha)$

$\Lambda 7$: $\forall x(\alpha \to \beta) \to \forall x \alpha \to \forall x \beta$, $\qquad \Lambda 8$: $\forall y \alpha \frac{y}{x} \to \forall x \alpha$ $(y \notin \text{var } \alpha)$,

$\Lambda 9$: $t = t$, $\qquad\qquad\qquad\qquad\qquad \Lambda 10$: $x = y \to \alpha \to \alpha \frac{y}{x}$ $(\alpha$ prime$)$.

It is easy to recognize $\Lambda 1$–$\Lambda 10$ as tautologies. For $\Lambda 1$–$\Lambda 4$ this is clear by **1.6**. For $\Lambda 5$–$\Lambda 8$ the reasoning proceeds straightforwardly by accounting for Corollary 2.3.6 on page 71 and the logical equivalences in **2.4**. For $\Lambda 9$ the claim is trivial, and $\Lambda 10$ is equivalent to $x = y, \alpha \vDash \frac{y}{x}$, and the latter is obviously the case.

Axiom $\Lambda 5$ corresponds to the rule $(\forall 1)$ of the calculus in **3.1**, while $\Lambda 6$ serves to deal with superfluous prefixes. The role of $\Lambda 7$ will become clear in the completeness proof for $\vdash$, and $\Lambda 8$ is part of bound renaming. $\Lambda 9$ and $\Lambda 10$ control the formal treatment of identity. If $\varphi$ is a tautology, then for any prefix block $\forall \vec{x}$, so too is $\forall \vec{x} \varphi$. Thus, $\Lambda$ consists solely of tautologies. The same holds for formulas derivable from $\Lambda$ using MP, simply because $\vDash \alpha, \alpha \to \beta$ implies $\vDash \beta$.

Let $X \vdash \alpha$ if there exists a *proof* $\Phi = (\varphi_0, \ldots, \varphi_n)$ of $\alpha$ *from* $X$, that is, $\alpha = \varphi_n$, and for all $k \leqslant n$ either $\varphi_k \in X \cup \Lambda$ or there exists some $\varphi$ such that $\varphi$ and $\varphi \to \varphi_k$ appear as members of $\Phi$ before $\varphi_k$. This definition and its consequences are the same as in **1.6**. As is the case there and proved in the same way, it holds that $X \vdash \alpha, \alpha \to \beta \Rightarrow X \vdash \beta$. Moreover, Theorem 1.6.1 also carries over unaltered, whose application will often be announced by the heading "proof by induction on $X \vdash \alpha$." For instance, the soundness of $\vdash$ is proved by induction on $X \vdash \alpha$, where soundness is as usual to mean $X \vdash \alpha \Rightarrow X \vDash \alpha$, for all $X$ and $\alpha$. In short, $\vdash\, \subseteq\, \vDash$. The proof runs exactly as on page 37.

The completeness of $\vDash$ can now be relatively easily be traced back to that of the rule calculus $\vdash$ of **3.1**. Indeed, much of the work was already undertaken in **1.6**, and we can immediately formulate the completeness of the calculus $\vDash$.

**Theorem 6.1 (Completeness theorem for $\vDash$).** $\vDash\ =\ \vDash$.

**Proof.** $\vDash\ \subseteq\ \vDash$ has already been verified. $\vDash\ \subseteq\ \vDash$ follows from the claim that $\vDash$ satisfies all nine basic rules of $\vdash$. This implies $\vdash\ \subseteq\ \vDash$, and since $\vdash\ =\ \vDash$ we have also $\vDash\ \subseteq\ \vDash$. For the rules $(\wedge 1)$ through $(\neg 2)$ the claim holds according to their proof for the Hilbert calculus in **1.6**. Lemmas 1.6.2 through 1.6.5 carry over word for word, because we have kept the four axioms on which the proofs are based and have taken no new rules into account. $(\forall 1)$ follows immediately from $\Lambda 5$ using MP, and (IR) is dealt with by $\Lambda 9$. Only $(\forall 2)$ and $(=)$ provide us with a little work, which, by the way, will clear up the role of axioms $\Lambda 6$, $\Lambda 7$, and $\Lambda 8$.

$(\forall 2)$: Suppose $x \notin \mathit{free}\, X$. We first prove $X \vDash \alpha \Rightarrow X \vDash \forall x \alpha$ by induction on $X \vDash \alpha$. *Initial step*: If $\alpha \in X$ then $x$ is not free in $\alpha$. Hence, $X \vDash \alpha \rightarrow \forall x \alpha$ using $\Lambda 6$, and MP yields $X \vDash \forall x \alpha$. If $\alpha \in \Lambda$ then also $\forall x \alpha \in \Lambda$, and likewise $X \vDash \forall x \alpha$. *Induction step*: Let $X \vDash \alpha, \alpha \rightarrow \beta$ and $X \vDash \forall x \alpha, \forall x (\alpha \rightarrow \beta)$ according to the induction hypothesis. This yields $X \vDash \forall x \alpha, \forall x \alpha \rightarrow \forall x \beta$ by Axiom $\Lambda 7$ and MP and, by another application of MP, the induction claim $X \vDash \forall x \beta$. Now, to verify $(\forall 2)$, let $X \vDash \alpha \frac{y}{x}$ and $y \notin \mathit{free}\, X \cup \mathit{var}\, \alpha$. By what we have just proved, we get $X \vDash \forall y \alpha \frac{y}{x}$. This, MP, and $X \vDash \forall y \alpha \frac{y}{x} \rightarrow \forall x \alpha$ (Axiom $\Lambda 8$) yield the conclusion of $(\forall 2)$, $X \vDash \forall x \alpha$. Thus, $\vDash$ indeed satisfies the rule $(\forall 2)$.

$(=)$: Let $\alpha$ be a prime formula and $X \vDash s = t, \alpha \frac{s}{x}$. Further, let $y$ be a variable $\neq x$ not in $s$ and $\alpha$. Then certainly $X \vDash \forall x \forall y (x = y \rightarrow \alpha \rightarrow \alpha \frac{y}{x})$, because the latter is a logical axiom in view of $\Lambda 10$. By the choice of $y$, rule $(\forall 1)$ shows that

$$X \vDash [\forall y (x = y \rightarrow \alpha \rightarrow \alpha \tfrac{y}{x})] \tfrac{s}{x}\ =\ \forall y (s = y \rightarrow \alpha \tfrac{s}{x} \rightarrow \alpha \tfrac{y}{x}).$$

Because of $y \notin \mathit{var}\, \alpha, s$ and $\alpha \frac{y}{x} \frac{t}{y} = \alpha \frac{t}{x}$, another application of $(\forall 1)$ yields

$$X \vDash [s = y \rightarrow \alpha \tfrac{s}{x} \rightarrow \alpha \tfrac{y}{x}] \tfrac{t}{y}\ =\ s = t \rightarrow \alpha \tfrac{s}{x} \rightarrow \alpha \tfrac{y}{x} \tfrac{t}{y}\ =\ s = t \rightarrow \alpha \tfrac{s}{x} \rightarrow \alpha \tfrac{t}{x}.$$

Since $X \vDash s = t, \alpha \frac{s}{x}$ by assumption, two applications of MP then leads to the desired conclusion $X \vDash \alpha \frac{t}{x}$. $\blacksquare$

A special case of the above completeness theorem is the following

**Corollary 6.2.** *For any $\alpha \in \mathcal{L}$, the following properties are equivalent:*

(i)   $\vdash \alpha$, *that is, $\alpha$ is derivable from $\Lambda$ by means of* MP *only,*

(ii)  *$\alpha$ is derivable from $\Lambda 1$–$\Lambda 10$ by means of* MP *and* MQ,

(iii) $\vDash \alpha$, *i.e., $\alpha$ is a tautology.*

The equivalence of (i) and (iii) renders especially intuitive the possibility to construct a "deduction machine" that effectively enumerates the set of all tautologies of $\mathcal{L}$. Here, we are dealing with just one rule of inference, modus ponens; hence we need the help of a machine to list the logical axioms, a "deducer" to check whether MP is applicable, and, if so, to apply it, and an output unit that emits the results and feeds them back into the deducer for further processing. However, similar to the case of a sequent calculus, such a procedure is not actually practicable; the distinction between significant and insignificant derivations is too involved to be taken into account. Who would be interested to find in the listing such a weird-looking tautology as for instance $\exists x(rx \rightarrow \forall y\, ry)$?

Next we want to show that the global consequence relation $\vDash^g$ defined in **2.5** can also be completely characterized by a Hilbert calculus. It is necessary only to adjoin the generalization rule MQ to the calculus $\vdash$. The resulting Hilbert calculus, denoted by $\vdash^g$, has two rules of inference, MP and MQ. Proofs in $\vdash^g$ have to be correspondingly redefined.

Like every Hilbert calculus, $\vdash^g$ is transitive: $X \vdash^g Y \ \& \ Y \vdash^g \alpha \Rightarrow X \vdash^g \alpha$. This was verified in **1.6** for propositional Hilbert calculi, but the same argument applies also to Hilbert calculi in first-order languages. With this remark, the completeness of $\vdash^g$ follows easily from that of $\vdash$:

**Theorem 6.3 (Completeness theorem for $\vdash^g$ ).** $\vdash^g \ = \ \vDash^g$.

**Proof.** Certainly $\vdash^g \subseteq \vDash^g$, since both MP and MQ are sound for $\vDash^g$. Now let $X \vDash^g \alpha$, so that $X^g \vDash \alpha$ by (1) of **2.5**. This yields $X^g \vdash \alpha$ by Theorem 6.1, and so $X^g \vdash^g \alpha$, since $\vdash \ \subseteq \ \vdash^g$ by definition of $\vdash^g$. Clearly, $X \vdash^g X^g$ in virtue of MQ; hence transitivity provides the desired $X \vdash^g \alpha$.   ☐

We now are going to discuss a notion of equal interest for both logic and computer science. $\alpha \in \mathcal{L}^0$ is called *generally valid in the finite* if $\mathcal{A} \vDash \alpha$ for all finite structures $\mathcal{A}$. Examples of such sentences $\alpha$ not being tautologies can be constructed in every first-order language $\mathcal{L}$ that contains at least a unary function or a binary relation symbol. For instance, consider

$\forall x \forall y (fx = fy \rightarrow x = y) \rightarrow \forall y \exists x\, y = fx$ from a language $\mathcal{L}$ containing the function symbol $f$. This sentence can be refuted only in an infinite $\mathcal{L}$-structure since an injection in a finite $\mathcal{L}$-structure is surjective. It holds in *all* finite $\mathcal{L}$-structures $\mathcal{A}$, but is not in $\mathsf{Taut}_{\mathcal{L}}$. Thus, $\mathsf{Taut}_{\mathcal{L}}$ is properly extended by the set of $\mathcal{L}$-sentences valid in the finite, $\mathsf{Tautfin}_{\mathcal{L}}$.

$\mathsf{Tautfin}$ ($= \mathsf{Tautfin}_{\mathcal{L}}$) is for *each* $\mathcal{L}$ a theory $T$ with the *finite model property*, i.e., every $\alpha \in \mathcal{L}^0$ compatible with $T$ has a finite $T$-model. More generally, the theory $T = Th\,\mathbf{K}$ has for any class $\mathbf{K}$ of finite $\mathcal{L}$-structures the finite model property. Indeed, if $T + \alpha$ is consistent, i.e., $\neg\alpha \notin T$, then $\mathcal{A} \nvDash \neg\alpha$ for some $\mathcal{A} \in \mathbf{K}$; hence $\mathcal{A} \vDash \alpha$. Examples are the theories $\mathsf{FSG}$ and $\mathsf{FG}$ of all finite semigroups and finite groups in $\mathcal{L}_\circ$, respectively. Both theories are undecidable. As regards $\mathsf{FSG}$, the proof is not particularly difficult; see **6.6**. Unlike $\mathsf{Taut}$, the set $\mathsf{Tautfin}$ is not axiomatizable for most languages $\mathcal{L}$. This is the claim of

**Theorem 6.4 (Trachtenbrot).** $\mathsf{Tautfin}_{\mathcal{L}}$ *is not (recursively) axiomatizable for any first-order language $\mathcal{L}$ containing at least one binary operation or a binary relation symbol.*

**Proof.** We restrict ourselves to the first case; for a binary relation symbol, the same follows easily by means of interpretation (Theorem 6.6.3). If $\mathsf{Tautfin}_{\mathcal{L}}$ were axiomatizable it would also be decidable because of the finite model property; Exercise 2. The same is true also for $\mathsf{Tautfin}_{\mathcal{L}_\circ}$, and by Exercise 1 in **3.5**, so too for $\mathsf{FSG}$, because $\mathsf{FSG}$ is the extension of $\mathsf{Tautfin}_{\mathcal{L}_\circ}$ by a single sentence, the law of associativity. But as already mentioned, $\mathsf{FSG}$ is undecidable. $\square$

The theorem is in fact a corollary of much stronger results that have been established in the meantime. For the newer literature on decision problems of this type consult [Id]. Unlike $\mathsf{FG}$, the theory of finite abelian groups, as well as of all abelian groups, is decidable, [Sz]. The former is a proper extension of the latter; for instance (as stated in Exercise 4),

$$\forall x \exists y\, y + y = x \rightarrow \forall x (x + x = 0 \rightarrow x = 0)$$

does not hold in all abelian groups, though it does in all finite ones.

As early as 1922 Behmann discovered by quantifier elimination that $\mathsf{Taut}$ possesses the finite model property provided the signature contains only unary predicate symbols; one can also prove this without difficulty by the Ehrenfeucht game of **5.3**. In this case, then, *Tautfin* = *Taut*, because $\alpha \notin$ *Taut* implies $\neg\alpha$ is satisfiable and therefore has a finite model. Thus, $\alpha \notin$ *Tautfin*. This proves *Tautfin* $\subseteq$ *Taut* and hence *Tautfin* = *Taut*. With the Ehrenfeucht game also a quite natural axiomatization of the theory FO of all finite ordered sets is obtained. See Exercise 3 in **5.3**.

### Exercises

1. Show that MQ is unprovable in $\vdash\!\!\!\sim$ (that is, $X \vdash\!\!\!\sim \alpha \Rightarrow X \vdash\!\!\!\sim \forall x\alpha$ does not hold, in general).

2. Suppose (i) a theory $T$ has the finite model property, (ii) the finite $T$-models are effectively enumerable (more precisely, a system of representatives thereof up to isomorphism). Show that (a) the sentences $\alpha$ refutable in $T$ are effectively enumerable, (b) if $T$ is axiomatizable then it is also decidable.

3. Let $T$ be a *finitely* axiomatizable theory with the finite model property. Show by working back to Exercise 2 that $T$ is decidable.

4. Show that $\forall x \exists y\, y + y = x \rightarrow \forall x(x + x = 0 \rightarrow x = 0)$ holds in all finite abelian groups. Moreover, provide an example of an infinite abelian group for which the above proposition fails.

## 3.7   First-Order Fragments

Subsequent to Gödel's completeness theorem it makes sense to investigate some fragments of first-order languages aiming at a formal characterization of deduction *inside* the fragment. In this section we present some results in this regard; in the next section we shall do the same for some extensions. First-order fragments are formalisms that come along without the full means of expression in a first-order language, for instance by the omission of some or all logical connectives, or restricted quantification. These formalisms are interesting for various reasons, partly because of the growing interest in automatic information processing with its more or less restricted user interface. The poorer a linguistic fragment, the more modest the possibilities for the formulation of sound rules. Therefore, the completeness problem for fragments is in general nontrivial.

A useful example dealt with more closely is the *language of equations*, whose only formulas are equations of a fixed algebraic signature. We think of the variables in the equations as being tacitly generalized and call these generalizations *identities*, though we often speak somewhat sloppily of equations. Theories with axiom systems of identities are called *equational theories* and their model classes *equational-defined classes* or *varieties*.

Let $\Gamma$ denote a set of identities defining an equational theory, $\gamma$ a single equation, and assume $\Gamma^g \vDash \gamma$. By Theorem 2.7 there is a formal proof for $\gamma$ from $\Gamma$. But because of the special form of the equations, it can be expected that one does not need the whole formalism to verify $\Gamma^g \vdash \gamma$. Indeed, Theorem 7.2 states that the *Birkhoff rules* (B0)–(B4) below, taken from [Bi], suffice. This result is so pleasing because when operating with (B0)–(B4), we remain completely inside the language of equations. The rules define a Hilbert-style calculus denoted by $\overset{B}{\vdash}$ and look as follows:

(B0) $/t=t$,  (B1) $s=t/t=s$,  (B2) $t=s, s=t'/t=t'$,

(B3) $t_1=t'_1,\ldots,t_n=t'_n/ft_1\cdots t_n=ft'_1\cdots t'_n$, (B4) $s=t/s^\sigma=t^\sigma$.

Here $\sigma$ is a global substitution, though as explained in **2.2** it would suffice to consider just simple $\sigma$. (B0) has no premise, which means that $t=t$ is derivable from any set of identities (or $t=t$ is added as an axiom to $\Gamma$). These rules are formally stated with respect to unquantified equations. However, we think of all variables as being generalized in a formal derivation sequence. We are forced to do this by the soundness requirement of (B4), because in general only $(s=t)^g \vDash s^\sigma = t^\sigma$. To verify $\Gamma \overset{B}{\vdash} \gamma \Rightarrow \Gamma^g \vDash \gamma$, we need only to show that the property $\Gamma^g \vDash \gamma$ is closed under (B0)–(B4), i.e., $\mathcal{A} \vDash t=t$ (which is trivial), $\mathcal{A} \vDash s=t \Rightarrow \mathcal{A} \vDash t=s$, etc. We have already come across the rules of $\overset{B}{\vdash}$ in **3.1**, stated there as Gentzen-style rules; they ensure that by $s \approx t :\Leftrightarrow \Gamma \overset{B}{\vdash} s=t$, a congruence in the term algebra $\mathcal{T}$ is defined as in Lemma 2.5. (B4) states the *substitution invariance* of $\approx$, which is to mean $s \approx t \Rightarrow s^\sigma \approx t^\sigma$.

Let $\mathcal{F}$ denote the factor structure of $\mathcal{T}$ with respect to $\approx$ (no distinction is made between the algebra $\mathcal{T}$ and its domain), and let $\bar{t}$ denote the congruence class modulo $\approx$ to which the term $t$ belongs, so that

(1)  $\overline{t_1} = \overline{t_2} \Leftrightarrow \Gamma \overset{B}{\vdash} t_1=t_2$.

Further, let $w: \mathrm{Var} \to \mathcal{F}$, say $x^w = \overline{t_x}$, with arbitrary $t_x \in x^w$. Any such choice determines a global substitution $\sigma_w : x \mapsto t_x$. Induction on $t$ yields

(2)  $t^{\mathcal{F},w} = \overline{t^\sigma}$  $(\sigma := \sigma_w)$.

**Lemma 7.1.** $\Gamma \overset{B}{\vdash} t_1 = t_2 \Leftrightarrow \mathcal{F} \vDash t_1 = t_2$.

**Proof.** Let $\Gamma \overset{B}{\vdash} t_1 = t_2$, $w \colon \text{Var} \to \mathcal{F}$, and $\sigma = \sigma_w$. By (B4) then also $\Gamma \overset{B}{\vdash} t_1^\sigma = t_2^\sigma$, so that $\overline{t_1^\sigma} = \overline{t_2^\sigma}$ by (1). Thus, $t_1^{\mathcal{F},w} = t_2^{\mathcal{F},w}$ using (2). Since $w$ was arbitrary, $\mathcal{F} \vDash t_1 = t_2$. Now suppose the latter and let $\varkappa$ be the so-called *canonical valuation* $x \mapsto \overline{x}$. Here we choose $\sigma_\varkappa = \iota$ (the identical substitution), hence $t_i^{\mathcal{F},\varkappa} = \overline{t_i}$ by (2). $\mathcal{F} \vDash t_1 = t_2$ implies $t_1^{\mathcal{F},\varkappa} = t_2^{\mathcal{F},\varkappa}$, and in view of $t_i^{\mathcal{F},\varkappa} = \overline{t_i}$, we get $\overline{t_1} = \overline{t_2}$ and so $\Gamma \overset{B}{\vdash} t_1 = t_2$ by (1). $\quad\square$

**Theorem 7.2 (Birkhoff's completeness theorem).** *Let $\Gamma$ be a set of identities and $t_1 = t_2$ an equation. Then $\Gamma \overset{B}{\vdash} t_1 = t_2 \Leftrightarrow \Gamma^g \vDash t_1 = t_2$.*

**Proof.** The direction $\Rightarrow$ is the soundness of $\overset{B}{\vdash}$. Now let $\Gamma^g \vDash t_1 = t_2$. Then certainly $\mathcal{F} \vDash \Gamma$ according to Lemma 7.1, or equivalently $\mathcal{F} \vDash \Gamma^g$. Thus, $\mathcal{F} \vDash t_1 = t_2$. Using Lemma 7.1 once again yields $\Gamma \overset{B}{\vdash} t_1 = t_2$. $\quad\square$

This proof is distinguished on the one hand by its simplicity and on the other by its highly abstract character. It has manifold variations and is valid in a corresponding sense, for example, for sentences of the form $\forall \vec{x}\pi$ with *arbitrary* prime formulas $\pi$ of any given first-order language. It is rather obvious how to strengthen the Birkhoff rules to cover this more general case: Keep (B0), (B1), and (B3) and replace the conclusions of (B3) and (B4) by arbitrary prime formulas of the language.

There is also a special calculus for sentences of the form

(3)  $\forall \vec{x}\,(\gamma_1 \wedge \cdots \wedge \gamma_n \to \gamma_0)$      $(n \geqslant 0,\ \text{all } \gamma_i \text{ equations}),$

called *quasi-identities*. Theories whose axioms are of the form (3) are called *quasi-equational theories* and their model classes *quasi-varieties*. The latter are important both for algebra and logic. (B0) is retained and (B1)–(B3) are replaced by the rules without premises (axioms)

$$/x = y \to y = x, \quad /x = y \wedge y = z \to x = z, \quad /\bigwedge_{i=1}^{n} x_i = y_i \to f\vec{x} = f\vec{y}.$$

Besides an adaptation of (B4), some rules are required for the formal handling of the premises $\gamma_1, \ldots, \gamma_n$ in (3), for instance their permutability (for details see e.g. [Se]). A highly important additional rule is here a variant of the cut rule, namely the binary Hilbert-style rule

$$\alpha \wedge \delta \to \gamma, \alpha \to \delta / \alpha \to \gamma \qquad (\alpha \text{ a conjunction of equations}).$$

The most interesting case for automated information processing, where Hilbert rules remaining inside the fragment still provide completeness, is that of universal Horn theories. Here, roughly speaking, the equations $\gamma_i$

in (3) may be *any* prime formulas. Horn theories are treated in Chapter **4**. But for enabling a real machine implementation, the calculus considered there, the resolution calculus, is different from a Hilbert- or a Gentzen-style calculus.

### Exercises

1. Show that a variety $K$ is closed with respect to homomorphic images, taking subalgebras, and forming arbitrary direct products of members of $K$; in short, $K$ has the properties **H**, **S**, and **P**.[7]

2. Develop a calculus for quasi-varieties as indicated in the text and prove its completeness. This exercise is a comprehensive task; we recommend to start with a study of [Se].

## 3.8 Extensions of First-Order Languages

Now we consider a few of the numerous possibilities for extending first-order languages to increase the power of expression: We say that a language $\mathcal{L}' \supseteq \mathcal{L}$ *of the same* signature as $\mathcal{L}$ is *more expressive* than $\mathcal{L}$ if for at least one sentence $\alpha \in \mathcal{L}'$, Md $\alpha$ is distinct from all Md $\beta$ for $\beta \in \mathcal{L}$. In $\mathcal{L}'$, some of the properties of first-order languages are lost. Indeed, the claim of the next theorem is that first-order languages are optimal in regard to the richness of their applications.

**Lindström's Theorem** (see [EFT] or [CK]). *There is no language of a given signature that is more expressive than the first-order language and for which both the compactness theorem and the Löwenheim–Skolem theorem hold.*

**Many-sorted languages.** In describing geometric facts it is convenient to use several variables, for points, lines, and, depending on dimension, also for geometrical objects of higher dimension. For every argument of a predicate or operation symbol of such a language, it is useful to fix its *sort*. For instance, the incidence relation of plane geometry has arguments for points and lines. For function symbols, the sort of their

---

[7] Conversely, if a class $K$ has these three properties then $K$ is a variety. This is Birkhoff's **HSP** theorem, a basic theorem of universal algebra; see e.g. [Mo].

values must additionally be given. If $\mathcal{L}$ is of sort $k$ and $v_0^s, v_1^s, \ldots$ are variables of sort $s$ $(1 \leqslant s \leqslant k)$ then every relation symbol $r$ is assigned a sequence $(s_1, \ldots, s_n)$; in a language without function symbols, prime formulas beginning with $r$ have the form $r x_1^{s_1} \cdots x_n^{s_n}$, where $x_i^{s_i}$ denotes a variable of the sort $s_i$.

Many-sorted languages represent only an inessential extension of the concept hitherto expounded, provided the sorts are given equal rights. Instead of a language $\mathcal{L}$ with $k$ sorts of variables, we can consider a one-sorted language $\mathcal{L}'$ with additional unary predicate symbols $P_1, \ldots, P_k$ and the adoption of certain new axioms: $\exists x P_i x$ for $i = 1, \ldots, k$ (no sort is empty, for otherwise it is dispensable) and $\neg \exists x (P_i x \wedge P_j x)$ for $i \neq j$ (sort disjunction). For example, plane geometry could also be described in a one-sorted language with the additional predicates $pt$ (to be a point) and $li$ (to be a line). Apart from a few differences in dealing with term insertion, many-sorted first-order languages behave almost exactly like one-sorted languages.

**Second-order languages.** Some frequently quoted axioms, e.g., the induction axiom IA, may be looked upon as second-order sentences. The simplest extension of a first-order language to one of higher order is the *monadic second-order language*, a two-sorted language. Let us consider such a language $\mathcal{L}$ with variables $x, y, z, \ldots$ for individuals and variables $X, Y, Z, \ldots$ for sets of these individuals, along with at least one binary relation symbol $\in$ but without function symbols. Prime formulas are $x = y$, $X = Y$, and $x \in X$. An $\mathcal{L}$-structure is generally of the form $(A, B, \in)$, where $\in \subseteq A \times B$. The goal is that by formulating additional axioms such as $\forall X Y [\forall x (x \in X \leftrightarrow x \in Y) \to X = Y]$ (which corresponds to the axiom of extensionality AE in **3.4**), the relation symbol $\in$ should be interpretable as the membership relation $\in$; hence $B$ should consist of the subsets of $A$. This goal is not fully attainable, but nearly so. Axioms on $A, B$ can be found such that $B$ is interpretable as a subset of $\mathfrak{P}A$, with $\in$ interpreted as $\in$. The same works by adding sort variables for members of $\mathfrak{P}\mathfrak{P}A$, $\mathfrak{P}\mathfrak{P}\mathfrak{P}A$, etc. This "completeness of the theory of types" plays a basic role, for instance, in the higher nonstandard analysis.

A more enveloping second-order language, $\mathcal{L}_{II}$, is won by adopting quantifiable variables for relations of each finite arity on the domains of individuals. But $\mathcal{L}_{II}$ then fails to satisfy both the finiteness and the

Löwenheim–Skolem theorem (Theorem 4.1), even for $\mathcal{L} = \mathcal{L}_{=}$. The former fails because a sentence $\alpha_{\text{fin}}$ can be given in $\mathcal{L}_{II}$ such that $\mathcal{A} \vDash \alpha_{\text{fin}}$ iff $A$ is finite. For note that $A$ is finite iff every injective $f \colon A \to A$ is bijective. This can effortlessly be formalized using a single universally quantified binary predicate variable characterizing the graph of $f$.

The Löwenheim–Skolem theorem is also easily refutable for $\mathcal{L}_{II}$; one need only write down in $\mathcal{L}_{II}$ the sentence 'there exists a continuous order on $A$ without smallest or largest element'. This sentence has no countable model. For if there were such a model, it would be isomorphic to the ordered set of rationals according to a theorem of Cantor (Example 2 in **5.2**) and therefore has gaps, contradicting our assumption.

There is still a more serious problem as regards $\mathcal{L}_{II}$: The ZFC-axioms, seen as axioms of the underlying set theory, do not suffice to establish what a tautology in $\mathcal{L}_{II}$ should actually be. For instance, the continuum hypothesis CH (see page 174) can be easily formulated as an $\mathcal{L}_{II}$-sentence, $\alpha_{\text{CH}}$. But CH is independent of ZFC. Thus, if CH is true, $\alpha_{\text{CH}}$ is an $\mathcal{L}_{II}$-tautology, otherwise not. It does not look as though mathematical intuition suffices to decide this question unambiguously.

**New quantifiers.** A simple syntactic extension $\mathcal{L}_{\eth}$ of a first-order language $\mathcal{L}$ is obtained by taking on a new quantifier denoted by $\eth$, which formally is to be handled as the $\forall$-quantifier. However, in a model $\mathcal{M} = (\mathcal{A}, w)$, a new interpretation of $\eth$ is provided by means of the satisfaction clause

(0)    $\mathcal{M} \vDash \eth x\alpha \Leftrightarrow$ there are infinitely many $a \in A$ with $\mathcal{M}_x^a \vDash \alpha$.

With this interpretation, we write $\mathcal{L}_{\eth}^0$ instead of $\mathcal{L}_{\eth}$, since yet another interpretation of $\eth$ will be discussed. $\mathcal{L}_{\eth}^0$ is more expressive than $\mathcal{L}$, as seen by the fact, for example, that the finiteness theorem for $\mathcal{L}_{\eth}^0$ no longer holds: Let $X$ be the collection of all sentences $\exists_n$ (there exist at least $n$ elements) plus $\alpha_{\text{fin}} := \neg \eth x\, x = x$ (there exist only finitely many elements). Every finite subset of $X$ has a model, but $X$ itself does not. All the same, $\mathcal{L}_{\eth}^0$ still satisfies the Löwenheim–Skolem theorem. This can be proved straightforwardly with the methods of **5.1**. Once again, because of the missing finiteness theorem there cannot be a complete rule calculus for $\mathcal{L}_{\eth}^0$. Otherwise, just as in **3.1**, one could prove the finiteness theorem after all. However, there are several nontrivial, correct Hilbert-style rules for $\mathcal{L}_{\eth}^0$, for instance the four rules

(Q1) $/\neg\mho x(x = y \lor x = z)$,     (Q2) $\mho x\alpha/\mho y\alpha \tfrac{y}{x}$ $(y \notin \text{free}\,\alpha)$,

(Q3) $\forall x(\alpha \to \beta)/\mho x\alpha \to \mho x\beta$,   (Q4) $\mho x\exists y\alpha, \neg\mho y\exists x\alpha/\exists y\mho x\,\alpha$.

In rule (Q1), which has no premises, clearly $x \neq y, z$. Intuitively, this rule tells us that the pair set $\{y, z\}$ is finite. (Q2) is bound renaming. (Q3) says that a set containing an infinite subset is itself infinite. (Q4) is rendered intuitive with $\alpha = \alpha(x, y)$ as follows:

Suppose that $\mathcal{A} \vDash \mho x\exists y\alpha, \neg\mho y\exists x\alpha$ and let $A_b = \{a \in A \mid \mathcal{A} \vDash \alpha(a, b)\}$. Then $\mathcal{A} \vDash \mho x\exists y\,\alpha$ states '$\bigcup_{b \in A} A_b$ is infinite', and $\mathcal{A} \vDash \neg\mho y\exists x\,\alpha$ says 'there exist only finitely many $b$ such that $A_b \neq \emptyset$'. The conclusion $\exists y\mho x\alpha$ tells us therefore '$A_b$ is infinite for at least one index $b$'. Hence, (Q4) expresses altogether that the union of a finite system of finite sets is finite.

Now let us replace the satisfaction clause (0) by

(1) $\mathcal{M} \vDash \mho x\alpha \Leftrightarrow$ there are uncountably many $a \in A$ with $\mathcal{M}_x^a \vDash \alpha$.

Also with this interpretation, (Q1)–(Q4) are sound for $\mathcal{L}_{\mho}^1$ $(= \mathcal{L}_{\mho}$ with the interpretation (1)). Rule (Q4) now evidently expresses that a countable union of countable sets is again countable. Moreover, the logical calculus $\overset{1}{\vdash}$ resulting from the basic rules of **3.1** by adjoining (Q1)–(Q4) is, surprisingly, complete for these semantics when restricted to countable sets $X$. Thus, $X \overset{1}{\vdash} \alpha \Leftrightarrow X \vDash \alpha$, for any countable $X \subseteq \mathcal{L}_{\mho}^1$, [CK]. This implies the following compactness theorem for $\mathcal{L}_{\mho}^1$: *If every finite subset of a countable set $X \subseteq \mathcal{L}_{\mho}^1$ has a model then so too does $X$.* For uncountable sets of formulas this is false in general; Exercise 1.

The above is a fairly incomplete listing of languages with modified quantifiers. There are also several other extensions of first-order languages, for instance languages with infinite formulas (containing infinitely long conjunctions and disjunctions). In this respect we refer to the literature on model theory.

**Programming languages.**   All languages hitherto discussed are of static character inasmuch as there are spatially and temporally independent truth values for given valuations $w$ in a structure $\mathcal{A}$. But one can also connect a first-order language $\mathcal{L}$ in various ways with a programming language having *dynamic* character and aiming at the description of certain types of information processing. The choice of $\mathcal{L}$ depends on what the programming language is aiming at. $\mathcal{L}$ can as a rule be reconstructed from the description of the programming language's syntax.

The theory of programming languages, both syntax and semantics, has its roots in mathematical logic. Nonetheless, it has assumed an independent status and belongs rather to computer science than to logic. Hence our considerations will be rather brief.

We describe here a simple example of such a language, $\mathcal{PL}$, where $\mathcal{L}$ is a fixed first-order language. Only the open formulas of $\mathcal{L}$ will be used in $\mathcal{PL}$, not the quantifiers. The elements of $\mathcal{PL}$ are certain strings, called *programs*, denoted by $\mathcal{P}, \mathcal{Q}, \ldots$, and defined below.

The dynamic character of $\mathcal{PL}$ arises by modifying traditional semantics as follows: A program $\mathcal{P}$ starts with a valuation $w \colon Var \to A$ where $A$ is the domain of a given $\mathcal{L}$-structure $\mathcal{A}$. The program $\mathcal{P}$ alters stepwise the values of the variables as a run of the program proceeds in time. If $\mathcal{P}$ terminates upon feeding in $w$, i.e., the calculation ends after finitely many steps of calculation, then the result is a new valuation $w^{\mathcal{P}}$. Otherwise we take $w^{\mathcal{P}}$ to be undefined. The precise description of this in general only partially defined operation $w \mapsto w^{\mathcal{P}}$ is called the *procedural semantics* of $\mathcal{PL}$. A closer description will be given below.

It is possible to meaningfully consider issues of completeness, say, for procedural semantics as well. For instance, if $\mathcal{L}$ speaks about natural numbers one may ask what conditions have to be posed on $\mathcal{L}$ such that each computable function is programmable in $\mathcal{PL}$.

The syntax of $\mathcal{PL}$ is specified as follows: The logical signature of $\mathcal{L}$ is extended by the symbols WHILE, DO, END, :=, and ; (the semicolon serves only as a separator for concatenated programs and could be omitted if programs are arranged 2-dimensionally). *Programs* on $\mathcal{L}$ are defined inductively as strings of symbols in the following manner:

- For any $x \in Var$ and term $t \in \mathcal{T}_{\mathcal{L}}$, the string $x := t$ is a program.

- If $\alpha$ is an open formula in $\mathcal{L}$ and $\mathcal{P}, \mathcal{Q}$ are programs, so too are the strings $\mathcal{P}; \mathcal{Q}$ and WHILE $\alpha$ DO $\mathcal{P}$ END.

No other strings are programs in this context. $\mathcal{P}; \mathcal{Q}$ is to mean that first $\mathcal{P}$ and then $\mathcal{Q}$ are executed. Let $\mathcal{P}^n$ denote the $n$-times repeated execution of $\mathcal{P}$, more precisely, $\mathcal{P}^0$ is the *empty* program $(w^{\mathcal{P}^0} = w)$ and $\mathcal{P}^{n+1} = \mathcal{P}^n; \mathcal{P}$.

The procedural semantics for the programming language $\mathcal{PL}$ is made precise by the following stipulations:

(a) $w^{x := t} = w\frac{t^w}{x}$ (i.e., $w$ alters at most the value of the variable $x$).

(b) If $w^{\mathcal{P}}$ and $(w^{\mathcal{P}})^{\mathcal{Q}}$ are defined, so too is $w^{\mathcal{P};\mathcal{Q}}$, and $w^{\mathcal{P};\mathcal{Q}} = (w^{\mathcal{P}})^{\mathcal{Q}}$.

(c) For $\mathcal{Q} :=$ WHILE $\alpha$ DO $\mathcal{P}$ END let $w^{\mathcal{Q}} = w^{\mathcal{P}^k}$ with $k$ specified below.

According to our intuition regarding the "WHILE loop," $k$ is the smallest number such that $\mathcal{A} \vDash \alpha\,[w^{\mathcal{P}^i}]$ for all $i < k$ and $\mathcal{A} \nvDash \alpha\,[w^{\mathcal{P}^k}]$, provided such a $k$ exists and all $w^{\mathcal{P}^i}$ for $i \leqslant k$ are well defined. Otherwise $w^{\mathcal{Q}}$ is considered to be undefined. If $k = 0$, that is, $\mathcal{A} \nvDash \alpha\,[w]$, then $w^{\mathcal{Q}} = w$, which amounts to saying that $\mathcal{P}$ is not executed at all, in accordance with the meaning of WHILE in standard programming languages.

**Example.** Let $\mathcal{L} = \mathcal{L}\{0, \mathsf{S}, \mathsf{Pd}\}$ and consider $\mathcal{A} = (\mathbb{N}, 0, \mathsf{S}, \mathsf{Pd})$, where $\mathsf{S}$ denotes the successor function and $\mathsf{Pd}$ the *predecessor function*, defined by $y = \mathsf{Pd}\,x \leftrightarrow y = 0 \vee x = \mathsf{S}y$ (so that $\mathsf{Pd}\,0 = 0$). Let $\mathcal{P}$ be the program

$$z := x\,;v := y\,;\text{WHILE}\,v \neq 0\,\text{DO}\,z := \mathsf{S}z\;;v := \mathsf{Pd}\,v\,\text{END}.$$

If $x$ and $y$ initially have the values $x^w = m$ and $y^w = n$, the program ends with $z^{w^{\mathcal{P}}} = m + n$. In other words, $\mathcal{P}$ terminates for every input $m, n$ for $x, y$ and computes the output $m + n$ in the variable $z$, while $x, y$ keep their initial values.

In $\mathcal{PL}$, the self-explanatory program schema IF $\alpha$ THEN $\mathcal{P}$ ELSE $\mathcal{Q}$ END is definable by the following composed program:

$$x := 0\,;\text{WHILE}\,\alpha \wedge x = 0\,\text{DO}\,\mathcal{P}\,;x := \mathsf{S}0\,\text{END}\,;\text{WHILE}\,x = 0\,\text{DO}\,\mathcal{Q}\,;x := \mathsf{S}0\,\text{END},$$

where $x$ is a variable not appearing in $\mathcal{P}$, $\mathcal{Q}$, and $\alpha$.

### Exercises

1. Show (a) both $\mathcal{L}_{II}$ and $\mathcal{L}^1_{\omega}$ violate the Löwenheim–Skolem theorem, (b) the finiteness theorem is false for uncountable sets of formulas in $\mathcal{L}^1_{\omega}$, in general (although it holds for countable sets of formulas).

2. Express the continuum hypothesis as a (possibly false) theorem of $\mathcal{L}_{II}$.

3. Verify the correctness of the definition of IF $\alpha$ THEN $\mathcal{P}$ ELSE $\mathcal{Q}$ END given at the end of the above text.

4. Define the DO $\mathcal{P}$ UNTIL $\alpha$ END loop in the programming language $\mathcal{PL}$. In this loop, $\mathcal{P}$ is executed *before* the test $\alpha$ is started. That is, $\mathcal{P}$ is executed at least once.

# Chapter 4

# Foundations of Logic Programming

Logic programming aims not so much at solving numerical problems in science and technology, as at treating information processing in general, in particular at the creation of expert systems of artificial intelligence. A distinction has to be made between logic programming as theoretical subject matter and the widely used programming language for practical tasks of this kind, PROLOG. In regard to the latter, we confine ourselves to a presentation of a somewhat simplified version, which nonetheless preserves the typical features.

The notions dealt with in **4.1** are of fairly general nature. Their origin lies in certain theoretical questions posed by mathematical logic, and they took shape before the invention of the computer. For certain sets of formulas, in particular for sets of universal Horn formulas, which are very important for logic programming, term models are obtained canonically. The newcomer need not understand all details of **4.1** at once, but should learn about Horn formulas in **4.2** and after a glance at the theorems may then continue with **4.3**.

The resolution method and its combination with unification applied in PROLOG were directly inspired by mechanical information processing. This method is also of significance for tasks of automated theorem proving, which extends beyond logic programming. We treat resolution first in the framework of propositional logic in **4.3**. Its highlight, the resolution theorem, is proved constructively, without recourse to the propositional compactness theorem. In **4.5**, unification is dealt with, and **4.6** presents the combination of resolution with unification and its application to logic programming. An introduction to this area is also offered by [Ll].

W. Rautenberg, *A Concise Introduction to Mathematical Logic*, Universitext, DOI 10.1007/978-1-4419-1221-3_4, © Springer Science+Business Media, LLC 2010

## 4.1   Term Models and Herbrand's Theorem

In the proof of Lemma 3.2.5 as well as in Lemma 3.7.1 we have come across models whose domains are equivalence classes of terms of a first-order language $\mathcal{L}$. In general, a *term model* is to mean an $\mathcal{L}$-model $\mathcal{F}$ whose domain $F$ is the set of congruence classes $\bar{t} = t/\approx$ of a congruence $\approx (= \approx_{\mathcal{F}})$ on the algebra $\mathcal{T} = \mathcal{T}_{\mathcal{L}}$. If $\approx$ is the identity in $\mathcal{T}$, one identifies $F$ with $\mathcal{T}$ so that then $\bar{t} = t$. Function symbols and constants are always interpreted canonically: $f^{\mathcal{F}}(\bar{t}_1, \ldots, \bar{t}_n) := \overline{ft_1 \cdots t_n}$ and $c^{\mathcal{F}} := \bar{c}$, while no particular condition is imposed on realizing the relation symbols in $\mathcal{F}$.

Further, let $\kappa : x \mapsto \bar{x}$ ($x \in \mathrm{Var}$). This is called the *canonical* valuation. In the terminology of **2.3**, $\mathcal{F} = (\mathfrak{F}, \kappa)$, where $\mathfrak{F} = \mathcal{T}/\approx_{\mathcal{T}}$ denotes the $\mathcal{L}$-structure belonging to the model $\mathcal{F}$ with the domain $F = \{\bar{t} \mid t \in \mathcal{T}\}$. We claim that independent of a specification of $\approx_{\mathcal{F}}$ and of the $r^{\mathcal{F}}$,

(1)   $t^{\mathcal{F}} = \bar{t}$ for all $t \in \mathcal{T}$,

(2)   $\mathcal{F} \models \forall \vec{x}\alpha \;\Leftrightarrow\; \mathcal{F} \models \alpha \frac{\vec{t}}{\vec{x}}$ for all $\vec{t} \in \mathcal{T}^n$ ($\alpha$ open).

(1) is verified by an easy term induction (cf. the proof of (d) page 101). (2) follows from left to right by Corollary 2.3.6. The converse runs as follows: $\mathcal{F} \models \alpha \frac{\vec{t}}{\vec{x}}$ for all $\vec{t} \in \mathcal{T}^n$ implies $\mathcal{F}^{\bar{t}_1 \cdots \bar{t}_n}_{x_1 \cdots x_n} = \mathcal{F}^{\vec{t}^{\mathcal{F}}}_{\vec{x}} \models \alpha$ for all $t_1, \ldots, t_n \in \mathcal{T}$ in view of (1) and of Theorem 2.3.5. But this means that $\mathcal{F} \models \forall \vec{x}\alpha$, because the $\bar{t}$ for $t \in \mathcal{T}$ exhaust the domain of $\mathcal{F}$.

Interesting for both theoretical logic and automated theorem proving including logic programming, is the question, for which consistent $X \subseteq \mathcal{L}$ can a term model be constructed in $\mathcal{L}$. An answer to this question is given by Theorems 1.1 and 2.1 below. First, we associate with each given $X \subseteq \mathcal{L}$ a special term model as follows.

**Definition.** The *term model* $\mathcal{F} = \mathcal{F}X$ (associated with a set $X \subseteq \mathcal{L}$) is the term model for which $\approx_{\mathcal{F}X}$ and $r^{\mathcal{F}X}$ are defined by

$$s \approx_{\mathcal{F}X} t :\Leftrightarrow X \vdash s = t; \quad r^{\mathcal{F}X} \bar{t}_1 \cdots \bar{t}_n :\Leftrightarrow X \vdash rt_1 \cdots t_n .$$

It is easily verified that $\approx_{\mathcal{F}X}$ is a congruence in $\mathcal{T}$ and that the definition of $r^{\mathcal{F}X}$ does not depend on the representatives. If $X$ is the axiom system of some theory $T$, then we write also $\mathcal{F}T$ for $\mathcal{F}X$, and $s \approx_T t$ for $s \approx_{\mathcal{F}X} t$. By (1), $\mathcal{F}X \models s = t \Leftrightarrow \bar{s} = \bar{t} \Leftrightarrow X \vdash s = t$. Similarly $\mathcal{F}X \models r\vec{t} \Leftrightarrow X \vdash r\vec{t}$. In general, $\mathcal{F}X$ is not a model for $X$. Our definition merely implies

(3)   $\mathcal{F}X \models \pi \;\Leftrightarrow\; X \vdash \pi$    ($\pi$ prime).

Here is a simple example in which the associated term model $\mathcal{F}T$ *is* a model for $T$. It will turn out to be a special case of Theorem 2.1.

**Example 1.** Let $T$ be the theory of semigroups and $\mathfrak{F}$ the algebra belonging to the term model $\mathcal{F}T$. Every term $t$ is equivalent in $T$ to a term in left association, denoted by $x_1 \cdots x_n$ (the operation symbol is not written and $x_1, \ldots, x_n$ is an enumeration of the variables of $t$ in order of appearance in $t$ from left to right, possibly with repetitions). Thus, $t \approx_T x_1 \cdots x_n$. For instance, $\boldsymbol{v}_0((\boldsymbol{v}_1\boldsymbol{v}_0)\boldsymbol{v}_1) \approx_T \boldsymbol{v}_0\boldsymbol{v}_1\boldsymbol{v}_0\boldsymbol{v}_1$. It is easy to see that $\imath\colon S \to \mathfrak{F}$ with $\imath(x_1, \ldots, x_n) = x_1 \ldots x_n$ is an isomorphism, where $S$ is the semigroup of strings over the alphabet $Var$, and $(x_1, \ldots, x_n)$ denotes here the string with the letters $x_1, \ldots, x_n$. Thus, with $S$ also $\mathfrak{F}$ is a semigroup, hence $\mathfrak{F} \vDash T$. This amounts to the same as saying $\mathcal{F} \vDash T$.

As already announced earlier, we slightly extend the concept of a model. Let $\mathcal{L}^k$ and $Var_k$ be defined as in **2.2**. Pairs $(\mathcal{A}, w)$ with $dom\, w \supseteq Var_k$ are called $\mathcal{L}^k$-*models*. Here $w$ need not be defined for $\boldsymbol{v}_k, \boldsymbol{v}_{k+1}, \ldots$ One may also say that an allocation to these variables has deliberately been "forgotten." In the case $k = 0$ Choose is $w = \emptyset$, so that an $\mathcal{L}^0$-model coincides with an $\mathcal{L}$-structure. Put $\mathcal{T}_k := \{t \in \mathcal{T} \mid var\, t \subseteq Var_k\}$. To ensure that the set $\mathcal{T}_0$ of ground terms is nonempty, we tacitly assume in this chapter that $\mathcal{L}$ contains at least one constant when considering $\mathcal{T}_0$. Clearly, $\mathcal{T}_k$ is a subalgebra of $\mathcal{T}$, since $t_1, \ldots, t_n \in \mathcal{T}_k \Rightarrow f\vec{t} \in \mathcal{T}_k$.

The concept of a term model can equally be related to $\mathcal{L}^k$: Let $\approx$ be a congruence in $\mathcal{T}_k$ and $\mathfrak{F}_k$ the factor structure $\mathcal{T}_k/\approx$ whose domain is $F_k = \{\bar{t} \mid t \in \mathcal{T}_k\}$ with $\bar{t} = t/\approx$, together with some interpretation of the relation symbols in $\mathcal{T}_k$. We extend $\mathfrak{F}_k$ canonically to an $\mathcal{L}^k$-model $\mathcal{F}_k$ by the (partial) valuation $x \mapsto \bar{x}$ for $x \in Var_k$, which is empty for $k = 0$ so that $\mathcal{F}_0$ and $\mathfrak{F}_0$ can be identified. For each $k$, the following conditions are verified similarly as with (1), (2), (3). The $\mathcal{L}^k$-model $\mathcal{F}_kX$ in $(3_k)$ is defined analogously to $\mathcal{F}X$ on the domain $\mathcal{T}_k/\approx$ with $s \approx t \Leftrightarrow X \vdash s = t$ for $s, t \in \mathcal{T}_k$; in particular, $\mathcal{F}_0X$ arises by factorizing $\mathcal{T}_0$.

$(1_k)$   $t^{\mathcal{F}_k} = \bar{t}$ for all $t \in \mathcal{T}_k$,

$(2_k)$   $\mathcal{F}_k \vDash \forall \vec{x}\alpha \;\; \Leftrightarrow \;\; \mathcal{F}_k \vDash \alpha\frac{\vec{t}}{\vec{x}}$ for all $\vec{t} \in \mathcal{T}_k^n$    ($\alpha$ open),

$(3_k)$   $\mathcal{F}_kX \vDash \pi \;\; \Leftrightarrow \;\; X \vdash \pi$    ($\pi$ a prime formula from $\mathcal{L}^k$).

Let $\varphi = \forall \vec{x}\alpha$ a universal formula ($\forall$-formula). Then $\alpha\frac{\vec{t}}{\vec{x}}$ is called an *instance* of $\varphi$. And if $\vec{t} \in \mathcal{T}_k^n$ then $\alpha\frac{\vec{t}}{\vec{x}}$ is called a $\mathcal{T}_k$-*instance*, for $k = 0$

also called a *ground instance* of $\varphi$. If $U$ is a set of universal formulas, let GI$(U)$ denote the set of ground instances of all $\varphi \in U$. Note that GI$(U) \neq \emptyset$ for $U \neq \emptyset$, provided $\mathcal{L}$ contains constants.

**Theorem 1.1.** *Let $U$ ($\subseteq \mathcal{L}$) be a set of $\forall$-formulas and $\tilde{U}$ the set of all instances of the formulas in $U$. Then the following are equivalent:*

*(i) $U$ is consistent,   (ii) $\tilde{U}$ is consistent,   (iii) $U$ has a term model in $\mathcal{L}$. The same holds if $U \subseteq \mathcal{L}^k$ (in particular for sets $U \subseteq \mathcal{L}^0$ of $\forall$-sentences), where $\tilde{U}$ now denotes the set of all $T_k$-instances of the formulas in $U$.*

**Proof.** (i)$\Rightarrow$(ii): Clear since $U \vdash \tilde{U}$. (ii)$\Rightarrow$(iii): Choose some maximally consistent $X \supseteq \tilde{U}$. Then $\mathcal{F}X \vDash \pi \Leftrightarrow X \vdash \pi$ for prime formulas $\pi$, by (3). Induction on $\wedge, \neg$ easily yields $\mathcal{F}X \vDash \alpha \Leftrightarrow X \vdash \alpha$, for all open $\alpha$. Since $X \supseteq \tilde{U}$ we obtain $\mathcal{F}X \vDash \tilde{U}$. But this yields $\mathcal{F}X \vDash U$ for the term model $\mathcal{F} = \mathcal{F}X$ according to (2). (iii)$\Rightarrow$(i): Trivial. For the case $U \subseteq \mathcal{L}^k$ the proof runs similarly using $(3_k)$, $(2_k)$, and $\mathcal{F}_k = \mathcal{F}_k X$. $\blacksquare$

By Theorem 1.1, a consistent set $U$ of universal *sentences* has a term model $\mathcal{F}_0$. For our purposes, the important case is that $U$ is $=$-free. Then $U$ has a model $\mathfrak{T}$ on the set $T_0$ of all ground terms, since $\mathcal{F}_0 X$ (that replaces $\mathcal{F}X$ in the proof of Theorem 1.1 for $k = 0$) is then constructed *without* a factorization of $T_0$. Such a model $\mathfrak{T} \vDash U$ is called a *Herbrand model* (cf. also Exercise 1 in **3.2**). Its domain $T_0$ is called the *Herbrand universe* of $\mathfrak{T}$. In general, $U$ has many Herbrand models on the same domain $T_0$ with the same canonical interpretation of constants and functions: $c^{\mathfrak{T}} = c$ and $f^{\mathfrak{T}}(t_1, \ldots, t_n) = f\vec{t}$ for all $\vec{t} \in T_0^n$. Only the relations may vary. If $U$ is a universal Horn theory (to be explained in **4.2**), then $U$ has a distinguished Herbrand model, the *minimal* Herbrand model. It will be defined on page 142.

**Example 2.** Let $U \subseteq \mathcal{L}\{0, \mathsf{S}, <\}$ consist of the $=$-free universal sentences

(a) $\forall x \, x < \mathsf{S}x$,   (b) $\forall x \forall y \forall z (x < y \wedge y < z \rightarrow x < z)$.

Here the Herbrand universe $T_0$ consists of all ground terms $\underline{n}$ ($= \mathsf{S}^n 0$). Obviously, $\mathcal{N} := (\mathbb{N}, 0, \mathsf{S}, <) \vDash \tilde{U}$. Since $\mathsf{S}^{\mathcal{N}} t = \mathsf{S}t$ for each $t \in T_0$ (canonical interpretation), $\mathcal{N}$ itself is then a Herbrand model. There are sever other Herbrand models for $U$, since $<$ may be interpreted in various ways as will be seen in Example 3 in **4.2**.

**Remark 1.** With Theorem 1.1 the problem of satisfiability for $X \subseteq \mathcal{L}$ can basically be reduced to a propositional satisfiability problem. By Exercise 5 in **2.6**, $X$ is—after adding new operation symbols—satisfiably equivalent to a set $U$ of $\forall$-formulas which, by Theorem 1.1, is in turn satisfiably equivalent to the set of open formulas $\tilde{U}$. Now replace the prime formulas $\pi$ occurring in the formulas of $\tilde{U}$ with propositional variables $p_\pi$, distinct variables for distinct prime formulas, as in **1.5**. In this way one obtains a satisfiably equivalent set of propositional formulas. This works immediately on $=$-free sets of $\forall$-formulas. By dealing with the congruence conditions for $=$ (page 141), this method can be generalized for sets of arbitrary $\forall$-formulas but is slightly more involved.

Although we will focus on a certain variant of the next theorem, its basic concern (the construction of explicit solutions of existential assertions) is the same in logic programming and related areas.

**Theorem 1.2 (Herbrand's theorem).** *Let $U \subseteq \mathcal{L}$ be a set of universal formulas, $\exists \vec{x}\alpha \in \mathcal{L}$, $\alpha$ open, and let $\tilde{U}$ be the set of all instances of members from $U$. Then the following properties are equivalent:*

*(i)* $\quad U \vdash \exists \vec{x}\alpha$,

*(ii)* $\quad U \vdash \bigvee_{i \leqslant m} \alpha \frac{\vec{t}_i}{\vec{x}}$ *for some $m$ and some $\vec{t}_0, \ldots, \vec{t_m} \in T^n$,*

*(iii)* $\quad \tilde{U} \vdash \bigvee_{i \leqslant m} \alpha \frac{\vec{t}_i}{\vec{x}}$ *for some $m$ and some $\vec{t}_0, \ldots, \vec{t_m} \in T^n$.*

*The same holds if $\mathcal{L}$ is replaced here by $\mathcal{L}^k$, $T$ by $T_k$, and $T^n$ by $T_k^n$, for each $k \geqslant 0$, where $\tilde{U}$ is now the set of all $T_k$-instances.*

**Proof.** Because $U \vdash \tilde{U}$, certainly (iii)$\Rightarrow$(ii)$\Rightarrow$(i). It remains to be shown that (i)$\Rightarrow$(iii): According to (i), $X = U \cup \{\forall \vec{x}\neg\alpha\}$ is inconsistent; hence also $\tilde{U} \cup \{\neg\alpha \frac{\vec{t}}{\vec{x}} \mid \vec{t} \in T^n\}$ by Theorem 1.1. Replacing here the prime formulas with propositional variables as indicated in Remark 1 above, (iii) follows already propositionally according to Exercise 1 in **1.4** (with $Y = \{\alpha \frac{t}{\vec{x}} \mid t \in T\}$). The proof for $\mathcal{L}^k$, $T_k$, and $T_k^n$ runs analogously. $\quad\blacksquare$

**Remark 2.** Herbrand's theorem was originally a proof-theoretic statement. It has several versions. The theorem's assumption that $\alpha$ is open is essential, as can be seen from the example $\vdash \exists x\alpha$ with $\alpha := \forall y(ry \rightarrow rx)$ and $U = \emptyset$. Indeed, $\vdash \exists x\alpha$ holds, for $\exists x\alpha$ is a tautology, Example 2 in **2.6**. But there are no terms $t_0, \ldots, t_m$ (variables in this case) such that $\vdash \bigvee_{i \leqslant m} \alpha \frac{t_i}{\vec{x}}$. The last formula can be falsified in a model with $n + 2$ elements in its domain as is readily seen.

### Exercises

1. Verify the conditions $(1_k)$, $(2_k)$, $(3_k)$ in detail.

2. Prove Herbrands theorem also for $\mathcal{L}^k$ instead of $\mathcal{L}$.

## 4.2   Horn Formulas

We will define *Horn formulas* (after [Hor]) for a given language $\mathcal{L}$ recursively. The following definition covers also the propositional case; simply omit everything that refers to quantification.

**Definition.** (a) Literals (i.e. prim formulas and their negations) are basic Horn formulas. If $\alpha$ is prime and $\beta$ a basic Horn formula, then $\alpha \to \beta$ is a basic Horn formula. (b) Basic Horn formulas are Horn formulas. If $\alpha, \beta$ are Horn formulas then so too are $(\alpha \wedge \beta)$, $\forall x \alpha$, and $\exists x \alpha$.

For instance, $\forall y(ry \to rx)$ and $\forall x(y \in x \to x \notin y)$ are Horn formulas. According to our definition, $\alpha_1 \to \cdots \to \alpha_n \to \beta$ $(n \geqslant 0)$ is the general form of a basic Horn formula, where $\alpha_1, \ldots, \alpha_n$ are prime and $\beta$ is a literal. Note that in the propositional case, the $\alpha_i$ are propositional variables and $\beta$ is a propositional literal. We also call a formula a (basic) Horn formula if it is equivalent to an original (basic) Horn formula. Thus, since

$$\alpha_1 \to \cdots \to \alpha_n \to \beta \equiv \beta \vee \neg\alpha_1 \vee \cdots \vee \neg\alpha_n$$

and by writing $\alpha_0$ for $\beta$ in case $\beta$ is prime, and $\beta = \neg\alpha_0$ if $\beta$ is negated, basic Horn formulas are up to logical equivalence of the type

$$\text{I: } \alpha_0 \vee \neg\alpha_1 \vee \cdots \vee \neg\alpha_n \quad \text{or} \quad \text{II: } \neg\alpha_0 \vee \neg\alpha_1 \vee \cdots \vee \neg\alpha_n$$

for prime formulas $\alpha_0, \ldots, \alpha_n$. I and II are disjunctions of literals of which at most one is a prime formula. Basic Horn formulas are often defined in this way. But our definition above has pleasant advantages in inductive proofs as we shall see, for instance, in the proof of Theorem 2.1. Basic Horn formulas of type I are called *positive* and those of type II *negative*.

A *propositional Horn formula*, i.e., a conjunction of propositional basic Horn formulas, can always be conceived of as a CNF whose disjunctions contain at most one nonnegated element. It is possible to think of an open Horn formula of $\mathcal{L}$ as resulting from replacing the propositional variables of some suitable propositional Horn formula by prime formulas of $\mathcal{L}$.

Each Horn formula is equivalent to a prenex Horn formula. If its prefix contains only $\forall$-quantifiers, then the formula is called a *universal Horn formula*. If the kernel of a Horn formula $\varphi$ in prenex form is a conjunction of positive basic Horn formulas, $\varphi$ is termed a *positive* Horn formula. Horn formulas without free variables are called *Horn sentences*. The universal Horn sentences in the following example are all positive.

**Example 1.** Identities and quasi-identities are universal Horn sentences, as are transitivity $(x \leqslant y \wedge y \leqslant z \to x \leqslant z)^g$, reflexivity, and irreflexivity, but not connexity $(x \leqslant y \vee y \leqslant x)^g$. Also the congruence conditions for $=$ are Horn sentences. Therein $\vec{x} = \vec{y}$ is to mean $\bigwedge_{i=1}^{n} x_i = y_i$:

$$(x = x)^g, \qquad\qquad (x = y \wedge x = z \to y = z)^g,$$
$$(\vec{x} = \vec{y} \to r\vec{x} \to r\vec{y})^g, \quad (\vec{x} = \vec{y} \to f\vec{x} = f\vec{y})^g.$$

$\forall x \exists y \, x \circ y = e$ is a Horn sentence, while $\forall x \exists y (x \neq 0 \to x \cdot y = 0)$ is not, and is even not equivalent to a Horn sentence in the theory $T_F$ of fields. Otherwise $\mathrm{Md}\, T_F$ would be closed under direct products; see Exercise 1. This is not the case: $\mathbb{Q} \times \mathbb{Q}$ is a ring that has zero-divisors, for example $(1, 0) \cdot (0, 1) = 0$. Thus, $\mathbb{Q} \times \mathbb{Q}$ cannot be a field.

A *Horn theory* is to mean a theory $T$ with an axiom system of Horn sentences. If these axioms are universal Horn sentences, then $T$ is called a *universal Horn theory*. Examples are the theories of groups in various languages, of rings, and all equational and quasi-equational theories.

**Theorem 2.1.** *Let $U$ be a consistent set of universal Horn formulas in a language $\mathcal{L}$ with term set $\mathcal{T}$. Then $\mathcal{F} := \mathcal{F}U$ is a model for $U$. In the case $U \subseteq \mathcal{L}^k$, $\mathcal{F}_k := \mathcal{F}_k U$ is a model for $U$ as well.*

**Proof.** $\mathcal{F} \vDash U$ follows from $(*)$: $U \vdash \alpha \Rightarrow \mathcal{F} \vDash \alpha$, for Horn formulas $\alpha$. $(*)$ is proved inductively on the construction of $\alpha$ in the definition of Horn formulas. For prime formulas $\pi$, $(*)$ is clear, for then (3) reads as $\binom{*}{*}$: $U \vdash \pi \Leftrightarrow \mathcal{F} \vDash \pi$. Now suppose that $U \vdash \neg\pi$. Then $U \nvdash \pi$, for $U$ is consistent. Hence $\mathcal{F} \nvDash \pi$ by $\binom{*}{*}$, and so $\mathcal{F} \vDash \neg\pi$. This confirms $(*)$ for all literals. Now let $\alpha$ be prime, $\beta$ a basic Horn formula, $U \vdash \alpha \to \beta$, and assume $\mathcal{F} \vDash \alpha$. Then $U \vdash \alpha$; hence $U \vdash \beta$, and so $\mathcal{F} \vDash \beta$ by the induction hypothesis. This proves $\mathcal{F} \vDash \alpha \to \beta$. Induction on $\wedge$ is clear. Finally, let $U \vdash \forall \vec{x}\alpha$ for some open Horn formula $\alpha$, and $\vec{t} \in \mathcal{T}^n$. Since then certainly $U \vdash \alpha\frac{\vec{t}}{\vec{x}}$, we get $\mathcal{F} \vDash \alpha\frac{\vec{t}}{\vec{x}}$ by the induction hypothesis. $\vec{t}$ was arbitrary and hence $\mathcal{F} \vDash \forall \vec{x}\alpha$ by (2) from **4.1**. This proves $(*)$. The case $U \subseteq \mathcal{L}^k$ is treated analogously. Consider $(2_k)$, $(3_k)$ and take $\mathcal{F}_k$ for $\mathcal{F}$. $\blacksquare$

Incidentally, $U$'s consistency in the theorem is always secured if $U$ consists of positive Horn formulas; Exercise 2. The most interesting case in Theorem 2.1 is that $U$ is the axiom system of a universal Horn theory. Then $\mathcal{F}U \vDash T$, and since $U \subseteq \mathcal{L}^k$ for each $k$, also $\mathcal{F}_k U \vDash T$.

**Example 2.** Let $T$ be the simple universal Horn theory from Example 1 in **4.1**. That $\mathfrak{F} \vDash T$ for the algebra $\mathfrak{F}$ underlying $\mathcal{F}T$ was shown there by proving that $\mathfrak{F}$ is isomorphic to the word-semigroup on the alphabet Var, while Theorem 2.1 yields $\mathfrak{F} \vDash T$ directly. It is easily seen that $\mathfrak{F}$ is generated by $\overline{v_0}, \overline{v_1}, \ldots$ $\mathfrak{F}$ is called *the free semigroup with the free generators* $\overline{v_0}, \overline{v_1}, \ldots$ Also $T_G^=$ is a universal Horn theory so that the free group generated from $\overline{v_0}, \overline{v_1}, \ldots$ is defined as well.

**Remark 1.** A universal Horn theory $T$ is said to be *nontrivial* if $\nvdash_T \forall xy\, x = y$. The generators $\overline{v_0}, \overline{v_1}, \ldots$ of $\mathfrak{F}$ are then distinct and $\mathcal{F}T$ is called the *free model of $T$ with the free generators* $\overline{v_i}$. The word "free" comes from the fact that if $\mathcal{M} = (\mathcal{A}, w)$ is any $T$-model, then the mapping $\overline{x} \mapsto x^{\mathcal{M}}$ ($x \in$ Var) generates a homomorphism $h \colon \mathfrak{F} \mapsto \mathcal{A}$; moreover, we can make "free use" of the values $h\overline{x}$ of the free generators $\overline{x}$, keeping $\mathcal{A}$ fixed and choosing suitable valuations in $\mathcal{M} = (\mathcal{A}, w)$. $h$ is given by $h\overline{t} = t^{\mathcal{M}}$. Note that $h$ is well defined, for $\overline{t_1} = \overline{t_2} \Rightarrow T \vdash t_1 = t_2 \Rightarrow \mathcal{M} \vDash t_1 = t_2 \Rightarrow t_1^{\mathcal{M}} = t_2^{\mathcal{M}} \Rightarrow h\overline{t_1} = h\overline{t_2}$, and it is a matter of routine to check the homomorphism conditions p. 50. E.g., if $f$ is $n$-ary then $hf(\overline{t_1}, \ldots, \overline{t_n}) = h\overline{ft_1 \ldots t_n} = (f\vec{t})^{\mathcal{M}} = f(\vec{t}^{\mathcal{M}}) = f(h\overline{t_1}, \ldots, h\overline{t_n})$. Similarly, $\mathcal{F}_k T$ is the *free model of $T$ with the free generators* $\overline{v_0}, \ldots, \overline{v_{k-1}}$. We will not make use of these remarks, made for the more advanced reader.

Let $U \subseteq \mathcal{L}^0$ be as in Theorem 2.1 but $=$-free, and $T$ be axiomatized by $U$. Let $\mathcal{F}_0 U$ be well defined, i.e. $\mathcal{L}$ contains constant symbols. Then $\mathcal{F}_0 U$ is a Herbrand model for $T$, called the *free* or *minimal Herbrand model for $T$*, and henceforth denoted by $\mathcal{C}_U$ or $\mathcal{C}_T$. The domain of $\mathcal{C}_U$ is the set of ground terms. A not too simple example of describing the minimal Herbrand model for a set $U$ of =-free universal Horn sentences is

**Example 3.** Let $U$ and $\mathcal{N}$ be as in Example 2 in **4.1**. Both (a) and (b) are universal Horn sentences. We determine the minimal Herbrand model $\mathcal{C}_U$ (whose domain consists of the terms $\underline{n}$) by proving $\mathcal{N} \simeq \mathcal{C}_U$, with the isomorphism $n \mapsto \underline{n}$. Since $\mathcal{C}_U \vDash \underline{m} < \underline{k} \Leftrightarrow U \vdash \underline{m} < \underline{k}$ by the definition page 136, it suffices to prove (*): $m < k \Leftrightarrow U \vdash \underline{m} < \underline{k}$. The direction $\Rightarrow$ is shown by induction on $k$, beginning with $k = \mathsf{S}m$. The initial step is clear since $U \vdash \underline{m} < \underline{\mathsf{S}m}$ by (a). Let $m < \mathsf{S}k$. By the induction hypothesis we then get $U \vdash \underline{m} < \underline{k}$, or $m = k$. In both cases, $U \vdash \underline{m} < \underline{\mathsf{S}k}$ by (a) and (b). The direction $\Leftarrow$ in (*) is obvious because $\mathcal{N}$ is a model of $U$.

**Remark 2.** The set $U$ in Example 2 in **4.1** has many models on the Herbrand universe $\mathbb{N}$. One may interpret $<$ by *any* transitive relation on $\mathbb{N}$ that extends $<^{\mathbb{N}}$, e.g., by $\leqslant^{\mathbb{N}}$. This interpretation will be excluded by adding $\forall x\, x \not< x$ to $U$, but the minimal Herbrand model remains the same if enlarging $U$ this way.

Most useful for logic programming is the following variant of Herbrand's theorem. The main difference is that for sets $U$ of universal Horn formulas we get a single solution $\gamma\frac{\vec{t}}{\vec{x}}$ whenever $U \vdash \exists\vec{x}\gamma$. The theorem does also hold with the same proof if $k$ is dropped throughout.

**Theorem 2.2.** *Let $U \subseteq \mathcal{L}^k$ ($k \geqslant 0$) be a consistent set of universal Horn formulas, $\gamma = \gamma_0 \wedge \cdots \wedge \gamma_m$, where all $\gamma_i$ are prime, and $\exists\vec{x}\gamma \in \mathcal{L}^k$. Then the following are equivalent:*

*(i) $\mathcal{F}_k U \vDash \exists\vec{x}\gamma$,   (ii) $U \vdash \gamma\frac{\vec{t}}{\vec{x}}$ for some $\vec{t} \in \mathcal{T}_k^n$,   (iii) $U \vdash \exists\vec{x}\gamma$.*

*In particular, for a consistent universal Horn theory $T$ of any $=$-free language with constants, $\mathcal{C}_T \vDash \exists\vec{x}\gamma$ is always equivalent to $\vdash_T \exists\vec{x}\gamma$.*

**Proof.** (i)⇒(ii): Let $\mathcal{F}_k U \vDash \exists\vec{x}\gamma$. Then $\mathcal{F}_k U \vDash \gamma\frac{\vec{t}}{\vec{x}}$ for some $\vec{t}$, because $\mathcal{F}_k U \vDash \neg\gamma\frac{\vec{t}}{\vec{x}}$ for all $\vec{t}$ implies $\mathcal{F}_k U \vDash \forall\vec{x}\neg\gamma$ by $(2_k)$, contradicting (i). Thus, for all $i \leqslant m$, $\mathcal{F}_k U \vDash \gamma_i\frac{\vec{t}}{\vec{x}}$ . Therefore $U \vdash \gamma_i\frac{\vec{t}}{\vec{x}}$ by $(3_k)$, and so $U \vdash \gamma\frac{\vec{t}}{\vec{x}}$. (ii)⇒(iii): Trivial. (iii)⇒(i): Theorem 2.1 states that $\mathcal{F}_k U \vDash U$. Hence (iii) implies $\mathcal{F}_k U \vDash \exists\vec{x}\gamma$. The particular case follows from (i)⇔(iii) when we choose $k = 0$ and observe that $\mathcal{C}_T = \mathcal{F}_0 T$ by definition. ◻

### Exercises

1. Show that $\mathrm{Md}\, T$ for a Horn theory $T$ is closed under direct products (i.e. $(\forall i \in I)\mathcal{A}_i \vDash T \;\Rightarrow\; \prod_{i\in I} \mathcal{A}_i \vDash T$), and if $T$ is a universal Horn theory, then also under substructures ($\mathcal{A}' \subseteq \mathcal{A} \vDash T \;\Rightarrow\; \mathcal{A}' \vDash T$).

2. Prove that a set of positive Horn formulas is always consistent.

3. Prove $\mathcal{C}_U \simeq (\mathbb{N}, 0, \mathsf{S}, \leqslant)$, where $U$ consists of the universal Horn sentences $\forall x\, x \leqslant x$, $\forall x\, x \leqslant \mathsf{S}x$, $\forall x \forall y \forall z(x \leqslant y \wedge y \leqslant z \to x \leqslant z)$.

## 4.3  Propositional Resolution

We recall the problem of quickly deciding the satisfiability of propositional formulas. This problem is of eminent practical importance, for many non-numerical (sometimes called "logical") problems can be reduced to this. The truth table method, practical for formulas with few variables, grows in terms of calculation effort exponentially with the number of variables; even the most powerful computers of the forseeable future will not be able to carry out the table method for propositional formulas with just 100 variables. As a matter of fact, no essentially better procedure is known, unless one is dealing with formulas of a particular shape, for instance with certain normal forms. The general case represents an unsolved problem of theoretic computer science, not discussed here, the so-called P=NP problem; see for instance [GJ] or look for progress on the Internet.

For conjunctive normal forms, the optimal procedure for contemporary computers is the *resolution procedure* introduced in the following. For the sake of a sparing presentation one switches from a disjunction $\lambda_1 \vee \cdots \vee \lambda_n$ of literals $\lambda_i$ to the set $\{\lambda_1, \ldots, \lambda_n\}$. In so doing, the order of the disjuncts and their possible repetition, inessential factors for the question of satisfiability, are eliminated. For instance, $\lambda_1 \vee \lambda_2 \vee \lambda_1$ is equally represented by $\{\lambda_1, \lambda_2\}$.

A finite, possibly empty set of literals is called a (propositional) *clause*. By a clause *in* $p_1, \ldots, p_n$ is meant a clause $K$ with $\operatorname{var} K \subseteq \{p_1, \ldots, p_n\}$. In the following, $K, H, G, L, P, N$ denote clauses and $\mathcal{K}, \mathcal{H}, \mathcal{P}, \mathcal{N}$ sets of clauses. $K = \{\lambda_1, \ldots, \lambda_n\}$ corresponds to the formula $\lambda_1 \vee \cdots \vee \lambda_n$. The *empty clause* (i.e., $n = 0$) is denoted by $\Box$. It corresponds to the empty disjunction (which is $\bot$, see the footnote page 13).

$K = \{q_1, \ldots, q_m, \neg r_1, \ldots, \neg r_k\}$ $(q_i, r_j \in PV)$ is called a *positive* clause if $m > 0$, for $m = 1$ also a *definite*, and for $m = 0$ a *negative* clause. These conventions will also be adopted when the $q_i$ and $r_i$ later denote prime formulas of a first-order language.

Write $w \vDash K$ (a propositional valuation $w$ *satisfies the clause* $K$) if $K$ contains some $\lambda$ with $w \vDash \lambda$. $K$ is termed *satisfiable* if there is some $w$ with $w \vDash K$. Note that the empty clause $\Box$, as the definition's wording suggests, is not satisfiable. $w$ is a *model* for a set $\mathcal{K}$ of clauses if $w \vDash K$ for all $K \in \mathcal{K}$. If $\mathcal{K}$ has a model then $\mathcal{K}$ is called *satisfiable*. In contrast to the empty clause $\Box$, the empty set of clauses is clearly satisfied by every valuation, again by the definition's wording.

$w$ satisfies a CNF $\alpha$ iff $w$ satisfies all its conjuncts, and hence all of the clauses corresponding to these conjuncts. Since every propositional formula can be transformed into a CNF, $\alpha$ is satisfiably equivalent to a corresponding finite set of clauses. For instance, the CNF $(p \vee q) \wedge (\neg p \vee q \vee r) \wedge (q \vee \neg r) \wedge (\neg q \vee s) \wedge \neg s$ is satisfiably equivalent to the corresponding set of clauses $\{\{p, q\}, \{\neg p, q, r\}, \{q, \neg r\}, \{\neg q, s\}, \{\neg s\}\}$.

It will turn out later that this set is not satisfiable. We write $\mathcal{K} \vDash H$ if every model of $\mathcal{K}$ also satisfies the clause $H$. A set of clauses $\mathcal{K}$ is accordingly unsatisfiable if and only if $\mathcal{K} \vDash \square$.

For $\lambda \notin K$ we frequently denote the clause $K \cup \{\lambda\}$ by $K, \lambda$. Moreover, let $\bar{\lambda} = \neg p$ for $\lambda = p$, $\bar{\lambda} = p$ for $\lambda = \neg p$ (hence $\bar{\bar{\lambda}} = \lambda$ in any case), and set $\bar{K} = \{\bar{\lambda} \mid \lambda \in K\}$. The *resolution calculus* operates with sets of clauses and individual clauses, and has a single rule working with these objects, the so-called *resolution rule*

$$\text{RR:} \quad \frac{K, \lambda \mid L, \bar{\lambda}}{K \cup L} \quad (\lambda, \bar{\lambda} \notin K \cup L).$$

RR may be read as follows: If the clauses $K, \lambda$ and $L, \bar{\lambda}$ are derivable, then also the clause $K \cup L$, called a *resolvent* of the clauses $K, \lambda$ and $L, \bar{\lambda}$.

A clause $H$ is said to be *derivable* from a set of clauses $\mathcal{K}$, in symbols $\mathcal{K} \vdash^{RR} H$, if $H$ can be obtained from $\mathcal{K}$ by the stepwise application of RR; equivalently, if $H$ belongs to the *resolution closure* $Rc\mathcal{K}$ of $\mathcal{K}$, which is the smallest set of clauses $\mathcal{H} \supseteq \mathcal{K}$ closed with respect to applications of RR. The definition of the resolution closure corresponds completely to the definition of an MP-closed set of formulas in **1.6**.

**Example 1.** Let $\mathcal{K} = \{\{p, \neg q\}, \{q, \neg p\}\}$. Application of RR leads to the resolvents $\{p, \neg p\}$ and $\{q, \neg q\}$, from which we see that a pair of clauses has several resolvents, in general. Every subsequent application of RR yields already available clauses, so that $Rc\mathcal{K}$ contains only the clauses $\{p, \neg q\}$, $\{q, \neg p\}$, $\{p, \neg p\}$, and $\{q, \neg q\}$.

Applying RR to $\{p\}, \{\neg p\}$ yields the empty clause $\square$. Hence $\mathcal{K} \vdash^{RR} \square$, with the unsatifiable set of clauses $\mathcal{K} = \{\{p\}, \{\neg p\}\}$. By the resolution theorem below, the derivability of the empty clause from a set of clauses $\mathcal{K}$ is characteristic of the nonsatisfiability of $\mathcal{K}$. To test this one needs only to check whether $\mathcal{K} \vdash^{RR} \square$, i.e., $\square \in Rc\mathcal{K}$. This question is effectively decidable for finite sets $\mathcal{K}$ because then $Rc\mathcal{K}$ is finite as well. Indeed, a resolvent that results from applying RR to clauses in $p_1, \ldots, p_n$ contains at most these very same variables. Further, there are only finitely many clauses in $p_1, \ldots, p_n$, exactly $2^{2n}$. But that is still an exponential increase as $n$ increases. Aside from this, the mechanical implementation of the resolution calculus mostly involves potentially infinite sets of clauses. We consider this problem more closely at the end of **4.6**.

The derivation of a clause $H$ from a set of clauses $\mathcal{K}$, especially the derivation of the empty clause, can best be graphically represented by a *resolution tree* as in Example 2. This is a tree that branches "upward" with an endpoint $H$ without edge exits, called the *root* of the tree. Points without entering edges are called *leaves*. A point that is not a leaf has two entrances, and the points leading to them are called their *predecessors*. The points of a resolution tree bear sets of clauses in the sense that a point not being a leaf is a resolvent of the two clauses above it.

**Example 2.** The following figure shows one of the many resolution trees for the already-mentioned collection of clauses

$$\mathcal{K}_0 = \{\{p, q\}, \{\neg p, q, r\}, \{q, \neg r\}, \{\neg q, s\}, \{\neg s\}\}.$$

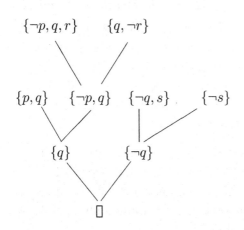

The leaves of this tree are all occupied by clauses in $\mathcal{K}_0$. It should be clear that an arbitrary clause $H$ belongs to the resolution closure of a set of clauses $\mathcal{K}$ just when there exists a resolution tree with leaves in $\mathcal{K}$ and root $H$. A resolution tree with leaves in $\mathcal{K}$ and the root $\square$ as shown in the figure on the left for $\mathcal{K} = \mathcal{K}_0$ is called a *resolution for $\mathcal{K}$*, or more precisely, a *successful resolution for $\mathcal{K}$*. Thus, because of $\square \in Rc\mathcal{K}_0$, the set of clauses $\mathcal{K}_0$ is unsatisfiable, and hence so is the conjunctive normal form that corresponds to the set $\mathcal{K}_0$, namely the formula

$$(p \vee q) \wedge (\neg p \vee q \vee r) \wedge (q \vee \neg r) \wedge (\neg q \vee s) \wedge \neg s.$$

**Remark 1.** If a resolution tree ends with a point $\neq \square$, either to which RR cannot be applied or where upon application the points are simply reproduced, then one talks of an *unsuccessful* resolution. In this case, most interpreters of the resolution calculus will "backtrack," which means the program searches backward along the tree for the first point where one of several resolution alternatives was chosen, and picks up another alternative. Some kind of selection strategy must in any case be implemented, since just as with any logical calculus, the resolution calculus is nondeterministic, that is, no natural preferences exist regarding the order of the derivations leading to a successful resolution, even if the existence of such a resolution is known for other reasons.

We remark that despite the derivability of the empty clause, for infinite unsatisfiable sets of clauses $\mathcal{K}$ there also may exist infinite resolution trees with nonrepeating points, where $\square$ never appears. Such a tree has no root. For example, the set of clauses

$$\mathcal{K} = \{\{p_1\}, \{\neg p_1\}, \{p_1, \neg p_2\}, \{p_2, \neg p_3\}, \dots\}$$

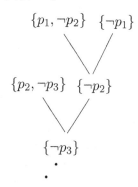

is not satisfiable. Here we obtain the infinite resolution tree in the figure on the right, occupied by leaves from $\mathcal{K}$, which has no root and does not reflect that $\square$ is derivable by just a single application of RR to the first two clauses of $\mathcal{K}$. In the diagram, the resolution calculus is running on $\mathcal{K}$ with a rather stupid strategy. This and similar examples indicate that the resolution calculus is incapable in general of deciding the satisfiability of infinite sets $\mathcal{K}$ of clauses. Indeed, this will be confirmed in **4.6**. Nonetheless, by Theorem 3.2 below there does exist—if $\mathcal{K}$ is in actual fact unsatisfiable—a successful resolution for $\mathcal{K}$ that can in principle be found in finitely many steps.

We commence the more detailed study of the resolution calculus with

**Lemma 3.1 (Soundness lemma).** *Let $\mathcal{K}$ be a set of clauses and $K$ a single clause. Then $\mathcal{K} \vdash^{RR} H \ \Rightarrow \ \mathcal{K} \vDash H$.*

**Proof.** As in the case of a Hilbert calculus, it suffices to confirm the soundness of the rule RR, that is, to prove that a model for $K, \lambda$ and $L, \overline{\lambda}$ is also one for $K \cup L$. Thus let $w \vDash K, \lambda$ and $w \vDash L, \overline{\lambda}$. **Case 1:** $w \nvDash \lambda$. Then there must be a literal $\lambda' \in K$ with $w \vDash \lambda'$. Hence $w \vDash K$ and therefore $w \vDash K \cup L$. **Case 2:** $w \vDash \lambda$. Then $w \nvDash \overline{\lambda}$. Similar to the above we get $w \vDash L$. Hence $w \vDash K \cup L$ as well.  $\blacksquare$

For the case $\mathcal{K} \vdash^{RR} \square$ the lemma shows $\mathcal{K} \vDash \square$, that is, the unsatisfiability of $\mathcal{K}$. The converse of Lemma 3.1 is in general not valid; for instance $\{\{p\}\} \vDash \{p, q\}$, but $\{\{p\}\} \nvdash^{RR} \{p, q\}$. It does hold, though, for $H = \square$. This follows from Theorem 3.2 below, often stated as "$\mathcal{K}$ *is unsatisfiable iff* $\mathcal{K} \vdash^{RR} \square$." In its proof we recursively construct a global valuation $w$ from partial valuations, defined only for $p_1, \dots, p_n$.

**Theorem 3.2 (Resolution theorem).** $\mathcal{K}$ *is satisfiable iff* $\mathcal{K} \nvdash^{RR} \square$.

**Proof.** Clearly $\mathcal{K} \nvDash \square$ if $\mathcal{K}$ is satisfiable, so $\mathcal{K} \nvdash^{RR} \square$ by Lemma 3.1. Now let $\mathcal{K} \nvdash^{RR} \square$, or equivalently, $\square \notin \mathcal{H}$ with $\mathcal{H} := Rc\mathcal{K}$. Let $\Lambda^{(n)}$ denote the set of all literals in $p_1, \ldots, p_n$, and $\mathcal{H}^{(n)}$ be the set of all $K \in \mathcal{H}$ with $K \subseteq \Lambda^{(n)}$ such that $p_n$ or $\neg p_n$ or both belong to $K$. Note that $\Lambda^{(0)} = \mathcal{H}^{(0)} = \emptyset$, $\mathrm{var}\,\mathcal{H}^{(n)} \subseteq \{p_1, \ldots, p_n\}$, and $\mathcal{H} = \bigcup_{n \in \mathbb{N}} \mathcal{H}^{(n)}$. We will construct a model for $\mathcal{H}$ (hence for $\mathcal{K}$) stepwise. $v_n := wp_n$ will be defined recursively on $n$ (more precisely, by naive course-of-value recursion, discussed e.g. in **6.1**) such that $w_n := (v_1, \ldots, v_n)$ has the property $(*)$: $w_n \vDash \bigcup_{i \leqslant n} \mathcal{H}^{(i)}$.

We agree to say that the "empty valuation" satisfies $\mathcal{H}^{(0)} = \emptyset$, hence $(*)$ holds trivially for $n = 0$. Let $v_1, \ldots, v_n$ be defined so that $(*)$ is true. We will define $v_{n+1} = wp_{n+1}$ such that $(+)$: $w_{n+1} \vDash \mathcal{H}^{(n+1)}$ is satisfied. This clearly implies $w_{n+1} \vDash \bigcup_{i \leqslant n+1} \mathcal{H}^{(i)}$, hence $w_n \vDash \mathcal{H}^{(n)}$ for all $n$, so that $w = (v_1, v_2, \ldots)$ is a model for the whole of $\mathcal{H}$. In order to verify $(+)$ we need to consider only those $K \in \mathcal{H}^{(n+1)}$ containing no $\lambda \in \Lambda^{(n)}$ with $w_n \vDash \lambda$ and not both $p_{n+1}$ and $\neg p_{n+1}$. These $K$ will be called *sensitive* clauses during this proof, since every other $K \in \mathcal{H}^{(n+1)}$ either contains some $\lambda \in \Lambda^{(n)}$ with $w_n \vDash \lambda$ or else both $p_{n+1}$ and $\neg p_{n+1}$, and hence is satisfied by *any* expansion of $w_n$ to $w_{n+1}$.[1] We may assume that there is a sensitive $K \in \mathcal{H}^{(n+1)}$ (otherwise put $v_{n+1} = 0$) and prove the following **claim:** Either $p_{n+1} \in K$ for all sensitive $K$—then put $v_{n+1} = 1$—or else $\neg p_{n+1} \in K$ for all sensitive $K$, in which case put $v_{n+1} = 0$, so that $(+)$ holds in either case. To prove this claim assume that there are sensitive $K, H \in \mathcal{H}^{(n+1)}$ with $p_{n+1} \in K$ and $\neg p_{n+1} \in H$, hence $\neg p_{n+1} \notin K$ and $p_{n+1} \notin H$. Applying RR to $H, K$, we then obtain either $\square$ (a contradiction to $\square \notin \mathcal{H}$), or else a clause from $\mathcal{H}^{(i)}$ for some $i \leqslant n$ whose literals are not satisfied by $w_n$, a contradiction to $(*)$, i.e. to $w_n \vDash \mathcal{H}^{(i)}$ for all $i \leqslant n$. This confirms the claim and completes the proof. $\square$

**Remark 2.** The foregoing proof is constructive, that is, if $\mathcal{K} \nvdash^{RR} \square$ and the $\mathcal{H}^{(n)}$ in the proof above are computable, then a valuation satisfying $\mathcal{K}$ is computable as well. Moreover, we incidentally proved the propositional compactness theorem for countable sets of formulas $X$ once again. Here is the argument: Every formula is equivalent to some KNF, and hence $X$ is satisfiably equivalent to a set of clauses $\mathcal{K}_X$. So if $X$ is not satisfiable, the same is true of $\mathcal{K}_X$. Consequently,

---

[1] The newcomer should write down all eight candidates for the subset $\mathcal{H}^{(1)}$ of $\Lambda^{(n)} = \{\{p_1\}, \{\neg p_1\}, \{p_1, \neg p_1\}\}$. Only $\{p_1\}$ and $\{\neg p_1\}$ are sensitive to $v_1 = wp_1$.

$\mathcal{K}_X \vdash^{RR} \square$ by Theorem 3.2. Therefore $\mathcal{K}_0 \vdash^{RR} \square$ for some finite subset $\mathcal{K}_0 \subseteq \mathcal{K}_X$, for there must be some successful resolution tree whose leaves are collected in $\mathcal{K}_0$. Having this, it is obvious that just a finite subset of $X$ is not satisfiable, namely the one that corresponds to the set of clauses $\mathcal{K}_0$.

## 4.4   Horn Resolution

A clause belonging to a propositional basic Horn formula is called a (propositional) *Horn clause*. It is called positive or negative if the corresponding Horn formula is positive or negative. Positive Horn clauses are of the form $\{\neg q_1, \dots, \neg q_n, p\}$ with $n \geqslant 0$, negative ones of the form $\{\neg q_1, \dots, \neg q_k\}$. The empty clause $(k = 0)$ is counted among the negative ones. It is important in practice that the resolution calculus can be formulated more specifically for Horn clauses. The empty clause, if it can be obtained from a set of Horn clauses at all, can also be obtained using a restricted resolution rule, which is applied only to pairs of Horn clauses in which one premise is positive (the left one in HR below) and the other one is negative. This is the *rule of Horn resolution*

HR : $\quad \dfrac{K, p \mid L, \neg p}{K \cup L} \quad$ ($K, L$ negative, $p, \neg p \notin K \cup L$).

The calculus operating with Horn clauses and rule HR is denoted by $\vdash^{HR}$. A positive Horn clause is clearly definite. Hence, the resolvent of an application of HR is unique and always negative. An *H*-resolution tree is therefore of the simple form illustrated by the figure on the right. Therein $P_0, \dots, P_\ell$ denote positive and $N_0, \dots, N_{\ell+1}$ negative Horn clauses. Such a tree is also called an *H-resolution for* $\mathcal{P}, N$—where $\mathcal{P}$ here and elsewhere is taken to mean a

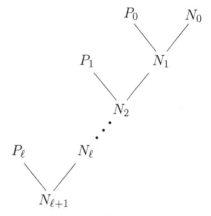

set of positive Horn clauses and $N$ is a negative clause $\neq \square$—if it satisfies (1) $P_i \in \mathcal{P}$ for all $i \leqslant \ell$, and (2) $N_0 = N$ & $N_{\ell+1} = \square$. It is evidently possible to regard an *H*-resolution for $\mathcal{P}, N$ simply as a sequence $(P_i, N_i)_{i \leqslant \ell}$ with the properties (0) $N_{i+1} = HR(P_i, N_i)$ for all $i \leqslant \ell$, (1), and (2). Here $HR(P, N)$ denotes the uniquely determined resolvent resulting from applying HR to the positive clause $P$ and the negative clause $N$.

Before proving the completeness of $\vdash^{HR}$ we require a little preparation. Let $\mathcal{P}$ be a set of positive Horn clauses. In order to gain an overview of all models $w$ of $\mathcal{P}$, consider the natural correspondence

$$w \longleftrightarrow V_w := \{p \in PV \mid w \vDash p\}$$

between valuations $w$ and subsets of $PV$. Put $w \leqslant w' :\Leftrightarrow V_w \subseteq V_{w'}$. Clearly, $\mathcal{P}$ is always satisfied by the "maximal" valuation $w$ with $V_w = PV$, for $w$ satisfies every positive clause and $\mathcal{P}$ contains only such clauses. It is obvious that $w \vDash \mathcal{P}$ iff $V = V_w$ satisfies the following two conditions:

(a) $p \in V$ provided $\{p\} \in \mathcal{P}$,

(b) $q_1, \ldots, q_n \in V \Rightarrow p \in V$, provided $\{\neg q_1, \ldots, \neg q_n, p\} \in \mathcal{P}$ $(n > 0)$.

Of all subsets $V \subseteq PV$ satisfying (a) and (b) there is clearly a smallest one, namely $V_{\mathcal{P}} := \bigcap\{V_w \mid w \vDash \mathcal{P}\}$. Let $w_{\mathcal{P}}$ be the $\mathcal{P}$-model corresponding to $V_{\mathcal{P}}$ and call it the *minimal $\mathcal{P}$-model*. We may define $V_{\mathcal{P}}$ for the minimal model $w_{\mathcal{P}}$ of $\mathcal{P}$ also as follows: Put $V_0 = \{p \in PV \mid \{p\} \in \mathcal{P}\}$ and

$$V_{k+1} = V_k \cup \{p \in PV \mid \{\neg q_1, \ldots, \neg q_n, p\} \in \mathcal{P} \text{ for some } q_1, \ldots, q_n \in V_k\}.$$

Then $V_{\mathcal{P}} = \bigcup_{k \in \mathbb{N}} V_k$. Indeed, $V_k \subseteq V_w$ for all $k$ and all $w \vDash \mathcal{P}$. Hence, $\bigcup_{k \in \mathbb{N}} V_k \subseteq V_{\mathcal{P}}$. Also $V_{\mathcal{P}} \subseteq \bigcup_{k \in \mathbb{N}} V_k$, since $w \vDash \mathcal{P}$ with $V_w = \bigcup_{k \in \mathbb{N}} V_k$.

The minimal $m$ with $p \in V_m$ is termed the $\mathcal{P}$-*rank* of $p$, denoted by $\rho_{\mathcal{P}} p$. Those $p$ with $\{p\} \in \mathcal{P}$ are of $\mathcal{P}$-rank 0. The variables arising from these by applying (b) have $\mathcal{P}$-rank 1 if not already in $V_0$, and so on.

**Lemma 4.1.** *Let $\mathcal{P}$ be a set of positive Horn clauses and $q_0, \ldots, q_k \in V_{\mathcal{P}}$. Then $\mathcal{P}, N \vdash^{HR} \square$, where $N = \{\neg q_0, \ldots, \neg q_k\}$.*

**Proof.** For $r_0, \ldots, r_n \in V_{\mathcal{P}}$ set $\rho_{\mathcal{P}}(r_0, \ldots, r_n) := \max\{\rho_{\mathcal{P}} r_0, \ldots, \rho_{\mathcal{P}} r_n\}$. Let $\mu(r_0, \ldots, r_n)$ be the number of $i \leqslant n$ such that $\rho_{\mathcal{P}} r_i = \rho_{\mathcal{P}}(r_0, \ldots, r_n)$. The claim is proved inductively on $\rho := \rho_{\mathcal{P}}(q_0, \ldots, q_k)$ and $\mu := \mu(q_0, \ldots, q_k)$. First suppose $\rho = 0$, i.e., $\{q_0\}, \ldots, \{q_k\} \in \mathcal{P}$. Then there is certainly an $H$-resolution for $\mathcal{P}, N$, namely the tree $(\{q_i\}, \{\neg q_i, \ldots, \neg q_k\})_{i \leqslant k}$. Now take $\rho > 0$ and w.l.o.g. $\rho = \rho_{\mathcal{P}} q_0$. Then there are $q_{k+1}, \ldots, q_m \in V_{\mathcal{P}}$ such that $P := \{\neg q_{k+1}, \ldots, \neg q_m, q_0\} \in \mathcal{P}$ and $\rho_{\mathcal{P}}(q_{k+1}, \ldots, q_m) < \rho$. Thus, $\rho_{\mathcal{P}}(q_1, \ldots, q_k, q_{k+1}, \ldots, q_m)$ is either $< \rho$, or it is $= \rho$, in which case $\mu(q_1, \ldots, q_m) < \mu$. By the induction hypothesis, in both cases $\mathcal{P}, N_1 \vdash^{HR} \square$ for $N_1 := \{\neg q_1, \ldots, \neg q_m\}$. Hence, an $H$-resolution $(P_i, N_i)_{1 \leqslant i \leqslant \ell}$ for $\mathcal{P}, N_1$ exists. But then $(P_i, N_i)_{i \leqslant \ell}$, with $P_0 := P$ and $N_0 := N$, is just an $H$-resolution for $\mathcal{P}, N$. $\square$

**Theorem 4.2 (the $H$-Resolution theorem).** *A set $\mathcal{K}$ of Horn clauses is satisfiable iff $\mathcal{K} \nvdash^{HR} \Box$.*

**Proof.** The condition $\mathcal{K} \nvdash^{HR} \Box$ is certainly necessary if $\mathcal{K}$ is satisfiable. For the converse assume that $\mathcal{K}$ is unsatisfiable, $\mathcal{K} = \mathcal{P} \cup \mathcal{N}$, all $P \in \mathcal{P}$ are positive and all $N \in \mathcal{N}$ negative. Since $w_\mathcal{P} \vDash \mathcal{P}$ but $w_\mathcal{P} \nvDash \mathcal{P} \cup \mathcal{N}$ there is some $N = \{\neg q_0, \ldots, \neg q_k\} \in \mathcal{N}$ such that $w_\mathcal{P} \nvDash N$. Consequently, $w_\mathcal{P} \vDash q_0, \ldots, q_k$ and therefore $q_0, \ldots, q_k \in V_\mathcal{P}$. By Lemma 4.1 we then obtain $\mathcal{P}, N \vdash^{HR} \Box$, and a fortiori $\mathcal{K} \vdash^{HR} \Box$. $\blacksquare$

**Corollary 4.3.** *Let $\mathcal{K} = \mathcal{P} \cup \mathcal{N}$ be a set of Horn clauses, all $P \in \mathcal{P}$ positive and all $N \in \mathcal{N}$ negative. Then the following conditions are equivalent:*

(i) $\mathcal{K}$ *is unsatisfiable,*     (ii) $\mathcal{P}, N$ *is unsatisfiable for some $N \in \mathcal{N}$.*

**Proof.** (i) implies $\mathcal{K} \vdash^{HR} \Box$ by Theorem 4.2. Hence, there is some $N \in \mathcal{N}$ and some $H$-Resolution for $\mathcal{P}, N$, whence $\mathcal{P}, N$ is unsatisfiable. (ii)$\Rightarrow$(i) is trivial because $\mathcal{P}, N$ is a subset of $\mathcal{K}$. $\blacksquare$

Thus, the investigation of sets of Horn clauses as regards satisfiability can completely be reduced to the case of just a single negative clause.

The hitherto illustrated techniques can without further ado be carried over to quantifier-free formulas of a first-order language $\mathcal{L}$, in that one thinks of the propositional variables to be replaced by prime formulas of $\mathcal{L}$. Clauses are then finite sets of literals in $\mathcal{L}$. By Remark 1 in **4.1** a set of $\mathcal{L}$-formulas is satisfiably equivalent to a set of open formulas, which w.l.o.g. are given in conjunctive normal form. Splitting these into their conjuncts provides a satisfiably equivalent set of disjunctions of literals. Converting these disjunctions into clauses, one obtains a set of clauses for which, by the remark just cited, a consistency condition can be stated propositionally. Now, because predicate-logical proofs are always reducible to the demonstration of certain inconsistencies by virtue of the equivalence of $X \vdash \alpha$ with the inconsistency of $X, \neg \alpha$, these proofs can basically also be carried out by resolution.

To sum up, resolution by Theorem 3.2 and 4.2 is not at all restricted to propositional logic but includes application to sets of literals of first-order languages. Theorem 7.3, the predicate logic version of Theorem 3.2, will essentially be reduced to the latter. Moreover, questions concerning resolution in first-order languages can basically be treated propositionally, as indicated by the exercises below.

Before elaborating on this, we consider an additional aid to automated proof procedures, namely unification. This will later be combined with resolution, and it is this combination that makes automated proof procedures fast enough for modern computers equipped with efficient interpreters of PROLOG.

## Exercises

1. Prove that the satisfiable set of clauses $\mathcal{P} = \{\{p_3\}, \{\neg p_3, p_1, p_2\}\}$ does not have a smallest model. The 2nd clause in $\mathcal{P}$ is not a Horn clause. Thus, in general only Horn clauses have a smallest model.

2. Let $p_{m,n,k}$ for $m, n, k \in \mathbb{N}$ be propositional variables, $\mathsf{S}$ the successor function, and $\mathcal{P}$ the set of all clauses belonging to the Horn formulas

   $$p_{m,0,m} \;\; ; \;\; p_{m,n,k} \to p_{m,\mathsf{S}n,\mathsf{S}k} \qquad (m, n, k \in \mathbb{N}).^2$$

   Let the standard model $w_s$ be defined by $w_s \vDash p_{m,n,k} \Leftrightarrow m+n = k$. Show that the minimal model $w_\mathcal{P}$ coincides with $w_s$.

3. Let $\mathcal{P}$ be the set of Horn clauses of Exercise 2. Prove that

   (a) $\mathcal{P}, \neg p_{n,m,n+m} \vdash^{HR} \Box$,   (b) $\mathcal{P}, \neg p_{n,m,k} \vdash^{HR} \Box \Rightarrow k = n + m$.

   (a) and (b) together are equivalent to the the single condition

   (c) $\mathcal{P}, \neg p_{n,m,k} \vdash^{HR} \Box \Leftrightarrow k = n + m$.

## 4.5   Unification

A decisive aid in logic programming is unification. This notion is meaningful for any set of formulas, but we confine ourself to $\neg$-*free* clauses $K \neq \Box$ of an identity-free language. $K$ contains only unnegated prime formulas, each starting with a relation symbol. Such a clause $K$ is called *unifiable* if a substitution $\sigma$ exists, a so-called *unifier* of $K$, such that $K^\sigma := \{\lambda^\sigma \mid \lambda \in K\}$ contains exactly one element; in other words, $K^\sigma$ is a *singleton*. Here $\sigma$ can most easily be understood as a simultaneous substitution, that is, $\sigma$ is globally defined and $x^\sigma = x$ for almost all variables $x$.

---

[2] In **4.6** these formulas will be interpreted as the ground instances of a logic program for computing the sum of two natural numbers.

Simultaneous substitutions form a semigroup with respect to composition, with the neutral element $\iota$, a fact we will heavily make use of.

**Example 1.** Consider $K = \{rxfxz, rfyzu\}$, $r$ and $f$ binary. Here $\omega = \frac{fyz}{x}\frac{ffyzz}{u}$ is a unifier: $K^\omega = \{rfyzffyzz\}$, as is readily confirmed. Clearly, $\omega$ as a composition of simple substitutions can be understood as a simultaneous substitution, see page 60.

Obviously, a clause containing prime formulas that start with distinct relation symbols is not unifiable. A further obstacle to unification is highlighted by

**Example 2.** Let $K = \{rx, rfx\}$ ($r, f$ unary). Assume $(rx)^\sigma = (rfx)^\sigma$. This clearly implies $rx^\sigma = rfx^\sigma$ and hence $x^\sigma = fx^\sigma$. This is impossible, for $x^\sigma$ and $fx^\sigma$ are clearly of different lengths. Hence, $K$ is not unifiable.

If $\sigma$ is a unifier then so too is $\sigma\tau$ for any substitution $\tau$. Call $\omega$ a *generic* or a *most general* unifier of $K$ if any other unifier $\tau$ of $K$ has a representation $\tau = \omega\sigma$ for some substitution $\sigma$. By Theorem 5.1 below, each unifiable clause has a generic unifier. For instance, it will turn out below that $\omega$ in Example 1 *is* generic.

A *renaming of variables*, a *renaming* for short, is for the sake of simplicity a substitution $\rho$ such that $\rho^2 = \iota$. This definition could be rendered more generally, but it suffices for our purposes. $\rho$ is necessarily bijective and maps variables to variables. Let $x_i^\rho = y_i$ ($\neq x_i$) for $i = 1, \ldots, n$ and $z^\rho = z$ otherwise. Then clearly $y_i^\rho = x_i$, that is, $\rho$ swaps the variables $x_i$ and $y_i$. In this case we shall write $\rho = \left(\begin{smallmatrix} x_1 \cdots x_n \\ y_1 \cdots y_n \end{smallmatrix}\right)$.

If $\omega$ is a generic unifier of $K$ then so too is $\omega' = \omega\rho$, for any renaming $\rho$. Indeed, for any given unifier $\tau$ of $K$ there is some $\sigma$ such that $\tau = \omega\sigma$. For $\sigma' := \rho\sigma$ then $\tau = \omega\rho^2\sigma = (\omega\rho)(\rho\sigma) = \omega'\sigma'$. Choosing in Example 1 for instance $\rho := \left(\begin{smallmatrix} y \ z \\ u \ v \end{smallmatrix}\right)$, we obtain the generic unifier $\omega' = \omega\rho$ for $K$, with $K^{\omega'} = \{rfuvffuvv\}$.

We now consider a procedure in the form of a flow diagram, the *unification algorithm*, denoted by $\mathfrak{U}$. It checks each nonempty clause $K$ of prime formulas of an identity-free language for unifiability, and in the positive case it produces a generic unifier. $\mathfrak{U}$ uses a variable $\sigma$ for substitutions with initial value $\iota$, and a variable $L$ for clauses with initial value $K$. Later on, $L$ contains $K^\sigma$ for the actual value of $\sigma$ that depends on the actual state of the procedure. Here the diagram of $\mathfrak{U}$:

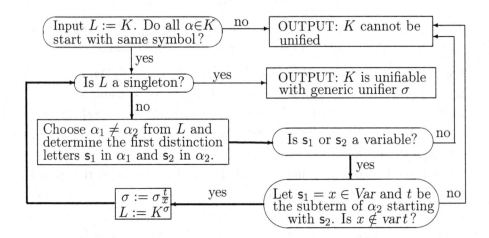

The *first distinction letters* of two strings are the first symbols from the left that distinguish the strings. The first letter of $\alpha \in L$ is a relation symbol. By Exercise 1 in **2.2**, any further symbol $s$ in $\alpha$ determines uniquely at each position of its occurrence a subterm of $\alpha$ whose initial symbol is $s$. The diagram has just one (thick-lined) loop that starts and ends in the test 'Is $L$ a singleton?'. The loop runs through the operation $\sigma := \sigma \frac{t}{x}$, $L := K^\sigma$, which assigns a new value to $\sigma$ and then the new value $K^\sigma$ to $L$. This reduces the number of variables in $L$, since $x \notin \mathrm{var}\, L$. Hence, $\mathfrak{U}$ necessarily leaves the loop after finitely many steps, and must stop and halt in one of the two OUTPUT boxes of $\mathfrak{U}$. But we do not yet know whether $\mathfrak{U}$ always ends up in the "right" box, i.e., whether $\mathfrak{U}$ answers correctly. The final value $\sigma$ is printed in the lower OUTPUT box.

**Example 3.** Let $\mathfrak{U}$ be executed on $K$ from Example 1. The first distinction letters of the two members $\alpha_1, \alpha_2 \in K$ are $s_1 = x$ and $s_2 = f$ at the second position. The subterm beginning with $s_2$ in $\alpha_2$ is $t = fyz$. Hence, after the first run through the loop with $\sigma := \iota \frac{fyz}{x} = \frac{fyz}{x}$ we get $K^\sigma = \{rfyzffyzz, rfyzu\}$. Here the first distinction letters are $f$ and $u$ at position 5. The term beginning with $f$ at this position is $t = ffyzz$. Since $u \notin \mathrm{var}\, ffyzz$, the loop is run through once again and we obtain $\sigma := \sigma \frac{ffyzz}{u} = \frac{fyz}{x} \frac{ffyzz}{u}$. This is a unifier, and $\mathfrak{U}$ comes to a halt with OUTPUT '$K$ is unifiable with the generic unifier $\sigma = \frac{fyz}{x} \frac{ffyzz}{u}$'.

We recommend a thorough study of this example. That the $\sigma$ at the end of the example is indeed generic follows from

**Theorem 5.1.** *The unification algorithm* $\mathfrak{U}$ *is sound, i.e., upon input of a negation-free clause* $K$ *it always answers correctly.*[3] $\mathfrak{U}$ *unifies with a generic unifier.*

**Proof.** This is obvious if two elements of $K$ are already distinguished by the first letter. Assume therefore that all $\alpha \in K$ begin with the same letter. If $\mathfrak{U}$ stops with the output '$K$ is unifiable ...', $K$ is in fact unifiable, since it must have been previously verified that $L = K^\sigma$ is a singleton. Conversely, we claim that $\mathfrak{U}$ also halts with the correct output, provided $K$ *is* unifiable. The latter will be our assumption till the end of the proof, along with the choice of an arbitrary but fixed unifier $\tau$ of $K$.

It has to be confirmed that both tests on the right side of the diagram do not end with the upper OUTPUT, i.e., the test questions are answered Yes. For the upper test this is clear, since substitutions preserve the symbols $\mathsf{s}_1, \mathsf{s}_2$ so that unification would be impossible; this contradicts our assumption. For the lower test ('Is $x \notin \mathrm{var}\, t$?') the correctness of the answer will be verified below. Let $i\ (= 0, \ldots, m)$ denote the moment after the $i$th run through the loop has been finished. $i = 0$ before the first run. Put $\sigma_0 := \iota$ and let $\sigma_i$ for $i > 0$ be the value of $\sigma$ after the $i$th run through the loop. Below we shall prove that

(∗) there exists a substitution $\tau_i$ with $\sigma_i \tau_i = \tau\quad (i = 0, \ldots, m)$.

Assume that $x \in \mathrm{var}\, t$ in the $(i + 1)$th run, with $\alpha_1, \alpha_2 \in K^{\sigma_i}$ chosen as in the diagram. Choose $\tau_i$ according to (∗). Since $\sigma_i \tau_i = \tau$ is a unifier, $\alpha_1^{\tau_i} = \alpha_2^{\tau_i}$. Hence $\binom{*}{*}: x^{\tau_i} = t^{\tau_i}$ (the terms $x^{\tau_i}$ and $t^{\tau_i}$ start at the same position and must be identical; see Exercise 1 in **2.2**). But then $x \in \mathrm{var}\, t$ is impossible, since otherwise $x^{\tau_i}$ and $t^{\tau_i}$ were of different length as in Example 2. This confirms the correctness of the diagram in the unifiable case as claimed above. (∗) says in particular that $\sigma_m \tau_m = \tau$. Since $\tau$ was arbitrary, and the $\sigma_i$ do not depend on the choice of $\tau$, it follows that $\sigma := \sigma_m$ is indeed a generic unifier of $K$.

It remains to prove (∗) by induction on $i \leqslant m$. This is trivial for $i = 0$: choose simply $\tau_0 = \tau$, so that $\sigma_0 \tau_0 = \iota \tau = \tau$. Suppose (∗) holds for $i < m$. As was shown, $\binom{*}{*}$ holds while running through the test '$x \notin \mathrm{var}\, t$?'

---

[3] The proof will be a paradigm for a so-called correctness proof of an algorithm. Such a proof is often fairly lengthy and has almost always to be carried out inductively on the number of runs through a loop occurring in the algorithm.

We set $\tau_{i+1} := \frac{t}{x}\,\tau_i$ and claim that $\frac{t}{x}\,\tau_{i+1} = \tau_i$. Indeed, for $y \neq x$ we obtain $y^{\frac{t}{x}\tau_{i+1}} = y^{\tau_{i+1}} = y^{\frac{t}{x}\tau_i} = y^{\tau_i}$, but in view of $x^{\tau_i} = t^{\tau_i}$ we have also

$$x^{\frac{t}{x}\,\tau_{i+1}} = t^{\tau_{i+1}} = t^{\frac{t}{x}\,\tau_i} = t^{\tau_i} \quad \text{(since } x \notin \mathrm{var}\,t\text{)}$$
$$= x^{\tau_i}.$$

$\frac{t}{x}\,\tau_{i+1} = \tau_i$ and $\sigma_{i+1} = \sigma_i\,\frac{t}{x}$ yield the induction claim

$$\sigma_{i+1}\tau_{i+1} = \sigma_i\,\frac{t}{x}\,\tau_{i+1} = \sigma_i\tau_i = \tau.$$

This completes the proof.  ∎

### Exercises

1. Let $\alpha, \beta$ be prime formulas without shared variables. Show that the properties (i) and (ii) are equivalent:

   (i) $\{\alpha, \beta\}$ is unifiable, (ii) there are substitutions $\sigma, \tau$ with $\alpha^\sigma = \beta^\tau$.

2. Show: $\sigma = \frac{\vec{t}}{\vec{x}}$ is *idempotent* (which is to mean $\sigma^2 = \sigma$) if and only if $x_i \notin \mathrm{var}\,t_j$, for all $i, j$ with $1 \leqslant i, j \leqslant n$.

3. A renaming $\rho$ is termed a *separator* of a pair of clauses $K_0, K_1$ if $\mathrm{var}\,K_0^\rho \cap \mathrm{var}\,K_1 = \emptyset$. Show that if $K_0 \cup K_1$ is unifiable then so is $K_0^\rho \cup K_1$, but not conversely, in general.

4. Assume $\mathsf{s}_1, \mathsf{s}_2 \notin \mathrm{Var}$ for the first distinction letters $\mathsf{s}_1, \mathsf{s}_2$ of the clauses $K_1 \neq K_2$. Show rigorously that $\{K_1, K_2\}$ is not unifiable.

## 4.6  Logic Programming

A rather general starting point in dealing with systems of artificial intelligence consists in using computers to draw consequences $\varphi$ from certain data and facts given in the form of a set of formulas $X$, that is, proving $X \vdash \varphi$ mechanically. That this is possible in theory was the subject of **3.5**. In practice, however, such a project is in general realizable only under certain limitations regarding the pattern of the formulas in $X, \varphi$. These limitations refer to any first-order language $\mathcal{L}$ adapted to the needs of a particular investigation. For logic programming the following restrictions on the set $X$ and the formula $\varphi$ are characteristic:

- $\mathcal{L}$ is identity-free and contains at least one constant symbol,

- each $\alpha \in X$ is a positive universal Horn sentence,

- $\varphi$ is a sentence of the form $\exists \vec{x}(\gamma_0 \wedge \cdots \wedge \gamma_k)$ with prime formulas $\gamma_i$.

Note that $\neg\varphi$ is equivalent to $\forall \vec{x}(\neg\gamma_0 \vee \cdots \vee \neg\gamma_k)$ and hence a negative universal Horn sentence. Because $\forall$-quantifiers can be distributed among conjunctions, we may assume that each $\alpha \in X$ is of the form

$(*) \quad (\beta_1 \wedge \cdots \wedge \beta_m \to \beta)^g \quad (\beta, \beta_1, \ldots, \beta_m \text{ prime formulas}, m \geqslant 0).$

A finite set of sentences of this type is called a *logic program* and will henceforth be denoted by $\mathcal{P}$. The availability of a constant symbol just ensures the existence of a Herbrand model for $\mathcal{P}$. In the programming language PROLOG, $(*)$ is formally written without quantifiers as follows and called a *program clause*:

$$\beta :- \beta_1, \ldots, \beta_m \quad (\text{or just } \beta :- \text{ in case } m = 0).$$

$:-$ symbolizes converse implication mentioned in **1.1**. For $m = 0$ such program clauses are called *facts*, and for $m > 0$ *rules*. In the sequel we make no distinction between a logic program $\mathcal{P}$ as a set of formulas and its transcript in PROLOG. The sentence $\varphi = \exists \vec{x}(\gamma_0 \wedge \cdots \wedge \gamma_k)$ in the last bulleted item above is also called a *query* to $\mathcal{P}$. In PROLOG, $\neg\varphi$ is mostly denoted by $:- \gamma_0, \ldots, \gamma_k.$[4] $\exists \vec{x}$ may be empty. This notation comes from the logical equivalence of the kernel of $\neg\varphi$ $(\equiv \forall \vec{x}(\neg\gamma_0 \vee \cdots \vee \neg\gamma_k))$ to the converse implication $\perp \leftarrow \gamma_0 \wedge \cdots \wedge \gamma_k$, omitting the writing of $\perp$.

Using rules one not only proceeds from given facts to new facts but may also arrive at answers to queries. The restriction as regards the abstinence from $=$ is not really essential. This will become clear in Examples 1 and 4 and in the considerations of this section. Whenever required, $=$ can be treated as an additional binary relation symbol by adjoining the Horn sentences from Example 1 in **4.2**.

Program clauses and negated queries can equally well be written as Horn clauses: $\beta :- \beta_0, \ldots, \beta_m$ as $\{\neg\beta_1, \ldots, \neg\beta_n, \beta\}$, and $:- \gamma_0, \ldots, \gamma_k$ as $\{\neg\gamma_0, \ldots, \neg\gamma_k\}$. For a logic program $\mathcal{P}$, let $\mathcal{P}$ denote the corresponding

---

[4] Sometimes also $?- \gamma_0, \ldots, \gamma_k$. Like many programming languages, PROLOG also has numerous "dialects." We shall therefore not consistently stick to a particular syntax. We also disregard many details, for instance that variables always begin with capital letters and that PROLOG recognizes certain unchanging predicates like *read*, $\ldots$, to provide a convenient user interface.

set of positive Horn clauses. $P$ and $\mathcal{P}$ can almost always be identified. To justify this semantically, let $\mathcal{A} \vDash K$ for a given $\mathcal{L}$-structure $\mathcal{A}$ and $K = \{\lambda_0, \ldots, \lambda_k\}$ simply mean $\mathcal{A} \vDash \bigvee_{i \leqslant k} \lambda_i$, which is clearly equivalent to $\mathcal{A} \vDash (\bigvee_{i \leqslant k} \lambda_i)^g$. Note that $\mathcal{A} \nvDash \Box$, since $\Box$ corresponds to the formula $\bot$. For $\mathcal{L}$-models $\mathcal{M}$, let $\mathcal{M} \vDash K$ have its ordinary meaning $\mathcal{M} \vDash \bigvee_{i \leqslant k} \lambda_i$.

If $\mathcal{A} \vDash K$ for all $K \in \mathcal{K}$, then $\mathcal{A}$ is called a *model* for a given set $\mathcal{K}$ of clauses, and $\mathcal{K}$ is called *satisfiable* or *consistent* if such an $\mathcal{A}$ exists. This is clearly equivalent to the consistency of the set of sentences corresponding to $\mathcal{K}$. Further, let $\mathcal{K} \vDash H$ if every model for $\mathcal{K}$ also satisfies $H$. Evidently $\mathcal{K} \vDash K^\sigma$ for $K \in \mathcal{K}$ and arbitrary substitutions $\sigma$, since $\mathcal{A} \vDash K \Rightarrow \mathcal{A} \vDash K^\sigma$. The clause $K^\sigma$ is also termed an *instance* of $K$, in particular a *ground instance* whenever $K^\sigma$ contains no variables.

A logic program $\mathcal{P}$, considered as a set of positive Horn formulas, is always consistent. All facts and rules of $\mathcal{P}$ are valid in the minimal Herbrand model $\mathcal{C}_\mathcal{P}$. This model should be thought of as the model of a domain of objects about which one wishes to express properties by means of $\mathcal{P}$. A logic program $\mathcal{P}$ is always written such that a real situation is modeled as precisely as possible by the minimal Herbrand model $\mathcal{C}_\mathcal{P}$.

Suppose that $\mathcal{P} \vdash \exists \vec{x} \gamma$, where $\gamma$ is a conjunction of prime formulas as at the beginning, i.e., $\exists \vec{x} \gamma$ is a query. Then a central goal of logic programming is to gain "solutions" of $\mathcal{P} \vdash \exists \vec{x} \gamma$ in $\mathcal{C}_\mathcal{P}$, which by Theorem 2.2 always exist. Here $\gamma \frac{\vec{t}}{\vec{x}}$ is called a *solution* of $\mathcal{P} \vdash \exists \vec{x} \gamma$ whenever $\mathcal{P} \vdash \gamma \frac{\vec{t}}{\vec{x}}$. One also speaks of the solution $\vec{x} := \vec{t}$, or an *answer* to the query $:- \gamma$.

Logic programming follows the strategy of proving $\mathcal{P} \vdash \varphi$ for a query $\varphi$ by establishing the inconsistency of $\mathcal{P}, \neg\varphi$. To verify this we know from Theorem 1.1 that an inconsistency proof of $\mathrm{GI}(\mathcal{P}, \neg\varphi)$ suffices. The resolution theorem shows that for this proof in turn, it suffices to derive the empty clause from the set of clauses $\mathrm{GI}(\mathcal{P}, N)$ corresponding to $\mathrm{GI}(\mathcal{P}, \neg\varphi)$. Here $\mathrm{GI}(\mathcal{K})$ generally denotes the set of all ground instances of members of a set $\mathcal{K}$ of clauses, and $N = \{\neg\gamma_1, \ldots, \neg\gamma_n\}$ is the negative clause corresponding to the query $\varphi$, the so-called *goal clause*.

As a matter of fact, we proceed somewhat more artfully and work not only with ground instances but also with arbitrary instances. Nor does the search for resolutions take place coincidentally or arbitrarily, but rather with the most sparing use of substitutions possible for the purpose of unification. Before the general formulation of Theorem 6.2, we exhibit this method of "unified resolution" by means of two easy examples. In the first of these, sum denotes the ternary relation $\text{graph}_+$ in $\mathbb{N}$.

**Example 1.** Consider the following program $\mathcal{P} = \mathcal{P}_+$ in $\mathcal{L}\{0, \mathsf{S}, \text{sum}\}$:

$$\forall x\, \text{sum}\, x0x \quad ; \quad \forall x \forall y \forall z (\text{sum}\, xyz \to \text{sum}\, x\mathsf{S}y\mathsf{S}z).$$

In PROLOG one may write this program somewhat more briefly as

$$\text{sum}\, x0x :\!- \quad ; \quad \text{sum}\, x\mathsf{S}y\mathsf{S}z :\!- \text{sum}\, xyz.$$

The first program clause is a "fact," the second one is a "rule." The set of Horn clauses that belongs to $\mathcal{P}$ is

$$\mathcal{P} = \{\{\text{sum}\, x0x\}, \{\neg \text{sum}\, xyz, \text{sum}\, x\mathsf{S}y\mathsf{S}z\}\}.$$

$\mathcal{P}$ describes sum $= \text{graph}_+$ in $\mathbb{N}$ together with $0, \mathsf{S}$; more precisely,

$$\mathcal{C}_\mathcal{P} \simeq \mathcal{N} := (\mathbb{N}, 0, \mathsf{S}, \text{sum}\,),$$

that is, $\mathcal{C}_\mathcal{P} \vDash \text{sum}\, \underline{m}\, \underline{n}\, \underline{k} \Leftrightarrow \mathcal{N} \vDash \text{sum}\, \underline{m}\, \underline{n}\, \underline{k}$ ($\Leftrightarrow m + n = k$). This is deduced in Example 3 on page 164, but more directly from Exercise 2 in **4.3**. By replacing therein $p_{m,n,k}$ with sum $\underline{m}\, \underline{n}\, \underline{k}$, the formulas of this exercise correspond precisely to the ground instances of $\mathcal{P}_+$.

Examples of queries to $\mathcal{P}$ are $\exists u \exists v\, \text{sum}\, u\underline{1}v$ and $\exists u\, \text{sum}\, uu\underline{6}$. Another example is sum $\underline{n}\, \underline{2}\, \underline{n+2}$ (here the $\exists$-prefix is empty). For each of these three queries $\varphi$, clearly $\mathcal{C}_\mathcal{P} \vDash \varphi$ holds. Hence, $\mathcal{P} \vdash \varphi$ by the last part of Theorem 2.2. But how can this be confirmed by a computer?

As an illustration, let $\varphi := \exists u \exists v\, \text{sum}\, u\underline{1}v$. Since $\mathcal{P} \vdash \text{sum}\, \underline{n}, \mathsf{S}\underline{n}$ we know that $(u, v) := (\underline{n}, \mathsf{S}\underline{n})$ is a solution of

$$(*) \quad \mathcal{P} \vdash \exists u \exists v\, \text{sum}\, u\underline{1}v.$$

We will show that $\mathcal{P} \vdash \text{sum}\, x\underline{1}\mathsf{S}x$, where $x$ occurs free in the last formula, is the general solution of $(*)$. The inconsistency proof of $\mathcal{P}, \neg\varphi$ results by deriving $\square$ from suitable instances of $\mathcal{P}, N$ that will be constructed by certain substitutions. $N := \{\neg \text{sum}\, u\underline{1}v\}$ is the goal clause corresponding to $\varphi$. The resolution rule is not directly applicable to $\mathcal{P}, N$. But with $\omega_0 := \frac{u}{x}\frac{0}{y}\frac{\mathsf{S}z}{v}$ it is applicable to $P^{\omega_0}, N^{\omega_0}$, with the Horn clause $P := \{\neg \text{sum}\, xyz, \text{sum}\, x\mathsf{S}y\mathsf{S}z\} \in \mathcal{P}$. Indeed, one easily confirms that $P^{\omega_0} = \{\neg \text{sum}\, u0z, \text{sum}\, u\underline{1}\mathsf{S}z\}$ and $N^{\omega_0} = \{\neg \text{sum}\, u\underline{1}\mathsf{S}z\}$. The resolvent of the pair of Horn clauses $P^{\omega_0}, N^{\omega_0}$ is $N_1 := \{\neg \text{sum}\, u0z\}$. This can be stated as follows: Resolution is becoming possible thanks to the unifiability of the clause $\{\text{sum}\, x\mathsf{S}y\mathsf{S}z, \text{sum}\, u\underline{1}v\}$, where sum $x\mathsf{S}y\mathsf{S}z$ belongs to $P$

and $\neg$sum $u\underline{1}v$ to $N$. But we have still to continue to try to get the empty clause. Let $P_1 := \{\text{sum } x0x\} \in \mathcal{P}$. Then $P_1, N_1$ can be brought to resolution by unification with $\omega_1 := \frac{x}{u}\frac{x}{z}$. For notice that $P_1^{\omega_1} = \{\text{sum } x0x\}$ and $N_1^{\omega_1} = \{\neg\text{sum } x0x\}$. Now simply apply RR to this pair of clauses to obtain $\square$. The diagram on the left makes this kind of a description more intuitive. Note that the set braces of the clauses have been omitted in the diagram. This resolution can certainly be produced by

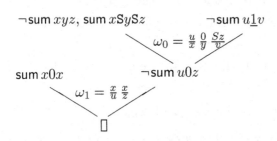

a computer; what the computer has to do is just to look for appropriate unifiers for pairs of clauses. In this way, $(*)$ is proved by Theorem 6.2(a) below. At the same time, by Theorem 6.2(b), applied to the resolution represented by the above diagram, we obtain a solution of $(*)$, namely $(\text{sum } u\underline{1}v)^{\omega_0\omega_1} = \text{sum } x\underline{1}Sx$. This solution is an example of a most general solution of $(*)$, because by substitution we obtain from sum $x\underline{1}Sx$ all individual solutions, namely the sentences sum $\underline{n}\,\underline{1}\,S\underline{n}$.

**Example 2.** The logic program $\mathcal{P} = \{\forall x(\text{hu } x \to \text{mt } x), \text{hu Socr}\}$, written hu Socr ; mt $x :-$ hu $x$ in PROLOG, formalizes the two premises of the old classical Aristotelian syllogism

*All humans are mortal; Socrates is a human. Hence, Socrates is mortal.*

Here $\mathcal{C}_\mathcal{P}$ is just the single-point model $\{\text{Socr}\}$, since Socr is the only constant and no functions occur. The figure on the right shows a resolution of the query $: - \text{ mt } x$ (in words,

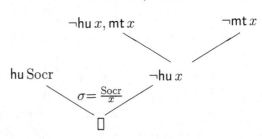

"Is there a mortal $x$ in the Herbrand model $\mathcal{C}_\mathcal{P}$?"), with the solution $x := \text{Socr}$. The familiar logic argument runs as follows: $\forall x(\text{hu } x \to \text{mt } x)$ implies hu Socr $\to$ mt Socr by specification of $x$. Thus, since hu Socr, MP yields mt Socr. We learn from this example among other things that proofs using MP can also be gained by resolution.

Of course, the above examples are far too simple to display the efficiency of logic programming in practice. Here we are interested only in illustrating the methods, which are essentially a combination of resolution and unification. Clearly, these methods concern basically the implementation of PROLOG and may be less interesting to the programmer who cares in the first line about successful programming, whereas the logician cares about the theory behind logical programming.

Following these preliminary considerations we will now generalize our examples and start with the following definition of the rules UR and UHR of *unified resolution* and *unified Horn resolution*, respectively. Therein, $K_0, K_1$ denote clauses and $\omega$ a substitution.

**Definition** Let $U_\omega R(K_0, K_1)$ be the set of all clauses $K$ such that there are clauses $H_0, H_1$ and negation-free clauses $G_0, G_1 \neq \Box$ such that after a possible swapping of the indices $0, 1$,

(a) $K_0 = H_0 \cup G_0$ and $K_1 = H_1 \cup \overline{G_1}$    $(\overline{G_1} = \{\overline{\lambda} \mid \lambda \in G_1\})$,

(b) $\omega$ is a (w.l.o.g. generic) unifier of $G_0 \cup G_1$ and $K = H_0^\omega \cup H_1^\omega$.

$K$ is called a *U-resolvent* of $K_0, K_1$ or an *application of the rule UR to* $K_0, K_1$ if $K \in U_\omega R(K_0^\rho, K_1)$ for some $\omega$ and some separator $\rho$ of $K_0, K_1$. The restriction of UR to Horn clauses $K_0, K_1$ ($K_0$ positive, $K_1$ negative) is denoted by UHR and $U_\omega R(K_0, K_1)$ by $U_\omega HR(K_0, K_1)$. The resolvent $K$ is then termed a *UH-resolvent* of $K_0, K_1$.

This definition becomes more lucid by some additional explanations. According to (b), $G_0^\omega = \{\pi\} = G_1^\omega$ for some prime formula $\pi$. Hence $K$ results from applying standard resolution on suitable premises. Applying UR or UHR to $K_0, K_1$ always includes a choice of $\omega$ and $\rho$ ($\rho$ may enlarge the set of resolvents, see Exercise 3 in **4.5**). In the examples we used UHR. In the first resolution step of Example 1 is $\neg\mathsf{sum}\, u0z \in U_{\omega_0}HR(P^\rho, N)$ (with $\rho = \iota$). The splitting of $K_0$ and $K_1$ according (a) above reads $H_0 = \{\neg\mathsf{sum}\, xyz\}$, $G_0 = \{\mathsf{sum}\, x\mathsf{S}y\mathsf{S}z\}$, and $H_1 = \emptyset$, $G_1 = \{\mathsf{sum}\, u\underline{1}v\}$. UHR was used again in the second resolution step, as well as in Example 2, strictly following the instruction of the above definition.

We write $\mathcal{K} \vdash^{UR} H$ if $H$ is derivable from the set of clauses $\mathcal{K}$ using UR. Accordingly, let $\mathcal{K} \vdash^{UHR} H$ be defined for sets of Horn clauses $\mathcal{K}$, where only UHR is used. As in **4.3**, derivations in $\vdash^{UR}$ or $\vdash^{UHR}$ can be visualized by means of trees. A (successful) *U-Resolution* for $\mathcal{K}$ is just a $U$-resolution tree with leaves in $\mathcal{K}$ and the root $\square$, where the applied substitutions are tied to the resolution nodes as in the diagram below.

A *UH-resolution* is defined similarly; it may as well be regarded as a sequence $(P_i^{\rho_i}, N_i, \omega_i)_{i \leqslant \ell}$ with $N_{i+1} \in U_{\omega_i} HR(P_i^{\rho_i}, N_i)$ for $i < \ell$ and $\square \in U_{\omega_\ell} HR(P_\ell^{\rho_\ell}, N_\ell)$. If $\mathcal{P}$ is a set of positive clauses and $N$ a negative clause, and if further $P_i \in \mathcal{P}$ holds for all $i \leqslant \ell$ and $N_0 = N$, one speaks of a *UH-resolution for* $\mathcal{P}, N$. In general, $\mathcal{P}$ consists of the clauses of some logic

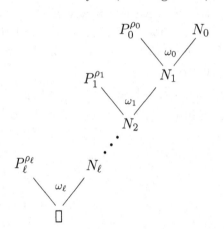

program and $N$ is any given goal clause. In place of *UH-resolution* one may also speak of *SLD-resolution* (**L**inear resolution with **S**election function for **D**efinite clauses). This name has nothing to do with some special strategy for searching a successful resolution, implemented in PROLOG. For details on this matter see for instance

[Ll]. The diagram illustrates a *UH*-resolution $(P_i^{\rho_i}, N_i, \omega_i)_{i \leqslant \ell}$ for $\mathcal{P}, N$. It generalizes the diagrams in Examples 1 and 2, which represent particularly simple examples of *UH*-resolutions.

First of all we prove the soundness of the calculus $\vdash^{UR}$ in Lemma 6.1. This clearly covers also the calculus $\vdash^{UHR}$ of unified Horn resolution, where one has to do with special clauses only.

**Lemma 6.1 (Soundness lemma).** $\mathcal{K} \vdash^{UR} H$ *implies* $\mathcal{K} \vDash H$.

**Proof.** It suffices to show that $K_0, K_1 \vDash H$ if $H$ is a $U$-resolvent of $K_0, K_1$. Let $H \in U_\omega R(K_0^\rho, K_1)$, $K_0^\rho = H_0 \cup G_0$, $K_1 = H_1 \cup \overline{G_1}$, $G_0^\omega = \{\pi\} = G_1^\omega$, $H = H_0^{\rho\omega} \cup H_1^\omega$, and $\mathcal{A} \vDash K_0, K_1$. Then $\mathcal{A} \vDash K_0^{\rho\omega}, K_1^\omega$ as well. Further, let $w \colon \mathrm{Var} \to A$, with $\mathcal{M} := (\mathcal{A}, w) \vDash K_0^{\rho\omega} = H_0^{\rho\omega} \cup \{\pi\}$, $\mathcal{M} \vDash H_1^\omega \cup \{\neg\pi\}$. If $\mathcal{M} \nvDash \pi$ then evidently $\mathcal{M} \vDash H_0^{\rho\omega}$. Otherwise $\mathcal{M} \nvDash \neg\pi$, hence $\mathcal{M} \vDash H_1^\omega$. So $\mathcal{M} \vDash H_0^\omega \cup H_1^\omega = H$ in any case. This states that $\mathcal{A} \vDash H$, because $w$ was arbitrary. $\square$

With respect to the calculus $\vdash^{UHR}$ this lemma serves the proof of (a) in

**Theorem 6.2 (Main theorem of logic programming).** *Let $\mathcal{P}$ be a logic program, $\exists \vec{x} \gamma$ a query, $\gamma = \gamma_0 \wedge \cdots \wedge \gamma_k$, and $N = \{\neg\gamma_0, \ldots, \neg\gamma_k\}$. Then the following hold:*

(a) $\mathcal{P} \vdash \exists \vec{x} \gamma$ *iff* $\mathcal{P}, N \vdash^{UHR} \square$ *(Adequacy).*

(b) *If* $(P_i^{\rho_i}, N_i, \omega_i)_{i \leqslant \ell}$ *is a UH-resolution for $\mathcal{P}, N$ and $\omega := \omega_0 \cdots \omega_\ell$, then* $\mathcal{P} \vdash \gamma^\omega$ *(Solution soundness).*

(c) *Suppose that $\mathcal{P} \vdash \gamma\frac{\vec{t}}{\vec{x}}$ $(\vec{t} \in \mathcal{T}_0^n)$. Then there is some UH-resolution $(K_i^{\rho_i}, N_i, \omega_i)_{i \leqslant \ell}$ and some $\sigma$ such that $x_i^{\omega\sigma} = t_i$ for $i = 1, \ldots, n$, where $\omega = \omega_0 \cdots \omega_\ell$ (Solution completeness).*

The proof of this theorem is undertaken in **4.7**. It has a highly technical character and uses a substantial amount of substitutions. Here are just a few comments. In view of $\neg\exists \vec{x} \gamma \equiv \forall \vec{x} \neg\gamma$ it is clear that

(∗) $\quad \mathcal{P} \vdash \exists \vec{x} \gamma$

is equivalent to the inconsistency of $\mathcal{P}, \forall \vec{x} \neg\gamma$, hence also to that of the corresponding set of Horn clauses $\mathcal{P}, N$. Theorem 6.2(a) states that this is equivalent to $\mathcal{P}, N \vdash^{UHR} \square$, which is not obvious. (b) tells us how to achieve a solution of (∗) by a successful resolution. Since $\gamma^\omega$ in (b) may still contain free variables (like $(\mathsf{sum}\, u\underline{1}v)^\omega$ for $\omega = \omega_1\omega_2$ in Example 1) and since $\mathcal{P} \vdash \gamma^\omega \Rightarrow \mathcal{P} \vdash \gamma^{\omega\tau}$ for any $\tau$, we often obtain whole families of solutions of (∗) in the Herbrand model $\mathcal{C}_\mathcal{P}$ by substituting ground terms. By (c), all solutions in $\mathcal{C}_\mathcal{P}$ are gained in this way, though not always with a single generic resolution as in Example 1. However, the theorem makes no claim as to whether and under what circumstances (∗) is solvable.

Logic programming is also expedient for purely theoretical purposes. For instance, it can be used to make the notion of computable functions on $\mathbb{N}$ entirely precise. The definition below provides just one of several similarly styled, intuitively illuminating possibilities. We will construct an undecidable problem in Theorem 6.3 below that explains the principal difficulties surrounding a general answer to the question $\mathcal{P} \vdash \exists \vec{x} \gamma$. Because in **6.1** computable functions are equated with recursive functions, we keep things fairly brief here.

**Definition.** $f \colon \mathbb{N}^n \to \mathbb{N}$ is called *computable* if there is a logic program $\mathcal{P}_f$ in a language that, in addition to 0 and $\mathsf{S}$, contains only relation symbols, including an $(n+1)$-ary relation symbol denoted by $r_f$ (to mean graph $f$), such that for all $\vec{k} = (k_1, \ldots, k_n)$ and $m$ the following is satisfied:

(1)   $\mathcal{P}_f \vdash r_f \underline{\vec{k}}\,\underline{m} \;\Leftrightarrow\; f\vec{k} = m$     $(\vec{k} = (\underline{k_1}, \ldots, \underline{k_n}))$.[5]

A function $f: \mathbb{N}^n \to \mathbb{N}$ satisfying (1) is certainly computable in the intuitive sense: a deduction machine is set to list all formulas provable from $\mathcal{P}_f$, and one simply has to wait until a sentence $r\vec{k}\,\underline{m}$ appears. Then the value $m = f\vec{k}$ is computed. By Theorem 6.2(a), the left-hand side of (1) is for $\mathcal{P} = \mathcal{P}_f$ equivalent to $\mathcal{P}, \{\neg r\underline{\vec{k}}\,\underline{m}\} \vdash^{UHR} \Box$. Therefore, $f$ is basically also computable with the Horn resolution calculus.

The domain of the Herbrand model $\mathcal{C}_{\mathcal{P}_f}$ is $\mathbb{N}$, and by Theorem 2.2,

$$\mathcal{P}_f \vdash r_f \underline{\vec{k}}\,\underline{m} \;\Leftrightarrow\; \mathcal{C}_{\mathcal{P}_f} \vDash r_f \underline{\vec{k}}\,\underline{m},$$

so that (1) holds when just the following claim has been proved:

(2)   $\mathcal{C}_{\mathcal{P}_f} \vDash r_f \underline{\vec{k}}\,\underline{m} \;\Leftrightarrow\; f\vec{k} = m$, for all $\vec{k}, m$.

**Example 3.** $\mathcal{P}_+$ in Example 1 computes $+$, more precisely graph$_+$, since $\mathcal{C}_{\mathcal{P}_+} \vDash \mathsf{sum}\,\underline{k}\,\underline{n}\,\underline{m} \Leftrightarrow k+n = m$ was shown there. So (2) holds and hence also (1). A logic program $\mathcal{P}_\times$ for computing $\mathsf{prd} := \mathsf{graph}\cdot$ arises from $\mathcal{P}_+$ by expanding the language to $\mathcal{L}\{0, \mathsf{S}, \mathsf{sum}, \mathsf{prd}\,\}$ and adding to $\mathcal{P}_+$ the program clauses $\mathsf{prd}\,x00 :-$ and $\mathsf{prd}\,x\mathsf{S}yu :- \mathsf{prd}\,xyz, \mathsf{sum}\,zxu$. Here we see that besides the graph of the target function some additional relations may be involved in the function's computation.

**Example 4.** The program $\mathcal{P}_\mathsf{S}$ in $\mathcal{L}\{0, \mathsf{S}, r_\mathsf{S}\,\}$, containing only $r_\mathsf{S}\,x\mathsf{S}x :-$ , computes the graph $r_\mathsf{S}$ of the successor function. Clearly, $\mathcal{P}_\mathsf{S} \vdash r_\mathsf{S}\,\underline{n}\mathsf{S}\underline{n}$, since $\mathcal{P}_\mathsf{S}, \{\neg r_\mathsf{S}\,\underline{n}\mathsf{S}\underline{n}\} \vdash^{UHR} \Box$ ($\Box$ is a resolvent of $\{r_\mathsf{S}\,x\mathsf{S}x\}^\sigma$ and $\{\neg r_\mathsf{S}\,\underline{n}\mathsf{S}\underline{n}\}^\sigma$, where $\sigma$ equals $\frac{\mathsf{S}^n 0}{x}$). Let $m \neq \mathsf{S}n$. Then $(\mathbb{N}, 0, \mathsf{S}, \mathsf{graph}\,\mathsf{S}) \vDash \mathcal{P}_\mathsf{S}, \neg r_\mathsf{S}\,\underline{n}\,\underline{m}$. Hence, $\mathcal{P}_\mathsf{S} \nvdash r_\mathsf{S}\,\underline{n}\,\underline{m}$. This confirms (1).

It is not difficult to recognize that each recursive function $f$ can be computed by a logic program $\mathcal{P}_f$ in the above sense in a language that in addition to some relation symbols contains only the operation symbols $0, \mathsf{S}$. Exercises 1, 2, and 3 provide the main steps in the proof, which proceeds by induction on the generating operations $\boldsymbol{Oc}$, $\boldsymbol{Op}$, and $\boldsymbol{O\mu}$ of recursive functions from **6.1**. Example 4 confirms the induction initiation for the initial primitive recursive function $\mathsf{S}$. The interested reader should study **6.1** to some extent in order to understand what is going on.

---

[5] By grounding computability in different terms one could $f$ provisionally call *LP-computable*. Our definition is related to the Herbrand–Gödel definition of computable functions but we will not step into further details in this respect.

Thus, the concept of logic programming is very comprehensive. On the other hand, this has the consequence that the question $\mathcal{P} \vdash \exists \vec{x}\gamma$ is, in general, not effectively decidable. Indeed, this undecidability is the assertion of our next theorem.

**Theorem 6.3.** *A logic program $\mathcal{P}$ exists whose signature contains at least a binary relation symbol $r$, but no operation symbols other than $0, \mathsf{S}$, such that no algorithm answers the question $\mathcal{P} \vdash \exists x\, r x \underline{k}$ for each $k$.*

**Proof.** Let $f : \mathbb{N} \to \mathbb{N}$ be recursive, but $\operatorname{ran} f = \{m \in \mathbb{N} \mid \exists k\, f k = m\}$ not recursive. Such a function $f$ exists; see Exercise 5 in **6.5**. Then we get for $\mathcal{P} := \mathcal{P}_f$ from the definition page 163,

$$
\begin{aligned}
\mathcal{P} \vdash \exists x\, r x \underline{m} &\Leftrightarrow \mathcal{C}_\mathcal{P} \vDash \exists x\, r x \underline{m} && (\text{Theorem 2.2, } r \text{ means } r_f) \\
&\Leftrightarrow \mathcal{C}_\mathcal{P} \vDash r \underline{k}\,\underline{m} \text{ for some } k && (\mathcal{C}_\mathcal{P} \text{ has the domain } \mathbb{N}) \\
&\Leftrightarrow f k = m \text{ for some } k && (\text{by } (2)) \\
&\Leftrightarrow m \in \operatorname{ran} f.
\end{aligned}
$$

Thus, if the question $\mathcal{P} \vdash \exists x\, r x \underline{m}$ were decidable then so too would be the question $m \in \operatorname{ran} f$, and this is a contradiction to the choice of $f$. $\quad\blacksquare$

### Exercises

1. Let $g : \mathbb{N}^n \to \mathbb{N}$ and $h : \mathbb{N}^{n+2} \to \mathbb{N}$ be computable by means of the logic programs $\mathcal{P}_g$ and $\mathcal{P}_h$, and let $f : \mathbb{N}^{n+1} \to \mathbb{N}$ arise from $g, h$ by primitive recursion. This is to mean that

   $$ f(\vec{a}, 0) = g\vec{a} \text{ and } f(\vec{a}, k+1) = h(\vec{a}, k, f(\vec{a}, k)) \text{ for all } \vec{a} \in \mathbb{N}^n. $$

   Provide a logic program for computing (the graph of) $f$.

2. Let $\mathcal{P}_h$ and $\mathcal{P}_{g_i}$ be logic programs for computing $h : \mathbb{N}^m \to \mathbb{N}$ and $g_i : \mathbb{N}^n \to \mathbb{N}$ ($i = 1, \ldots, m$). Further, let the function $f$ be defined by $f\vec{a} = h(g_1\vec{a}, \ldots, g_m\vec{a})$ for all $\vec{a} \in \mathbb{N}^n$. Give a logic program for computing $f$.

3. Let $g : \mathbb{N}^{n+1} \to \mathbb{N}$ and $\mathcal{P}_{g_i}$ be logic program for computing $g$. Further assume that to each $\vec{a} \in \mathbb{N}^n$ there is some $b \in \mathbb{N}$ with $g(\vec{a}, b) = 0$ and let $h\vec{a}$ for $\vec{a} \in \mathbb{N}^n$ be the smallest $m$ such that $g(\vec{a}, m) = 0$. Give a logic program for computing $h : \mathbb{N}^n \to \mathbb{N}$.

   These exercises and the examples show that the *LP*-computable functions coincide with the general recursive functions.

## 4.7   A Proof of the Main Theorem

While we actually require the following lemmas and Theorem 7.3 below only for the unified Horn resolution, the proofs are carried out here for the more general U-resolution. These proofs are not essentially more difficult. As a matter of fact, the only difficulty in the proofs is the handling of substitutions. The calculi $\vdash^{RR}$ and $\vdash^{HR}$ from **4.3** now operate with variable-free clauses of a fixed identity-free first-order language with at least one constant. The presence of constants assures that variable-free clauses are available. $\rho, \sigma, \tau$ denote simultaneous substitutions throughout.

**Lemma 7.1.** *Let $K_0, K_1$ be clauses with separator $\rho$ and let $K_0^{\sigma_0}$, $K_1^{\sigma_1}$ be variable-free. Suppose that $K$ is a resolvent of $K_0^{\sigma_0}, K_1^{\sigma_1}$. Then there are substitutions $\omega, \tau$ and some $H \in U_\omega R(K_0^\rho, K_1)$ such that $H^\tau = K$, i.e., $K$ is a ground instance of some U-resolvent of $K_0, K_1$. Further, for a given finite set $V$ of variables, $\omega, \tau$ can be selected in such a way that $x^{\omega\tau} = x^{\sigma_1}$ for all $x \in V$. The same holds for Horn resolution.*

**Proof.** Suppose w.l.o.g. that $K_0^{\sigma_0} = L_0, \pi$ and $K_1^{\sigma_1} = L_1, \neg\pi$ for some prime formula $\pi$, and $K = L_0 \cup L_1$. Put $H_i := \{\alpha \in K_i \mid \alpha^{\sigma_i} \in L_i\}$, $G_0 := \{\alpha \in K_0 \mid \alpha^{\sigma_0} = \pi\}$, and $G_1 := \{\beta \in \overline{K_1} \mid \beta^{\sigma_1} = \pi\}$, $i = 0, 1$. Then $K_0 = H_0 \cup G_0$, $K_1 = H_1 \cup \overline{G_1}$, $H_i^{\sigma_i} = L_i$, $G_i^{\sigma_i} = \{\pi\}$. Let $\rho$ be a separator of $K_0, K_1$ and define $\sigma$ by $x^\sigma = x^{\rho\sigma_0}$ in case $x \in \text{var}\,K_0^\rho$, and $x^\sigma = x^{\sigma_1}$ else, so that e.g. $K_1^\sigma = K_1^{\sigma_1}$. Note that also $\rho\sigma = \rho\rho\sigma_0 = \sigma_0$, hence $K_0^{\rho\sigma} = K_0^{\sigma_0}$. Therefore $(G_0^\rho \cup G_1)^\sigma = G_0^{\rho\sigma} \cup G_1^\sigma = G_0^{\sigma_0} \cup G_1^{\sigma_1} = \{\pi\}$, that is, $\sigma$ unifies $G_0^\rho \cup G_1$. Let $\omega$ be a generic unifier of this clause, so that $\sigma = \omega\tau$ for suitable $\tau$. Then $H := H_0^{\rho\omega} \cup H_1^\omega \in U_\omega R(K_0^\rho, K_1)$ by definition of the rule UR, and $H_0^{\rho\sigma} = H_0^{\sigma_0}$, $H_1^\sigma = H_1^{\sigma_1}$ yield the desired

$$H^\tau = H_0^{\rho\omega\tau} \cup H_1^{\omega\tau} = H_0^{\rho\sigma} \cup H_1^\sigma = H_0^{\sigma_0} \cup H_1^{\sigma_1} = L_0 \cup L_1 = K.$$

The second part of our lemma is easily confirmed. Since $V$ is finite, $\rho$ can clearly be chosen such that $V \cap \text{var}\,K_0^\rho = \emptyset$, hence $x^\sigma = x^{\sigma_1}$ for all $x \in V$. By definition of $\sigma$ and in view of $\omega\tau = \sigma$ it then obviously follows that $x^{\omega\tau} = x^{\sigma_1}$ whenever $x \in V$. $\quad\blacksquare$

The next lemma claims that if the $\square$ is derivable from $GI(\mathcal{K})$ with resolution only (i.e., without unification and separation), then $\square$ is also directly derivable from $\mathcal{K}$ in the calculi $\vdash^{UR}$ and $\vdash^{UHR}$, respectively.

**Lemma 7.2 (Lifting lemma).** *Suppose that* $\mathrm{GI}(\mathcal{K}) \vdash^{RR} \square$ *for some set of clauses* $\mathcal{K}$. *Then* $\mathcal{K} \vdash^{UR} \square$. *And if* $\mathcal{K}$ *consists of Horn clauses only, then also* $\mathcal{K} \vdash^{UHR} \square$.

**Proof.** We shall verify the more general claim

(∗) If $\mathrm{GI}(\mathcal{K}) \vdash^{RR} K$ then $\mathcal{K} \vdash^{UR} H$ and $K = H^\tau$ for some $H$ and $\tau$.

For $K = \square$, (∗) is our claim since $\square^\tau = \square$. (∗) follows straightforwardly by induction on $\mathrm{GI}(\mathcal{K}) \vdash^{RR} K$. It is trivial for $K \in \mathrm{GI}(\mathcal{K})$, by definition of $\mathrm{GI}(\mathcal{K})$. For the inductive step let $\mathrm{GI}(\mathcal{K}) \vdash^{RR} K_0^{\sigma_0}, K_1^{\sigma_1}$, with $K_0, K_1 \in \mathcal{K}$ for suitable $\sigma_0, \sigma_1$ according to the induction hypotheses, and let $K$ be a resolvent of $K_0^{\sigma_0}, K_1^{\sigma_1}$. That then $H \in U_\omega R(K_0^\rho, K_1)$ and $K = H^\tau$ for suitable $H, \omega, \tau$ is exactly the first claim of Lemma 7.1. This proves (∗). The case for Horn clauses is completely similar. $\square$

**Theorem 7.3 (U-Resolution theorem).** *A set of clauses* $\mathcal{K}$ *is inconsistent iff* $\mathcal{K} \vdash^{UR} \square$; *a set of Horn clauses* $\mathcal{K}$ *is inconsistent iff* $\mathcal{K} \vdash^{UHR} \square$.

**Proof.** If $\mathcal{K} \vdash^{UR} \square$ then $\mathcal{K} \vDash \square$ by Lemma 6.1; hence $\mathcal{K}$ is inconsistent. Suppose now the latter, so that the set $U$ of ∀-sentences corresponding to $\mathcal{K}$ is inconsistent as well. Then $\mathrm{GI}(U)$ is inconsistent according to Theorem 1.1; hence $\mathrm{GI}(\mathcal{K})$ as well. Thus, $\mathrm{GI}(\mathcal{K}) \vdash^{RR} \square$ by Theorem 3.2 and so $\mathcal{K} \vdash^{UR} \square$ by the Lifting lemma. For sets of Horn clauses the proof runs analogously, using Theorem 4.2 instead of Theorem 3.2. $\square$

**Proof of Theorem 6.2.** (a): $\mathcal{P} \vdash \exists \vec{x} \gamma$ is equivalent to the inconsistency of $\mathcal{P}, \forall \vec{x} \neg \gamma$ or of $\mathcal{P}, N$. But the inconsistency of $\mathcal{P}, N$ is, by the U-Resolution theorem, precisely the same as saying $\mathcal{P}, N \vdash^{UHR} \square$.

(b): Proof by induction on the length $\ell$ of a (successful) $UH$-resolution $(P_i^{\rho_i}, N_i, \omega_i)_{i \leqslant \ell}$ for $\mathcal{P}, N$. Let $\ell = 0$, so that $\square \in U_\omega HR(P_0^\rho, N)$ for suitable $\rho, \omega \ (= \omega_0)$, and $P_0 \in \mathcal{P}$. Then $\omega$ unifies $P_0^\rho \cup N$, i.e., $P_0^{\rho\omega} = \{\pi\} = \gamma_i^\omega$ for some prime formula $\pi$ and all $i \leqslant k$. Hence trivially $\mathcal{P}_0^{\rho\omega} \vdash \gamma_i^\omega \ (= \pi)$ for each $i \leqslant k$, and so $\mathcal{P} \vdash \gamma_0^\omega \wedge \cdots \wedge \gamma_k^\omega = \gamma^\omega$ as claimed in Theorem 6.2(b) for the case $\ell = 0$. Now let $\ell > 0$. Then $(P_i^{\rho_i}, N_i, \omega_i)_{1 \leqslant i \leqslant \ell}$ is a $UH$-resolution for $\mathcal{P}, N_1$ as well. By the induction hypothesis,

(1)    $\mathcal{P} \vdash \alpha^{\omega_1 \cdots \omega_\ell}$ whenever $\neg \alpha \in N_1$.

It suffices to verify $\mathcal{P} \vdash \gamma_i^\omega$ for all $i \leqslant k$ with $\omega = \omega_0 \ldots \omega_n$ in agreement with the notation in Theorem 6.2. To this end we distinguish two cases for given $i$: if $\neg \gamma_i^{\omega_0} \in N_1$ then $\mathcal{P} \vdash (\gamma_i^{\omega_0})^{\omega_1 \cdots \omega_\ell}$ by (1), hence $\mathcal{P} \vdash \gamma_i^\omega$. Now suppose $\neg\gamma_i^{\omega_0} \notin N_1$. Then $\gamma_i^{\omega_0}$ disappears in the resolution step from $P_0^{\rho_0}, N_0 \,(= N)$ to $N_1$. So $P_0$ takes the form $P_0 = \{\neg\beta_1, \ldots, \neg\beta_m, \beta\}$, where $\beta^{\rho_0\omega_0} = \gamma_i^{\omega_0}$ and $\neg\beta_j^{\rho_0\omega_0} \in N_1$ for $j = 1, \ldots, m$. Thus (1) evidently yields $\mathcal{P} \vdash (\beta_j^{\rho_0\omega_0})^{\omega_1\cdots\omega_\ell}$ and therefore $\mathcal{P} \vdash \bigwedge_{j=1}^m \beta_j^{\rho_0\omega}$. At the same time, it holds $\mathcal{P} \vdash \bigwedge_{j=1}^m \beta_j^{\rho_0\omega} \to \beta^{\rho_0\omega}$ because of $\mathcal{P} \vdash \neg\beta_1^{\rho_0\omega} \vee \ldots \vee \neg\beta_m^{\rho_0\omega} \vee \beta^{\rho_0\omega}$ (the latter holds since $P_0 = \{\neg\beta_1, \ldots, \neg\beta_m, \beta\}$). Using MP we then obtain $\mathcal{P} \vdash \beta^{\rho_0\omega}$. From $\beta^{\rho_0\omega_0} = \gamma_i^{\omega_0}$, applying $\omega_1 \cdots \omega_\ell$ on both sides, we obtain $\beta^{\rho_0\omega} = \gamma_i^{\omega}$. Hence $\mathcal{P} \vdash \gamma_i^{\omega}$ also in the second case. Thus, $\beta^{\rho_0\omega} = \gamma_i^{\omega}$ for all $i \leqslant n$, independent on our case distinction. This proves (b).

(c): Let $\mathcal{P} \vdash \gamma^\tau$ with $\tau := \frac{\vec{t}}{\vec{x}}$, and $\mathcal{P}' = \mathrm{GI}(\mathcal{P})$. Then $\mathcal{P}, \neg\gamma^\tau$ is inconsistent, and so too is $\mathcal{P}', \neg\gamma^\tau$ by Theorem 1.1 (consider $\mathrm{GI}(\neg\gamma^\tau) = \{\neg\gamma^\tau\}$). According to the $H$-resolution theorem 4.2, there is some $H$-resolution $\boldsymbol{B} = (P_i', Q_i)_{i \leqslant \ell}$ for $\mathcal{P}', N^\tau$ with $Q_0 = N^\tau$. Here let, say, $P_i' = P_i^{\sigma_i}$ for appropriate clauses $P_i \in \mathcal{P}$ and $\sigma_i$. From this facts we will derive

(2) *for finite $V \subseteq \mathrm{Var}$ there exist $\rho_i, N_i, \omega_i, \tau$ such that $(P_i^{\rho_i}, N_i, \omega_i)_{i \leqslant \ell}$ is a UH-resolution for $\mathcal{P}, N$. Moreover, $x^{\omega\tau} = x^\sigma$ for $\omega := \omega_0 \cdots \omega_\ell$ and all $x \in V$.*

This completes our reasoning, because (2) yields (for $V = \{x_1, \ldots, x_n\}$) $x_i^{\omega\tau} = x_i^\sigma = t_i$ for $i = 1, \ldots, n$, whence (c). For the inductive proof of (2) look at the first resolution step $Q_1 = HR(P_0', Q_0)$ in $\boldsymbol{B}$, with $P_0' = P_0^{\sigma_0}$, $Q_0 = N^{\sigma_1}$, $\sigma_1 := \tau$. By Lemma 7.1 with $K_0 := P_0$, $K_1 := N_0 := N$, we choose $\omega_0, \rho_0, \tau_0, H$ such that $H \in U_\omega HR(P_0^{\rho_0}, N_0)$ and $H^{\tau_0} = Q_1$, as well as $x^{\omega_0\tau_0} = x^\sigma$ for all $x \in V$. If $\ell = 0$, that is, if $Q_1 = \Box$, then also $H = \Box$ and (2) is proved with $\tau = \tau_0$. Now suppose $\ell > 0$. For the $H$-resolution $(P_i', Q_i)_{1 \leqslant i \leqslant \ell}$ for $\mathcal{P}', Q_1$ and for $V' := \mathrm{var}\{x^{\omega_0} \mid x \in V\}$ there exist by the induction hypothesis $\rho_i, N_i, \omega_i$ for $i = 1, \ldots, \ell$ and some $\tau$, such that $(P_i^{\rho_i}, N_i, \omega_i)_{1 \leqslant i \leqslant \ell}$ is a UH-resolution for $\mathcal{P}, H$ and simultaneously $y^{\omega_1\cdots\omega_\ell\tau} = y^{\tau_0}$ for all $y \in V'$ (instead of $Q_0 = N^\sigma$ we have now to consider $Q_1 = H^{\tau_0}$). Because of $\mathrm{var}\, x^{\omega_0} \subseteq V'$ and $x^{\omega_0\tau_0} = x^\sigma$ for $x \in V$ we get

(3) $\quad x^{\omega\tau} = (x^{\omega_0})^{\omega_1\cdots\omega_\ell\tau} = x^{\omega_0\tau_0} = x^\sigma$, for all $x \in V$.

$(P_i^{\rho_i}, N_i, \omega_i)_{i \leqslant \ell}$ is certainly a UH-resolution. Moreover, by virtue of (3), and by choosing $V = \{x_1, \ldots, x_n\}$, it holds $x_i^{\omega\tau} = x_i^\sigma = t_i$ for $i = 1, \ldots, n$. This proves (2), hence (c), and completes the proof of Theorem 6.2.

# Chapter 5
# Elements of Model Theory

Model theory is a main branch of applied mathematical logic. Here the techniques developed in mathematical logic are combined with construction methods of other areas (such as algebra and analysis) to their mutual benefit. The following demonstrations can provide only a first glimpse in this respect, a deeper understanding being gained, for instance, from [CK] or [Ho]. For further-ranging topics, such as saturated models, stability theory, and the model theory of languages other than first-order, we refer to the special literature, [Bue], [Mar], [Pz], [Rot], [Sa], [She].

The theorems of Löwenheim and Skolem were first formulated in the generality given in **5.1** by Tarski. These and the compactness theorem form the basis of model theory, a now wide-ranging discipline that arose around 1950. Key concepts of model theory are elementary equivalence and elementary extension. These not only are interesting in themselves but also have multiple applications to model constructions in set theory, nonstandard analysis, algebra, geometry and elsewhere.

Complete axiomatizable theories are decidable; see **3.5**. The question of decidability and completeness of mathematical theories and the development of well-honed methods that solve these questions have always been a driving force for the further development of mathematical logic. Of the numerous methods, we introduce here the most important: Vaught's test, Ehrenfeucht's game, Robinson's method of model completeness, and quantifier elimination. For more involved cases, such as the theories of algebraically closed and real closed fields, model-theoretic criteria are developed and applied. For a complete understanding of the material in **5.5** the reader should to some extent be familiar with some basic algebraic constructions, mainly concerning the theory of fields.

W. Rautenberg, *A Concise Introduction to Mathematical Logic*,
Universitext, DOI 10.1007/978-1-4419-1221-3_5,
© Springer Science+Business Media, LLC 2010

## 5.1 Elementary Extensions

In **3.3** nonstandard models were obtained using a method that we now generalize. For given $\mathcal{L}$ and a set $A$ let $\mathcal{L}A$ denote the language resulting from $\mathcal{L}$ by adjoining new constant symbols $\boldsymbol{a}$ for all $a \in A$. The symbol $\boldsymbol{a}$ should depend only on $a$, not on $A$, so that $\mathcal{L}A \subseteq \mathcal{L}B$ whenever $A \subseteq B$. To simplify notation we shall write from Theorem 1.3 onward just $a$ rather than $\boldsymbol{a}$; there will be no risk of misunderstanding.

Let $\mathcal{B}$ be an $\mathcal{L}$-structure and $A \subseteq B$ (the domain of $\mathcal{B}$). Then the $\mathcal{L}A$-expansion in which $\boldsymbol{a}$ is interpreted by $a \in A$ will be denoted by $\mathcal{B}_A$. According to Exercise 3 in **2.3** we have for arbitrary $\alpha = \alpha(\vec{x}) \in \mathcal{L}$ and arbitrary $\vec{a} \in A^n$,

(1) $\quad \mathcal{B} \vDash \alpha\,[\vec{a}] \Leftrightarrow \mathcal{B}_A \vDash \alpha(\vec{a}) \quad (\alpha(\vec{a}) := \alpha \frac{a_1}{x_1} \cdots \frac{a_n}{x_n})$.

It is important to notice that every sentence from $\mathcal{L}A$ is of the form $\alpha(\vec{a})$ for suitable $\alpha(\vec{x}) \in \mathcal{L}$ and $\vec{a} \in A^n$. Instead of $\mathcal{B}_A \vDash \alpha(\vec{a})$ (which is equivalent to $\mathcal{B} \vDash \alpha\,[\vec{a}]$) we later will write just $\mathcal{B}_A \vDash \alpha(\vec{a})$ or even $\mathcal{B} \vDash \alpha(\vec{a})$, as in Theorem 1.3. Thus, $\mathcal{B}$ may also denote a constant expansion of $\mathcal{B}$ if it is not the distinction that is to be emphasized. This notation is somewhat sloppy but points up the ideas behind the constructions.

Note that for an $\mathcal{L}$-structure $\mathcal{A}$, the $\mathcal{L}A$-expansion $\mathcal{A}_A$ receives a new constant symbol for *every* $a \in A$, even if some elements of $\mathcal{A}$ already possess names in $\mathcal{L}$. The set of all variable-free literals $\lambda \in \mathcal{L}A$ such that $\mathcal{A}_A \vDash \lambda$ is called the *diagram of* $\mathcal{A}$, denoted by $D\mathcal{A}$. For instance, $D(\mathbb{R}, <)$ contains for all $a, b \in \mathbb{R}$ the literals $\boldsymbol{a} = \boldsymbol{b}$, $\boldsymbol{a} \neq \boldsymbol{b}$, $\boldsymbol{a} < \boldsymbol{b}$, or $\boldsymbol{a} \nless \boldsymbol{b}$, depending on whether indeed $a{=}b$, $a{\neq}b$, $a{<}b$, or $a{\nless}b$ for the reals $a, b$. Diagrams are important for various constructions in model theory.

The notion of an embedding $\imath : \mathcal{A} \to \mathcal{B}$ as defined in **2.1** (that is, the image of $\mathcal{A}$ under $\imath$ is an isomorphic copy of $\mathcal{A}$) embraces the notion of a substructure. Indeed, $\mathcal{A} \subseteq \mathcal{B}$ iff $\imath = \mathrm{id}_A$, i.e. if $\imath$ is the *identical* or *trivial embedding* of $\mathcal{A}$ into $\mathcal{B}$.

Let $\mathcal{L}_0 \subseteq \mathcal{L}$. In this chapter, the embeddability of an $\mathcal{L}_0$-structure $\mathcal{A}$ into a given $\mathcal{L}$-structure $\mathcal{B}$ often means the embeddability of $\mathcal{A}$ into the $\mathcal{L}_0$-reduct $\mathcal{B}_0$ of $\mathcal{B}$, and we shall write $\mathcal{A} \subseteq \mathcal{B}$ also in such a situation. In this sense the group $\mathbb{Z}$, for example, is embeddable into the field $\mathbb{Q}$.

**Theorem 1.1.** *Let $\mathcal{L}_0 \subseteq \mathcal{L}$, $\mathcal{A}$ be an $\mathcal{L}_0$-structure, and $\mathcal{B}$ an $\mathcal{L}\mathcal{A}$-structure. Then $\mathcal{B} \vDash D\mathcal{A}$ iff $\imath \colon a \mapsto a^{\mathcal{B}}$ is an embedding of $\mathcal{A}$ in $\mathcal{B}$.*

**Proof.** $\Rightarrow$: Let $\mathcal{B} \vDash D\mathcal{A}$ and $a, b \in A$, $a \neq b$. Then $\boldsymbol{a} \not\boldsymbol{=} \boldsymbol{b} \in D\mathcal{A}$. Hence $\mathcal{B} \vDash \boldsymbol{a} \neq \boldsymbol{b}$, or equivalently, $a^{\mathcal{B}} \neq b^{\mathcal{B}}$. Thus $\imath$ is injective. For a relation symbol $r$ from $\mathcal{L}_0$ and $\vec{a} \in A^n$ we have in view of $\mathcal{B} \vDash D\mathcal{A}$,

$$r^{\mathcal{A}}\vec{a} \Leftrightarrow r\vec{a} \in D\mathcal{A} \Leftrightarrow \mathcal{B} \vDash r\vec{a} \Leftrightarrow r^{\mathcal{B}}\imath\vec{a} \quad (\imath\vec{a} := (\imath a_1, \ldots, \imath a_n)).$$

Similarly $\imath f^{\mathcal{A}}\vec{a} = f^{\mathcal{B}}\imath\vec{a}$ is obtained, for note that whenever $\vec{a} \in A^n$ and $b \in A$ then $f^{\mathcal{A}}\vec{a} = b \Leftrightarrow f\vec{a} \boldsymbol{=} \boldsymbol{b} \in D\mathcal{A} \Leftrightarrow \mathcal{B} \vDash f\vec{a} \boldsymbol{=} \boldsymbol{b} \Leftrightarrow f^{\mathcal{B}}\imath\vec{a} = \imath b$. Thus, $\imath$ is indeed an embedding. $\Leftarrow$: For variable-free terms $t$ in $\mathcal{L}_0 A$ one easily verifies $\imath t^{\mathcal{A}} = t^{\mathcal{B}}$, where here and elsewhere $t^{\mathcal{A}}$ means more precisely $t^{\mathcal{A}_A}$. Since $\imath$ is injective, it follows for variable-free equations $t_1 \boldsymbol{=} t_2$ in $\mathcal{L}_0 A$,

$$t_1 \boldsymbol{=} t_2 \in D\mathcal{A} \Leftrightarrow \imath_1^{\mathcal{A}} = t_2^{\mathcal{A}} \Leftrightarrow \imath t_1^{\mathcal{A}} = \imath t_2^{\mathcal{A}} \Leftrightarrow t_1^{\mathcal{B}} = t_2^{\mathcal{B}} \Leftrightarrow \mathcal{B} \vDash t_1 \boldsymbol{=} t_2.$$

In the same way we get $t_1 \not\boldsymbol{=} t_2 \in D\mathcal{A} \Leftrightarrow \mathcal{B} \vDash t_1 \not\boldsymbol{=} t_2$. Sentences of the form $r\vec{t}$ and their negations are dealt with analogously. Thus, $\mathcal{B} \vDash D\mathcal{A}$. ◻

**Corollary 1.2.** *Let $\mathcal{A}, \mathcal{B}$ be $\mathcal{L}$-structures and $\mathcal{B}'$ an $\mathcal{L}\mathcal{A}$-expansion of $\mathcal{B}$. Then $\mathcal{B}' \vDash D\mathcal{A}$ iff $\mathcal{A}$ is embeddable into $\mathcal{B}$. Moreover, if $A \subseteq B$ then $\mathcal{B}_A \vDash D\mathcal{A} \Leftrightarrow \mathcal{A} \subseteq \mathcal{B}$.*

Indeed, by the theorem with $\mathcal{L}_0 = \mathcal{L}$, the mapping $\imath \colon a \mapsto a^{\mathcal{B}'}$ realizes the embedding, and also the converse of the first claim is obvious. $\imath$ is the identical mapping in case $A \subseteq B$, which verifies the "Moreover" part with $\mathcal{B}' = \mathcal{B}_A$. Frequent use will be made of this corollary, without mentioning it explicitly. Taking a *prime model* for a theory $T$ to mean a model embeddable into every $T$-model, the corollary states that $\mathcal{A}_A$ is a prime model for $D\mathcal{A}$, understood as a theory. We are using the concept of a prime model only in this sense. It must be distinguished from the concept of an *elementary* prime model for $T$ as defined in Exercise 2.

Probably the most important concept in model theory, for which a first example appears on the next page, is given by the following

**Definition.** Let $\mathcal{L}$ be a first-order language and let $\mathcal{A}, \mathcal{B}$ be $\mathcal{L}$-structures. $\mathcal{A}$ is called an *elementary substructure* of $\mathcal{B}$, and $\mathcal{B}$ an *elementary extension* of $\mathcal{A}$, in symbols $\mathcal{A} \preccurlyeq \mathcal{B}$, if $A \subseteq B$ and

(2) $\quad \mathcal{A} \vDash \alpha\,[\vec{a}] \;\Leftrightarrow\; \mathcal{B} \vDash \alpha\,[\vec{a}]$, for all $\alpha = \alpha(\vec{x}) \in \mathcal{L}$ and $\vec{a} \in A^n$.

Clearly, $\mathcal{A} \preccurlyeq \mathcal{B} \Rightarrow \mathcal{A} \subseteq \mathcal{B}$. Terming $D_{el}\mathcal{A} := \{\alpha \in \mathcal{L}A^0 \mid \mathcal{A}_A \vDash \alpha\}$ the *elementary diagram* of $\mathcal{A}$, $\mathcal{A} \preccurlyeq \mathcal{B}$ is obviously equivalent to $\mathcal{A} \subseteq \mathcal{B}$ and $\mathcal{B}_A \vDash D_{el}\mathcal{A}$. Indeed, (2) already holds given only $\mathcal{A} \vDash \alpha\,[\vec{a}] \Rightarrow \mathcal{B} \vDash \alpha\,[\vec{a}]$, for all $\alpha = \alpha(\vec{x}) \in \mathcal{L}$ and $\vec{a} \in A^n$.

(2) is equivalent to $\mathcal{A}_A \vDash \alpha(\vec{a}) \Leftrightarrow \mathcal{B}_A \vDash \alpha(\vec{a})$, by (1). And since every $\alpha \in \mathcal{L}A$ is of the form $\alpha(\vec{a})$ for appropriate $\alpha(\vec{x}) \in \mathcal{L}$, $\vec{a} \in A^n$, and $n \geqslant 0$, the property $\mathcal{A} \preccurlyeq \mathcal{B}$ is also characterized by $\mathcal{A} \subseteq \mathcal{B}$ and $\mathcal{A}_A \equiv \mathcal{B}_A$ (elementary equivalence in $\mathcal{L}A$).

In general, $\mathcal{A} \preccurlyeq \mathcal{B}$ means much more than $\mathcal{A} \subseteq \mathcal{B}$ and $\mathcal{A} \equiv \mathcal{B}$. For instance, let $\mathcal{A} = (\mathbb{N}_+, <)$ and $\mathcal{B} = (\mathbb{N}, <)$. Then certainly $\mathcal{A} \subseteq \mathcal{B}$, and since $\mathcal{A} \simeq \mathcal{B}$, we have also $\mathcal{A} \equiv \mathcal{B}$. But $\mathcal{A} \preccurlyeq \mathcal{B}$ is false. For example, $\exists x\, x < 1$ is true in $\mathcal{B}_A$, but obviously not in $\mathcal{A}_A$. The following theorem will prove to be very useful for, among other things, the provision of nontrivial examples for $\mathcal{A} \preccurlyeq \mathcal{B}$:

**Theorem 1.3 (Tarski's criterion).** *For arbitrary $\mathcal{L}$-structures $\mathcal{A}, \mathcal{B}$ with $\mathcal{A} \subseteq \mathcal{B}$ the following conditions are equivalent:*

  (i) $\mathcal{A} \preccurlyeq \mathcal{B}$,

  (ii) *For all $\varphi(\vec{x}, y) \in \mathcal{L}$ and $\vec{a} \in A^n$ holds the implication*
    $\mathcal{B} \vDash \exists y \varphi(\vec{a}, y) \Rightarrow \mathcal{B} \vDash \varphi(\vec{a}, a)$ *for some $a \in A$.*

**Proof.** (i)$\Rightarrow$(ii): Let $\mathcal{A} \preccurlyeq \mathcal{B}$ and $\mathcal{B} \vDash \exists y \varphi(\vec{a}, y)$, so that also $\mathcal{A} \vDash \exists y \varphi(\vec{a}, y)$. Then $\mathcal{A} \vDash \varphi(\vec{a}, a)$ for some $a \in A$. But $\mathcal{A} \preccurlyeq \mathcal{B}$; hence $\mathcal{B} \vDash \varphi(\vec{a}, a)$. (ii)$\Rightarrow$(i): Since $\mathcal{A} \subseteq \mathcal{B}$, (2) certainly holds for prime formulas. The induction steps for $\wedge, \neg$ are obvious. Only the quantifier step needs a closer look:

$$\mathcal{A} \vDash \forall y \varphi(\vec{a}, y) \Leftrightarrow \mathcal{A} \vDash \varphi(\vec{a}, a) \text{ for all } a \in A$$
$$\Leftrightarrow \mathcal{B} \vDash \varphi(\vec{a}, a) \text{ for all } a \in A \quad \text{(induction hypothesis)}$$
$$\Leftrightarrow \mathcal{B} \vDash \forall y \varphi(\vec{a}, y) \quad \text{(see below).}$$

We prove the direction $\Rightarrow$ in the last equivalence indirectly: Assume that $\mathcal{B} \nvDash \forall y \varphi(\vec{a}, y)$. Then $\mathcal{B} \vDash \exists y \neg \varphi(\vec{a}, y)$. Hence $\mathcal{B} \vDash \neg \varphi(\vec{a}, a)$ for some $a \in A$ according to (ii). Thus, $\mathcal{B} \vDash \varphi(\vec{a}, a)$ cannot hold for all $a \in A$. $\square$

Interesting examples for $\mathcal{A} \preccurlyeq \mathcal{B}$ are provided in a surprisingly simple way by the following theorem, which, unfortunately, is applicable only if $\mathcal{B}$ has "many automorphisms," as is the case in the example below, and in geometry, for instance.

**Theorem 1.4.** *Let $\mathcal{A} \subseteq \mathcal{B}$. Suppose that for all $n$, all $\vec{a} \in A^n$, and all $b \in B$ there is an automorphism $\imath \colon \mathcal{B} \to \mathcal{B}$ such that $\imath\vec{a} = \vec{a}$, and $\imath b \in A$. Then $\mathcal{A} \preccurlyeq \mathcal{B}$.*

**Proof.** It suffices to verify (ii) in Theorem 1.3. Let $\mathcal{B} \vDash \exists y \varphi(\vec{a}, y)$, or equivalently $\mathcal{B} \vDash \varphi(\vec{a}, b)$ for some $b \in B$. Then $\mathcal{B} \vDash \varphi(\imath\vec{a}, \imath b)$ according to Theorem 2.3.4, and since $\imath\vec{a} = \vec{a}$, we obtain $\mathcal{B} \vDash \varphi(\vec{a}, a)$ with $a := \imath b \in A$. This proves (ii). $\quad\blacksquare$

**Example.** It is readily shown that for given $a_1, \ldots, a_n \in \mathbb{Q}$ and $b \in \mathbb{R}$ there exists an automorphism of $(\mathbb{R}, <)$ that maps $b$ to a rational number and leaves $a_1, \ldots, a_n$ fixed (Exercise 3). Thus, $(\mathbb{Q}, <) \preccurlyeq (\mathbb{R}, <)$. In particular $(\mathbb{Q}, <) \equiv (\mathbb{R}, <)$.

Here is a look at some less simple examples of elementary extensions, considered more closely in **5.5**. Let $\mathcal{A} = (\mathbb{A}, 0, 1, +, \cdot)$ denote the *field of algebraic numbers* and $\mathcal{C}$ the field of complex numbers. The domain $\mathbb{A}$ consists of all complex numbers that are zeros of (monic) polynomials with rational coefficients. Then $\mathcal{A} \preccurlyeq \mathcal{C}$. Similarly, $\mathcal{A}_r \preccurlyeq \mathcal{R}$ where $\mathcal{A}_r$ denotes the field of all *real algebraic numbers* and $\mathcal{R}$ is the field of all reals. The claim $\mathcal{A} \preccurlyeq \mathcal{C}$ follows from the model completeness of the theory ACF proved on page 198. Similarly $\mathcal{A}_r \preccurlyeq \mathcal{R}$ will be shown.

Before continuing we will acquaint ourselves somewhat with transfinite cardinal numbers. It is possible to assign a set-theoretic object denoted by $|M|$ not only to finite sets but to arbitrary sets $M$ such that

(3) $\quad M \sim N \Leftrightarrow |M| = |N| \quad$ ($\sim$ means equipotency; see page 111).

$|M|$ is called the *cardinal number* or *cardinality* of $M$. This is just the number of elements in $M$ for a finite $M$; for an infinite set $M$, $|M|$ is called a *transfinite cardinal number*, or briefly a *transfinite cardinal*.

At this stage it is unimportant just how $|M|$ is defined in detail. The interested reader will find some definition in every textbook on set theory. Significant are (4) and (5), taken as granted, from which (6) and (7) straightforwardly follow.

(4) The cardinal numbers are well-ordered according to size, i.e., each nonempty collection of them possesses a smallest element. Here let $|N| \leqslant |M|$ if there is an injection from $N$ to $M$. The smallest transfinite cardinality is $|\mathbb{N}|$, i.e., $|\mathbb{N}| \leqslant |M|$ for all infinite sets $M$.

(5) $|M \cup N| = |M \times N| = \max\{|M|, |N|\}$ for arbitrary sets $M$ and $N$ of which at least one is infinite.

**Remark.** With this definition of $\leqslant$ it follows that $|M| \leqslant |N|$ & $|N| \leqslant |M|$ implies $|M| = |N|$ (without AC). This is called the *Cantor–Bernstein theorem*. Actually, the first proof of this theorem without AC (even more elegant than Bernstein's) is due to Dedekind, who left it unpublished in his diary from 1887. This theorem holds under surprisingly weak assumptions; see [De].

We first derive from (4) and (5) that $M^* := \bigcup_{n>0} M^n$ has the same cardinality as $M$ for infinite $M$, where $M^*$ denotes the set of all nonempty finite sequences of elements of $M$. In short,

(6) $|M^*| = |M|$     ($M$ infinite).

Indeed, $|M^1| = |M|$, and the hypothesis $|M^n| = |M|$ obviously yields $|M^{n+1}| = |M^n \times M| = |M|$ by (5). Thus $|M^n| = |M|$ for all $n$. Therefore $|M^*| = |\bigcup_{n>0} M^n| = |M \times \mathbb{N}| = |M|$. One similarly obtains from (4), (5) for every transfinite cardinal $\kappa$ the property

(7) If $A_0, A_1 \ldots$ are sets and $|A_n| \leqslant \kappa$ for all $n \in \mathbb{N}$ then $|\bigcup_{n \in \mathbb{N}} A_n| \leqslant \kappa$.

The smallest transfinite cardinal number (i.e., $|\mathbb{N}|$) is that of the countably infinite sets, denoted by $\aleph_0$. The next one is $\aleph_1$. Then follows $\aleph_2$, $\aleph_3, \ldots$ There is a smallest cardinal larger than all $\aleph_n$, denoted by $\aleph_\omega$, etc. The Cantor–Bernstein theorem shows that the power set $\mathfrak{P}\mathbb{N}$ and the set $\mathbb{R}$ have the same cardinality, denoted by $2^{\aleph_0}$. Certainly $\aleph_0 < 2^{\aleph_0}$, hence $\aleph_1 \leqslant 2^{\aleph_0}$. Cantor's *continuum hypothesis* (CH) states that $\aleph_1 = 2^{\aleph_0}$.

CH is independent in ZFC; see e.g. [Ku]. While there are axioms extending beyond ZFC that decide CH one way or another, none of these is sufficiently plausible to be regarded as "true." In the last decades some evidence has been collected that suggests that $2^{\aleph_0} = \aleph_2$, but this seemingly does not yet convince the majority of mathematicians.

The cardinality of a structure $\mathcal{A}$ is always that of its domain, that is, $|\mathcal{A}| := |A|$. Theorem 1.5 below, essentially due to Tarski and therefore sometimes called the Löwenheim–Skolem–Tarski theorem, generalizes Theorem 3.4.1 (page 112) essentially. The additive "downward" prevents a mix-up of these theorems. For $|\mathcal{B}| \geqslant |\mathcal{L}|$, Theorem 1.5 ensures the existence of some $\mathcal{A} \preccurlyeq \mathcal{B}$ (in particular $\mathcal{A} \equiv \mathcal{B}$) such that $|\mathcal{A}| \leqslant |\mathcal{L}|$.

**Theorem 1.5 (Löwenheim–Skolem theorem downward).** *Suppose that $\mathcal{B}$ is an $\mathcal{L}$-structure such that $|\mathcal{L}| \leqslant |\mathcal{B}|$ and let $A_0 \subseteq B$ be arbitrary. Then $\mathcal{B}$ has an elementary substructure $\mathcal{A}$ of cardinality $\leqslant \max\{|A_0|, |\mathcal{L}|\}$ such that $A_0 \subseteq A$.*

**Proof.** We construct a sequence $A_0 \subseteq A_1 \subseteq \cdots \subseteq B$ as follows. Let $A_k$ be given. For every $\alpha = \alpha(\vec{x}, y)$ and $\vec{a} \in A_k^n$ such that $\mathcal{B} \vDash \exists y \alpha(\vec{a}, y)$ we select some $b \in B$ with $\mathcal{B} \vDash \alpha(\vec{a}, b)$ and adjoin $b$ to $A_k$, thus getting $A_{k+1}$. In particular, if $\alpha$ is $f\vec{x} = y$ then certainly $\mathcal{B} \vDash \exists y \, f\vec{a} = y$. Since $\mathcal{B} \vDash \exists! y \, f\vec{a} = y$, there is no alternative selection; hence $f^{\mathcal{B}} \vec{a} \in A_{k+1}$. Thus, $A := \bigcup_{k \in \mathbb{N}} A_k$ is closed under the operations of $\mathcal{B}$, and therefore defines a substructure $\mathcal{A} \subseteq \mathcal{B}$. We shall prove $\mathcal{A} \preccurlyeq \mathcal{B}$ by Tarski's criterion. Let $\mathcal{B} \vDash \exists y \alpha(\vec{a}, y)$ for $\alpha = \alpha(\vec{x}, y)$ and let $\vec{a} \in A^n$. Then $\vec{a} \in A_k^n$ for some $k$. Therefore, there is some $a \in A_{k+1}$ (hence $a \in A$) such that $\mathcal{B} \vDash \alpha(\vec{a}, a)$. This proves (ii) in Theorem 1.3 and so $\mathcal{A} \preccurlyeq \mathcal{B}$. It remains to show that $|A| \leqslant \kappa := \max\{|A_0|, |\mathcal{L}|\}$. There are at most $\kappa$ formulas and $\kappa$ finite sequences of elements in $A_0$. Thus, by definition of $A_1$, at most $\kappa$ new elements are adjoined to $A_0$. Hence $|A_1| \leqslant \kappa$. Similarly, $|A_n| \leqslant \kappa$ is verified for each $n > 0$. By (7) we thus get $|\bigcup_{n \in \mathbb{N}} A_n| \leqslant \kappa$. ◻

Combined with the compactness theorem, the above theorem yields

**Theorem 1.6 (Löwenheim–Skolem theorem upward).** *Let $\mathcal{C}$ be any infinite $\mathcal{L}$-structure and $\kappa \geqslant \max\{|\mathcal{C}|, |\mathcal{L}|\}$. Then there exists an $\mathcal{A} \succcurlyeq \mathcal{C}$ with $|\mathcal{A}| = \kappa$.*

**Proof.** Choose some $D \supseteq C$ with $|D| = \kappa$. From (6) it follows that $|\mathcal{L}D| = \kappa$, because the alphabet of $\mathcal{L}D$ has cardinality $\kappa$. Since $|C| \geqslant \aleph_0$, by the compactness theorem, $D_{el}\mathcal{C} \cup \{c \neq d \mid c, d \in D, \ c \neq d\}$ has a model $\mathcal{B}$. Since $d \mapsto d^{\mathcal{B}}$ $(d \in D)$ is injective, we may assume $d^{\mathcal{B}} = d$ for all $d \in D$, i.e., $D \subseteq B$. By Theorem 1.5 with $\mathcal{L}D$ for $\mathcal{L}$ and $D$ for $A_0$, there is some $\mathcal{A} \preccurlyeq \mathcal{B}$ with $D \subseteq A$ and $\kappa \leqslant |D| \leqslant |A| \leqslant \max\{|\mathcal{L}D|, |D|\} = \kappa$. Hence $|A| = \kappa$. From $C \subseteq D$ and $\mathcal{A} \equiv_{\mathcal{L}D} \mathcal{B} \vDash D_{el}\mathcal{C}$ it follows that $\mathcal{A} \vDash D_{el}\mathcal{C}$. Since $C \subseteq D \subseteq A$ in addition, the $\mathcal{L}$-reduct of $\mathcal{A}$ is an elementary extension of the given structure $\mathcal{C}$. ◻

These theorems show in particular that a countable theory $T$ with at least one infinite model also has models in every infinite cardinality. Further, $\vdash_T \alpha$ already holds when merely $\mathcal{A} \vDash \alpha$ for all $T$-models $\mathcal{A}$ of a *single* infinite cardinal number $\kappa$, as long as $T$ has only infinite models, because under this assumption every $T$-model is elementarily equivalent to a $T$-model of cardinality $\kappa$.

## Exercises

1. Let $\mathcal{A} \preccurlyeq \mathcal{C}$ and $\mathcal{B} \preccurlyeq \mathcal{C}$, where $A \subseteq B$. Prove that $\mathcal{A} \preccurlyeq \mathcal{B}$.

2. An embedding $\imath \colon \mathcal{A} \to \mathcal{B}$ is termed *elementary* if $\imath \mathcal{A} \preccurlyeq \mathcal{B}$, where $\imath \mathcal{A}$ denotes the image of $\mathcal{A}$ under $\imath$. Show similarly to Theorem 1.1 that an $\mathcal{L}A$-structure $\mathcal{B}$ is a model of $D_{el}\mathcal{A}$ iff $\mathcal{A}$ is elementarily embeddable into $\mathcal{B}$.

3. Let $a_1, \ldots, a_n \in \mathbb{Q}$ and $b \in \mathbb{R}$. Show that there is an automorphism of $(\mathbb{R}, <)$ that maps $b$ to a rational number and leaves all $a_i$ fixed.

4. Let $\mathcal{A} \equiv \mathcal{B}$. Construct a structure $\mathcal{C}$ in which $\mathcal{A}, \mathcal{B}$ are both elementarily embeddable.

5. Let $\mathcal{A}$ be an $\mathcal{L}$-structure generated from $G \subseteq A$ and $\mathcal{T}_G$ the set of ground terms in $\mathcal{L}G$. Prove that (a) for every $a \in A$ there is some $t \in \mathcal{T}_G$ such that $a = t^{\mathcal{A}}$, (b) if $\mathcal{A} \vDash T$ and $D\mathcal{A} \vdash_T \alpha \ (\in \mathcal{L}G)$ then $D_G\mathcal{A} \vdash_T \alpha$. Here $D_G\mathcal{A} := D\mathcal{A} \cap \mathcal{L}G$.

## 5.2 Complete and $\kappa$-Categorical Theories

According to the definition on page 105, a theory $T \subseteq \mathcal{L}^0$ is complete if it is consistent and each extended theory $T' \supset T$ in $\mathcal{L}^0$ is inconsistent. A complete theory need not be maximally consistent in the whole of $\mathcal{L}$. For instance, in general neither $\vdash_T x = y$ nor $\vdash_T x \neq y$, even if $T$ is complete. Some equivalent formulations of completeness, whose usefulness depends on the situation at hand, are presented by the following

**Theorem 2.1.** *For a consistent theory $T$ the following conditions are equivalent:*[1]

---

[1] All these conditions are also equivalent (they all hold) if the inconsistent theory is taken to be complete, which is not the case here, as we agreed upon in **3.3**.

(i)   $T$ is complete,
(ii)  $T = Th\,\mathcal{A}$ for every $\mathcal{A} \vDash T$,
(iii) $\mathcal{A} \equiv \mathcal{B}$ for all $\mathcal{A}, \mathcal{B} \vDash T$,
(iv)  $\vdash_T \alpha \lor \beta$ implies $\vdash_T \alpha$ or $\vdash_T \beta$  $(\alpha, \beta \in \mathcal{L}^0)$,
(v)   $\vdash_T \alpha$ or $\vdash_T \neg\alpha$  $(\alpha \in \mathcal{L}^0)$.

**Proof.** (i) $\Rightarrow$ (ii): Since $T \subseteq Th\,\mathcal{A}$ for each model $\mathcal{A} \vDash T$, it must be that $T = Th\,\mathcal{A}$. (ii) $\Rightarrow$ (iii): For $\mathcal{A}, \mathcal{B} \vDash T$ we have by (ii) $Th\,\mathcal{A} = T = Th\,\mathcal{B}$, and therefore $\mathcal{A} \equiv \mathcal{B}$. (iii) $\Rightarrow$ (iv): Let $\vdash_T \alpha \lor \beta$, $\mathcal{A} \vDash T$, and $\mathcal{A} \vDash \alpha$, say. Then $\mathcal{B} \vDash \alpha$ for all $\mathcal{B} \vDash T$ by (iii), hence $\vdash_T \alpha$. (v) is a special case of (iv) because $\vdash_T \alpha \lor \neg\alpha$, for arbitrary $\alpha \in \mathcal{L}^0$. (v) $\Rightarrow$ (i): Let $T' \supset T$ and $\alpha \in T' \setminus T$. Then $\vdash_T \neg\alpha$ by (v); hence also $\vdash_{T'} \neg\alpha$. But then $T'$ is inconsistent. Hence, by the above definition, $T$ is complete. $\blacksquare$

We now present various methods by which conjectured completeness can be confirmed. The completeness question is important for many reasons. For example, according to Theorem 3.5.2, a complete axiomatizable theory is decidable whatever the means of proving completeness might have been.

An elementary theory with at least one infinite model, even if it is complete, has many different infinite models. For instance, according to Theorem 1.6, the theory possesses models of arbitrarily high cardinality. However, sometimes it happens that all of its models of a given finite or infinite cardinal number $\kappa$ are isomorphic. The following definition bears this circumstance in mind.

**Definition.** A theory $T$ is $\kappa$-*categorical* if there exists up to isomorphism precisely one $T$-model of cardinality $\kappa$.

**Example 1.** The theory $\mathsf{Taut}_=$ of tautological sentences in $\mathcal{L}_=$ is $\kappa$-categorical for every cardinal $\kappa$. Indeed, here models $\mathcal{A}, \mathcal{B}$ of cardinality $\kappa$ are naked sets and these are trivially isomorphic under any bijection from $A$ onto $B$.

The theory $\mathsf{DO}$ of densely ordered sets results from the theory of ordered sets (formalized in **2.3**; see also **2.1**) by adjoining the axioms

$$\exists x \exists y\, x \neq y\,; \quad \forall x \forall y \exists z (x < y \to x < z \land z < y).$$

It is obvious that a densely ordered set is infinite. $\mathsf{DO}$ can be extended by the axioms $\mathsf{L} := \exists x \forall y\, x \leqslant y$ and $\mathsf{R} := \exists x \forall y\, y \leqslant x$ to the theory $\mathsf{DO}_{11}$ of densely ordered sets with edge elements. Replacing $\mathsf{R}$ by $\neg\mathsf{R}$ results in the theory $DO_{10}$ of densely ordered sets with left but without right edge elements. Accordingly $DO_{01}$ denotes the theory with right but without left, and $DO_{00}$ that of dense orders without edge elements. The paradigm of a model for $DO_{00}$ is $(\mathbb{Q}, <)$. Another model is $(\mathbb{Q}_+, <)$.

**Example 2.** $DO_{00}$ is $\aleph_0$-categorical (Exercise 1 treats the other $DO_{ij}$). The following proof is due to Cantor. A function $f$ with $\text{dom} f \subseteq M$ and $\text{ran} f \subseteq N$ is said to be a *partial function* from $M$ to $N$. Now let $A = \{a_0, a_1, \dots\}$ and $B = \{b_0, b_1, \dots\}$ be countable $DO_{00}$-models. Define $f_0$ by $f_0 a_0 = b_0$ so that $\text{dom} f_0 = \{a_0\}$, $\text{ran} f_0 = \{b_0\}$ (step 0). Assume that in the $n$th step a partial function $f_n$ *from* $A$ to $B$ with finite domain was constructed with $a < a' \Leftrightarrow f_n a < f_n a'$, for all $a, a' \in \text{dom} f_n$ (a so-called *partial isomorphism*), and that $\{a_0, \dots, a_n\} \subseteq \text{dom} f_n$ and $\{b_0, \dots, b_n\} \subseteq \text{ran} f_n$. These conditions are trivially satisfied for $f_0$. Let $m$ be minimal with $a_m \in A \setminus \text{dom} f_n$. Choose $b \in B \setminus \text{ran} f_n$ such that $g_n := f_n \cup \{(a_m, b)\}$ is also a partial isomorphism. This is possible thanks to the denseness of $B$. Now let $m$ be minimal with $b_m \in B \setminus \text{ran} g_n$. Choose a suitable $a \in A \setminus \text{dom} g_n$ such that $f_{n+1} := g_n \cup \{(a, b_m)\}$ is a partial isomorphism too. This "to and fro" construction clearly provides for both $a_{n+1} \in \text{dom} f_{n+1}$ and $b_{n+1} \in \text{ran} f_{n+1}$. Claim: $f = \bigcup_{n \in \mathbb{N}} f_n$ is an isomorphism from $A$ onto $B$. Indeed, $f$ is a function. Moreover, $\text{dom} f = A$ and $\text{ran} f = B$. The isomorphism condition $x < y \Leftrightarrow fx < fy$ is clear, since any $x, y \in A$ belong already to $\text{dom} f_n$ for suitable $n$.

**Example 3.** The *successor theory* $T_{\text{suc}}$ in $\mathcal{L}\{0, S\}$ has the axioms

$$\forall x\, 0 \neq Sx, \quad \forall xy(Sx = Sy \to x = y), \quad (\forall x \neq 0)\exists y\, x = Sy,$$

$$\forall x_0 \cdots x_n (\bigwedge_{i < n} Sx_i = x_{i+1} \to x_0 \neq x_n) \quad (n = 1, 2, \dots).$$

The last axiom says *there are no "circles."* $T_{\text{suc}}$ is not $\aleph_0$-categorical, but it is $\aleph_1$-categorical. Indeed, each model $\mathcal{A} \vDash T_{\text{suc}}$ with $|\mathcal{A}| = \aleph_1$ consists up to isomorphism of the (countable) standard model $(\mathbb{N}, 0, S)$ and $\aleph_1$ many "threads" of isomorphism type $(\mathbb{Z}, S)$, where $S : z \mapsto z + 1$. For if there were only countably many such threads then the entire model would be countable. Hence any two $T_{\text{suc}}$-models of cardinality $\aleph_1$ are isomorphic.

**Example 4.** The theory $ACF_p$ of a.c. fields of given characteristic $p$ (page 105) is $\aleph_1$-categorical. We sketch here a proof very briefly because $ACF_p$ is analyzed in **5.5** in a different way. The claim follows from the facts that each field is embeddable into an a.c. field (cf. Example 1 in **5.5**) and that a *transcendental extension* $\mathcal{K}'$ of a field $\mathcal{K}$ (that is, every $a \in K' \backslash K$ is transcendental over $K$) has a *transcendence basis* $B$. This is a maximal system of algebraically independent elements in $K' \backslash K$. The isomorphism type of $\mathcal{K}'$ is completely determined by the cardinality of $B$.

It is fairly plausible that in Examples 3 and 4 $\kappa$-categoricity holds for every cardinal $\kappa > \aleph_0$. This observation is no coincidence. It is explained by the following theorem.

**Morley's theorem.** *If a countable theory $T$ is $\kappa$-categorical for some $\kappa > \aleph_0$ then it is $\kappa$-categorial for all $\kappa > \aleph_0$.*

The proof makes use of extensive methods and must be passed over here. On the other hand, the proof of the following theorem requires but little effort.

**Theorem 2.2 (Vaught's test).** *A countable consistent theory $T$ without finite models is complete provided $T$ is $\kappa$-categorical for some $\kappa$.*

**Proof.** Note first that $\kappa \geqslant \aleph_0$ because $T$ possesses no finite models. Assume that $T$ is incomplete. Choose some $\alpha \in \mathcal{L}^0$ with $\nvdash_T \alpha$ and $\nvdash_T \neg\alpha$. Then $T, \alpha$ and $T, \neg\alpha$ are consistent. These sets have countable infinite models by Theorem 1.5, and according to Theorem 1.6 there are also models $\mathcal{A}$ and $\mathcal{B}$ of cardinal $\kappa$. Since $\mathcal{A}, \mathcal{B} \vDash T$, by hypothesis $\mathcal{A} \simeq \mathcal{B}$; hence $\mathcal{A} \equiv \mathcal{B}$, which contradicts $\mathcal{A} \vDash \alpha$ and $\mathcal{B} \vDash \neg\alpha$. $\square$

**Example 5.** (a) $\mathsf{DO}_{00}$ has only infinite models and is $\aleph_0$-categorical by Example 2. Hence $\mathsf{DO}_{00}$ is complete by Vaught's test. This fact confirms $(\mathbb{Q}, <) \equiv (\mathbb{R}, <)$ once again. In fact, each $\mathsf{DO}_{ij}$ is complete (Exercise 1). Thus, $\mathcal{A} \equiv \mathcal{B}$ for $\mathcal{A}, \mathcal{B} \vDash \mathsf{DO}$ iff $\mathcal{A}, \mathcal{B}$ have "the same edge configuration," which tells us that the $\mathsf{DO}_{ij}$ are the only completions of $\mathsf{DO}$. Since $\mathsf{DO}$ is axiomatizable, it follows by Exercise 3 in **3.5** that $\mathsf{DO}$ is decidable. The same applies to each of its finite extensions $\mathsf{DO}_{ij}$.

(b) The successor theory $T_{\mathrm{suc}}$ is $\aleph_1$-categorical (Example 3) and has only infinite models. Hence it is complete by Vaught's test and as an axiomatizable theory thus decidable (Theorem 3.5.2).

(c) $\mathsf{ACF}_p$ is $\aleph_1$-categorical by Example 4. Each a.c. field $\mathcal{A}$ is infinite. For assume the converse, $A = \{a_0, \ldots, a_n\}$, say. Then the polynomial $1 + \prod_{i \leqslant n}(x - a_i)$ would have no root, a contradiction. Thus, by Vaught's test, $\mathsf{ACF}_p$ is complete and decidable (since it is axiomatizable). This result will be derived by quite different methods in **5.5**.

The model classes of first-order sentences are called *elementary classes*. These clearly include the model classes of finitely axiomatizable theories. For each such theory $T$, $\operatorname{Md} T = \bigcap_{\alpha \in T} \operatorname{Md} \alpha$ is an intersection of elementary classes, also termed a $\Delta$-*elementary class*. Thus, the class of all fields is elementary, and that of all a.c. fields is $\Delta$-elementary. On the other hand, the class of all finite fields is not $\Delta$-elementary because its theory evidently has infinite models. An algebraic characterization of elementary and $\Delta$-elementary classes will be provided in **5.7**.

The model classes of complete first-order theories are called *elementary types*. $\operatorname{Md} T$ is the union of the elementary types belonging to the completions of a theory $T$. For instance, DO has just the four completions $\mathsf{DO}_{ij}$ determined by the edge configuration, that is, by those of the sentences $\mathsf{L}, \mathsf{R}, \neg\mathsf{L}, \neg\mathsf{R}$, valid in the respective completion. For this case, the next theorem provides more information on $T$, in particular on $\equiv_T$.

Let $X \subseteq \mathcal{L}$ be nonempty and $T$ a theory. Take $\langle X \rangle$ to denote the set (still dependent on $T$) of all formulas equivalent in $T$ to Boolean combinations of formulas in $X$. Clearly, $\top \in \langle X \rangle$ since $\top \equiv_T \alpha \vee \neg \alpha$ for $\alpha \in X$. Therefore, $T \subseteq \langle X \rangle$, because $\alpha \equiv_T \top$ whenever $\alpha \in T$. Call $X \subseteq \mathcal{L}^0$ a *Boolean basis for* $\mathcal{L}^0$ *in* $T$ if *every* $\alpha \in \mathcal{L}^0$ belongs to $\langle X \rangle$, i.e., every sentence in $\mathcal{L}$ is a Boolean combination of sentences from $X$. Example 6(b) below indicates how useful a Boolean base for decision problems can be. $\mathcal{A} \equiv_X \mathcal{B}$ is to mean $\mathcal{A} \vDash \alpha \Leftrightarrow \mathcal{B} \vDash \alpha$, for all $\alpha \in X$.

**Theorem 2.3 (Basis theorem for sentences).** *Let $T$ be a theory and $X \subseteq \mathcal{L}^0$ a set of sentences with $\mathcal{A} \equiv_X \mathcal{B} \Rightarrow \mathcal{A} \equiv \mathcal{B}$, for all $\mathcal{A}, \mathcal{B} \vDash T$.[2] Then $X$ is a Boolean basis for $\mathcal{L}^0$ in $T$.*

**Proof.** Let $\alpha \in \mathcal{L}^0$ and $Y_\alpha := \{\beta \in \langle X \rangle \mid \alpha \vdash_T \beta\}$. We claim $(*)$: $Y_\alpha \vdash_T \alpha$. Otherwise let $\mathcal{A} \vDash T, Y_\alpha, \neg\alpha$. Then $T_X \mathcal{A} := \{\gamma \in \langle X \rangle \mid \mathcal{A} \vDash \gamma\} \vdash \neg\alpha$; indeed, for any $\mathcal{B} \vDash T_X \mathcal{A}$ we have $\mathcal{B} \equiv_X \mathcal{A}$ and hence $\mathcal{B} \equiv \mathcal{A}$. Therefore $\gamma \vdash_T \neg\alpha$ for some $\gamma \in T_X \mathcal{A}$, because $\langle X \rangle$ is closed under conjunctions. This yields $\alpha \vdash_T \neg\gamma$, i.e., $\neg\gamma \in Y_\alpha$. Thus $\mathcal{A} \vDash \neg\gamma$, in contradiction to $\mathcal{A} \vDash \gamma$. So $(*)$ holds. Hence there are $\beta_0, \ldots, \beta_m \in Y_\alpha$ such that $\beta := \bigwedge_{i \leqslant m} \beta_i \vdash_T \alpha$. We know that $\alpha \vdash_T \beta_i$ and so that $\alpha \vdash_T \beta$ as well. This and $\beta \vdash_T \alpha$ confirms $\alpha \equiv_T \beta$, and since $\beta \in \langle X \rangle$, also $\alpha \in \langle X \rangle$. ∎

---

[2] This assumption is equivalent to the assertion that $\{\gamma \in \langle X \rangle \mid \mathcal{A} \vDash \gamma\}$ is complete; see the subsequent proof. For refinements of the theorem we refer to [HR].

**Example 6.** (a) Let $T = \mathsf{DO}$ and $X = \{\mathsf{L}, \mathsf{R}\}$. Then $\mathcal{A} \equiv_X \mathcal{B} \Rightarrow \mathcal{A} \equiv \mathcal{B}$, for all $\mathcal{A}, \mathcal{B} \vDash T$. Indeed, $\mathcal{A} \equiv_X \mathcal{B}$ states that $\mathcal{A}, \mathcal{B}$ possess the same edge configuration. But then $\mathcal{A} \equiv \mathcal{B}$, because the $\mathsf{DO}_{ij}$ are all complete; see Example 5(a). Therefore, $\mathsf{L}$ and $\mathsf{R}$ form a Boolean basis for $\mathcal{L}_<^0$ in $\mathsf{DO}$. This theory has four completions, and so by Exercise 5 in **3.5**, exactly $15 \,(= 2^4 - 1)$ consistent extensions.

(b) Let $T = \mathsf{ACF}$ and $X = \{\mathtt{char}_p \mid p \text{ prime}\}$. Again, $\mathcal{A} \equiv_X \mathcal{B} \Rightarrow \mathcal{A} \equiv \mathcal{B}$, for all $\mathcal{A}, \mathcal{B} \vDash T$, because by Example 5(c), $\mathsf{ACF}_p$ is complete for each $p$ (including $p = 0$). Hence, by Theorem 2.3, the $\mathtt{char}_p$ constitute a Boolean basis for sentences modulo $\mathsf{ACF}$. This implies the decidability of $\mathsf{ACF}$: let $\alpha \in \mathcal{L}^0$ be given; just wait in an enumeration process of the theorems of $\mathsf{ACF}$ until a sentence of the form $\alpha \leftrightarrow \beta$ appears, where $\beta$ is a Boolean combination of the $\mathtt{char}_p$. Such a sentence definitely appears. Then test whether $\beta \equiv_{\mathsf{ACF}} \top$, for example by converting $\beta$ into a CNF.

**Corollary 2.4.** *Let $T \subseteq \mathcal{L}^0$ be a theory with arbitrarily large finite models, such that all finite $T$-models with the same number of elements and all infinite $T$-models are elementarily equivalent. Then*

(a) *the sentences $\exists_n$ form a Boolean basis for $\mathcal{L}^0$ in $T$,*

(b) *$T$ is decidable provided $T$ is finitely axiomatizable.*

**Proof.** Let $X := \{\exists_k \mid k \in \mathbb{N}\}$. Then by hypothesis, $\mathcal{A} \equiv_X \mathcal{B} \Rightarrow \mathcal{A} \equiv \mathcal{B}$, for all $\mathcal{A}, \mathcal{B} \vDash T$. Thus, (a) follows by Theorem 2.3. By (a) and Exercise 4 in **2.3** each $\alpha \in \mathcal{L}^0$ is equivalent in $T$ to $\bigvee_{\nu \leqslant n} \exists_{=k_\nu}$ with $k_0 < \cdots < k_n$ or to $\exists_k \vee \bigvee_{\nu \leqslant n} \exists_{=k_\nu}$ for some $k$. Hence a sentence $\alpha$ that has a $T$-model has also a finite $T$-model by the first assumption on $T$, i.e., $T$ has the finite model property. Thus, (b) holds by Exercise 3 in **3.6**. $\qquad \square$

An easy example of application is the theory $\mathsf{Taut}_=$ of tautologies in $\mathcal{L}_=$. The formulas constructed from the Boolean base $\{\exists_n \mid n \in \omega\}$ in the proof also permit a simple description of the elementary classes of $\mathcal{L}_=$. These are finite unions of classes determined by sentences $\exists_k$ and $\exists_{=m}$. Another example is the theory $\mathsf{FO}$ of finite ordered sets. We prove in the next section that $\mathsf{FO}$ satisfies the assumptions of Corollary 2.4. Hence, the elementary classes of $\mathsf{FO}$ have the same simple description.

These examples illustrate the following: If we know the elementary types of a theory $T$—these correspond to the completions of $T$—then we also know their elementary classes. As a rule, the type classification, that is, finding an appropriate set $X$ satisfying the hypothesis of Theorem 2.3, is successful only in particular cases. The required work tends to be extensive. We mention in this regard the theories of abelian groups, of Boolean algebras, and of other locally finite varieties; see for instance [MV]. The above examples are just the simplest ones.

Easy to deal with is the case of an incomplete theory $T$ that has finitely many completions. Example 6(a) is just a special case. According to Exercise 5 in **3.5**, $T$ then has finitely many extensions. Moreover, all these are *finite* extensions. Indeed, if $T + \{\alpha_i \mid i \in \mathbb{N}\}$ is a nonfinite extension then w.l.o.g. $\bigwedge_{i<n} \alpha_i \nvdash_T \alpha_n$, which obviously implies that $T$ has infinitely many completions, contradicting our hypothesis. Thus, we may assume that $T_1, \ldots, T_m$ are the completions of $T$ and that $T_i = T + \alpha_i$ for some $\alpha_i \in \mathcal{L}^0$. Then $\{\alpha_1, \ldots, \alpha_m\}$ is a Boolean basis for $\mathcal{L}^0$ in $T$. Exercise 4 provides a canonical axiomatization of all consistent extensions of $T$.

## Exercises

1. Prove that also $\mathsf{DO}_{10}$, $\mathsf{DO}_{11}$, and $\mathsf{DO}_{01}$ are $\aleph_0$-categorical and hence complete. In addition, verify that these theories and $\mathsf{DO}_{00}$ are the only completions of $\mathsf{DO}$.

2. Prove that $T_{\mathrm{suc}}$ (page 178) is also completely axiomatized by the first two given axioms plus IS: $\varphi \frac{0}{x} \wedge \forall x(\varphi \rightarrow \varphi \frac{\mathsf{S}x}{x}) \rightarrow \forall x\varphi$; here $\varphi$ runs over all formulas of $\mathcal{L}\{0, \mathsf{S}\}$ (the "induction schema" for $\mathcal{L}\{0, \mathsf{S}\}$).

3. Show that the theory $T$ of torsion-free divisible abelian groups is $\aleph_1$-categorical and complete (hence decidable). This shows, e.g., that the groups $(\mathbb{R}, 0, +)$ and $(\mathbb{Q}, 0, +)$ are elementarily equivalent.

4. Let $T$ be incomplete and let $T+\alpha_0, \ldots, T+\alpha_m$ be all the completions of $T$. Prove that $T + \bigvee_{\nu \leqslant n} \alpha_{i_\nu}$ are all consistent extensions of $T$. Here $n \leqslant m$ and $i_0 < \cdots < i_n \leqslant m$. (Note that $T = T + \bigvee_{i \leqslant m} \alpha_i$.)

5. Show that an $\aleph_0$-categorical theory $T$ with no finite models has an elementary prime model. Example: $(\mathbb{Q}, <)$ is an elementary prime model for $\mathsf{DO}_{00}$.

## 5.3   The Ehrenfeucht Game

Unfortunately, Vaught's criterion has only limited applications because many complete theories are not categorical in any transfinite cardinality. Let SO denote the *theory of discretely ordered sets*, i.e., of all $(M, <)$ such that every $a \in M$ has an immediate successor provided $a$ is not the right edge element, and likewise an immediate predecessor provided $a$ is not a left edge element. "SO" is intended to recall "step order," because the word "discrete" in connection with orders often has the stronger sense "each cut is a jump." $SO_{ij}$ ($i.j \in \{0,1\}$) is defined analogously to $DO_{ij}$ (see page 177). For instance, $SO_{10}$ is the theory of discretely ordered sets with left and without right edge element. Clearly, $(\mathbb{N}, <)$ is a prime model for $SO_{10}$. The models of $SO_{10}$ arise from arbitrary orders $(M, <)$ with a left edge element by replacing the latter by $(\mathbb{N}, <)$ and every other element of $M$ by a specimen of $(\mathbb{Z}, <)$. From this it follows that $SO_{10}$ cannot be $\kappa$-categorical for any $\kappa \geqslant \aleph_0$. Yet this theory is complete, as will be shown, and the same applies to $SO_{00}$ and $SO_{01}$. Only $SO_{11}$ is incomplete and is the only one of the four theories that has finite models. It coincides with the elementary theory of all finite ordered sets; Exercise 3.

We prove the completeness of $SO_{10}$ game-theoretically using a two-person game with players I and II, *Ehrenfeucht's game* $\Gamma_k(\mathcal{A}, \mathcal{B})$, which is played in $k$ rounds, $k \geqslant 0$. Here $\mathcal{A}, \mathcal{B}$ are given $\mathcal{L}$-structures and $\mathcal{L}$ is a *relational language*, i.e., $\mathcal{L}$ does not contain constants or operation symbols. With regard to our goal this presents no real loss of generality because each structure can be converted into a relational one by replacing its operations by the corresponding graphs. Another advantage of relational structures used in the sequel is that there is a bijective correspondence between subsets and substructures.

We now describe the game $\Gamma_k(\mathcal{A}, \mathcal{B})$. Player I chooses in each of the $k$ rounds one of the two structures $\mathcal{A}$ and $\mathcal{B}$. If this is $\mathcal{A}$, he selects some $a \in A$. Then player II has to answer with some element $b \in B$. If player I chooses $\mathcal{B}$ and some $b$ from $B$ then player II must answer with some element $a \in A$. This is the entire game. Clearly, it has still to be explained who wins. After $k$ rounds, elements $a_1, \ldots, a_k \in A$ and $b_1, \ldots, b_k \in B$ have been selected, where $a_i, b_i$ denote the elements selected in round $i$. Player II wins if the mapping $a_i \mapsto b_i$ ($i = 1, \ldots, k$) is a partial isomorphism from $\mathcal{A}$ to $\mathcal{B}$; in other words, if the substructure of $\mathcal{A}$ with the domain $\{a_1, \ldots, a_k\}$ is isomorphic to the substructure of $\mathcal{B}$ with the domain $\{b_1, \ldots, b_k\}$. Otherwise, player I is the winner.

We write $\mathcal{A} \sim_k \mathcal{B}$ if player II has a winning strategy in the game $\Gamma_k(\mathcal{A}, \mathcal{B})$, that is, in every round player II can answer any move from player I such that at the end player II is the winner. For the "zero-round game" let $\mathcal{A} \sim_0 \mathcal{B}$ by definition.

**Example.** Let $\mathcal{A} = (\mathbb{N}, <)$ be a proper initial segment of $\mathcal{B} \vDash SO_{10}$. We show that $\mathcal{A} \sim_k \mathcal{B}$ for arbitrary $k > 0$. Player II plays as follows: If player I chooses some $b_1$ in $\mathcal{B}$ in the first round then player II answers with $a_1 = 2^{k-1}-1$ if $d(0, b_1) \geqslant 2^{k-1}-1$, otherwise with $a_1 = d(0, b_1)$.[3] The procedure is similar if player I begins with $\mathcal{A}$. If player I now selects some $b_2 \in \mathcal{B}$ such that $d(0, b_2), d(b_1, b_2) \geqslant 2^{k-2}-1$, then player II answers with $a_2 = a_1 \pm 2^{k-2}$ depending on whether $b_2 > b_1$ or $b_2 < b_1$, and otherwise with the element of the same distance from 0 or $a_1$ as that of $b_2$ from 0 or $b_1$ in $\mathcal{B}$. Similarly in the third round, etc.

$$\bullet \ \bullet \ \bullet \ \bullet \cdots \mathcal{B} \cdots \bullet \ \bullet \ \bullet \quad \bullet \quad \bullet \ \bullet \ \bullet \cdots$$
$$b_2 \qquad\qquad b_3 \qquad b_1$$

$$\bullet \ \bullet \ \bullet \ \bullet \cdots \mathcal{A}$$
$$a_2 \ a_3 \ a_1 \qquad a_1 = 2^2 - 1 = 3, \ a_2 = a_1 - 2^1 = 1$$

The figure shows the course of a 3-round game played in the described way. Player I has chosen from $\mathcal{B}$ only for simplicity. With this strategy player II wins every game, as can be shown by induction on $k$. The reader should play a few rounds before proving this rigorously.

In contrast to the example, for $\mathcal{A} = (\mathbb{N}, <)$ and $\mathcal{B} = (\mathbb{Z}, <)$ player II's chances have already dropped in $\Gamma_2(\mathcal{A}, \mathcal{B})$ if player I selects $0 \in A$ in the first round. Player II will lose already in the second round. This has to do with the fact that the existence of an edge element is expressible by a sentence of quantifier rank 2. We write $\mathcal{A} \equiv_k \mathcal{B}$ for $\mathcal{L}$-structures $\mathcal{A}, \mathcal{B}$ if $\mathcal{A} \vDash \alpha \Leftrightarrow \mathcal{B} \vDash \alpha$, for all $\alpha \in \mathcal{L}^0$ with $\mathrm{qr}\,\alpha \leqslant k$. It is always the case that $\mathcal{A} \equiv_0 \mathcal{B}$ for all $\mathcal{A}, \mathcal{B}$, because in relational languages there are no sentences of quantifier rank 0. Below we will prove the following remarkable

**Theorem 3.1.** $\mathcal{A} \sim_k \mathcal{B}$ *implies* $\mathcal{A} \equiv_k \mathcal{B}$. *Hence,* $\mathcal{A} \equiv \mathcal{B}$ *provided* $\mathcal{A} \sim_k \mathcal{B}$ *for all* $k$.

---

[3] The "distance" $d(a, b)$ between elements $a, b$ of some $SO$-model is 0 for $a = b$, $1 +$ the number of elements between $a$ and $b$ if it is finite, and $d(a, b) = \infty$ otherwise.

For finite signatures a somewhat weaker version of the converse of the theorem is valid as well, though we do not discuss this here. Before proving Theorem 3.1 we demonstrate its applicability. The theorem and the above example yield $(\mathbb{N}, <) \equiv_k \mathcal{B}$ for all $k$ and hence $(\mathbb{N}, <) \equiv \mathcal{B}$ for every $\mathcal{B} \vDash \mathsf{SO}_{10}$, because $(\mathbb{N}, <)$ is a prime model for $\mathsf{SO}_{10}$ and hence can always be regarded as an initial segment of $\mathcal{B}$. Therefore $\mathsf{SO}_{10}$ is complete. For reasons of symmetry the same holds for $\mathsf{SO}_{01}$, and likewise for $\mathsf{SO}_{00}$. On the other hand, $\mathsf{SO}_{11}$ has the finite model property according to Exercise 3 and coincides with the theory $\mathsf{FO}$ of all finite ordered sets.

For the proof of Theorem 3.1 we first consider a minor generalization of $\Gamma_k(\mathcal{A}, B)$, the game $\Gamma_k(\mathcal{A}, \mathcal{B}, \vec{a}, \vec{b})$ with prior moves $\vec{a} \in A^n, \vec{b} \in B^n$. In the first round player I selects some $a_{n+1} \in A$ or $b_{n+1} \in B$ and player II answers with $b_{n+1}$ or $a_{n+1}$, etc. The game protocol consists of sequences $(a_1, \ldots, a_{n+k})$ and $(b_1, \ldots, b_{n+k})$ at the end. Player II has won if $a_i \mapsto b_i$ $(i = 1, \ldots, n + k)$ is a partial isomorphism. Clearly, for $n = 0$ we obtain precisely the original game $\Gamma_k(\mathcal{A}, \mathcal{B})$.

This adjustment brings about an inductive characterization of a winning strategy for player II independent of more general concepts as follows:

**Definition.** Player II has a winning strategy in $\Gamma_0(\mathcal{A}, \mathcal{B}, \vec{a}, \vec{b})$ provided $a_i \mapsto b_i$ for $i = 1, \ldots, n$ is a partial isomorphism. Player II has a winning strategy in $\Gamma_{k+1}(\mathcal{A}, \mathcal{B}, \vec{a}, \vec{b})$ if for every $a \in A$ there is some $b \in B$, and for every $b \in B$ some $a \in A$, such that player II has a winning strategy in $\Gamma_k(\mathcal{A}, \mathcal{B}, \vec{a}\_a, \vec{b}\_b)$. Here $\vec{c}\_c$ denotes the operation of appending the element $c$ to the sequence $\vec{c}$.

We shall write $(\mathcal{A}, \vec{a}) \sim_k (\mathcal{B}, \vec{b})$ if player II has a winning strategy in $\Gamma_k(\mathcal{A}, \mathcal{B}, \vec{a}, \vec{b})$. In particular, $\mathcal{A} \sim_k \mathcal{B}$ (this represents the choice $\vec{a} = \vec{b} = \emptyset$) is now precisely defined.

**Lemma 3.2.** *Let $(\mathcal{A}, \vec{a}) \sim_k (\mathcal{B}, \vec{b})$, where $\vec{a} \in A^n$ and $\vec{b} \in B^n$. Then $(*) \colon \mathcal{A} \vDash \varphi(\vec{a}) \Leftrightarrow \mathcal{B} \vDash \varphi(\vec{b})$, for all $\varphi = \varphi(\vec{x})$ such that $\mathrm{qr}\, \varphi \leqslant k$.*

**Proof** by induction on $k$. Let $k = 0$. Since $a_i \mapsto b_i$ $(i = 1, \ldots, n)$ is a partial isomorphism, $(*)$ is valid for prime formulas, and since the induction steps in the proof of $(*)$ for $\neg$, $\wedge$ are obvious, it is valid also for all formulas $\varphi$ with $\mathrm{qr}\, \varphi = 0$. Now let $(\mathcal{A}, \vec{a}) \sim_{k+1} (\mathcal{B}, \vec{b})$. The only interesting case is $\varphi = \forall y \alpha(\vec{x}, y)$ such that $\mathrm{qr}\, \varphi = k + 1$, because it is easily seen that every other formula of quantifier rank $k+1$ is a Boolean combination of such formulas and of formulas of quantifier rank $\leqslant k$. Induction over $\neg$ and $\wedge$ in proving $(*)$ is harmless. Let $\mathcal{A} \vDash \forall y \alpha(\vec{a}, y)$ and $b \in B$. Then Player II chooses some $a \in A$ with $(\mathcal{A}, \vec{a}\smile a) \sim_k (\mathcal{B}, \vec{b}\smile b)$, so that according to the induction hypothesis, $\mathcal{A} \vDash \alpha(\vec{a}, a) \Leftrightarrow \mathcal{B} \vDash \alpha(\vec{b}, b)$. Clearly, the latter is supposed to hold for sequences $\vec{a}, \vec{b}$ of elements of arbitrary length. Because of $\mathcal{A} \vDash \alpha(\vec{a}, a)$, also $\mathcal{B} \vDash \alpha(\vec{b}, b)$. Since $b$ was arbitrary, we obtain $\mathcal{B} \vDash \forall y \alpha(\vec{b}, y)$. For reasons of symmetry, $\mathcal{B} \vDash \forall y \alpha(\vec{b}, y) \Rightarrow \mathcal{A} \vDash \forall y \beta(\vec{a}, y)$ holds as well. ◻

Theorem 3.1 is just the application of this lemma for the case $n = 0$ and has therefore been proved. The method illustrated is wide-ranging and has many generalizations.

### Exercises

1. Let $\mathcal{A}, \mathcal{B}$ be two infinite discretely ordered sets with the same edge configuration. Prove that $\mathcal{A} \sim_k \mathcal{B}$ for all $k$. Hence $\mathcal{A}, \mathcal{B}$ are elementarily equivalent.

2. Let $\mathcal{A}, \mathcal{B} \vDash \mathsf{SO}_{11}$, $k > 0$, and $|A|, |B| \geqslant 2^k - 1$. Prove that $\mathcal{A} \sim_k \mathcal{B}$, so that $\mathcal{A} \equiv_k \mathcal{B}$ according to Theorem 3.1.

3. Infer from Exercise 2 that $\mathsf{SO}_{11}$ has the finite model property and coincides with the elementary theory $\mathsf{FO}$ of all finite ordered sets.

4. Show that $\mathsf{L}, \mathsf{R}, \exists_1, \exists_2, \ldots$ constitute a Boolean basis modulo $\mathsf{SO}$ and use this to prove the decidability of $\mathsf{SO}$.[4]

## 5.4  Embedding and Characterization Theorems

Many of the foregoing theories, for instance those of orders, of groups in $\cdot, e, ^{-1}$, and of rings, are universal or $\forall$-theories, considered already on page 83. We also know that for every theory $T$ of this kind $\mathcal{A} \subseteq \mathcal{B} \vDash T$ implies $\mathcal{A} \vDash T$; in short, $T$ is $\mathsf{S}$-*invariant*. $\mathsf{DO}$ obviously does not have

---

[4] Moreover, the theory of all linear orders is decidable (Ehrenfeucht), and thus each of its finite extensions; but the proof is incomparably more difficult than for $\mathsf{DO}$ or $\mathsf{SO}$.

this property, and so there cannot exist an axiom system of $\forall$-sentences for it. According to Theorem 4.3 the $\forall$-theories are completely characterized by the property of **S**-invariance. This fact presents a particularly simple example of the model-theoretic characterization of certain syntactic forms of axiom systems.

$T^\forall := \{\alpha \in T \mid \alpha \text{ is an } \forall\text{-sentence}\}$ is called the *universal part* of a theory $T$. Note the distinction between the set $T^\forall$ and the $\forall$-theory $T^\forall$, which of course contains more than just $\forall$-sentences. For $\mathcal{L}_0 \subseteq \mathcal{L}$ put $T_0^\forall := \mathcal{L}_0 \cap T^\forall$. If $\mathcal{A}$ is an $\mathcal{L}_0$-structure and $\mathcal{B}$ an $\mathcal{L}$-structure then $\mathcal{A} \subseteq \mathcal{B}$ or '$\mathcal{A}$ is a substructure of $\mathcal{B}$' will often mean in this section that $\mathcal{A}$ is a substructure of the $\mathcal{L}_0$-reduct of $\mathcal{B}$. The phrase '$\mathcal{A}$ is embeddable into $\mathcal{B}$' introduced in **5.1** is to be understood correspondingly. Examples will be found below. First we state the following

**Lemma 4.1.** *Every $T_0^\forall$-model $\mathcal{A}$ is embeddable into some $T$-model.*

**Proof.** It is enough to prove $(*)$: $T + D\mathcal{A}$ is consistent, because $\mathcal{A}$ is embeddable into each $\mathcal{B} \vDash T + D\mathcal{A}$ by Theorem 1.1. Assume that $(*)$ is false. Then there is a conjunction $\varkappa(\vec{a})$ of sentences in $D\mathcal{A}$ such that $\varkappa(\vec{a}) \vdash_T \bot$, or equivalently, $\vdash_T \neg\varkappa(\vec{a})$. Here let $\vec{a}$ embrace all the constants of $\mathcal{L}\mathcal{A}$ that appear in the members of $\varkappa$ but not in $T$. By the rule $(\forall 3)$ of constant quantification from **3.2**, $\vdash_T \forall\vec{x}\neg\varkappa(\vec{x})$. Hence $\forall\vec{x}\neg\varkappa(\vec{x}) \in T_0^\forall$ and thus $\mathcal{A} \vDash \forall\vec{x}\neg\varkappa(\vec{x})$, a contradiction to $\mathcal{A} \vDash \varkappa(\vec{a})$. $\quad\blacksquare$

**Lemma 4.2.** $\operatorname{Md} T^\forall$ *consists of just the substructures of all $T$-models.*

**Proof.** Every substructure of a $T$-model is of course a $T^\forall$-model. Furthermore, each $\mathcal{A} \vDash T^\forall$ is (by Lemma 4.1 for $\mathcal{L}_0 = \mathcal{L}$) embeddable into some $\mathcal{B} \vDash T$, and this is surely equivalent to $\mathcal{B}' \simeq \mathcal{B}$ and $\mathcal{A} \subseteq \mathcal{B}'$ for some $\mathcal{B}' \vDash T$, because $\operatorname{Md} T$ is always closed under isomorphic images. $\quad\blacksquare$

**Example.** (a) Let AG be the theory of abelian groups in $\mathcal{L}\{\circ\}$. A substructure of $\mathcal{A} \vDash$ AG is obviously a commutative regular semigroup. Conversely, it is not hard to prove that every such semigroup is embeddable into an abelian group. Therefore, the theory $\mathsf{AG}^\forall$ coincides with the theory of the commutative regular semigroups. *Warning*: noncommutative regular semigroups need not be embeddable into groups.

(b) Substructures of fields in $\mathcal{L}\{0, 1, +, -, \cdot\}$ are integral domains. Conversely, according to a basic algebraic construction, every integral domain (not every ring) is embeddable into a field, its *quotient field*. It is constructed similarly to the field $\mathbb{Q}$ from the integral domain $\mathbb{Z}$. Hence, by Lemma 4.2, the theory $T_J$ of integral domains, axiomatized by the axioms for commutative rings with 1 and without zero-divisors, has the same universal part as the theory $T_F$ of fields. Also, ACF has the same universal part, because every field is embeddable into some algebraically closed field, its algebraic closure; see [Wae] and Example 1 in **5.5**.

**Theorem 4.3.** $T$ *is a universal theory iff* $T$ *is* **S**-*invariant*.

**Proof.** This follows immediately from Lemma 4.2, since for an **S**-invariant theory $T$, clearly $\mathrm{Md}\,T = \mathrm{Md}\,T^\forall$. In other words, $T$ is axiomatized by its universal part $T^\forall$. ∎

This theorem is reminiscent of the **HSP** theorem cited on page 129. However, the latter concerns identities only. It has a proof that is akin to the proof of the following remarkable theorem, which presents an elegant model-theoretic characterization of universal Horn theories introduced in **4.2**. Call a theory $T$ **SP**-*invariant* if $\mathrm{Md}\,T$ is closed under direct products and substructures. Always remember that a statement like $\mathcal{A} \models \varphi(\vec{a})$ with $\vec{a} \in A^n$ is to mean either $\mathcal{A} \models \varphi(\vec{x})\,[\vec{a}]$ or $\mathcal{A}_A \models \varphi(\vec{a})$.

**Theorem 4.4.** $T$ *is a universal Horn theory iff* $T$ *is* **SP**-*invariant*.

**Proof.** $\Rightarrow$: Exercise 1 in **4.2**. $\Leftarrow$: Trivial if $\vdash_T \forall xy\, x = y$, for then $T$ is axiomatized by $\forall xy\, x = y$. Otherwise let $U$ be the set of all universal Horn sentences of $T$. We prove $\mathrm{Md}\,T = \mathrm{Md}\,U$. Only $\mathrm{Md}\,U \subseteq \mathrm{Md}\,T$ is not obvious. Let $\mathcal{A} \models U$. To verify $\mathcal{A} \models T$ it suffices to show (∗): $T \cup D\mathcal{A} \nvdash \bot$, since for $\mathcal{B} \models T, D\mathcal{A}$ w.l.o.g. $\mathcal{A} \subseteq \mathcal{B}$, so $\mathcal{A} \models T$ thanks to **S**-invariance. Let $P := \{\pi \in D\mathcal{A} \mid \pi \text{ prime}\}$, so that $D\mathcal{A} = P \cup \{\neg \pi_i \mid i \in I\}$ for some $I \neq \emptyset$, all $\pi_i$ prime. We first show $\binom{*}{*}$: $P \nvdash_T \pi_i$ for all $i \in I$. Indeed, otherwise $\vdash_T \varkappa(\vec{a}) \to \pi_i(\vec{a})$ for some conjunction $\varkappa(\vec{a})$ of sentences in $P$, with the tuple $\vec{a}$ of constants not in $T$. Therefore $\vdash_T \alpha := \forall\vec{x}(\varkappa(\vec{x}) \to \pi_i(\vec{x}))$. Hence $\alpha \in U$, for $\alpha$ is a universal Horn sentence in the language of $T$, whence $\mathcal{A} \models \alpha$. But this contradicts $\mathcal{A} \models \varkappa(\vec{a}) \wedge \neg \pi_i(\vec{a})$ and confirms $\binom{*}{*}$. Choose $\mathcal{A}_i \models T, P, \neg \pi_i$. Then $\mathcal{B} := \prod_{i \in I} \mathcal{A}_i \models T \cup P \cup \{\neg \pi_i \mid i \in I\} = T \cup D\mathcal{A}$ (note that $\mathcal{B} \models \pi_i$ is impossible since $\mathcal{A}_i \models \neg \pi_i$). This verifies (∗). ∎

The following application of Lemma 4.1 aims in a somewhat different direction.

**Theorem 4.5.** *Let $\mathcal{L}_0 \subseteq \mathcal{L}$ and let $\mathcal{A}$ be an $\mathcal{L}_0$-structure. For $T \subseteq \mathcal{L}^0$ the following are equivalent:*

   (i)    *$\mathcal{A}$ is embeddable into some $T$-model,*

   (ii)   *every finitely generated $\mathcal{B} \subseteq \mathcal{A}$ is embeddable into a $T$-model,*

   (iii)  *$\mathcal{A} \vDash T_0^\forall \ (= \mathcal{L}_0 \cap T^\forall)$.*

**Proof.** (i)$\Rightarrow$(ii): Trivial. (ii)$\Rightarrow$(iii): Let $\forall \vec{x}\alpha \in T_0^\forall$ with $\alpha = \alpha(\vec{x})$ open, w.l.o.g. $\vec{x} = (x_1, \ldots, x_n) \neq \emptyset$. Let $\mathcal{A}_0$ for $\vec{a} = (a_1, \ldots, a_n) \in A^n$ be the substructure in $\mathcal{A}$ generated from $a_1, \ldots, a_n$. By (ii), $\mathcal{A}_0 \subseteq \mathcal{B}$ for some model $\mathcal{B} \vDash T$. Since $\mathcal{B} \vDash \forall \vec{x}\alpha$, it holds that $\mathcal{A}_0 \vDash \forall \vec{x}\alpha$; therefore $\mathcal{A}_0 \vDash \alpha(\vec{a})$, so that $\mathcal{A} \vDash \alpha(\vec{a})$ by Theorem 2.3.2. Since $\vec{a} \in A^n$ was chosen arbitrarily, $\mathcal{A} \vDash \forall \vec{x}\alpha$, and since $\forall \vec{x}\alpha$ was arbitrarily taken from $T_0^\forall$, it follows that $\mathcal{A} \vDash T_0^\forall$. (iii)$\Rightarrow$(i): This is exactly the claim of Lemma 4.1. $\square$

**Examples of applications.** (a) Let $T$ be the theory of ordered abelian groups in $\mathcal{L} = \mathcal{L}\{0, +, -, <\}$. Such a group is clearly torsion-free, which is expressed by a schema of $\forall$-sentences in $\mathcal{L}_0 = \mathcal{L}\{0, +, -\}$. Conversely, Theorem 4.5 implies that a torsion-free abelian group (the $\mathcal{A}$ in the theorem) is orderable, or what amounts to the same thing, is embeddable into an ordered abelian group. One needs to show only that every finitely generated torsion-free abelian group $G$ is orderable. By a well-known result from group theory, $G \simeq \mathbb{Z}^n$ for some $n > 0$. But $\mathbb{Z}^n$ can be ordered lexicographically, as is easily seen by induction on $n$. For nonabelian groups, the conditions corresponding to torsion-freeness are somewhat more involved. (b) Without needing algebraic methods we know that there exists a set of universal sentences in $0, 1, +, -, \cdot$, whose adoption to the theory of fields characterizes the orderable fields. Sufficient for this, by Theorem 4.5, is the set of all $\forall$-sentences in $0, 1, +, -, \cdot$ provable from the axioms for ordered fields. Indeed, even the schema of sentences '$-1$ is not a sum of squares' is enough to characterize the orderable fields (see [Wae]).

Not just $\forall$-theories but also $\forall$-formulas can be characterized model-theoretically. Call $\alpha(\vec{x})$ **S**-*persistent* or simply *persistent* in $T$ provided all $\mathcal{A}, \mathcal{B} \vDash T$ have the property

        (sp)   $\mathcal{A} \subseteq \mathcal{B} \vDash \alpha(\vec{a}) \Rightarrow \mathcal{A} \vDash \alpha(\vec{a})$, for all $\vec{a} \in A^n$.

According to the next theorem this property is characteristic for the $\forall$-formulas up to equivalence in $T$.

**Theorem 4.6.** *If $\alpha = \alpha(\vec{x})$ is persistent in $T$ then $\alpha$ is equivalent to some $\forall$-formula $\alpha'$ in $T$, which can be chosen such that free $\alpha' \subseteq$ free $\alpha$.*

**Proof.** Let $Y$ be the set of all formulas $\forall \vec{y} \beta(\vec{x}, \vec{y})$ with $\alpha \vdash_T \forall \vec{y} \beta(\vec{x}, \vec{y})$, where $\beta$ is open and $\vec{x}$ and $\vec{y}$ are of length $n \geqslant 0$ and $m \geqslant 0$, respectively. We shall prove (a): $Y \vdash_T \alpha(\vec{x})$. This would complete the proof because then, thanks to free $Y \subseteq \{x_1, \ldots, x_n\}$, there is a conjunction $\varkappa = \varkappa(\vec{x})$ of formulas from $Y$ with $\varkappa \vdash_T \alpha$. Since also $\alpha \vdash_T \varkappa$, we have $\alpha \equiv_T \varkappa$. Moreover, $\alpha' := \varkappa \in Y$, since $Y$ is closed under conjunction according to Exercise 3 in **2.4**. This proves the claim. For proving (a) we assume (b) $(\mathcal{A}, \vec{a}) \vDash T, Y$ with $\vec{a} \in A^n$. We need to show that $(\mathcal{A}, \vec{a}) \vDash \alpha$. This follows from (c): $T, \alpha(\vec{a}), D\mathcal{A}$ is consistent, for if $\mathcal{B} \vDash T, \alpha(\vec{a}), D\mathcal{A}$, then w.l.o.g. $\mathcal{A} \subseteq \mathcal{B}$; and also $\mathcal{A} \vDash \alpha(\vec{a})$ since $\alpha$ is persistent. If (c) were false then $\alpha(\vec{a}) \vdash_T \neg \varkappa(\vec{a}, \vec{b})$ for some conjunction $\varkappa(\vec{a}, \vec{b})$ of sentences from $D\mathcal{A}$ with the $m$-tuple $\vec{b}$ of constants of $\varkappa$ from $A \backslash \{a_1, \ldots, a_n\}$. Thus $\alpha(\vec{a}) \vdash_T \forall \vec{y} \neg \varkappa(\vec{a}, \vec{y})$. Since the $a_1, \ldots, a_n$ do not appear in $T$, we get $\alpha(\vec{x}) \vdash_T \forall \vec{y} \neg \varkappa(\vec{x}, \vec{y}) \in Y$. Therefore, and by (b), $(\mathcal{A}, \vec{a}) \vDash \forall \vec{y} \neg \varkappa(\vec{x}, \vec{y})$, or equivalently $\mathcal{A} \vDash \forall \vec{y} \neg \varkappa(\vec{a}, \vec{y})$, in contradiction to $\mathcal{A} \vDash \varkappa(\vec{a}, \vec{b})$. $\qquad \square$

**Remark.** Let $T$ be countable and all $T$-models infinite. Then $\alpha$ is already equivalent in $T$ to an $\forall$-formula, provided $\alpha$ is $\kappa$-*persistent*; this means that (sp) holds for all $T$-models $\mathcal{A}, \mathcal{B}$ of some fixed cardinal $\kappa \geqslant \aleph_0$. For in this case each $T$-model is elementarily equivalent to a model of cardinality $\kappa$ by the Löwenheim–Skolem theorems. Hence, it suffices to verify (a) in the above proof by considering only models $\mathcal{A}, \mathcal{B}$ of cardinality $\kappa$.

Sentences of the form $\forall \vec{x} \exists \vec{y} \alpha$ with kernel $\alpha$ are called $\forall \exists$-*sentences*. Many theories, for instance of real or of algebraically closed fields and of divisible groups, are $\forall \exists$-*theories*, i.e., they possess axiom systems of $\forall \exists$-sentences. We shall characterize the $\forall \exists$-theories semantically.

A *chain $K$ of structures* is a set $K$ of $\mathcal{L}$-structures such that $\mathcal{A} \subseteq \mathcal{B}$ or $\mathcal{B} \subseteq \mathcal{A}$ for all $\mathcal{A}, \mathcal{B} \in K$. Chains are very often given as sequences $\mathcal{A}_0 \subseteq \mathcal{A}_1 \subseteq \mathcal{A}_2 \subseteq \cdots$. No matter how $K$ is given, a structure $\mathcal{C} := \bigcup K$ can be defined in a natural way: Let $C := \bigcup \{A \mid \mathcal{A} \in K\}$ be its domain. Further, let $r^{\mathcal{C}} \vec{a} \Leftrightarrow r^{\mathcal{A}} \vec{a}$ for $\vec{a} \in C^n$, where $\mathcal{A} \in K$ is chosen such that $\vec{a} \in A^n$. Such an $\mathcal{A} \in K$ exists: Let $\mathcal{A}$ simply be the maximum of the chain members containing $a_1, \ldots, a_n$, respectively. The definition of $r^{\mathcal{C}}$ is independent of the choice of $\mathcal{A}$. Indeed, let $\mathcal{A}' \in K$ and $a_1, \ldots, a_n \in A'$. Since $\mathcal{A} \subseteq \mathcal{A}'$ or $\mathcal{A}' \subseteq \mathcal{A}$, it holds that $r^{\mathcal{A}} \vec{a} \Leftrightarrow r^{\mathcal{A}'} \vec{a}$ in either case. Finally, for function symbols $f$, let $f^{\mathcal{C}}\vec{a} = f^{\mathcal{A}}\vec{a}$, where $\mathcal{A} \in K$ is chosen such that $\vec{a} \in A^n$. Here too the choice of $\mathcal{A} \in K$ is irrelevant. $\mathcal{C}$ was just defined in such a way that each $\mathcal{A} \in K$ is a substructure of $\mathcal{C}$.

**Example 1.** Let $\mathcal{D}_n$ be the additive group of $n$-place decimal numbers (with at most $n$ decimals after the decimal point). Since $\mathcal{D}_n \subseteq \mathcal{D}_{n+1}$, the $\mathcal{D}_n$ form a chain. Here $\mathcal{D} = \bigcup_{n \in \mathbb{N}} \mathcal{D}_n$ is just the additive group of finite decimal numbers. The corresponding holds if the $\mathcal{D}_n$ are understood as ordered sets. Because then $\mathcal{D} \vDash \mathsf{DO}$, while $\mathcal{D}_n \vDash \mathsf{SO}$ for all $n$, Md $\mathsf{SO}$ is not closed under union of chains. Therefore $\mathsf{SO}$ is not an $\forall\exists$-theory (in contrast to $\mathsf{DO}$), as follows from a simple observation in the next paragraph.

It is easy to see that an $\forall\exists$-sentence $\alpha = \forall x_1 \cdots x_n \exists y_1 \cdots y_m \beta(\vec{x}, \vec{y})$ valid in all members $\mathcal{A}$ of a chain $K$ of structures is also valid in $\mathcal{C} = \bigcup K$. For let $\vec{a} \in C^n$. Then clearly $\vec{a} \in A^n$ for some $\mathcal{A} \in K$, hence $\mathcal{A} \vDash \exists \vec{y}(\vec{a}, \vec{y})$. Since $\mathcal{A} \subseteq \mathcal{C}$, it follows that $\mathcal{C} \vDash \exists \vec{y}(\vec{a}, \vec{y})$ according to Corollary 2.3.3. Now, $\vec{a}$ was arbitrarily be chosen, hence indeed $\mathcal{C} \vDash \forall \vec{x} \exists \vec{y} \beta(\vec{x}, \vec{y})$. Thus, if $T$ is an $\forall\exists$-theory then Md $T$ is always closed under union of chains, or as is said, $T$ is *inductive*.

This property is characteristic of $\forall\exists$-theories, Theorem 4.9. However, the proof is no longer simple. It requires the notion of an *elementary chain*. This is a set $K$ of $\mathcal{L}$-structures such that $\mathcal{A} \preccurlyeq \mathcal{B}$ or $\mathcal{B} \preccurlyeq \mathcal{A}$, for all $\mathcal{A}, \mathcal{B} \in K$. Clearly, $K$ is then also a chain with respect to $\subseteq$.

**Lemma 4.7 (Tarski's chain lemma).** *Let $K$ be an elementary chain and put $\mathcal{C} = \bigcup K$. Then $\mathcal{A} \preccurlyeq \mathcal{C}$ for every $\mathcal{A} \in K$.*

**Proof.** We have to show that $\mathcal{A} \vDash \alpha(\vec{a}) \Leftrightarrow \mathcal{C} \vDash \alpha(\vec{a})$, with $\vec{a} \in A^n$. This follows by induction on $\alpha = \alpha(\vec{x})$ and is clear for prime formulas. The induction steps over $\wedge, \neg$ are also straightforward. Let $\mathcal{A} \vDash \forall y \alpha(y, \vec{a})$ and $a_0 \in C$ arbitrary. There is certainly some $\mathcal{B} \in K$ such that $a_0, \ldots, a_n \in B$ and $\mathcal{A} \preccurlyeq \mathcal{B}$. Thus, $\mathcal{B} \vDash \forall y \alpha(y, \vec{a})$ and hence $\mathcal{B} \vDash \alpha(a_0, \vec{a})$. By the induction hypothesis (which is supposed to hold for *any* chain member) so too $\mathcal{C} \vDash \alpha(a_0, \vec{a})$. Since $a_0 \in C$ was arbitrary, $\mathcal{C} \vDash \forall y \alpha(y, \vec{a})$. The converse $\mathcal{C} \vDash \forall y \alpha(y, \vec{a}) \Rightarrow \mathcal{A} \vDash \forall y \alpha(y, \vec{a})$ follows similarly. $\square$

We require yet another useful concept, found in many of the examples in **5.5**. Let $\mathcal{A} \subseteq \mathcal{B}$. Then $\mathcal{A}$ is termed *existentially closed in $\mathcal{B}$*, in symbols $\mathcal{A} \subseteq_{ec} \mathcal{B}$, provided

$(*)$   $\mathcal{B} \vDash \exists \vec{x} \varphi(\vec{x}, \vec{a}) \Rightarrow \mathcal{A} \vDash \exists \vec{x} \varphi(\vec{x}, \vec{a})$   $(\vec{a} \in A^n)$,

where $\varphi = \varphi(\vec{x}, \vec{a})$ runs through all conjunctions of literals from $\mathcal{L}A$. $(*)$ then holds automatically for all open $\varphi \in \mathcal{L}A$. One sees this straight away by converting $\varphi$ into a disjunctive normal form and distributing $\exists \vec{x}$ over the disjuncts. Clearly $\mathcal{A} \preccurlyeq \mathcal{B} \Rightarrow \mathcal{A} \subseteq_{ec} \mathcal{B} \Rightarrow \mathcal{A} \subseteq \mathcal{B}$. Moreover, $\subseteq_{ec}$ satisfies a readily proved chain lemma as well: If $K$ is a chain of structures such that $\mathcal{A} \subseteq_{ec} \mathcal{B}$ or $\mathcal{B} \subseteq_{ec} \mathcal{A}$ for all $\mathcal{A}, \mathcal{B} \in K$, then $\mathcal{A} \subseteq_{ec} \bigcup K$ for every $\mathcal{A} \in K$. This is an easy exercise.

The next lemma presents various characterizations of $\mathcal{A} \subseteq_{ec} \mathcal{B}$. Let $D_\forall \mathcal{A}$ denote the *universal diagram* of $\mathcal{A}$, which is the set of all $\forall$-sentences of $\mathcal{L}A$ valid in $\mathcal{A}$. Clearly $D_\forall \mathcal{A} \subseteq D_{el}\mathcal{A}$. In (iii) the indexing of $\mathcal{B}$ with $A$ is omitted to ease legibility.

**Lemma 4.8.** *Let $\mathcal{A}, \mathcal{B}$ be $\mathcal{L}$-structures and $\mathcal{A} \subseteq \mathcal{B}$. Then are equivalent* (i) $\mathcal{A} \subseteq_{ec} \mathcal{B}$,   (ii) *there is an* $\mathcal{A}' \supseteq \mathcal{B}$ *such that* $\mathcal{A} \preccurlyeq \mathcal{A}'$,   (iii) $\mathcal{B} \vDash D_\forall \mathcal{A}$.

**Proof.** (i)$\Rightarrow$(ii): Let $\mathcal{A} \subseteq_{ec} \mathcal{B}$. We obtain some $\mathcal{A}' \supseteq \mathcal{B}$ such that $\mathcal{A} \preccurlyeq \mathcal{A}'$ as a model of $D_{el}\mathcal{A} \cup D\mathcal{B}$ (more precisely, as the $\mathcal{L}$-reduct of such a model), so that it remains only to show the consistency. Suppose the opposite, so that $D_{el}\mathcal{A} \vdash \neg \varkappa(\vec{b})$ for some conjunction $\varkappa(\vec{b})$ of members from $D\mathcal{B}$ with the $n$-tuple $\vec{b}$ of all constants of $B \setminus A$ in $\varkappa$. Since $b_1, \ldots, b_n$ do not occur in $D_{el}\mathcal{A}$, we get $D_{el}\mathcal{A} \vdash \forall \vec{x} \neg \varkappa(\vec{x})$. Thus $\mathcal{A} \vDash \forall \vec{x} \neg \varkappa(\vec{x})$. On the other hand, $\mathcal{B} \vDash \varkappa(\vec{b})$; hence $\mathcal{B} \vDash \exists \vec{x} \varkappa(\vec{x})$. With (i) and $\varkappa(\vec{x}) \in \mathcal{L}A$ also $\mathcal{A} \vDash \exists \vec{x} \varkappa(\vec{x})$, in contradiction to $\mathcal{A} \vDash \forall \vec{x} \neg \varkappa(\vec{x})$. (ii)$\Rightarrow$(iii): Since $\mathcal{A} \preccurlyeq \mathcal{A}'$, we have $\mathcal{A}' \vDash D_{el}\mathcal{A} \supseteq D_\forall \mathcal{A}$. Because of $\mathcal{B} \subseteq \mathcal{A}' \vDash D_\forall \mathcal{A}$, evidently $\mathcal{B} \vDash D_\forall \mathcal{A}$. (iii)$\Rightarrow$(i): By (iii), $\mathcal{A} \vDash \alpha \Rightarrow \mathcal{B} \vDash \alpha$, for all $\forall$-sentences $\alpha$ of $\mathcal{L}A$. The latter is equivalent to $\mathcal{B} \vDash \alpha \Rightarrow \mathcal{A} \vDash \alpha$, for all $\exists$-sentences $\alpha \in \mathcal{L}A$, and hence to property (i).   ◻

**Theorem 4.9.** *A theory $T$ is an $\forall \exists$-theory if and only if $T$ is inductive.*

**Proof.** As already shown, an $\forall \exists$-theory $T$ is inductive. Conversely let $T$ be inductive. We show that $\operatorname{Md} T = \operatorname{Md} T^{\forall \exists}$, where $T^{\forall \exists}$ denotes the set of all $\forall \exists$-theorems provable in $T$. The nontrivial part is the verification of $\operatorname{Md} T^{\forall \exists} \subseteq \operatorname{Md} T$. So let $\mathcal{A} \vDash T^{\forall \exists}$. **Claim:** $T \cup D_\forall \mathcal{A}$ is consistent. Otherwise $\vdash_T \neg \varkappa$ for some conjunction $\varkappa = \varkappa(\vec{a})$ of sentences of $D_\forall \mathcal{A}$ with the tuple $\vec{a}$ of constants in $A$ appearing in $\varkappa$ but not in $T$. Hence $\vdash_T \forall \vec{x} \neg \varkappa(\vec{x})$. Now, $\varkappa(\vec{x})$ is equivalent to an $\forall$-formula, and so $\neg \varkappa(\vec{x})$ to an

∃-formula. Thus, $\forall \vec{x} \neg \varkappa(\vec{x})$ belongs up to equivalence to $T^{\forall \exists}$. Therefore $\mathcal{A} \vDash \forall \vec{x} \neg \varkappa(\vec{x})$, which contradicts $\mathcal{A} \vDash \varkappa(\vec{a})$. This proves the claim.

Now let $\mathcal{A}_1 \vDash T \cup D_\forall \mathcal{A}$ and w.l.o.g. $\mathcal{A}_1 \supseteq \mathcal{A}$. Then also $\mathcal{A} \subseteq_{ec} \mathcal{A}_1$ in view of Lemma 4.8. By the same lemma there exists an $\mathcal{A}_2 \supseteq \mathcal{A}_1$ with $\mathcal{A}_0 := \mathcal{A} \preccurlyeq \mathcal{A}_2$, so that $\mathcal{A}_2 \vDash T^{\forall \exists}$ as well. We now repeat this construction with $\mathcal{A}_2$ in place of $\mathcal{A}_0$ and obtain structures $\mathcal{A}_3, \mathcal{A}_4$ such that $\mathcal{A}_2 \subseteq_{ec} \mathcal{A}_3 \vDash T$, $\mathcal{A}_3 \subseteq \mathcal{A}_4$, and $\mathcal{A}_2 \preccurlyeq \mathcal{A}_4$. Continuing this construction produces a sequence $\mathcal{A}_0 \subseteq \mathcal{A}_1 \subseteq \mathcal{A}_2 \subseteq \cdots$ of structures with the inclusion relation illustrated in the following figure:

$$\mathcal{A} = \mathcal{A}_0 \; \underbrace{\overset{\subseteq}{\rule{1.5cm}{0.4pt}} \; \mathcal{A}_1 \; \overset{\subseteq}{\rule{1.5cm}{0.4pt}} \; \mathcal{A}_2}_{\preccurlyeq} \; \underbrace{\overset{\subseteq}{\rule{1.5cm}{0.4pt}} \; \mathcal{A}_3 \; \overset{\subseteq}{\rule{1.5cm}{0.4pt}} \mathcal{A}_4}_{\preccurlyeq} \subseteq \cdots \subseteq \mathcal{C}$$

Let $\mathcal{C} := \bigcup_{i \in \mathbb{N}} \mathcal{A}_i$. Clearly, also $\mathcal{C} = \bigcup_{i \in \mathbb{N}} \mathcal{A}_{2i}$, and since by construction $\mathcal{A} = \mathcal{A}_0 \preccurlyeq \mathcal{A}_2 \preccurlyeq \cdots$, we get $\mathcal{A} \preccurlyeq \mathcal{C}$ from the chain lemma. At the same time we also have $\mathcal{C} = \bigcup_{i \in \mathbb{N}} \mathcal{A}_{2i+1}$, and since by construction $\mathcal{A}_{2i+1} \vDash T$ for all $i$, it holds that $\mathcal{C} \vDash T$, for $T$ is inductive. But then too $\mathcal{A} \vDash T$ because $\mathcal{A} \preccurlyeq \mathcal{C}$. This is what we had to prove. ☐

A decent application of the theorem is that $\mathsf{SO}_{10}$ cannot be axiomatized by $\forall \exists$-axioms, for $\mathsf{SO}_{10}$ is not inductive according to Example 1. $\mathsf{SO}_{10}$ is an $\forall \exists \forall$-theory, and we see now that at least one $\forall \exists \forall$-axiom is needed in its axiomatization.

The "sandwich" construction in the proof of Theorem 4.9 can still be generalized. We will not elaborate on this but rather add some words about so-called model compatibility. Let $T_0 + T_1$ be the smallest theory containing $T_0$ and $T_1$. From the consistency of $T_0$ and $T_1$ we cannot infer that $T_0, T_1$ are compatible, i.e., $T_0 + T_1$ is consistent, even if $T_0$ and $T_1$ are *model compatible* in the following sense: every $T_0$-model is embeddable into some $T_1$-model and vice versa. This property is equivalent to $T_0^\forall = T_1^\forall$ by Theorem 4.5, hence it is an equivalence relation. Thus, the class of consistent $\mathcal{L}$-theories splits into disjoint classes of pairwise model compatible theories. That model compatible theories need not be compatible in the ordinary sense is shown by the following

**Example 2.** $\mathsf{DO}$ and $\mathsf{SO}$ are model compatible (Exercise 2) but $\mathsf{DO} + \mathsf{SO}$ is clearly inconsistent. Since $\mathsf{DO}$ is inductive, we get another argument that SO is not inductive: if it were inductive, $DO + SO$ would be consistent according to Exercise 3.

### Exercises

1. Let $X$ be a set of *positive* sentences, i.e., the $\alpha \in X$ are constructed from prime formulas by means of $\wedge, \vee, \forall, \exists$ only. Prove that if $\mathcal{A} \vDash X$ then also $\mathcal{B} \vDash X$, whenever $\mathcal{B}$ is a homomorphic image of $\mathcal{A}$, that is, $\mathrm{Md}\, X$ is closed under homomorphic images. Once again the converse holds (Lyndon's theorem; see [CK]).

2. Show that the theories $DO$ and $SO$ are model compatible.

3. Suppose $T_0$ and $T_1$ are model compatible and inductive. Show that $T_0 + T_1$ is an inductive theory that, in addition is model compatible with $T_0$ and $T_1$.

4. For inductive $T$ show that of all inductive extensions model compatible with $T$ there exists a largest one, the *inductive completion* of $T$. For instance, this is $ACF$ for the theory $T_F$ of fields.

## 5.5   Model Completeness

A theory $T$ is called *model complete* if for every model $\mathcal{A} \vDash T$ the theory $T + D\mathcal{A}$ is complete in $\mathcal{L}\mathcal{A}$. This notion was introduced in [Ro1]. For $\mathcal{A}, \mathcal{B} \vDash T$ where $\mathcal{A} \subseteq \mathcal{B}$ (hence $\mathcal{B}_A \vDash D\mathcal{A}$), the completeness of $T + D\mathcal{A}$ obviously means the same as $\mathcal{A}_A \equiv \mathcal{B}_A$, or equivalently, $\mathcal{A} \preccurlyeq \mathcal{B}$. In short, a model complete theory $T$ has the property

$$(*) \quad \mathcal{A} \subseteq \mathcal{B} \Rightarrow \mathcal{A} \preccurlyeq \mathcal{B}, \text{ for all } \mathcal{A}, \mathcal{B} \vDash T.$$

Conversely, if $(*)$ is satisfied then $T + D\mathcal{A}$ is also complete. Indeed, let $\mathcal{B} \vDash T, D\mathcal{A}$ so that w.l.o.g. $\mathcal{A} \subseteq \mathcal{B}$ and hence $\mathcal{A} \preccurlyeq \mathcal{B}$. But then all these $\mathcal{B}$ are elementarily equivalent in $\mathcal{L}\mathcal{A}$ to $\mathcal{A}_A$ and therefore to each other, which tells us that $T + D\mathcal{A}$ is complete. $(*)$ is therefore an equivalent definition of model completeness, and this definition, which is easy to remember, will be preferred in the sequel.

It is clear that if $T \subseteq \mathcal{L}$ is model complete then so too is every theory that extends it in $\mathcal{L}$. Furthermore, $T$ is then inductive. Indeed, a chain

$K$ of $T$-models is always elementary, by $(*)$. By Lemma 4.7, we obtain that $\mathcal{A} \preccurlyeq \bigcup K$ for any $\mathcal{A} \in K$, and therefore $\bigcup K \vDash T$ thanks to $\mathcal{A} \vDash T$, which confirms the claim. Hence, by Theorem 4.9, only an $\forall\exists$-theory can be model complete.

An example of an $\forall\exists$-theory that is not model complete is $\mathsf{DO}$. Let $\mathbb{Q}_a$ be $\{x \in \mathbb{Q} \mid a \leqslant x\}$ for $a \in \mathbb{Q}$. Then $(\mathbb{Q}_1, <) \subseteq (\mathbb{Q}_0, <)$ but $(\mathbb{Q}_1, <) \npreccurlyeq (\mathbb{Q}_0, <)$ so that $(*)$ does not hold. These two models also show that the complete theory $\mathsf{DO}_{10}$ is not model complete. Another example of a complete but not model complete theory is $\mathsf{SO}_{10}$, since as was noticed on page 193, $\mathsf{SO}_{10}$ is not an $\forall\exists$-theory. Conversely, a model complete theory need not be complete: A prominent example is $\mathsf{ACF}$, which will be treated in Theorem 5.4. Nonetheless, with the following theorem the completeness of a theory can often be obtained more easily than with other methods.

**Theorem 5.1.** *If a theory $T$ is model complete and has a prime model then $T$ is complete.*

**Proof.** Suppose that $\mathcal{A} \vDash T$ and let $\mathcal{P} \vDash T$ be a prime model. Then $\mathcal{P} \subseteq \mathcal{A}$ up to isomorphism, and so $\mathcal{P} \preccurlyeq \mathcal{A}$ by $(*)$, in particular $\mathcal{P} \equiv \mathcal{A}$. Hence, all $T$-models are elementarily equivalent to each other so that $T$ is in fact complete. ❏

The following theorem states additional characterizations of model completeness, of which (ii) is as a rule more easily verifiable than the definition. The implication (ii)$\Rightarrow$(i) carries the name *Robinson's test* for model completeness.

**Theorem 5.2.** *For any theory $T$ the following items are equivalent:*

(i)  *$T$ is model complete,*

(ii)  *$\mathcal{A} \subseteq \mathcal{B} \Rightarrow \mathcal{A} \subseteq_{ec} \mathcal{B}$, for all $\mathcal{A}, \mathcal{B} \vDash T$,*

(iii)  *each $\exists$-formula $\alpha$ is equivalent in $T$ to an $\forall$-formula $\beta$ such that $\mathrm{free}\,\beta \subseteq \mathrm{free}\,\alpha$,*

(iv)  *each formula $\alpha$ is equivalent in $T$ to an $\forall$-formula $\beta$ such that $\mathrm{free}\,\beta \subseteq \mathrm{free}\,\alpha$.*

**Proof.** (i)$\Rightarrow$(ii): evident, since $\mathcal{A} \subseteq \mathcal{B} \Rightarrow \mathcal{A} \preccurlyeq \mathcal{B} \Rightarrow \mathcal{A} \subseteq_{ec} \mathcal{B}$. (ii)$\Rightarrow$(iii): According to Theorem 4.6 it is enough to verify that every $\exists$-formula $\alpha = \alpha(\vec{x}) \in \mathcal{L}$ is persistent in $T$. Let $\mathcal{A}, \mathcal{B} \vDash T$, $\mathcal{A} \subseteq \mathcal{B}$, $\vec{a} \in A^n$, and

$\mathcal{B} \vDash \alpha(\vec{a})$. Then $\mathcal{A} \vDash \alpha(\vec{a})$, because $\mathcal{A} \subseteq_{ec} \mathcal{B}$ thanks to (ii). (iii)$\Rightarrow$(iv): induction on $\alpha$. (iii) is used only in the $\neg$-step: Let $\alpha \equiv \beta$, $\beta$ some $\forall$-formula (induction hypothesis). Then $\neg\beta \equiv \gamma$ for some $\forall$-formula $\gamma$, hence $\neg\alpha \equiv \gamma$. (iv)$\Rightarrow$(i): let $\mathcal{A}, \mathcal{B} \vDash T$, $\mathcal{A} \subseteq \mathcal{B}$, and $\mathcal{B} \vDash \alpha(\vec{a})$ with $\vec{a} \in A^n$. Then $\mathcal{A} \vDash \alpha(\vec{a})$, since by (iv), $\alpha(\vec{x}) \equiv_T \beta$ for some $\forall$-formula $\beta$. This shows that $\mathcal{A} \preccurlyeq \mathcal{B}$, hence (i). $\qquad \square$

**Remark 1.** If $T$ is countable and has infinite models only, then it is possible to restrict the criterion (ii) to models $\mathcal{A}, \mathcal{B}$ of any chosen infinite cardinal number $\kappa$. Then we can prove that an $\exists$-formula is $\kappa$-persistent as defined in the remark on page 190, which by the same remark suffices to prove the claim of Theorem 5.2 and hence (iii). Once we have obtained (iii) we have also (i). This remark is significant for Lindström's criterion, Theorem 5.7.

A relatively simple example of a model complete theory is $T_{V\mathbb{Q}}$, the theory of (nontrivial) $\mathbb{Q}$-vector spaces $\mathcal{V} = (V, +, 0, \mathbb{Q})$, where $0$ denotes the zero vector and each $r \in \mathbb{Q}$ is taken to be a unary operation on the set of vectors $V$. $T_{V\mathbb{Q}}$ formulates the familiar vector axioms, where, for example, the axiom $r(a + b) = ra + rb$ is reproduced as a schema of sentences, namely $\forall a \forall b\, r(a + b) = ra + rb$ for all $r \in \mathbb{Q}$. Let $\mathcal{V}, \mathcal{V}' \vDash T_{V\mathbb{Q}}$ with $\mathcal{V} \subseteq \mathcal{V}'$. We claim that $\mathcal{V} \subseteq_{ec} \mathcal{V}'$. By Theorem 5.2(iii), $T_{V\mathbb{Q}}$ is then model complete. For the claim let $\mathcal{V}' \vDash \exists \vec{x} \alpha$, with a conjunction $\alpha$ of literals in $x_1, \ldots, x_n$ and constants $a_1, \ldots, a_m, b_1, \ldots, b_k \in V$. Then $\alpha$ is essentially a system of the form

$$
\text{(s)} \quad
\begin{cases}
r_{11}x_1 + \cdots + r_{1n}x_n = a_1 \qquad s_{11}x_1 + \cdots + s_{1n}x_n \neq b_1 \\
\qquad\qquad \vdots \qquad\qquad\qquad\qquad\qquad\qquad \vdots \\
r_{m1}x_1 + \cdots + r_{mn}x_n = a_m \qquad s_{k1}x_1 + \cdots + s_{kn}x_n \neq b_k
\end{cases}
$$

Indeed, the only prime formulas are term equations, and every term in $x_1, \ldots, x_n$ is equivalent in $T_{V\mathbb{Q}}$ to some term of the form $r_1 x_1 + \cdots + r_n x_n$. Without going into detail, it is plausible by the properties of linear systems that the system (s) has already a solution in $\mathcal{V}$, provided it is solvable at all; see for instance [Zi].

For the rest of this section we assume some knowledge of classical algebra, where *closure constructions* are frequently undertaken. For instance, a torsion-free abelian group has a *divisible closure*, a field $\mathcal{A}$ has an *algebraic closure* (a minimal a.c. extension of $\mathcal{A}$), and an ordered field has a *real closure*; see Example 2 below. Generally speaking, we start from a theory $T$ and $\mathcal{A} \models T^\forall$. By a *closure of $\mathcal{A}$ in $T$* we mean a $T$-model $\bar{\mathcal{A}} \supseteq \mathcal{A}$ such that $\mathcal{A} \subseteq \mathcal{B} \Rightarrow \bar{\mathcal{A}} \subseteq \mathcal{B}$, for every $\mathcal{B} \models T$. More precisely, if $\mathcal{A} \subseteq \mathcal{B}$ then there is an embedding of $\bar{\mathcal{A}}$ into $\mathcal{B}$ leaving $A$ pointwise fixed. In this case we say that $T$ *permits a closure operation*.

Supposing this, let $\mathcal{A}, \mathcal{B} \models T$, $\mathcal{A} \subset \mathcal{B}$, and $b \in B \backslash A$. Then there is a smallest submodel of $\mathcal{B}$ containing $A \cup \{b\}$, the $T^\forall$-model generated in $\mathcal{B}$ by $A \cup \{b\}$, denoted by $\mathcal{A}(b)$. Its closure in $T$ is denoted by $\mathcal{A}^b$. In view of $\mathcal{A} \subset \mathcal{A}^b \subseteq \mathcal{B}$, it is called an *immediate extension* of $\mathcal{A}$ in $T$.

**Example 1.** Let $T := \mathsf{ACF}$. A $T^\forall$-model $\mathcal{A}$ is here an integral domain. $T$ permits a closure operation: $\bar{\mathcal{A}}$ is the so-called *algebraic closure* of the quotient field of $\mathcal{A}$. That there exists an a.c. field $\bar{\mathcal{A}} \supseteq \mathcal{A}$ that in addition is embeddable into every a.c. field $\mathcal{B} \supseteq \mathcal{A}$, is Steinitz's theorem regarding a.c. fields, [Wae, p. 201]. Let now $\mathcal{A}, \mathcal{B} \models T$ with $\mathcal{A} \subset \mathcal{B}$ and $b \in B \backslash A$. Then $b$ is transcendental over $\mathcal{A}$, because $\mathcal{A}$ is a.c. Thus, $a_0 + a_1 b + \cdots + a_n b^n \neq 0$, for all $a_0, \ldots, a_n \in A$ with $a_n \neq 0$. For this reason $\mathcal{A}(b)$ is isomorphic to the ring $\mathcal{A}(x)$ of polynomials $\sum_{i \leqslant n} a_i x^i$ with the "unknown" $x$ (the image of $b$). Hence, $\mathcal{A}(b) \simeq \mathcal{A}(x) \simeq \mathcal{A}(c)$ provided $\mathcal{A}, \mathcal{B}, \mathcal{C} \models T$, with $\mathcal{A} \subset \mathcal{B}, \mathcal{C}$ and $b \in B \backslash A$, $c \in C \backslash A$. The isomorphism of $\mathcal{A}(b), \mathcal{A}(c)$ extends in a natural way to their quotient fields (represented by the field of rational functions over $\mathcal{A}$) and hence to their closures $\mathcal{A}^b$ and $\mathcal{A}^c$. Thus, a $T$-model has up to isomorphism only one immediate extension in $T$. Not so in the next, more involved, example.

**Example 2.** A *real closed field* is an ordered field $\mathcal{A}$ (such as $\mathbb{R}$) in which every polynomial of odd degree has a zero and every $a \geqslant 0$ is a square in $A$. These properties will turn out to be equivalent to the continuity scheme CS (p. 110). Let RCF denote the theory of these fields. Although $\leqslant$ is definable in RCF by $x \leqslant y \leftrightarrow \exists z \, y - x = z^2$, order should here be a basic relation. Let $T := \mathsf{RCF}$. A $T^\forall$-model $\mathcal{A}$ is an ordered integral domain that determines the order of its quotient field $\mathcal{Q}$. According to Artin's theorem for real closed fields [Wae, p. 244], some $\bar{\mathcal{A}} = \bar{\mathcal{Q}} \models T$ can be constructed, called the *real closure of $\mathcal{A}$* or of its quotient field $\mathcal{Q}$ in $T$. Let $\mathcal{A}, \mathcal{B} \models \mathsf{RCF}$, $\mathcal{A} \subset \mathcal{B}$, and $b \in B \backslash A$. Then $b$ is transcendental over $\mathcal{A}$, because no algebraic extension of $\mathcal{A}$ is orderable—this is another characterization of real closed fields. Here $\mathcal{A}(b)$ is isomorphic to the *ordered* ring $\mathcal{A}(x)$ of polynomials over $\mathcal{A}$ and determines the isomorphism type of its quotient field $\mathcal{Q}(b)$ and of its real closure $\mathcal{A}^b = \overline{\mathcal{Q}(b)}$. Actually, $<^{\mathcal{A}^b}$ is determined by its restriction to $A \cup \{b\}$, or by the partition $A = \{a \in A \mid a <^{\mathcal{A}^b} b\} \cup \{a \in A \mid b <^{\mathcal{A}^b} a\}$. To see this, note that it is provable in RCF that a polynomial $p(x)$ with the zeros $a_1, \ldots, a_n \in A$ decomposes in $\mathcal{A} \vDash$ RCF as $c \cdot q(x) \cdot \prod_{i=1}^{n}(x - a_i)$ with $c \in A$, $n \geqslant 0$, and $q(x)$ a product of irreducible polynomials of degree 2 or $q(x) = 1$. In $\mathcal{Q}(b)$ (and $\mathcal{A}^b$) one has $q(b) > 0$. Indeed, each irreducible factor $b^2 + db + e$ of $q(b)$ is $> 0$ since $b^2 + db + e = (b + \frac{d}{2})^2 + e - \frac{d^2}{4} > 0$ $(d, e \in A)$. Thus we know whether $p(b) > 0$ if we know the signs of $b - a_i$ for all zeros $a_i$ of $p(x)$ in $A$. This suffices to fix the order in $\mathcal{Q}(b)$ as is easily seen, and hence in $\mathcal{A}^b$ by Artin's theorem.

For inductive theories $T$ that permit a closure operation, Robinson's test for model completeness can still be simplified as follows:

**Lemma 5.3.** *Let $T$ be inductive, and suppose $T$ permits a closure operation. Assume further that $\mathcal{A} \subseteq_{ec} \mathcal{A}'$ for all $\mathcal{A}, \mathcal{A}' \vDash T$ in the case that $\mathcal{A}'$ is an immediate extension of $\mathcal{A}$ in $T$. Then $T$ is model complete.*

**Proof.** Let $\mathcal{A}, \mathcal{B} \vDash T$, $\mathcal{A} \subseteq \mathcal{B}$. By Theorem 5.2(ii) it suffices to show that $\mathcal{A} \subseteq_{ec} \mathcal{B}$. Let $H$ be the set of all $\mathcal{C} \subseteq \mathcal{B}$ such that $\mathcal{A} \subseteq_{ec} \mathcal{C} \vDash T$. Trivially $\mathcal{A} \in H$. Since $T$ is inductive, a chain $K \subseteq H$ satisfies $\bigcup K \vDash T$. One easily verifies $\mathcal{A} \subseteq_{ec} \bigcup K$ as well, so that $\bigcup K \in H$. By Zorn's lemma there is a maximal element $\mathcal{A}_m \in H$. **Claim:** $\mathcal{A}_m = \mathcal{B}$. Assume $\mathcal{A}_m \subset \mathcal{B}$. Then there is an immediate extension $\mathcal{A}'_m \vDash T$ of $\mathcal{A}_m$ such that $\mathcal{A}_m \subset \mathcal{A}'_m \subseteq \mathcal{B}$. Since $\mathcal{A} \subseteq_{ec} \mathcal{A}_m$, and by hypothesis $\mathcal{A}_m \subseteq_{ec} \mathcal{A}'_m$, we get $\mathcal{A} \subseteq_{ec} \mathcal{A}'_m$. This, however, contradicts the maximality of $\mathcal{A}_m$ in $H$. Therefore, it must be the case that $\mathcal{A}_m = \mathcal{B}$. Consequently, $\mathcal{A} \subseteq_{ec} \mathcal{B}$. ∎

**Theorem 5.4.** ACF *is model complete and thus so too* $\mathsf{ACF}_p$, *the theory of a.c. fields of given characteristic $p$ ($= 0$ or a prime). Moreover, $\mathsf{ACF}_p$ is complete.*

**Proof.** Let $\mathcal{A}, \mathcal{B} \vDash$ ACF, $\mathcal{A} \subset \mathcal{B}$, and $b \in B \backslash A$. By Lemma 5.3 it suffices to show that $\mathcal{A} \subseteq_{ec} \mathcal{A}^b$. Here $\mathcal{A}^b$ is an immediate extension of $\mathcal{A}$ in ACF. Let $\alpha := \exists \vec{x} \beta(\vec{x}, \vec{a}) \in \mathcal{L}A$, $\beta$ quantifier-free, and $\mathcal{A}^b \vDash \alpha$. We shall prove $\mathcal{A} \vDash \alpha$ and for this we consider

$$X := \mathsf{ACF} \cup D\mathcal{A} \cup \{p(x) \neq 0 \mid p(x) \text{ a monic polynomial on } A\}.$$

With $b$ for $x$ one sees that $(\mathcal{A}^b, b) \vDash X$ (for $b$ is trancendental over $\mathcal{A}$). Let $(\mathcal{C}, c) \vDash X$, with $c$ for $x$. Since $\mathcal{C} \vDash D\mathcal{A}$, w.l.o.g. $\mathcal{A} \subseteq \mathcal{C}$. By Example 1 $\mathcal{A}^b \simeq \mathcal{A}^c$, and so $\mathcal{A}^c \vDash \alpha$. $\mathcal{A}^c \subseteq \mathcal{C}$ implies $\mathcal{C} \vDash \alpha$, for $\alpha$ is an $\exists$-sentence. Since $(\mathcal{C}, c)$ has been chosen arbitrarily we obtain $X \vdash \alpha$, and from this by the finiteness theorem evidently

$$D\mathcal{A}, \bigwedge_{i \leqslant k} p_i(x) \neq 0 \vdash_{\mathsf{ACF}} \alpha,$$

for some $k$ and monic polynomials $p_0, \dots, p_k$. Particularization and the deduction theorem show that $D\mathcal{A} \vdash_{\mathsf{ACF}} \exists x \bigwedge_{i \leqslant k} p_i(x) \neq 0 \rightarrow \alpha$. Every a.c. field is infinite (Example 5(c) in **5.2**), and a polynomial has only finitely many zeros in a field. Thus, $D\mathcal{A} \vdash_{\mathsf{ACF}} \exists x \bigwedge_{i \leqslant k} p_i(x) \neq 0$. Hence, $D\mathcal{A} \vdash_{\mathsf{ACF}} \alpha$ and so $\mathcal{A} \vDash \alpha$. This proves $\mathcal{A} \subseteq_{ec} \mathcal{A}^b$ and in view of Lemma 5.3 the first part of the theorem. The algebraic closure of the prime field of characteristic $p$ is obviously a prime model for $\mathsf{ACF}_p$. Therefore, by Theorem 5.1, $\mathsf{ACF}_p$ is complete. $\quad\Box$

The following significant theorem is won similarly. It was originally proved in [Ta3] by means of quantifier elimination. Incidentally, the claim of completeness is not obtainable by means of Vaught's criterion, in contrast to the case of $\mathsf{ACF}$.

**Theorem 5.5.** *The theory* $\mathsf{RCF}$ *of real closed fields is model complete and complete. It is thus identical to the theory of the ordered field of real numbers, and as a complete axiomatizable theory it is also decidable.*

**Proof.** Let $\mathcal{A} \vDash \mathsf{RCF}$. It once again suffices to show that $\mathcal{A} \subseteq_{ec} \mathcal{A}^b$ for an immediate extension $\mathcal{A}^b$ of $\mathcal{A}$ in $\mathsf{RCF}$. Let $U := \{a \in A \mid a <^{\mathcal{B}} b\}$, $V := \{a \in A \mid b <^{\mathcal{B}} a\}$, with $\mathcal{B} := \mathcal{A}^b$. Then $U \cup V = A$. Now let $\mathcal{A}^b \vDash \exists \vec{x} \beta(\vec{x}, \vec{a})$, $\beta$ quantifier-free, $\vec{a} \in A^m$. The model $(\mathcal{B}, b)$ with $b$ for $x$ then clearly satisfies the set of formulas

$$X := \mathsf{RCF} \cup D\mathcal{A} \cup \{a < x \mid a \in U\} \cup \{x < a \mid a \in V\}.$$

Suppose $(\mathcal{C}, c) \vDash X$, interpreting $x$ as $c$. We may assume $\mathcal{A} \subseteq \mathcal{C}$ because $\mathcal{C} \vDash D\mathcal{A}$. Since $c \notin U \cup V = A$, $c$ is transcendental over $\mathcal{A}$ (see Example 2). Hence, the quotient field $\mathcal{Q}(c)$ of $\mathcal{A}(c)$ is isomorphic to the field of rational functions over $\mathcal{A}$ with the unknown $x$. The order of $\mathcal{Q}(c)$ is fixed by the partition $A = U \cup V$ coming from $\mathcal{Q}(b)$. Thus, $\mathcal{Q}(b) \simeq \mathcal{Q}(x) \simeq \mathcal{Q}(c)$. The isomorphism $\mathcal{Q}(b) \simeq \mathcal{Q}(c)$ extends to one between the real closures $\mathcal{A}^b$ and $\mathcal{A}^c$. As in Theorem 5.4 we thus obtain $X \vdash \alpha$, and so for some

$a_1, \ldots, a_k, b_1, \ldots, b_l \in A$ (where $k, l \geqslant 0$ but $k + l > 0$),

$$D\mathcal{A} \vdash_{\mathsf{RCF}} \exists x (\textstyle\bigwedge_{i=1}^{k} a_i < x \wedge \bigwedge_{i=1}^{l} x < b_i) \to \alpha \qquad (a_i \in U, \; b_i \in V).$$

Now, an ordered field is densely ordered without edge elements, hence is infinite. Therefore, $\vdash_{\mathsf{RCF}} \exists x (\bigwedge_{i=1}^{k} a_i < x \wedge \bigwedge_{i=1}^{l} x < b_i)$ which results in $D\mathcal{A} \vdash_{\mathsf{RCF}} \alpha$. Thus, $\mathcal{A} \vDash \alpha$, and $\mathcal{A} \subseteq_{ec} \mathcal{A}^b$ is proved. To verify completeness observe that $\mathsf{RCF}$ has a prime model, namely the real closure of $\mathbb{Q}$, the ordered field of all real algebraic numbers. Applying Theorem 5.1 once again confirms the completeness of $\mathsf{RCF}$. $\quad\square$

A theory $T$ is called the *model completion* of a theory $T_0$ of the same language if $T_0 \subseteq T$ and $T + D\mathcal{A}$ is complete for every $\mathcal{A} \vDash T_0$. Clearly, $T$ is then model complete; moreover, $T$ is model compatible with $T_0$ ($\mathcal{A} \vDash T_0$ implies $(\exists \mathcal{C} \in \mathrm{Md}\, T)\mathcal{A} \subseteq \mathcal{C}$, since $T + D\mathcal{A}$ is consistent). The existence of a model complete extension is necessary for the existence of a model completion of $T_0$, but not sufficient. See Exercise 1.

**Remark 2.** A somewhat surprising fact is that a model completion of $T$ is uniquely determined provided it exists. Indeed, let $T, T'$ be model completions of $T_0$. Both theories are model compatible with $T_0$, and hence with each other. $T, T'$ are model complete and therefore inductive, so that $T + T'$ is model compatible with $T$ (Exercise 3 in **5.4**). Thus, if $\mathcal{A} \vDash T$ then there exist some $\mathcal{B} \vDash T + T'$ with $\mathcal{A} \subseteq \mathcal{B}$, and since $T$ is model complete we obtain $\mathcal{A} \preccurlyeq \mathcal{B}$. This implies $\mathcal{A} \equiv \mathcal{B} \vDash T'$, and consequently $\mathcal{A} \vDash T'$. For reasons of symmetry, $\mathcal{A} \vDash T' \Rightarrow \mathcal{A} \vDash T$ as well. Therefore $T = T'$.

**Example 3.** $\mathsf{ACF}$ is the model completion of the theory $T_J$ of integral domains, hence also of the theory $T_F$ of fields. Indeed, let $\mathcal{A} \vDash T_J$. By Theorem 5.4, $\mathsf{ACF}$ is model complete, hence also $T := \mathsf{ACF} + D\mathcal{A}$ (in $\mathcal{L}A$). Moreover, $T$ is complete, since by Example 1, $T$ has a prime model, the closure $\bar{A}$ of $\mathcal{A}$ in $\mathsf{ACF}$. Using Theorem 5.5, one shows analogously that $\mathsf{RCF}$ is the model completion of the theory of ordered commutative rings with unit element. Each such ring is embeddable into an ordered field (an algebraic standard construction) and hence into a real closed field.

$\mathcal{A} \vDash T$ is called *existentially closed in* $T$, or *$\exists$-closed in* $T$ for short, if $\mathcal{A} \subseteq_{ec} \mathcal{B}$ for each $\mathcal{B} \vDash T$ with $\mathcal{A} \subseteq \mathcal{B}$. For instance, every a.c. field $\mathcal{A}$ is $\exists$-closed in the theory of fields. For let $\mathcal{B} \supseteq \mathcal{A}$ be any field and $\mathcal{C}$ be any a.c. extension of $\mathcal{B}$. Then $\mathcal{A} \preccurlyeq \mathcal{C}$ thanks to the model completeness of $\mathsf{ACF}$. Hence $\mathcal{A} \subseteq_{ec} \mathcal{B}$ by Lemma 4.8(ii). The following lemma generalizes in some sense the fact that every field is embeddable into an a.c. field.

Similarly, a group, for instance, is embeddable into a group that is ∃-closed in the theory of groups.

**Lemma 5.6.** *Let $T$ be an $\forall\exists$-theory of some countable language $\mathcal{L}$. Then every infinite model $\mathcal{A}$ of $T$ can be extended to a model $\mathcal{A}^*$ of $T$ such that $|\mathcal{A}^*| = |\mathcal{A}|$, which is $\exists$-closed in $T$.*

**Proof.** For the proof we assume, for simplicity, that $\mathcal{A}$ is countable. Then $\mathcal{L}\mathcal{A}$ is also countable. Let $\alpha_0, \alpha_1, \ldots$ be an enumeration of the $\exists$-sentences of $\mathcal{L}\mathcal{A}$ and $\mathcal{A}_0 = \mathcal{A}_\mathcal{A}$.[5] Let $\mathcal{A}_{n+1}$ be an extension of $\mathcal{A}_n$ in $\mathcal{L}\mathcal{A}$ such that $\mathcal{A}_{n+1} \vDash T + \alpha_n$, as long as such an extension exists; otherwise simply put $\mathcal{A}_{n+1} = \mathcal{A}_n$. Since $T$ is inductive, $\mathcal{B}_0 = \bigcup_{n\in\mathbb{N}} \mathcal{A}_n \vDash T$. If $\alpha = \alpha_n$ is an $\exists$-sentence in $\mathcal{L}\mathcal{A}$ valid in some extension $\mathcal{B} \vDash T$ of $\mathcal{B}_0$, then already $\mathcal{A}_{n+1} \vDash \alpha$ and thus also $\mathcal{B}_0 \vDash \alpha$. Now we repeat this construction with $\mathcal{B}_0$ in place of $\mathcal{A}_0$ with respect to an enumeration of all $\exists$-sentences in $\mathcal{L}\mathcal{B}_0$ and obtain an $\mathcal{L}\mathcal{B}_0$-structure $\mathcal{B}_1 \vDash T$. Subsequent reiterations produce a sequence $\mathcal{B}_1 \subseteq \mathcal{B}_2 \subseteq \cdots$ of $\mathcal{L}\mathcal{B}_n$-structures $\mathcal{B}_{n+1} \vDash T$. Let $\mathcal{A}^*$ ($\vDash T$) be the $\mathcal{L}$-reduct of $\bigcup_{n\in\mathbb{N}} \mathcal{B}_n \vDash T$ and $\mathcal{A}^* \subseteq \mathcal{B} \vDash T$. Assume $\mathcal{B} \vDash \exists \vec{x} \beta(\vec{a}, \vec{x})$, $\vec{a} \in (\mathcal{A}^*)^n$. Then $\mathcal{B}_m \vDash \beta(\vec{a}, \vec{b})$ for suitable $m$. Hence $\bigcup_{n\in\mathbb{N}} \mathcal{B}_n \vDash \beta(\vec{a}, \vec{b})$, and so $\mathcal{A}^* \vDash \exists \vec{x} \beta(\vec{a}, \vec{x})$. $\blacksquare$

With this lemma one readily obtains the following highly applicable criterion for proving the model completeness of countable $\kappa$-categorical theories, which, by Vaught's criterion, are always complete.

**Theorem 5.7 (Lindström's criterion).** *A countable $\kappa$-categorical $\forall\exists$-theory $T$ without finite models is model complete.*

**Proof.** Since all $T$-models are infinite, $T$ has a model of cardinality $\kappa$, and by Lemma 5.6 also one that is $\exists$-closed in $T$. But then all $T$-models of cardinality $\kappa$ are $\exists$-closed in $T$, because all these are isomorphic. Thus $\mathcal{A} \subseteq \mathcal{B} \Rightarrow \mathcal{A} \subseteq_{ec} \mathcal{B}$, for all $\mathcal{A}, \mathcal{B} \vDash T$ of cardinality $\kappa$. Therefore, $T$ is model complete according to Remark 1 on page 196. $\blacksquare$

**Examples of applications.**

(a) The $\aleph_0$-categorical theory of atomless Boolean algebras.

(b) The $\aleph_1$-categorical theory of nontrivial $\mathbb{Q}$-vector spaces.

(c) The $\aleph_1$-categorical theory of a.c. fields of given characteristic.

---

[5] For uncountable $\mathcal{A}$ we have $|\mathcal{L}\mathcal{A}| = |\mathcal{A}|$. In this case one proceeds with an ordinal enumeration of $\mathcal{L}\mathcal{A}$ rather than an ordinary one. But the proof is almost the same.

A few comments: A Boolean algebra $\mathcal{B}$ is called *atomless* if for each $a \neq 0$ in $B$ there is some $b \neq 0$ in $B$ with $b < a$ ($<$ is the partial lattice order of $\mathcal{B}$). The proof of (a) is similar to that for densely ordered sets. Also (b) is easily verified. Observe that a $\mathbb{Q}$-vector space of cardinality $\aleph_1$ has a base of cardinality $\aleph_1$. From (c) the model completeness of ACF follows in a new way: If $\mathcal{A}, \mathcal{B} \vDash$ ACF and $\mathcal{A} \subseteq \mathcal{B}$ then both fields have the same characteristic $p \geqslant 0$. Since $\mathsf{ACF}_p$ is model complete by (c), $\mathcal{A} \preccurlyeq \mathcal{B}$ follows. This obviously implies that ACF itself is model complete.

### Exercises

1. Prove that of the four theories $\mathsf{DO}_{ij}$ only $\mathsf{DO}_{00}$ is model complete. Moreover, show that $\mathsf{DO}_{00}$ is the model completion of both DO and the theory $T_{ord}$ of all orders, but not of the theory of all irreflexive relations which is not model-compatible with $T_{ord}$.

2. Let $T$ be the theory of divisible torsion-free abelian groups. Show that $T$ is the model completion of the theory $T_0$ of torsion-free abelian groups.

3. $T^*$ is called the *model companion* of $T$ provided $T, T^*$ are model compatible and $T^*$ is model complete. Show that if $T^*$ exists then $T^*$ is uniquely determined provided it exists. Moreover, show that each $\mathcal{A} \in \operatorname{Md} T^*$ is $\exists$-closed in $T$.

4. Prove that an $\forall\exists$-sentence valid in all finite fields is valid also in all a.c. fields. This fact is highly useful in algebraic geometry.

## 5.6    Quantifier Elimination

Because $\exists x(y < x \wedge x < z) \equiv_{\mathsf{DO}} y < z$, in the theory of densely ordered sets the quantifier in the left-hand formula can be eliminated. In fact, in some theories, including the theory $\mathsf{DO}_{00}$ (see **5.2**), the quantifiers can be eliminated from every formula. One says that $T$ ($\subseteq \mathcal{L}^0$) *allows quantifier elimination* if for every $\varphi \in \mathcal{L}$ there exists some open formula $\varphi' \in \mathcal{L}$ such that $\varphi \equiv_T \varphi'$. Quantifier elimination is the oldest method of showing that certain theories are decidable and occasionally also complete. Some presentations demand additionally $\text{free}\,\varphi' = \text{free}\,\varphi$, but one is not obliged to to do so. A theory $T$ that allows quantifier elimination is model complete by Theorem 5.2(iv), because open formulas are $\forall$-formulas. Hence, $T$ is an $\forall\exists$-theory, which is a remarkable necessary condition for quantifier eliminability.

In order to confirm quantifier elimination for a theory $T$ it suffices to eliminate the prefix $\exists x$ from every formula of the form $\exists x\alpha$, where $\alpha$ is open. Indeed, think of all subformulas of the form $\forall x\alpha$ in a formula $\varphi$ as being equivalently replaced by $\neg\exists x\neg\alpha$, so that only the $\exists$-quantifier appears in $\varphi$. Looking at the farthest-right prefix $\exists x$ in $\varphi$ one can write $\varphi = \cdots\exists x\alpha\cdots$ with some quantifier-free $\alpha$. Now, if $\exists x\alpha$ is replaceable by an open formula $\alpha'$ then this process can be iterated no matter how long it takes for all $\exists$-quantifiers in $\varphi$ to disappear.

Thanks to the $\vee$-distributivity of $\exists$-quantifiers we may moreover assume that the quantifier-free part $\alpha$ of $\exists x\alpha$ from which $\exists x$ has to be eliminated is a conjunction of literals, and that $x$ explicitly occurs in each of these literals: simply convert $\alpha$ into a DNF and distribute $\exists x$ over the disjuncts such that $\exists x$ stands in front of a conjunction of literals only. If $x$ does not appear in any of these literals, $\exists x$ can simply be discarded. Otherwise remove the literals not containing $x$ beyond the scope of $\exists x$, observing that $\exists x(\alpha \wedge \beta) \equiv \exists x\alpha \wedge \beta$ if $x \notin \text{var}\,\beta$.

Furthermore, it can be supposed that none of the conjuncts is of the form $x = t$ with $x \notin \text{var}\,t$. Indeed, since $\exists x(x = t \wedge \alpha) \equiv \alpha\frac{t}{x}$, the quantifier has then already been eliminated. We may also assume $x \neq \boldsymbol{v}_0$ (using bound renaming) and that neither $x = x$ nor $x \neq x$ is among the conjuncts. For $x = x$ can equivalently be replaced by $\top$, as can $x \neq x$ by $\bot$. Here one may define $\top$ and $\bot$ as $\boldsymbol{v}_0 = \boldsymbol{v}_0$ and $\boldsymbol{v}_0 \neq \boldsymbol{v}_0$, respectively. Replacement will then introduce $\boldsymbol{v}_0$ as a possible new free variable, but that is harmless. If the language contains a constant $c$ one may replace $\boldsymbol{v}_0$ by $c$ in the above consideration. If not, one may add a constant or even $\bot$ as a new prime formula to the language, similar to what is proposed below for $\mathsf{DO}_{00}$.

Call an $\exists$-formula *simple* if it is of the form $\exists x \bigwedge_i \alpha_i$, where every $\alpha_i$ is a literal with $x \in \text{var}\,\alpha_i$. Then the above considerations result in the following theorem.

**Theorem 6.1.** *T allows quantifier elimination if every simple $\exists$-formula $\exists x \bigwedge_i \alpha_i$ is equivalent in $T$ to some open formula. Here w.l.o.g., none of the literals $\alpha_i$ is $x = x$, $x \neq x$, or of the form $x = t$ with $x \notin \operatorname{var} t$.*

**Example 1.** $T = \mathsf{DO}_{00}$ allows quantifier elimination. Because of

$$y \not< z \equiv_T z < y \vee z = y \quad \text{and} \quad z \neq y \equiv_T z < y \vee y < z$$

and since in general $(\alpha \vee \beta) \wedge \gamma \equiv (\alpha \wedge \gamma) \vee (\beta \wedge \gamma)$, we may suppose that the conjunction of the $\alpha_i$ in Theorem 6.1 does not contain the negation symbol. We are therefore dealing with a formula of the form

$$\exists x (y_1 < x \wedge \cdots \wedge y_m < x \ \wedge \ x < z_1 \wedge \cdots \wedge x < z_k),$$

which is equivalent to $\bot$ ($\equiv v_0 \neq v_0$) if $x$ is one of the variables $y_i, z_j$, or to $\top$ whenever $m = 0$ or $k = 0$, and in the remaining case to $\bigwedge_{i,j=1}^{n} y_i < z_j$. Who wants to avoid the use of $v_0$ as an extra variable may also regard at $\bot$ as an additional 0-ary relation symbol.

$\mathsf{DO}$ itself does not allow quantifier elimination, since in $\alpha(y) := \exists x \, x < y$ the quantifier is not eliminable. Indeed, if $\alpha(y)$ were equivalent in $\mathsf{DO}$ to an open formula then $\mathcal{A}, \mathcal{B} \vDash \mathsf{DO}$, $\mathcal{A} \subseteq \mathcal{B}$, $a \in A$, and $\mathcal{B} \vDash \alpha(a)$ would imply $\mathcal{A} \vDash \alpha(a)$. But this is not so for the densely ordered sets $\mathcal{A}, \mathcal{B}$ with $A = \{x \in \mathbb{Q} \mid 1 \leqslant x\}$ and $B = \mathbb{Q}$. Choose $a = 1$. Quantifier elimination does however become possible if the signature $\{<\}$ is expanded by considering $\mathsf{L}, \mathsf{R}$ as 0-ary predicate symbols. The fact that $\{\mathsf{L}, \mathsf{R}\}$ forms a Boolean basis for sentences in $\mathsf{DO}$ is not yet sufficient for quantifier eliminability if looking at $\mathsf{L}$ and $\mathsf{R}$ as formulas, for these contain quantifiers.

Also the theory $\mathsf{SO}$ does not allow quantifier elimination in the original language, simply because it is not an $\forall\exists$-theory as was noticed earlier. The same is true for the expansions $\mathsf{SO}_{ij}$ of $\mathsf{SO}$.

**Example 2.** A classical and nontrivial result of quantifier elimination by Presburger [Pr] refers to $Th(\mathbb{N}, 0, 1, +, <)$, with the additional unary predicate symbols $m|$ ($m = 2, 3, \dots$), defined by $m|x \leftrightarrow \exists y \, my = x$, where $my$ denotes the $m$-fold sum $y + \cdots + y$ of $y$. We shall prove a related result with respect to the group $\mathbb{Z}$ in $\mathcal{L}\{0, 1, +, -, <, 2|, 3|, \dots\}$. Denote the $k$-fold sum $1 + \cdots + 1$ by $k$, and set $(-k)x := -kx$.

Let $\mathsf{ZGE}$ be the elementary theory in $\mathcal{L}\{0, 1, +, -, <, 2|, 3|, \dots\}$ whose axioms subsume those for ordered abelian groups plus the axioms

$$\forall x (0 < x \leftrightarrow 1 \leqslant x), \ \forall x (m|x \leftrightarrow \exists y \, my = x), \text{ and } \vartheta_m := \forall x \bigvee_{k < m} m|x + k$$

for $m = 2, 3, \ldots$ The reducts of ZGE-models to $\mathcal{L} := \mathcal{L}\{0, 1, +, -, <\}$ are called $\mathbb{Z}$-*groups*. These are ordered with the smallest positive element 1. The $\vartheta_m$ state for a $\mathbb{Z}$-group $G$ that the factor groups $G/mG$ are cyclic of order $m$. Here $mG := \{mx \mid x \in G\}$. Let ZG denote the reduct theory of ZGE in $\mathcal{L}$. Its models are precisely the $\mathbb{Z}$-groups. ZGE is a definitorial, hence conservative extension of ZG (cf. **2.6**). It will turn out that $\mathbb{Z}$-groups are just the ordered abelian groups elementarily equivalent to the paradigm of a $\mathbb{Z}$-group, $(\mathbb{Z}, 0, 1, +, -, <)$. Let us notice that $\vdash_{\mathsf{ZG}} \forall x \eta_n$ for each $n$, where $\eta_n$ is the formula $0 \leqslant x < n \rightarrow \bigvee_{k<n} x = k$.

We are now going to prove that ZGE allows quantifier elimination. Observe first that since $t \neq s \equiv_{\mathsf{ZGE}} s < t \vee t < s$, $m \nmid t \equiv_{\mathsf{ZGE}} \bigvee_{i=1}^{m-1} m \mid t + i$, and $m \mid t \equiv_{\mathsf{ZGE}} m \mid -t$, we may assume that the kernel of a simple $\exists$-formula is a conjunction of formulas of the form $n_i x = t_i^0$, $n_i' x < t_i^1$, $t_i^2 < n_i'' x$, and $m_i \mid n_i''' x + t_i^3$, where $x \notin \mathrm{var}\, t_i^j$. By multiplying these formulas by a suitable integer and using $t < s \equiv_{\mathsf{ZGE}} nt < ns$ and $m \mid t \equiv_{\mathsf{ZGE}} nm \mid nt$ for $n \neq 0$, one sees that all the $n_i, n_i', n_i'', n_i'''$ can be made equal to some number $n > 1$. Clearly, in doing so, $t_i^j$ and the "modules" $m_i$ all change. But the problem of elimination is thus reduced to formulas of the following form, where the $j$th conjunct disappears whenever $k_j = 0$ ($j \leqslant 3$):

(1) $\exists x \big( \bigwedge_{i=1}^{k_0} nx = t_i^0 \wedge \bigwedge_{i=1}^{k_1} t_i^1 < nx \wedge \bigwedge_{i=1}^{k_2} nx < t_i^2 \wedge \bigwedge_{i=1}^{k_3} m_i \mid nx + t_i^3 \big).$

With $y$ for $nx$ and $m_0 = n$, (1) is certainly equivalent in ZGE to

(2) $\exists y \big( \bigwedge_{i=1}^{k_0} y = t_i^0 \wedge \bigwedge_{i=1}^{k_1} t_i^1 < y \wedge \bigwedge_{i=1}^{k_2} y < t_i^2 \wedge \bigwedge_{i=1}^{k_3} m_i \mid y + t_i^3 \wedge m_0 \mid y \big).$

According to Theorem 6.1 we can at once assume that $k_0 = 0$, so that the elimination problem, after renaming $y$ back to $x$, reduces to formulas of the following form, where $x \notin \mathrm{var}\, t_i^j$:

(3) $\exists x \big( \bigwedge_{i=1}^{k_1} t_i^1 < x \wedge \bigwedge_{i=1}^{k_2} x < t_i^2 \wedge \bigwedge_{i=0}^{k_3} m_i \mid x + t_i^3 \big).$

Let $m$ be the least common multiple of $m_0, \ldots, m_{k_3}$.

**Case 1:** $k_1, k_2 = 0$. Then (3) is equivalent in ZGE to $\bigvee_{j=1}^{m} \bigwedge_{i=0}^{k_3} m_i \mid j + t_i^3$. Indeed, if an $x$ such that $\bigwedge_{i=0}^{k_3} m_i \mid x + t_i^3$ exists at all, then so does some $x = j \in \{1, \ldots, m\}$. For let $j$ with $m \mid x + (m - j)$ be given by axiom $\vartheta_m$. Then also $m \mid x - j$ and consequently $m_i \mid x - j$ for all $i \leqslant k_3$. Therefore $m_i \mid x + t_i^3 - (x - j) = j + t_i^3$ also holds for $i = 0, \ldots, k_3$ as was claimed.

**Case 2:** $k_1 \neq 0$ and $j$ as above. Then (3) is equivalent to

(4) $\bigvee_{\mu=1}^{k_1} [\bigwedge_{i=1}^{k_1} t_i^1 \leqslant t_\mu^1 \wedge \bigvee_{j=1}^{m} (\bigwedge_{i=1}^{k_2} t_\mu^1 + j < t_i^2 \wedge \bigwedge_{i=0}^{k_3} m_i \mid t_\mu^1 + j + t_i^3)].$

This is a case distinction according to the maximum among the values of the $t_i^1$. From each disjunct in (4) certainly (3) follows in ZGE (consider $t_i^1 < t_\mu^1 + j$). Now suppose conversely that $x$ is a solution of (3). Then in the case $\bigwedge_{i=1}^{k_1} t_i^1 \leqslant t_\mu^1$ the $\mu$th disjunct of (4) is also valid. To prove this we need only confirm $t_\mu^1 + j < t_i^2$, which comes down to $t_\mu^1 + j \leqslant x$. Were $x < t_\mu^1 + j$, i.e., $0 < x - t_\mu^1 < j$, then $x - t_\mu^1 = k$ follows for some $k < j$ by $\eta_j$, that is, $x = t_\mu^1 + k$. Thus, $m_i | t_\mu^1 + j - x = j - k$ for all $i \leqslant k_3$. But this yields the contradiction $m \, | \, j - k < m$.

**Case 3:** $k_1 = 0$ and $k_2 \neq 0$. The argument is analogous to Case 2 but with a distinction according to the smallest term among the $t_i^2$.

From this remarkable example we obtain the following

**Corollary 6.2.** ZGE *is model complete. Moreover,* ZGE *and* ZG *are both complete and decidable.*

**Proof.** Since $\mathbb{Z}$ determines a prime model for ZGE, completeness follows from model completeness, which in turn follows from quantifier eliminability. Clearly, along with ZGE also its reduct theory ZG is complete. Hence, as complete axiomatizable theories, both these theories are decidable. $\square$

**Remark 1.** Also ZG is model complete; Exercise 1. Actually, ZG is the model completion of the theory of discretely ordered abelian groups because every such group is embeddable into some $\mathbb{Z}$-group (which is not quite easy to prove). This is a main reason for the interest in ZG. Although model complete, ZG does not allow quantifier elimination.

We now intend to show that theories ACF and RCF of algebraically and real closed fields respectively allow quantifier elimination, even without any expansion of their signatures. Attacking the elimination problem in a direct manner as above would fill a separate chapter. We therefore undertake a model-theoretic proof in Theorem 6.4, applying thereby a variant of Theorem 2.3. Call $X \subseteq \mathcal{L}$ a *Boolean basis for* $\mathcal{L}$ *in* $T$ if *every* $\varphi \in \mathcal{L}$ belongs to $\langle X \rangle$, the set of Boolean combinations of formulas from $X$. Let $\mathcal{M}, \mathcal{M}'$ be $\mathcal{L}$-models. Write $\mathcal{M} \equiv_X \mathcal{M}'$ whenever $\mathcal{M} \vDash \varphi \Leftrightarrow \mathcal{M}' \vDash \varphi$, for all $\varphi \in X$, and $\mathcal{M} \equiv \mathcal{M}'$ whenever $\mathcal{M} \vDash \varphi \Leftrightarrow \mathcal{M}' \vDash \varphi$, for *all* $\varphi \in \mathcal{L}$.

**Theorem 6.3 (Basis theorem for formulas).** *Let* $T \subseteq \mathcal{L}^0$ *be a theory,* $X \subseteq \mathcal{L}$, *and suppose that* $\mathcal{M} \equiv_X \mathcal{M}' \Rightarrow \mathcal{M} \equiv \mathcal{M}'$, *for all* $\mathcal{M}, \mathcal{M}' \vDash T$. *Then* $X$ *is a Boolean basis for* $\mathcal{L}$ *in* $T$.

**Proof.** Let $\alpha \in \mathcal{L}$ and $Y_\alpha := \{\gamma \in \langle X \rangle \mid \alpha \vdash_T \gamma\}$, where $\langle X \rangle$ is defined as on page 180. One then shows that $Y_\alpha \vdash_T \alpha$ as in the proof of Theorem 2.3

by arguing with a model $\mathcal{M}$ rather than a structure $\mathcal{A}$. The remainder of the proof proceeds along the lines of the proof of the mentioned theorem and is therefore left to the reader. ∎

A theory $T$ is called *substructure complete* if for all $\mathcal{A}, \mathcal{B}$ with $\mathcal{A} \subseteq \mathcal{B} \vDash T$ the theory $T + \mathcal{D}\mathcal{A}$ is complete. This generalizes model completeness and is basically only a reformulation of '$T$ is the model completion of $T^\forall$'. Indeed, let $T$ be substructure complete and $\mathcal{A} \vDash T^\forall$. Then by Lemma 4.1, $\mathcal{A} \subseteq \mathcal{B}$ for some $\mathcal{B} \vDash T$, hence $T + \mathcal{D}\mathcal{A}$ is complete. Thus, $T$ is the model completion of $T^\forall$. Conversely, let $T$ be the model completion of $T^\forall$ and $\mathcal{A} \subseteq \mathcal{B} \vDash T$. Then $\mathcal{A} \vDash T^\forall$, so that $T + \mathcal{D}\mathcal{A}$ is complete. Thus, $T$ is substructure complete. The criterion (ii) in the next theorem may therefore also be reformulated. There exist yet other criteria, in particular the amalgamability of models of $T^\forall$; see e.g. [CK].

**Theorem 6.4.** *For every theory $T$ in $\mathcal{L}$ the following are equivalent:*

(i) *$T$ allows quantifier elimination,*     (ii) *$T$ is substructure complete.*

**Proof.** (i)⇒(ii): Let $\mathcal{A}$ be a substructure of a $T$-model, $\varphi(\vec{x}) \in \mathcal{L}$, and $\vec{a} \in A^n$ such that $\mathcal{A} \vDash \varphi[\vec{a}]$. Further, let $\mathcal{B} \vDash T, \mathcal{D}\mathcal{A}$ so that w.l.o.g. $\mathcal{B} \supseteq \mathcal{A}$. Then also $\mathcal{B} \vDash \varphi(\vec{a})$, because in view of (i) we may suppose that $\varphi$ is open. Since $\mathcal{B}$ was arbitrary, $\mathcal{D}\mathcal{A} \vdash_T \varphi(\vec{a})$. Hence $T + \mathcal{D}\mathcal{A}$ is complete.
(ii)⇒(i): Let $X$ denote the set all of literals of $\mathcal{L}$. It suffices to prove

$$(*) \quad \mathcal{M} \equiv_X \mathcal{M}' \Rightarrow \mathcal{M} \equiv \mathcal{M}', \text{ for all } \mathcal{M}, \mathcal{M}' \vDash T,$$

for then $X$ is a Boolean basis for $\mathcal{L}$ in $T$ according to Theorem 6.3. This obviously amounts to saying that $T$ allows quantifier elimination. Let $\mathcal{M}, \mathcal{M}' \vDash T$, $\mathcal{M} = (\mathcal{A}, w)$, $\mathcal{M} \vDash \varphi(\vec{x})$, $\vec{x} \neq \emptyset$, and $a_1 = x_1^w, \ldots, a_n = x_n^w$. Let $\mathcal{A}^E$ be the substructure generated in $\mathcal{A}$ from $E := \{a_1, \ldots, a_n\}$. By (ii), $T + \mathcal{D}\mathcal{A}^E$ is complete and consistent with $\varphi(\vec{a})$, since $\mathcal{A}_A$ satisfies $T + \mathcal{D}\mathcal{A}^E + \varphi(\vec{a})$. Hence $\mathcal{D}\mathcal{A}^E \vdash_T \varphi(\vec{a})$. Moreover, $\mathcal{D}\mathcal{A}^E \cap \mathcal{L}E \vdash_T \varphi(\vec{a})$ by Exercise 5 in **5.1**. Thus, by the finiteness theorem, there are literals $\lambda_0(\vec{x}), \ldots, \lambda_k(\vec{x}) \in \mathcal{L}$ with $\lambda_i(\vec{a}) \in \mathcal{D}\mathcal{A}^E$ and $\bigwedge_{i \leqslant k} \lambda_i(\vec{a}) \vdash_T \varphi(\vec{a})$. Therefore $\bigwedge_{i \leqslant k} \lambda_i(\vec{x}) \vdash_T \varphi(\vec{x})$, because $a_1, \ldots, a_n$ do not appear in $T$. Since $\mathcal{M} \vDash \bigwedge_{i \leqslant k} \lambda_i(\vec{x})$ and $\mathcal{M} \equiv_X \mathcal{M}'$, also $\mathcal{M}' \vDash \bigwedge_{i \leqslant k} \lambda_i(\vec{x})$ and so $\mathcal{M}' \vDash \varphi(\vec{x})$. The above holds for arbitrary formulas $\varphi(\vec{x}) \in \mathcal{L}$ provided $\vec{x} \neq \emptyset$. These include sentences as well which completes the proof of $(*)$. ∎

**Corollary 6.5.** *An ∀-theory $T$ permits quantifier elimination if and only if $T$ is model complete.*

**Proof.** Due to $\mathcal{A} \subseteq \mathcal{B} \vDash T \Rightarrow \mathcal{A} \vDash T$, (ii) in Theorem 6.4 is satisfied provided only that $T + D\mathcal{A}$ is complete for all $\mathcal{A} \vDash T$. But this is granted if $T$ is model complete. ❏

**Example 3.** Let $T$ be the ∀-theory with two unary function symbols $f, g$ whose axioms state that $f$ and $g$ are injective, $f$ and $g$ are mutually inverse ($\forall x\, fgx = x$ and $\forall x\, gfx = x$), and there are no circles, i.e., no sequences $x_0, \ldots, x_n$ ($n > 0$) such that $x_{i+1} = fx_i$ and $x_0 = x_n$. This implies in particular $\forall x\, x \neq fx$. Note that $\forall y \exists x fx = y$ is provable (choose $x = gy$). Hence, $f$ and $g$ are bijective. The $T$-models consist of disjoint countable infinite "threads," which occurred also in Example 3 in **5.2**. Hence, $T$ is $\aleph_1$-categorical and thus model complete by Lindström's criterion. By the corollary, $T$ permits the elimination of quantifiers.

**Theorem 6.6.** ACF *and* RCF *allow quantifier elimination.*

**Proof.** By Theorem 6.4 it is enough to show that ACF and RCF are substructure complete, or put another way, ACF and RCF are the model completions of ACF$^\forall$ and RCF$^\forall$, respectively. Both claims are clear from Example 3 in **5.5** according to which ACF$^\forall$ coincides with the theory of integral domains, and RCF$^\forall$ with the theory of ordered commutative rings with a unit element. ❏

Theorem 6.6 was originally proved by Tarski in [Ta3]. While thanks to a host of model-theoretic methods the above proof is significantly shorter than Tarski's original, the latter is still of import in many algorithmic questions. Decidability and eliminability of quantifiers in RCF have great impact also on other fields of research, in particular on the foundations of geometry, which are not treated in this book. Let us mention that both Euclidean and non-Euclidean geometry can entirely be based on RCF and hence are decidable as well.

**Remark 2.** Due to the completeness of RCF, one may also say that the first-order theory of the ordered field $\mathbb{R}$ allows quantifier elimination. Incidentally, the quantifiers in RCF are not eliminable if the order, which is definable in RCF, is not considered as a basic relation. Also $T := Th(\mathbb{R}, <, 0, 1, +, -, \cdot, \exp)$, a (complete) theory with the exponential function exp in the language, does not allow quantifier elimination. Nonetheless, $T$ is model complete, as was shown in [Wi]. Because of its completeness, the decision problem for $T$ reduces to the still unsolved axiomatization problem, whose solution hinges on the unanswered problem concerning transcendental numbers, *Schanuel's conjecture*, which lies outside the scope of logic (consult the Internet). A particular question related to the conjecture is whether $e^e$ is transcendental.

### Exercises

1. Show that the theory ZG is model complete in its language, and even in the language $\mathcal{L}\{0, 1, +, -\}$.

2. A structure elementarily equivalent to $(\mathbb{N}, 0, 1, +, <)$ is called an $\mathbb{N}$-*semigroup*. Axiomatize the theory of $\mathbb{N}$-semigroups and show (by tracing back to ZG) that it allows quantifier elimination in $\mathcal{L}\{0, 1, +, <, 1|, 2|, \dots \}$.

3. Let RCF$^\circ$ be the theory of real closed fields without order as a basic notion. Prove that $\exists y$ is not eliminable in $\alpha(x) = \exists y\, y \cdot y = x$ in the frame of RCF$^\circ$.

4. Show that RCF is axiomatized alternatively by the axioms for ordered fields and the continuity scheme in **3.3** page 110.

5. Show that the theory $T$ of divisible ordered abelian groups allows quantifier elimination.

## 5.7   Reduced Products and Ultraproducts

In order to merely indicate the usefulness of the following constructions consider for instance $\mathbb{Z}^n$, a direct power of the additive group $\mathbb{Z}$. By componentwise verification of the axioms it can be shown that $\mathbb{Z}^n$ is itself an abelian group ($n \geqslant 2$). But in this and similar examples we can save ourselves the bother, because by Theorem 7.5 below a Horn sentence valid in all $\mathcal{A}_i$ is also valid in the direct product $\prod_{i \in I} \mathcal{A}_i$, and the group axioms are Horn sentences in each reasonable signature for groups.

Let $(\mathcal{A}_i)_{i \in I}$ be a family of $\mathcal{L}$-structures and $F$ a proper filter on a nonempty index set $I$ (see page 34). We define a relation $\approx_F$ on the domain $B$ of the product $\mathcal{B} := \prod_{i \in I} \mathcal{A}_i$ by

$$a \approx_F b \;\Leftrightarrow\; \{i \in I \mid a_i = b_i\} \in F.$$

This is an equivalence relation on $B$. Indeed, let $I_{a=b} := \{i \in I \mid a_i = b_i\}$. $\approx_F$ is reflexive (since $I_{a=a} = I \in F$) and trivially symmetric, but also transitive, because $I_{a=b}, I_{b=c} \in F \Rightarrow I_{a=c} \in F$, thanks to the obvious fact $I_{a=b} \cap I_{b=c} \subseteq I_{a=c}$.

Furthermore, $\approx_F$ is a *congruence* in the algebraic reduct of $\mathcal{B}$. To see this let $f$ be an $n$-ary function symbol and $\vec{a} \approx_F \vec{b}$ (which for $\vec{a} = (a^1, \ldots, a^n)$, $\vec{b} = (b^1, \ldots, b^n)$ in $B^n$ abbreviates $a^1 \approx_F b^1, \ldots, a^n \approx_F b^n$). Then clearly $I_{\vec{a}=\vec{b}} := \bigcap_{\nu=1}^{n} I_{a^\nu = b^\nu}$ belongs to $F$. Since certainly $I_{\vec{a}=\vec{b}} \subseteq I_{f\vec{a}=f\vec{b}}$, we get $I_{f\vec{a}=f\vec{b}} \in F$ and hence $f^{\mathcal{B}} \vec{a} \approx_F f^{\mathcal{B}} \vec{b}$.

Let $C := \{a/F \mid a \in B\}$, where $a/F$ denotes the congruence class of $\approx_F$ to which $a \in B$ belongs. Thus, $a/F = b/F \Leftrightarrow I_{a=b} \in F$. This $C$ becomes the domain of some $\mathcal{L}$-structure $\mathcal{C}$ in which first the operations $f^{\mathcal{C}}$ are defined in a canonical way. With $\vec{a}/F := (a^1/F, \ldots, a^n/F)$ we set $f^{\mathcal{C}}(\vec{a}/F) := (f^{\mathcal{B}} \vec{a})/F$. This definition is sound because $\approx_F$ is a congruence. For constant symbols $c$ let $c^{\mathcal{C}}$ of course be $c^{\mathcal{B}}/F$.

Similar to the identity, the relation symbols are interpreted in $\mathcal{C}$ in a completely natural way as follows:

$$r^{\mathcal{C}} \vec{a}/F :\Leftrightarrow I_{r\vec{a}} \in F \quad \left( I_{r\vec{a}} := \{i \in I \mid r^{\mathcal{A}_i} \vec{a}_i\}, \ \vec{a}_i := (a_i^1, \ldots, a_i^n) \right).$$

Also this definition is sound, since $I_{r\vec{a}} \in F$ and $\vec{a} \approx_F \vec{b}$ imply $I_{r\vec{b}} \in F$. Indeed, $\vec{a} \approx_F \vec{b}$ is equivalent to $I_{\vec{a}=\vec{b}} \in F$ and it is readily verified that $I_{r\vec{a}} \cap I_{\vec{a}=\vec{b}} \subseteq I_{r\vec{b}}$.

The $\mathcal{L}$-structure $\mathcal{C}$ so defined is called a *reduced product of the $\mathcal{A}_i$ by the filter $F$* and is denoted by $\prod_{i \in I}^{F} \mathcal{A}_i$ (some authors denote it by $\prod_{i \in I} \mathcal{A}_i/F$). Imagining a filter $F$ as a system of subsets of $I$ each of which contains "almost all indices," one may think of $\prod_{i \in I}^{F} \mathcal{A}_i$ as arising from $\mathcal{B} = \prod_{i \in I} \mathcal{A}_i$ by identification of those $a, b \in B$ for which the $i$th projections are the same for almost all indices $i$.

Let $\mathcal{C} = \prod_{i \in I}^{F} \mathcal{A}_i$. For $w : \mathrm{Var} \to B \ (= \prod_{i \in I} A_i)$ the valuation $x \mapsto (x^w)_i$ to $A_i$ is denoted by $w_i$, so that $x^w = (x^{w_i})_{i \in I}$. Induction on $t$ yields $t^w = (t^{w_i})_{i \in I}$. Define the valuation $w/F : \mathrm{Var} \to C$ by $x^{w/F} = x^w/F$. This setting generalizes inductively to

(1) $t^{w/F} = t^w/F$, for all terms $t$ and valuations $w : \mathrm{Var} \to B$.

(1) follows from $(f\vec{t})^{w/F} = f^{\mathcal{C}}(\vec{t}^{\,w/F}) = f^{\mathcal{C}}(\vec{t}^{\,w}/F) = f^{\mathcal{B}}(\vec{t}^{\,w})/F = (f\vec{t})^w/F$. It is easily seen that each $w' : \mathrm{Var} \to C$ is of the form $w/F$ for a suitable valuation $w : \mathrm{Var} \to B$.

Let $w \colon \mathit{Var} \to B$ and $\alpha \in \mathcal{L}$. Define $I_\alpha^w := \{i \in I \mid \mathcal{A}_i \vDash \alpha\,[w_i]\}$. Then

(2) $I_{\exists x \beta}^w \subseteq I_\beta^{w'}$ for some $a \in B$ and $w' = w_x^a$.

Indeed, let $i \in I_{\exists x \beta}^w$, i.e., $\mathcal{A}_i \vDash \exists x \beta\,[w_i]$. Choose some $a_i \in A_i$ with $\mathcal{A}_i \vDash \beta\,[w_i\,{}_x^{a_i}]$. For $i \notin I_{\exists x \beta}^w$ pick up *any* $a_i \in A_i$. Then clearly (2) holds with $a = (a_i)_{i \in I}$ and $w' = w_x^a$.

The case that $F$ is an ultrafilter on $I$ is of particular interest. By Theorem 7.1, all elementary properties valid in almost all factors carry over to the reduced product, which in this case is called an *ultraproduct*. If $\mathcal{A}_i = \mathcal{A}$ for all $i \in I$ then $\prod_{i \in I}^F \mathcal{A}_i$ is termed an *ultrapower of* $\mathcal{A}$, usually denoted by $\mathcal{A}^I/F$.

The importance of ultrapowers is underlined by Shelah's theorem (not proved here) that $\mathcal{A} \equiv \mathcal{B}$ iff $\mathcal{A}$ and $\mathcal{B}$ have isomorphic ultrapowers. The proof of Theorem 7.1 uses mainly filter properties; the specific ultrafilter property is applied only for confirming $I_{\neg \alpha}^w \in F \Leftrightarrow I_\alpha^w \notin F$.

**Theorem 7.1 (Łoś's ultraproduct theorem).** *Let* $\mathcal{C} = \prod_{i \in I}^F \mathcal{A}_i$ *be an ultraproduct of the* $\mathcal{L}$-*structures* $\mathcal{A}_i$. *Then for all formulas* $\alpha \in \mathcal{L}$ *and all* $w \colon \mathit{Var} \to \prod_{i \in I} A_i$,

$$(\ast) \quad \mathcal{C} \vDash \alpha[w/F] \Leftrightarrow I_\alpha^w \in F.$$

**Proof** by induction on $\alpha$. $(\ast)$ is obtained for equations $t_1 = t_2$ as follows:

$$
\begin{aligned}
\mathcal{C} \vDash t_1 = t_2\,[w/F] \quad &\Leftrightarrow \quad t_1^{w/F} = t_2^{w/F} \Leftrightarrow t_1^w/F = t_2^w/F \quad \text{(by (1))} \\
&\Leftrightarrow \quad \{i \in I \mid t_1^{w_i} = t_2^{w_i}\} \in F \qquad \left(t^w = (t^{w_i})_{i \in I}\right) \\
&\Leftrightarrow \quad \{i \in I \mid \mathcal{A}_i \vDash t_1 = t_2\,[w_i]\} \in F \ \Leftrightarrow \ I_{t_1 = t_2}^w \in F.
\end{aligned}
$$

One similarly proves $(\ast)$ for prime formulas $r\vec{t}$. Induction steps:

$$
\begin{aligned}
\mathcal{C} \vDash \alpha \wedge \beta\,[w/F] \quad &\Leftrightarrow \quad \mathcal{C} \vDash \alpha, \beta\,[w/F] \\
&\Leftrightarrow \quad I_\alpha^w, I_\beta^w \in F \qquad \text{(induction hypothesis)} \\
&\Leftrightarrow \quad I_\alpha^w \cap I_\beta^w \in F \qquad \text{(filter property)} \\
&\Leftrightarrow \quad I_{\alpha \wedge \beta}^w \in F \qquad (\text{since } I_{\alpha \wedge \beta}^w = I_\alpha^w \cap I_\beta^w).
\end{aligned}
$$

Further, $\mathcal{C} \vDash \neg \alpha\,[w/F] \Leftrightarrow \mathcal{C} \nvDash \alpha\,[w/F] \Leftrightarrow I_\alpha^w \notin F \Leftrightarrow I \setminus I_\alpha^w \in F \Leftrightarrow I_{\neg \alpha}^w \in F$. Now let $I_{\forall x \alpha}^w \in F$, $a \in \prod_{i \in I} A_i$, and $w' := w_x^a$. Since $I_{\forall x \alpha}^w \subseteq I_\alpha^{w'}$, also $I_\alpha^{w'} \in F$. Hence, $\mathcal{C} \vDash \alpha\,[w'/F]$ by the induction hypothesis. $a$ was arbitrary, therefore $\mathcal{C} \vDash \forall x \alpha\,[w/F]$. The converse is, with $\beta := \neg \alpha$, equivalent to $I_{\exists x \beta}^w \in F \Rightarrow \mathcal{C} \vDash \exists x \beta\,[w/F]$. This follows from (2), since $(\ast)$ holds by the induction hypothesis for $\alpha$, hence also for $\neg \alpha$. $\blacksquare$

**Corollary 7.2.** *A sentence $\alpha$ is valid in the ultraproduct $\prod_{i \in I}^{F} \mathcal{A}_i$ iff $\alpha$ is valid in "almost all" $\mathcal{A}_i$, that is, $\{i \in I \mid \mathcal{A}_i \vDash \alpha\} \in F$. In particular, $\mathcal{A}^I/F \vDash \alpha \Leftrightarrow \mathcal{A} \vDash \alpha$, for all $\alpha \in \mathcal{L}^0$. In words, an ultrapower of $\mathcal{A}$ is elementarily equivalent to $\mathcal{A}$.*

The last claim is clear, since the validity of $\alpha$ in a structure does not depend on the valuation chosen. The ultrapower case can be further strengthened to $\mathcal{A} \preccurlyeq \mathcal{A}^I/F$ (Exercise 2), useful for the construction of special nonstandard models, for instance. From countless applications of ultraproducts, we present here a very short proof of the compactness theorem for arbitrary first-order languages. The proof is tricky, but it is undoubtedly the most elegant proof of the compactness theorem.

**Theorem 7.3.** *Let $X \subseteq \mathcal{L}$ and let $I$ be the set of all finite subsets of $X$. Assume that every $i \in I$ has a model $(\mathcal{A}_i, w_i)$. Then there exists an ultrafilter $F$ on $I$ such that $\prod_{i \in I}^{F} \mathcal{A}_i \vDash X [w/F]$, where $x^w = (x^{w_i})_{i \in I}$ for $x \in \mathrm{Var}$. In short, if every finite subset of $X \subseteq \mathcal{L}$ has a model then the same applies to the whole of $X$.*

**Proof.** Let $J_\alpha := \{i \in I \mid \alpha \in i\}$ for $\alpha \in X$. The intersection of finitely many members of $E := \{J_\alpha \mid \alpha \in X\}$ is $\neq \emptyset$; for instance $\{\alpha_0, \ldots, \alpha_n\}$ belongs to $J_{\alpha_0} \cap \cdots \cap J_{\alpha_n}$. By the ultrafilter theorem (page 35), there exists an ultrafilter $F \supseteq E$. If $\alpha \in X$ and $i \in J_\alpha$ (that is, $\alpha \in i$) then $\mathcal{A}_i \vDash \alpha [w_i]$. Consequently, $J_\alpha \subseteq I_\alpha^w$; hence $I_\alpha^w \in F$. Therefore, $\prod_{i \in I}^{F} \mathcal{A}_i \vDash \alpha [w/F]$ by Theorem 7.1, as claimed. ☐

A noteworthy consequence of these results is the following theorem in which $\boldsymbol{K}_{\mathcal{L}}$ denotes the class of all $\mathcal{L}$-structures; by Shelah's theorem mentioned above, condition (a) can be converted into a purely algebraic one.

**Theorem 7.4.** *Let $\boldsymbol{K} \subseteq \boldsymbol{K}_{\mathcal{L}}$. Then*

   (a) *$\boldsymbol{K}$ is $\Delta$-elementary iff $\boldsymbol{K}$ is closed under elementary equivalence and under ultraproducts,*

   (b) *$\boldsymbol{K}$ is elementary iff $\boldsymbol{K}$ is closed under elementary equivalence and both $\boldsymbol{K}$ and $\setminus \boldsymbol{K}$ ($= \boldsymbol{K}_{\mathcal{L}} \setminus \boldsymbol{K}$) are closed under ultraproducts.*

**Proof.** (a): A $\Delta$-elementary class is trivially closed under elementary equivalence. The rest of the direction $\Rightarrow$ holds by Theorem 7.1. $\Leftarrow$: Let $T := Th\,\boldsymbol{K}$ and $\mathcal{A} \vDash T$, and let $I$ be the set of all finite subsets of $Th\,\mathcal{A}$.

For each $i = \{\alpha_1, \ldots, \alpha_n\} \in I$ there exists some $\mathcal{A}_i \in \mathbf{K}$ such that $\mathcal{A}_i \vDash i$, for otherwise $\bigvee_{\nu=1}^n \neg\alpha_\nu \in T$ which contradicts $i \subseteq Th\,\mathcal{A}$. According to Theorem 7.3 (with $X = Th\,\mathcal{A}$) there is a $\mathcal{C} := \prod_{i \in I}^F \mathcal{A}_i \vDash Th\,\mathcal{A}$, and if $\mathcal{A}_i \in \mathbf{K}$ then so too $\mathcal{C} \in \mathbf{K}$. Because of $\mathcal{C} \vDash Th\,\mathcal{A}$ we know that $\mathcal{C} \equiv \mathcal{A}$, and therefore $\mathcal{A} \in \mathbf{K}$. This shows that $\mathcal{A} \vDash T \Rightarrow \mathcal{A} \in \mathbf{K}$. Hence $\mathcal{A} \vDash T \Leftrightarrow \mathcal{A} \in \mathbf{K}$, i.e., $\mathbf{K}$ is $\Delta$-elementary. (b): $\Rightarrow$ is obvious by (a), because for $\mathbf{K} = \mathrm{Md}\,\alpha$ we have $\backslash \mathbf{K} = \mathrm{Md}\,\neg\alpha$. $\Leftarrow$: By (a), $\mathbf{K} = \mathrm{Md}\,S$ for some $S \subseteq \mathcal{L}^0$. Let $I$ be the set of all finite nonempty subsets of $S$. **Claim:** There is some $i = \{\alpha_0, \ldots, \alpha_n\} \in I$ with $\mathrm{Md}\,i \subseteq \mathbf{K}$. Otherwise let $\mathcal{A}_i \vDash i$ such that $\mathcal{A}_i \in \backslash \mathbf{K}$ for all $i \in I$. Then there exists an ultraproduct $\mathcal{C}$ of the $\mathcal{A}_i$ such that $\mathcal{C} \in \backslash \mathbf{K}$ and $\mathcal{C} \vDash i$ for all $i \in I$; hence $\mathcal{C} \vDash S$. This is a contradiction to $\mathrm{Md}\,S \subseteq \mathbf{K}$. Thus, the claim holds. Since also $\mathbf{K} = \mathrm{Md}\,S \subseteq \mathrm{Md}\,i$, we obtain $\mathbf{K} = \mathrm{Md}\,i = \mathrm{Md}\,\bigwedge_{\nu \leqslant n} \alpha_\nu$. $\blacksquare$

**Application.** Let $\mathbf{K}$ be the ($\Delta$-elementary) class of all fields of characteristic 0. We show that $\mathbf{K}$ is not elementary, and thus in a new way that $Th\,\mathbf{K}$ is not finitely axiomatizable. Let $\mathcal{P}_i$ denote the prime field of characteristic $p_i$ ($p_0{=}2$, $p_1{=}3$, $\ldots$) and let $F$ be a nontrivial ultrafilter on $\mathbb{N}$. We claim that the field $\prod_{i \in \mathbb{N}}^F \mathcal{P}_i$ has characteristic 0. Indeed, $\{i \in I \mid \mathcal{P}_i \vDash \neg\mathrm{char}_p\}$ is for a given prime $p$ certainly cofinite and belongs to $F$, so that $\prod_{i \in \mathbb{N}}^F \mathcal{P}_i \vDash \neg\mathrm{char}_p$ for all $p$. Hence $\backslash \mathbf{K}$ is not closed under ultraproducts, and so by Theorem 7.4(b), $\mathbf{K}$ cannot be elementary.

We now turn to reduced products. Everything said below on reduced products remains valid for direct products; these are the special case with the minimal filter $F = \{I\}$. More precisely, $\prod_{i \in I}^{\{I\}} \mathcal{A}_i \simeq \prod_{i \in I} \mathcal{A}_i$. Filters are always supposed to be proper in the sequel.

**Theorem 7.5.** *Let $\mathcal{C} = \prod_{i \in I}^F \mathcal{A}_i$ be a reduced product, $w \colon \mathrm{Var} \to \prod_{i \in I} \mathcal{A}_i$, and $\alpha$ a Horn formula from the corresponding first-order language. Then*

$$(\star) \quad I_\alpha^w \in F \Rightarrow \mathcal{C} \vDash \alpha\,[w/F].$$

*In particular, a Horn sentence valid in almost all $\mathcal{A}_i$ is also valid in $\mathcal{C}$.*

**Proof** by induction on the construction of Horn formulas. For prime formulas the converse of $(\star)$ is also valid, because in the proof of $(*)$ from Theorem 7.1 for prime formulas no specific ultrafilter property was used. Moreover, $I_{\neg\alpha}^w \in F \Rightarrow I_\alpha^w \notin F \Rightarrow \mathcal{C} \nvDash \alpha\,[w/F] \Rightarrow \mathcal{C} \vDash \neg\alpha\,[w/F]$, provided $\alpha$ is prime. Hence, $(\star)$ is correct for all literals. Now suppose $(\star)$ for a prime formula $\alpha$ and a basic Horn formula $\beta$, and let $I_{\alpha \to \beta}^w \in F$. We show that $\mathcal{C} \vDash \alpha \to \beta \, [w/F]$. Let $\mathcal{C} \vDash \alpha \, [w/F]$. Then $I_\alpha^w \in F$ since $\alpha$ is prime. $I_\alpha^w \cap I_{\alpha \to \beta}^w \subseteq I_\beta^w$ leads to $I_\beta^w \in F$; hence $\mathcal{C} \vDash \beta \, [w/F]$ by the induction hypothesis. This shows that $\mathcal{C} \vDash \alpha \to \beta \, [w/F]$ and proves $(\star)$ for all basic Horn formulas. Induction on $\wedge$ and $\forall$ proceeds as in Theorem 7.1 and the $\exists$-step easily follows with the help of (2) above. $\qquad \square$

According to this theorem the model classes of Horn theories are always closed under reduced products, in particular under direct products. This result strengthens Exercise 1 in **4.2** significantly. We mention finally that also the converse holds: every theory with a model class closed with respect to reduced products is a Horn theory. But the proof of this claim, presented in [CK], is essentially more difficult than that for the similar-sounding Theorem 4.4.

## Exercises

1. Show that $\prod_{i \in I}^F \mathcal{A}_i$ is isomorphic to $\mathcal{A}_{i_0}$ for some $i_0 \in I$ if $F$ is a trivial ultrafilter. This applies, for instance, to ultraproducts on a finite index set (Exercise 3 in **1.5**). Thus, ultraproducts are interesting only if the index set $I$ is infinite.

2. Prove that $\mathcal{A}$ is elementarily embeddable into an ultrapower $\mathcal{A}^I/F$.

3. (Basic in nonclassical logics). Let $\vDash_{\boldsymbol{K}} := \bigcap \{\vDash_{\mathcal{A}} | \; \mathcal{A} \in \boldsymbol{K}\}$ be the consequence relation defined by a class $\boldsymbol{K}$ of $L$-matrices (page 49). Show that $\vDash_{\boldsymbol{K}}$ is finitary provided $\boldsymbol{K}$ is closed under ultraproducts (which is the case, for instance, if $\boldsymbol{K} = \{\mathcal{A}\}$ with finite $\mathcal{A}$). Thus, $\vDash_{\mathcal{A}}$ is finitary for each finite logical matrix.

4. Let $\mathcal{A}, \mathcal{B}$ be Boolean algebras. Prove that $\mathcal{A} \vDash \alpha \Leftrightarrow \mathcal{B} \vDash \alpha$ for all universal Horn sentences $\alpha$. This holds in particular for identities and quasi-identities. Every sentence of this kind valid in $\mathcal{2}$ is therefore valid in all Boolean algebras.

5. Let $\mathcal{A}_i'$ for each $i \in I \, (\neq \emptyset)$ be an expansion of the $\mathcal{L}$-structure $\mathcal{A}_i$ to $\mathcal{L}' \supseteq \mathcal{L}$. Prove that the reduced product $\prod_{i \in I}^F \mathcal{A}_i'$ is an $\mathcal{L}'$-expansion of the reduced product $\prod_{i \in I}^F \mathcal{A}_i$ (the "expansion theorem").

# Chapter 6

# Incompleteness and Undecidability

Gödel's fundamental results concerning the incompleteness of formal systems sufficiently rich in content, along with Tarski's on the nondefinability of the notion of truth and Church's on the undecidability of logic, as well as other undecidability results, are all based on essentially the same arguments. A widely known popularization of Gödel's first incompleteness theorem runs as follows:

Consider a formalized axiomatic theory $T$ that describes a given domain of objects $\mathcal{A}$ in a manner that we hope is complete. Moreover, suppose that $T$ is capable of talking in its language $\mathcal{L}$ about its own syntax and proofs from its axioms. This is often possible if $T$ has actually been devised to investigate other things (numbers or sets, say), namely by means of an internal encoding of the syntax of $\mathcal{L}$. Then the sentence $\gamma$: "I am unprovable in $T$" belongs to $\mathcal{L}$, where "I" refers precisely to the sentence $\gamma$ (clearly, this possibility of self-reference has to be laid down in detail, which was the main work in [Gö2]). *Then $\gamma$ is true in $\mathcal{A}$ but unprovable in $T$.*

Indeed, if we assume that $\gamma$ is provable, then, like any other provable sentence in $T$, $\gamma$ would be true in $\mathcal{A}$ and so unprovable, since this is just what $\gamma$ claims. Thus, our assumption leads to a contradiction. Hence, $\gamma$ is indeed unprovable, that is, $\gamma$'s assertion conforms with truth; moreover, $\gamma$ belongs to the sentences from $\mathcal{L}$ true in $\mathcal{A}$, as it will turn out.

Put together, our goal of exhaustively capturing all theorems valid in $\mathcal{A}$ by means of the axioms of $T$ has not been achieved and is in fact not achievable, as we will see. No matter how strong our axiomatic theory $T$ is, there are always sentences true in $\mathcal{A}$ but unprovable in $T$.

W. Rautenberg, *A Concise Introduction to Mathematical Logic*,
Universitext, DOI 10.1007/978-1-4419-1221-3_6,
© Springer Science+Business Media, LLC 2010

Clearly, the above is just a rough simplification of Gödel's theorem that does not speak at all about a domain of objects, but is rather a *proof-theoretic* assertion the proof of which can be carried out in the framework of Hilbert's finitistic metamathematics. This in turn means about the same as being formalizable and provable in Peano arithmetic PA, introduced in **3.3**.

This result was a decisive point for a wellfounded criticism of *Hilbert's program*, which aimed to justify infinitistic methods by means of a finitistic understanding of metamathematics. For a detailed description of what Hilbert was aiming at, see [Kl2] or consult [HB, Vol. 1]. The paradigm of a domain of objects in the above sense is, for a variety of reasons, the structure $\mathcal{N} = (\mathbb{N}, 0, \mathsf{S}, +, \cdot)$. Gödel's theorem states that even for $\mathcal{N}$ a complete axiomatic characterization in its language is impossible, a result with far-reaching consequences. In particular, PA, which aims at telling us as much as possible about $\mathcal{N}$, is shown to be incomplete. PA is the center point of Chapter **7**. It is of special importance because most of classical number theory and of discrete mathematics can be developed in it. In addition, known methods for investigating mathematical foundations can be formalized and proved in PA. These methods have stood firm against all kinds of criticism, leaving aside some objections concerning the unrestricted use of two-valued logic, not to be discussed here.

Some steps in Gödel's proof require only modest suppositions regarding $T$, namely the numeralwise representability of relevant syntactic predicates and functions in $T$ in the sense of **6.3**. It was one of Gödel's decisive discoveries that *all* the predicates required in $\gamma$'s construction above are primitive recursive[1] and that all predicates and functions of this type are indeed representable in $T$. As remarked by Tarski and Mostowski, the latter works even in certain finitely axiomatizable, highly incomplete theories $T$ and, in addition, covers all recursive functions. This yields not only the recursive undecidability of $T$ and all its subtheories in $\mathcal{L}$ (in particular the theory $\mathsf{Taut}_{\mathcal{L}}$), but also of all consistent extensions of $T$.

---

[1] All these predicates are also elementary in the recursion-theoretic sense, see e.g. [Mo], although it requires much more effort to verify this. Roughly speaking, the elementary functions are the "not too rapidly growing" primitive recursive functions. The exponential function $(m, n) \mapsto m^n$ is still elementary; however, the hyperexponential function defined on page 239 is not.

From this it follows that the first incompleteness theorem as well as Church's and Tarski's results can all be obtained in one go, making essential use of the *fixed point lemma* in **6.5**, also called the *diagonalization lemma* because it is shown by some kind of diagonalization on the primitive recursive substitution function. Its basic idea can even be recognized in the ancient liar paradox, and is also used in the foregoing popularization of the first incompleteness theorem.

In **6.1** we develop the theory of recursive and primitive recursive functions to the required extent. **6.2** deals with the arithmetization of syntax and of formal proofs. **6.3** and **6.4** treat the representability of recursive functions in axiomatic theories. In **6.5** all the aforementioned results are proved, while the deeper-lying second incompleteness theorem is dealt with in Chapter **7**. Section **6.6** concerns the transferability of decidability and undecidability by interpretation, and **6.7** describes the first-order arithmetical hierarchy, which vividly illustrates the close relationship between logic and recursion theory. At the end we consider special $\Sigma_1$-formulas, important for Chapter **7**.

# 6.1 Recursive and Primitive Recursive Functions

From now on, along with $i, \ldots, n$ we take $a, \ldots, e$ to denote natural numbers, unless stated otherwise. Let $\mathbf{F}_n$ denote the set of all $n$-ary functions with arguments and values in $\mathbb{N}$, and put $\mathbf{F} := \bigcup_{n \in \mathbb{N}} \mathbf{F}_n$. For $f \in \mathbf{F}_m$ and $g_1, \ldots, g_m \in \mathbf{F}_n$, we call $h : \vec{a} \mapsto f(g_1\vec{a}, \ldots, g_m\vec{a})$ the (canonical) *composition* of $f$ and the $g_i$ and write $h = f[g_1, \ldots, g_m]$. The arity of $h$ is $n$. Analogously, let $P[g_1, \ldots, g_m]$ for a given predicate $P \subseteq \mathbb{N}^m$ ($m > 0$) denote the $n$-ary predicate $\{\vec{a} \in \mathbb{N}^n \mid P(g_1\vec{a}, \ldots, g_m\vec{a})\}$.

In an intuitive sense $f \in \mathbf{F}_n$ is computable if there is an algorithm for computing $f\vec{a}$ for every $\vec{a}$ in finitely many steps. Sum and product are simple examples. There are uncountably many unary functions on $\mathbb{N}$, and because of the finiteness of every set of computation instructions, only countably many of these can be computable. Thus, there must be noncomputable functions. This existence proof brings to mind the one for transcendental real numbers, based on the countability of the set of algebraic numbers. Coming up with concrete examples is, in both cases, much less simple.

The computable functions in the intuitive sense obviously have the following properties:

**Oc**: If $h \in \mathbf{F}_m$ and $g_1, \ldots, g_m \in \mathbf{F}_n$ are computable, so too is the composition $f = h[g_1, \ldots, g_m]$.

**Op**: If $g \in \mathbf{F}_n$ and $h \in \mathbf{F}_{n+2}$ are computable then so is $f \in \mathbf{F}_{n+1}$, uniquely determined by the equations

$$f(\vec{a}, 0) = g\vec{a}; \quad f(\vec{a}, \mathsf{S}b) = h(\vec{a}, b, f(\vec{a}, b)).$$

These are called the *recursion equations* for $f$, and $f$ is said *to result from $g, h$ by primitive recursion*, or $f = \mathbf{Op}(g, h)$ for short.

**Oμ**: Let $g \in \mathbf{F}_{n+1}$ be such that $\forall \vec{a} \, \exists b \, g(\vec{a}, b) = 0$. If $g$ is computable then so is $f$, given by $f\vec{a} = \mu b[g(\vec{a}, b) = 0]$. Here the right-hand term denotes the smallest $b$ with $g(\vec{a}, b) = 0$. $f$ is said to result from $g$ by the so-called *μ-operation*.

Considering **Oc**, **Op**, and **Oμ** as generating operations for obtaining new functions from already-constructed ones, we state the following definition due to Kleene:

**Definition.** The set of p.r. (*primitive recursive*) functions consists of all functions on $\mathbb{N}$ that can be obtained by finitely many applications of **Oc** and **Op** starting with the following *initial functions*: the constant 0, the successor function $\mathsf{S}$, and the projection functions $\mathrm{I}_\nu^n : \vec{a} \mapsto a_\nu \ (1 \leqslant \nu \leqslant n, n = 1, 2, \ldots)$. With the additional generating schema **Oμ** one obtains the set of all *recursive* or *μ-recursive* functions. A predicate $P \subseteq \mathbb{N}^n$ is called p.r. or *recursive* (also *decidable*) provided the *characteristic function* $\chi_P$ of $P$ has the respective property, defined by

$$\chi_P \vec{a} = \begin{cases} 1 & \text{in case } P\vec{a}, \\ 0 & \text{in case } \neg P\vec{a}. \end{cases}$$

**Remark 1.** It should at least be noticed that it was Dedekind who first proved that **Op** defines exactly one function $f \in \mathbf{F}_n$ in the sense of set theory. Note also that for $n = 0$ the recursion equations reduce to $f0 = c$ and $f\mathsf{S}b = h(b, fb)$, where $c \in \mathbf{F}_0$ and $h \in \mathbf{F}_2$. If the condition $\forall \vec{a} \, \exists b \, g(\vec{a}, b) = 0$ in **Oμ** is omitted, then $f$ is regarded as undefined for those $\vec{a}$ for which there is no $b$ with $g(\vec{a}, b) = 0$. In this way the so-called partially recursive functions are defined. These are very important for recursion theory. However, we will not require them.

The following examples make it clear that by means of the functions $I_\nu^n$ our stipulations concerning arity in $\mathbf{Oc}$ and $\mathbf{Op}$ can be extensively relaxed. In the examples, however, we will still adjoin the normalized notation each time in parentheses.

**Examples.** Let $\mathsf{S}^0 = I_1^1$ and $\mathsf{S}^{k+1} = \mathsf{S}[\mathsf{S}^k]$, so that clearly $\mathsf{S}^k : a \mapsto a + k$. By $\mathbf{Oc}$ these functions are all p.r. The $n$-ary *constant functions* $\mathsf{K}_c^n : \vec{a} \mapsto c$ can be seen to be p.r. as follows: $\mathsf{K}_c^0 = \mathsf{S}^c[0]$ for arbitrary $c \in \mathbb{N}$, while $\mathsf{K}_c^1 0 = c \ (= \mathsf{K}_c^0)$ and $\mathsf{K}_c^1 \mathsf{S}b = c \ (= I_2^2(b, \mathsf{K}_c^1 b))$. Thus, $\mathsf{K}_c^1 = \mathbf{Op}(\mathsf{K}_c^0, I_1^2)$. For $n > 1$ we have $\mathsf{K}_c^n = \mathsf{K}_c^1[I_1^n]$. Further, the recursion equations

$$a + 0 = a \ (= I_1^1(a)); \quad a + \mathsf{S}b = \mathsf{S}(a + b) \ (= \mathsf{S}I_3^3(a, b, a + b))$$

show addition to be a p.r. function. Since

$$a \cdot 0 = 0 \ (= \mathsf{K}_0^1 a) \text{ and } a \cdot \mathsf{S}b = a \cdot b + a \ (= I_3^3(a, b, a \cdot b) + I_1^3(a, b, a \cdot b)),$$

it follows that $\cdot$ is p.r. and entirely analogously so is $(a, b) \mapsto a^b$. Also the predecessor function $\mathsf{Pd}$ is p.r. since $\mathsf{Pd}\, 0 = 0$ and $\mathsf{Pd}(\mathsf{S}b) = b \ (= I_1^2(b, \mathsf{Pd}\, b))$.

"Cut-off subtraction" $\dot{-}$, defined by $a \dot{-} b = a - b$ for $a \geqslant b$, and $a \dot{-} b = 0$ otherwise, is p.r. because

$$a \dot{-} 0 = a \ (= I_1^1(a)) \text{ and } a \dot{-} \mathsf{S}b = \mathsf{Pd}(a \dot{-} b) \ (= \mathsf{Pd}\, I_3^3(a, b, a \dot{-} b)).$$

The absolute difference $|a - b|$ is p.r., for $|a - b| = (a \dot{-} b) + (b \dot{-} a)$.

One sees easily that if $f$ is p.r. (resp. recursive) then so too is every function that results from $f$ by swapping, equating, or adjoining fictional arguments. For example, let $f \in \mathbf{F}_2$ and $f_1 := f[I_2^2, I_1^2]$. Then clearly $f_1(a, b) = f(b, a)$. For $f_2 := f[I_1^1, I_1^1]$ we have $f_2 a = f(a, a)$, and for $f_3 := f[I_1^3, I_2^3]$ we get $f_3(a, b, c) = f(a, b)$, for all $a, b, c$.

From now on we will be more relaxed in writing down applications of $\mathbf{Oc}$ or $\mathbf{Op}$; the $I_\nu^n$ will no longer explicitly appear. If $f \in \mathbf{F}_{n+1}$ is p.r. then so is the function $(\vec{a}, b) \mapsto \prod_{k<b} f(\vec{a}, k)$, since $\prod_{k<0} f(\vec{a}, k) = 1$, and $\prod_{k<\mathsf{S}b} f(\vec{a}, k) = \prod_{k<b} f(\vec{a}, k) \cdot f(\vec{a}, b)$. Also $(\vec{a}, b) \mapsto \sum_{k<b} f(\vec{a}, k)$, defined by $\sum_{k<0} f(\vec{a}, k) = 0$ and $\prod_{k<\mathsf{S}b} f(\vec{a}, k) = \sum_{k<b} f(\vec{a}, k) + f(\vec{a}, b)$ is p.r.

The $\delta$-*function* is defined by $\delta 0 = 1$, $\delta \mathsf{S}n = 0$ and hence is p.r. With $\delta$ we easily obtain the characteristic function of the identity relation: $\chi_=(a, b)$ equals $\delta|a - b|$. This in turn implies that every finite subset $E$ of $\mathbb{N}$ is p.r. because $\chi_\emptyset = \mathsf{K}_0^1$ and for $E = \{a_1, \ldots, a_n\} \neq \emptyset$ we have

$$\chi_E(a) = \chi_=(a, a_1) + \cdots + \chi_=(a, a_n).$$

The inequality relation $\neq$ is p.r. because $X_{\neq}(a,b) = \sigma|a-b|$ with the *signum function* $\sigma$, defined by $\sigma 0 = 0$, $\sigma S n = 1$. Also $\leqslant$ is p.r. because $X_{\leqslant}(a,b) = \sigma(Sb \div a)$, as is easily verified.

Almost all functions considered in number theory are p.r., in particular the *prime enumeration* $n \mapsto p_n$ (with $p_0 = 2$, $p_1 = 3, \ldots$). The same is true for standard predicates like $|$ (divides) and prim (to be a prime number). This will all be verified after some additional remarks.

Important is the closure of the set of p.r. functions with respect to *definition by* p.r. (resp. recursive) *case distinction*: If $P, g, h$ are p.r. (resp. recursive) then so is $f$, given by $f\vec{a} = g\vec{a} \cdot X_P \vec{a} + h\vec{a} \cdot \delta X_P \vec{a}$, or

$$f\vec{a} = \begin{cases} g\vec{a} & \text{in case } P\vec{a}, \\ h\vec{a} & \text{in case } \neg P\vec{a}. \end{cases}$$

A simple example is $(a,b) \mapsto \max(a,b)$, defined by $\max(a,b) = b$ if $a \leqslant b$ and $\max(a,b) = a$ otherwise. Case distinction easily generalizes to more than two cases. Sometimes equations in $\boldsymbol{Op}$ are given by p.r. case distinction as in a defining equation for $\text{rem}(a,b)$, *the remainder of dividing a by* $b$ ($\neq 0$): $\text{rem}(Sa, b) = 0$ if $b|Sa$, and $\text{rem}(Sa, b) = S\,\text{rem}(a,b)$ otherwise. Case distinction yields also that $a \mapsto b_i a$ is p.r., where $b_i a$ denotes the $i$th digit of the binary representation of $a \in \mathbb{N}$, $a = \sum_{i \geqslant 0} b_i a \cdot 2^i$. Indeed, $b_i a = 0$ if $\text{rem}(a, 2^{i+1}) < 2^i$ and $b_i a = 1$ otherwise.

Of fundamental importance is the hypothesis that recursive functions exhaust all the computable functions over $\mathbb{N}$. This hypothesis is called *Church's thesis*; all undecidability results are based on it. Though it is not at all obvious from looking at the definition of recursive functions that these functions exhaust all computable functions no matter what the computation procedures look like, all the variously defined computability concepts turned out to be equivalent, providing evidence in favor of the thesis. One of these concepts is computability by means of a *Turing machine* [Tu], a particularly simple model of automated information processing. Also, programming languages may be used to define the concept of computability, for instance PROLOG, as was seen in **4.6**.

Below we compile a list of the easily provable basic facts about p.r. and recursive predicates needed in the following. Further insights, above all concerning the form of their defining formulas, will emerge in **6.3** and thereafter. $P, Q, R$ now denote exclusively predicates of $\mathbb{N}$. In order to simplify the notation of properties of such predicates, we use as abbreviations in our metatheory the prefixes $(\exists a{<}b)$, $(\exists a{\leqslant}b)$, $(\forall a{<}b)$, and $(\forall a{\leqslant}b)$ as in (B) below. Their meaning is self-explanatory.

(A) The set of p.r. (resp. recursive) predicates is closed under forming the complement, union, and intersection of predicates of the same arity, as well as under insertion of p.r. (resp. recursive) functions, and finally under swapping, equating, and adjoining fictional arguments. This is proved as follows: for $P \subseteq \mathbb{N}^n$, $\delta[\chi_P]$ is exactly the characteristic function of $\neg P := \mathbb{N}^n \setminus P$; furthermore, $\chi_{P \cap Q} = \chi_P \cdot \chi_Q$ and $\chi_{P \cup Q} = \mathrm{sg}[\chi_P + \chi_Q]$ as well as $\chi_{P[g_1,\dots,g_m]} = \chi_P[g_1,\dots,g_m]$. Since $\chi_{\mathrm{graph}\,f}(\vec{a},b)$ is the same as $\chi_=(f\vec{a},b)$, graph $f$ is p.r. provided $f$ is (though the converse need not hold, see the end of this section). All the other above-mentioned closure properties are simply obtained from the corresponding properties of the characteristic functions.

(B) Let $P, Q, R \subseteq \mathbb{N}^{n+1}$. Suppose that $Q(\vec{a}, b) \Leftrightarrow (\forall k{<}b)P(\vec{a}, k)$ and $R(\vec{a}, b) \Leftrightarrow (\exists k{<}b)P(\vec{a}, k)$. Then we say that $Q, R$ result from $P$ by *bounded quantification*. The same will be said if $<$ in these definitions is replaced by $\leqslant$. If $P$ is p.r. so too are all these predicates, because $\chi_Q(\vec{a}, b) = \prod_{k<b}\chi_P(\vec{a}, k)$, $\chi_R(\vec{a}, b) = \mathrm{sg}(\sum_{k<b}\chi_P(\vec{a}, k))$, and similarly if $<$ is replaced by $\leqslant$. The proofs of these equations are so simple that we may pass over them. Briefly, the set of p.r. (resp. recursive) predicates is closed under bounded quantification. For instance, since $a\,|\,b \Leftrightarrow (\exists k{\leqslant}b)[a \cdot k = b]$, also $|$ is p.r. So too is the predicate prim, because $\mathrm{prim}\,p \Leftrightarrow p \neq 0,1 \ \& \ (\forall a{<}p)[a|p \Rightarrow a{=}1]$. Note that $a|p \Rightarrow a = 1$ is equivalent (at the metatheoretical level) to $a{\not|}\,p \vee a = 1$ and is therefore the union of p.r. predicates. Hence, this predicate is indeed p.r.

(C) Suppose $P \subseteq \mathbb{N}^{n+1}$ satisfies $\forall \vec{a}\,\exists b\,P(\vec{a}, b)$ and let $f\vec{a} = \mu k[P(\vec{a}, k)]$ be the smallest $k$ such that $P(\vec{a}, k)$. Then by $\boldsymbol{O\mu}$, if $P$ is recursive so too is the function $f$, because $f\vec{a} = \mu k[\delta\chi_P(\vec{a}, k) = 0]$. This important generalization of $\boldsymbol{O\mu}$ will henceforth likewise be denoted by $\boldsymbol{O\mu}$. On the other hand, $f$ need no longer be p.r., provided $P$ is. This does hold, though, for the *bounded $\mu$-operation*: if $P \subseteq \mathbb{N}^{n+1}$ is p.r. so too is $f$ defined by $f(\vec{a}, m) = \mu k{\leqslant}m[P(\vec{a}, k)]$. Here let

$$\mu k{\leqslant}m[P(\vec{a}, k)] = \begin{cases} \text{the smallest } k \leqslant m \text{ with } P(\vec{a}, k), \text{ if such } k \text{ exists,} \\ m \ \text{ otherwise.} \end{cases}$$

Clearly $f(\vec{a},0) = 0$, and $f(\vec{a},Sm) = f(\vec{a},m)$ if $(\exists k\leqslant m)P(\vec{a},k)$, and $f(\vec{a},Sm) = Sm$ otherwise. To convert this into a p.r. case distinction we define a p.r. function $g$ by

$$g(\vec{a},m,b) = \begin{cases} b \text{ if } (\exists k\leqslant m)P(\vec{a},k), \\ Sm \text{ otherwise.} \end{cases}$$

Then $f(\vec{a},Sm) = g(\vec{a},m,f(\vec{a},m))$ is readily confirmed. Therefore, $f$ is indeed a p.r. function.

Let $h$ and $P$ be p.r. and $\mu k\leqslant h\vec{a}[P(\vec{a},k)] := \mu k\leqslant m[P(\vec{a},k)\,\&\,m{=}h\vec{a}]$. Then also $\vec{a} \mapsto \mu k\leqslant h\vec{a}[P(\vec{a},k)]$ is p.r. A first application is the *pairing function* $\wp$, a bijective mapping from $\mathbb{N}^2$ to $\mathbb{N}$, defined by $\wp(a,b) = \sum_{i\leqslant a+b} i + a$. It enumerates the pairs $(a,b)$ as in the figure (cf. Exercise 2). Using the formula (∗) : $\sum_{i\leqslant n} i = \frac{1}{2}n(n+1)$ we obtain $\wp(a,b) = \frac{1}{2}(a+b)(a+b+1) + a$. From this equation one easily obtains a def-

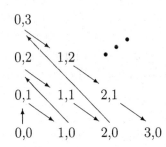

inition of $\wp$ by means of the bounded $\mu$-operation, for instance

$$\wp(a,b) = \mu k\leqslant(a+b)(a+b+1) + 2a\,[2k = (a+b)(a+b+1) + 2a].$$

The famous formula (∗) was probably first considered by Pythagoras, who counted the number $\mathbf{t}_n$ of points of a triangle with $n$ base points as in the right-hand figure in two ways. $\mathbf{t}_n$ equals $1+2+\cdots+n$ if one counts vertically. But two such triangles, put together appropriately, form a rectangle with $n\cdot(n+1)$ points. Thus, in fact $\mathbf{t}_n = \frac{1}{2}n(n+1)$.

$$1+2+3+4$$

Clearly, the bounded $\mu$-operation is not really needed in order to see that $\wp$ is p.r. A more convincing application of this operation is a rigorous proof that the prime number enumeration is p.r. If $p$ is prime then $p! + 1$ is certainly not divisible by a prime $q \leqslant p$. Indeed, if $q|p! + 1$ and $q|p!$, we obtain $q|p! + 1 - p! = 1$ and hence the contradiction $q|1$. Thus, a prime divisor of $p! + 1$ is necessarily a new prime. Therefore, the function $n \mapsto p_n$ is uniquely characterized by the equations

$$(\star) \quad p_0 = 2\,; \quad p_{n+1} = \mu q\leqslant p_n!+1[\text{prim } q\,\&\,q > p_n].$$

(⋆) is an application of **Op**, because with $f:(a,b) \mapsto \mu q\leqslant b[\text{prim } q\,\&\,q > a]$, $g : a \mapsto f(a,a! + 1)$ is p.r. as well, and the second equation in (⋆) can be written $p_{n+1} = g(p_n)$, as is easily verified. Hence, $n \mapsto p_n$ is indeed p.r.

**Remark 2.** Unlike the set of p.r. functions, the set of $\mu$-recursive functions can no longer be effectively enumerated, indeed, not even all unary ones: if $(f_n)_{n\in\mathbb{N}}$ were such an effective enumeration then $f: n \mapsto f_n(n) + 1$ would be computable and hence recursive by Church's thesis. Thus, $f = f_m$ for some $m$, so that $f_m(m) = f(m) = f_m(m) + 1$, a contradiction. While this seemingly speaks against the thesis, it can in fact be eliminated from the argument using some basic recursion theory. (C) clarifies the distinction between p.r. and recursive functions to some extent. The former can be computed with an effort that can in principle be estimated in advance, whereas the existence condition in the unbounded $\mu$-operation may be nonconstructive, so that even crude estimates of the effort required for computation are impossible. It is not very hard to construct a computable unary function that each p.r. unary function eventually overtakes. Nonetheless, also a p.r. function may grow extremely fast. For instance, it is physically impossible to compute the digits of $f6$ for the p.r. function $f: n \mapsto \underbrace{2^{2^{\cdot^{\cdot^{\cdot^2}}}}}_{n}$. While $f5$ has "only" $19\,729$ digits, the number of digits of $f6$ is already astronomical.

The following considerations are required in **6.2**. They concern the encoding of finite sequences of numbers of arbitrary length. There are basically several possibilities for doing this. One of these is to use the pairing function $\wp$ (or a similar one, cf. [Shoe]) repeatedly. Here we choose the particularly intuitive encoding from [Gö2], based on the prime enumeration $n \mapsto p_n$ and the unique prime factorization.

**Definition.** $\langle a_0, \ldots, a_n \rangle := p_0^{a_0+1} \cdots p_n^{a_n+1}$ $\left( = \prod_{i\leqslant n} p_i^{a_i+1} \right)$ is called the *Gödel number* of $(a_0, \ldots, a_n)$. The empty sequence has the Gödel number 1, also denoted by $\langle \rangle$. Let *GN* denote the set of all Gödel numbers.

Clearly, $\langle a_0, \ldots, a_n \rangle = \langle b_0, \ldots, b_m \rangle$ implies $m = n$ and $a_i = b_i$ for $i = 1, \ldots, n$. Also, $(a_0, \ldots, a_n) \mapsto \langle a_0, \ldots, a_n \rangle$ is certainly p.r. and by (A), (B) above, so is *GN*, since

$$a \in GN \Leftrightarrow a \neq 0 \ \& \ (\forall p\leqslant a)(\forall q\leqslant p)[\mathsf{prim}\, p, q \ \& \ p|a \ \Rightarrow \ q|a].$$

We now create a small provision of p.r. functions useful for the encoding of syntax in **6.2**. Using (C) we define a p.r. function $a \mapsto \ell a$ as follows:

$$\ell a = \mu k\leqslant a[p_k \nmid a].$$

We call $\ell a$ for a Gödel number $a$ the "length" of $a$, since clearly $\ell 1 = 0$, and for $a = \langle a_0, \ldots, a_n \rangle = \prod_{i\leqslant n} p_i^{a_i+1}$ is $\ell a = n + 1$, because $k = n + 1$ is the smallest index such that $p_k \nmid a$. Note that always $\ell a \leqslant a$ and for

$a \neq 0$ even $\ell a < a$, because $p_{a-1} \nmid a$ in view of $p_{a-1} > a$. Also the binary operation $(a, i) \mapsto (a)_i$ is p.r., where the term $(a)_i$ is defined by

$$(a)_i = \mu k \leqslant a[p_i^{k+2} \nmid a \, pd].$$

This is the "component-recognition function." $p_i^{k+1} | a$ and $p_i^{k+2} \nmid a$ imply $k = (a)_i$, hence $(\langle a_0, \ldots, a_n \rangle)_i = a_i$ for all $i \leqslant n$. This function, printed bold in order to catch the eye, always begins counting the components of a Gödel number with $i = 0$. Therefore, $(a)_{last} := (a)_{\ell a - 1}$ is the last component of a Gödel number $a \neq 1$, while $(1)_{last} = 0$. Which values $(a)_i$ and $\ell$ have for arguments outside $GN$ is irrelevant.

From the above definitions it follows that $a = \prod_{i < \ell a} p_i^{(a)_i + 1}$ for Gödel numbers $a$ including $a = 1$ (because the empty product equals 1). Next we define the *arithmetical concatenation* $* \in \mathbf{F}_2$ by

$$a * b = a \cdot \prod_{i < \ell b} p_{\ell a + i}^{(b)_i + 1} \text{ for } a, b \in GN \text{ and } a * b = 0 \text{ otherwise.}$$

Obviously, $\langle a_1, \ldots, a_n \rangle * \langle b_1, \ldots, b_m \rangle = \langle a_1, \ldots, a_n, b_1, \ldots, b_m \rangle$, so that $GN$ is closed under $*$. Moreover, $a, b \in GN \Rightarrow a, b \leqslant a * b$, as immediately follows from the definition. Note also that $a * b \in GN \Rightarrow a, b \in GN$, for all $a, b$. Clearly, $*$ is p.r., for its definition is based on p.r. case distinction.

The arithmetical function $*$ is useful for, among other things, a powerful generalization of $\mathbf{Op}$, the course-of-values recursion explained now. To every $f \in \mathbf{F}_{n+1}$ corresponds a function $\bar{f} \in \mathbf{F}_{n+1}$ given by

$$\bar{f}(\vec{a}, 0) = \langle \rangle \ (= 1); \quad \bar{f}(\vec{a}, b) = \langle f(\vec{a}, 0), \ldots, , f(\vec{a}, b - 1) \rangle \text{ for } b > 0.$$

$\bar{f}$ encodes the course of values of $f$ in the last argument. Let $F \in \mathbf{F}_{n+2}$. Then just as for $\mathbf{Op}$ there is one and only one $f \in \mathbf{F}_{n+1}$ that satisfies

$$\mathbf{Oq}: \quad f(\vec{a}, b) = F(\vec{a}, b, \bar{f}(\vec{a}, b)).$$

Namely, $f(\vec{a}, 0) = F(\vec{a}, 0, \langle \rangle) = F(\vec{a}, 0, 1)$, $f(\vec{a}, 1) = F(\vec{a}, 1, \langle f(\vec{a}, 0) \rangle)$, $f(\vec{a}, 2) = F(\vec{a}, 2, \langle f(\vec{a}, 0), f(\vec{a}, 1) \rangle)$, etc. In $\mathbf{Oq}$, $f(\vec{a}, b)$ in general depends for $b > 0$ on all values $f(\vec{a}, 0), \ldots, f(\vec{a}, b - 1)$, not just on $f(\vec{a}, b - 1)$ as in $\mathbf{Op}$. Hence $\mathbf{Oq}$ is called the schema of *course-of-values recursion*. A simple example is the Fibonacci sequence $(fn)_{n \in \mathbb{N}}$, defined by $f0 = 0$, $f1 = 1$, and $fn = f(n - 1) + f(n - 2)$ for $n \geqslant 2$. The $F$ in "normal form" $\mathbf{Oq}$ is given here by $F(b, c) = b$ for $b \leqslant 1$ and $F(b, c) = (c)_{b-1} + (c)_{b-2}$ otherwise. Indeed, $f0 = 0 = F(0, \bar{f}0)$, $f1 = 1 = F(1, \bar{f}1)$, and for $n \geqslant 2$, we have $fn = f(n - 1) + f(n - 2) = (\bar{f}n)_{n-1} + (\bar{f}n)_{n-2} = F(n, \bar{f}n)$.

**Op** is a special case of **Oq**. If $f = \mathbf{Op}(g, h)$ and $F$ is defined by the equations $F(\vec{a}, 0, c) = g(\vec{a})$ and $F(\vec{a}, \mathsf{S}b, c) = h(\vec{a}, b, (c)_b)$, then $f$ satisfies **Oq** with this $F$, as may straightforwardly be checked while observing that $f(\vec{a}, b)$ equals $(\bar{f}(\vec{a}, \mathsf{S}b))_b$.

**Theorem 1.1.** *Let $f$ satisfy **Oq**. If $F$ is p.r. then so too is $f$.*

**Proof.** Since $\langle c_0, \ldots, c_b \rangle = \langle c_0, \ldots, c_{b-1} \rangle * \langle c_b \rangle$ for $b > 0$, our $\bar{f}$ satisfies

$$\bar{f}(\vec{a}, 0) = 1; \quad \bar{f}(\vec{a}, \mathsf{S}b) = \bar{f}(\vec{a}, b) * \langle f(\vec{a}, b) \rangle = \bar{f}(\vec{a}, b) * \langle F(\vec{a}, b, \bar{f}(\vec{a}, b)) \rangle.$$

The second equation can be written $\bar{f}(\vec{a}, \mathsf{S}b) = h(\vec{a}, b, \bar{f}(\vec{a}, b))$, where $h$ is defined by $h(\vec{a}, b, c) = c * \langle F(\vec{a}, b, c) \rangle$. With $F$ also the function $h$ is p.r. Hence, by **Op**, $\bar{f}$ is p.r. But then so is $f$, because in view of **Oq**, $f$ is a composition of p.r. functions. ∎

We now make precise the intuitive notion of recursive (or effective) enumerability. $M \subseteq \mathbb{N}$ is called r.e. (*recursively enumerable*) if there is some recursive $R \subseteq \mathbb{N}^2$ such that $M = \{b \in \mathbb{N} \mid (\exists a \in \mathbb{N}) Rab\}$. In short, $M$ is the *range* of some recursive relation. Since $a \in M \Leftrightarrow (\exists b \in \mathbb{N}) R'ab$, where $R'ab \Leftrightarrow Rba$, $M$ is at the same time the *domain* of some recursive relation. More generally, $P \subseteq \mathbb{N}^n$ is called r.e. if $P\vec{a} \Leftrightarrow (\exists x \in \mathbb{N}) Q(x, \vec{a})$ for some $(n + 1)$-ary recursive predicate $Q$. Note that a recursive predicate $P$ is r.e. Indeed, $P\vec{a} \Leftrightarrow (\exists b \in \mathbb{N}) P'(b, \vec{a})$; here $P'(b, \vec{a}) :\Leftrightarrow P\vec{a}$ (adjoining a fictional variable). It is not quite easy to present an ad hoc example of an r.e. predicate that is not recursive. But such examples arise naturally in **6.5**, where we prove the undecidability of several axiomatic theories.

It is readily shown that $M \neq \emptyset$ is r.e. if and only if $M = \operatorname{ran} f$ for some recursive $f \in \mathbf{F}_1$; Exercise 5. This characterization corresponds perfectly to our intuition: stepwise computation of $f0, f1, \ldots$ provides an effective enumeration of $M$ in the intuitive sense. This enumeration can be carried out by a computer that puts out $f0, f1, \ldots$ successively and does not stop its execution by itself.

The empty set is r.e. because it is the domain of the empty binary relation, which is recursive, and even p.r., since its characteristic function is the constant function $\mathrm{K}_0^2$. In view of the above characterization of r.e. sets $M \neq \emptyset$, one could have defined these from the outset as the ranges of unary recursive functions. But the first definition has the advantage of immediately expanding to the $n$-dimensional case.

It is easily seen that a function $f \in \mathbf{F}_n$ is recursive provided graph $f$ is, simply because $f\vec{a} = \mu b[\text{graph } f(\vec{a}, b)]$ (or in strict terms of property $O\mu$, $f\vec{a} = \mu b[\delta \chi_{\text{graph } f}(\vec{a}, b) = 0]$), that is, $f$ can immediately be isolated from graph $f$ with the $\mu$-operator. Conversely, if $f$ is recursive then so is graph $f$, because $\chi_{\text{graph } f}(\vec{a}, b) = \chi_=(f\vec{a}, b)$. This equation also shows that graph $f$ is p.r. whenever $f$ is p.r. The converse need not be true. There are functions $f$ whose graph is p.r. although $f$ itself is not. A famous example is the (modified) *Ackermann function* $\circ \in \mathbf{F}_2$, defined by

$$0 \circ b = Sb \quad ; \quad Sa \circ 0 = a \circ 1 \quad ; \quad Sa \circ Sb = a \circ (Sa \circ b)$$

(see e.g. [Fel1, pp. 76–84]). Thus, not every recursion is primitive recursive.

### Exercises

1. Let $a \leqslant fa$ for all $a$. Prove that if $f$ is p.r. (resp. recursive) then so is $\operatorname{ran} f$. The same holds for $f \in \mathbf{F}_n$ if $a_1, \ldots, a_n \leqslant f\vec{a}$ for $\vec{a} \in \mathbb{N}^n$.

2. Prove in detail that the pairing function $\wp : \mathbb{N}^2 \to \mathbb{N}$ is bijective and that its diagram in the figure on page 222 is correct.

3. Since $\wp : \mathbb{N}^2 \to \mathbb{N}$ is bijective, there are functions $\varkappa_1, \varkappa_2 \in \mathbf{F}_1$ with $\wp(\varkappa_1 n, \varkappa_2 n) = n$, for all $n$. Prove that $\varkappa_1, \varkappa_2$ are p.r. (One need not exhibit explicit terms for $\varkappa_1, \varkappa_2$, although this is not difficult.)

4. Let $\operatorname{lcm}\{f\nu \mid \nu \leqslant n\}$ be the least common multiple of $f0, \ldots, fn$ with $f \in \mathbf{F}_1$. Show that $n \mapsto \operatorname{lcm}\{f\nu \mid \nu \leqslant n\}$ is p.r. provided $f$ is.

5. Let $M \subseteq \mathbb{N}$ be nonempty. Show that $M$ is r.e. iff $M = \operatorname{ran} f$ for some recursive $f \in \mathbf{F}_1$.

## 6.2　Arithmetization

Roughly put, arithmetization (or Gödelization) is the description of the syntax of a formal language $\mathcal{L}$ and of formal proofs from an axiom system by means of arithmetical operations and relations on natural numbers. It presupposes the encoding of strings from the alphabet of $\mathcal{L}$ by natural numbers. Syntactic functions and predicates correspond in this way to welldefined functions and predicates on $\mathbb{N}$.

Thus many goals at once become attainable. First of all, the intuitive idea of a computable word function can be made more precise using the notion of recursive functions. Second, syntactic predicates such as, for instance '$x \in var\,\alpha$', can be replaced by corresponding predicates of $\mathbb{N}$. Third, using encoding, statements about syntactic functions, predicates, and formal proofs can be formulated in theories $T \subseteq \mathcal{L}$ able to speak about arithmetic, and perhaps be proved in $T$.

We demonstrate the arithmetization of syntax using as an example the language $\mathcal{L} = \mathcal{L}_{ar}$, whose extralogical symbols are $0, \mathsf{S}, +, \cdot$. This is the language of Peano arithmetic PA. However, the same procedure can be carried out analogously for other formal languages, as will be apparent in the course of our considerations.

The first step is to assign uniquely to every basic symbol $\mathsf{s}$ of $\mathcal{L}$ a number $\sharp\mathsf{s}$, its *symbol code*. The following table provides an example for $\mathcal{L} = \mathcal{L}_{ar}$:

| s | $=$ | $\neg$ | $\wedge$ | $\forall$ | ( | ) | 0 | S | + | $\cdot$ | $v_0$ | $v_1$ | |
|---|---|---|---|---|---|---|---|---|---|---|---|---|---|
| $\sharp$s | 1 | 3 | 5 | 7 | 9 | 11 | 13 | 15 | 17 | 19 | 21 | 23 | $\cdots$ |

Next we encode the string $\xi = \mathsf{s}_0 \cdots \mathsf{s}_n$ by its *Gödel number*, which is the number $\langle \sharp\mathsf{s}_0, \ldots, \sharp\mathsf{s}_n \rangle = p_0^{1+\sharp\mathsf{s}_0} \cdot \ldots \cdot p_n^{1+\sharp\mathsf{s}_n}$. The empty string gets Gödel number 1. This is Gödel's original encoding, but there are several other possibilities to encode syntax by natural numbers.

**Example.** The term $0$ and the prime formula $0 = 0$ have the still comparatively small Gödel numbers $2^{1+\sharp 0} = 2^{14}$ and $2^{14} \cdot 3^2 \cdot 5^{14}$, respectively. The term $\underline{1}$ has Gödel number $2^{16} \cdot 3^{14}$. This encoding is not particularly economical, but that need not concern us here. Nor is it a problem that the symbol code of $=$ is just the Gödel number of the empty string. For note that $=$ , considered as an atomic string or a string of length 1, has Gödel number $2^{1+1} = 4$.

Let $\dot{\xi}$ be the Gödel number of $\xi \in \mathcal{S}_{\mathcal{L}}$, and $\dot{t}$ and $\dot{\alpha}$ therefore those of the term $t$ and the formula $\alpha$, respectively. If we write $\xi\eta$ for the concatenation of $\xi, \eta \in \mathcal{S}_{\mathcal{L}}$, then obviously $(\xi\eta)^{\cdot} = \dot{\xi} * \dot{\eta}$, where $*$ is the arithmetical concatenation from **6.1**. $\dot{\mathcal{S}}_{\mathcal{L}} = \{\dot{\xi} \mid \xi \in \mathcal{S}_{\mathcal{L}}\}$ is a p.r. subset of the set of all Gödel numbers. Indeed, since $\mathcal{L}$-symbols are encoded by odd numbers, $n \in \dot{\mathcal{S}}_{\mathcal{L}} \Leftrightarrow n \in GN$ & $(\forall k < \ell n)\, 2 \nmid (n)_k$.

When arithmetizing the syntax one has to carefully distinguish a symbol s from its corresponding atomic string, although these are normally denoted identically (see also the *Notation*). $\sharp$s is the symbol code of the symbol s, while $\dot{s} = 2^{1+\sharp s}$ is the Gödel number of the atomic string s. For example, the *symbol* 0 has the symbol code 13, while the *prime term* 0 has Gödel number $\dot{0} = 2^{1+\sharp 0} = 2^{14}$. We must also distinguish between the *symbol* $\boldsymbol{v}_i$ and the *prime term* $\boldsymbol{v}_i$. Thus, the set of *prime terms* $\boldsymbol{v}_1, \boldsymbol{v}_2, \ldots,$ denoted by $\mathcal{V}$, cannot simply be identified with Var.

**Remark 1.** Nonetheless, one could, right from the beginning, identify symbols with their codes and strings with their Gödel numbers, so that $\dot{\varphi} = \varphi$ and $\dot{t} = t$ for formulas $\varphi$ and terms $t$. Then syntactic predicates are arithmetical from the outset. This would even alleviate some of the following considerations in technical respect. However, we postpone this until we have convinced ourselves that syntax can indeed adequately be encoded in arithmetic. Further, the alphabet of $\mathcal{L}_{ar}$ could easily be replaced by a finite one, consisting, say, of the symbols $=, \neg, \ldots, \cdot, \boldsymbol{v}$, in that $\boldsymbol{v}_0$ is replaced by the string $\boldsymbol{v}0$, $\boldsymbol{v}_1$ by $\boldsymbol{v}S0$, etc. Other encodings found in the literature arise from the identification of the letters in such alphabets with the digits of a suitable number base.

In the following, let $\dot{W} = \{\dot{\xi} \mid \xi \in W\}$ for sets $W \subseteq \mathcal{S}_{\mathcal{L}}$ of words. A corresponding notation will be used for many-place word predicates $P$. We call $P$ p.r. or recursive whenever $\dot{P}$ is p.r. or recursive, respectively. So, for example, if we talk about a recursive axiom system $X \subseteq \mathcal{L}$, it is always understood that $\dot{X}$ is recursive. Other properties, such as recursively enumerable or representable, can be transferred to word predicates by means of the above or a similar arithmetization.

All these remarks refer not just to $\mathcal{L} = \mathcal{L}_{ar}$, but to an arbitrary *arithmetizable* (or *Gödelizable*) language $\mathcal{L}$, by which we simply mean that $\mathcal{L}$ possesses finitely or countably many specified basic symbols, so that each string can be assigned a number code in a computable way. In this way, the concepts of an axiomatizable or decidable theory, already used in **3.3**, obtain a precise meaning. Of course, one must clearly distinguish between the axioms and theorems of an axiomatic theory; the axiom systems of familiar theories like PA and ZFC are readily seen to be p.r., while these theories considered as sets of theorems are shown in **6.5** to be undecidable and cannot even be extended in any way to decidable theories.

The main goal now is the arithmetization of the formal proof method. We use $\vdash$ from now on to denote the Hilbert calculus of **3.6** consisting of the axiom system $\Lambda$ with the axiom schemata $\Lambda 1$–$\Lambda 10$ given there and the Modus Ponens MP as the only rule of inference. Here everything refers to a fixed arithmetizable language $\mathcal{L}$, which, as a rule, will be the arithmetical language $\mathcal{L}_{ar}$. Just as for strings, for a finite sequence $\Phi = (\varphi_0, \ldots, \varphi_n)$ of $\mathcal{L}$-formulas we call $\dot{\Phi} := \langle \dot{\varphi}_0, \ldots, \dot{\varphi}_n \rangle$ its *Gödel number*. This includes in particular the case that $\Phi$ is a proof from $X$ ($\subseteq \mathcal{L}$) in the sense of **3.6**, which in the general case also contains formulas from $\Lambda$. Note that $\dot{\Phi} \neq \dot{\xi}$ for all $\xi \in \mathcal{S}_{\mathcal{L}}$, because $2 | (\dot{\Phi})_0$ (a proof is not empty), whereas $2 \nmid (\dot{\xi})_0$ because the symbol codes are odd. This is the case in our example language $\mathcal{L}_{ar}$ and may actually be presupposed throughout. Thus, we can comfortably distinguish the Gödel numbers of formulas and terms from the Gödel numbers of finite sequences of formulas.

Now let $T$ ($\subseteq \mathcal{L}^0$) be a theory axiomatized by some fixed axiom system $X \subseteq T$. Examples are PA and ZFC.[2] A proof $\Phi = (\varphi_0, \ldots, \varphi_n)$ from $X$ is also called *a proof in $T$*. Here and elsewhere the axiom system $X$ is tacitly understood to be an essential part of $T$, which is originally understood as a set of sentences. First define the p.r. functions $\dot{\neg}$, $\dot{\wedge}$, $\dot{\rightarrow}$ as follows: $\dot{\neg} a := \dot{\neg} * a$, $a \dot{\wedge} b := \dot{(} * a * \dot{\wedge} * b * \dot{)}$ and $a \dot{\rightarrow} b := \dot{\neg}(a \dot{\wedge} \dot{\neg} b)$ (argument parentheses in the last expression should not be mixed up with parentheses belonging to the alphabet of $\mathcal{L}$). Clearly, both $\dot{\mathcal{S}}_{\mathcal{L}}$ and $\dot{\mathcal{L}}$ are closed under these operations.

Let *proof$_T$* denote the unary arithmetical predicate that corresponds to the syntactic predicate '$\Phi$ is a proof in $T$'. We denote the arithmetical predicates corresponding to '$\Phi$ is a proof in $T$ for $\varphi$' (the last component of $\Phi$) and to 'there is a proof for $\varphi$ in $T$' by *bew$_T$* and *bwb$_T$*, respectively (coming from beweis = proof and beweisbar = provable). Precise definitions of these predicates look as follows:

(1)  $proof_T(b) :\Leftrightarrow b \in GN \ \& \ b \neq 1$

$$\& \ (\forall k < \ell b)[(b)_k \in \dot{X} \cup \dot{\Lambda} \ \vee \ (\exists i, j < k)(b)_i = (b)_j \dot{\rightarrow} (b)_k],$$

(2)  $bew_T(b, a) :\Leftrightarrow proof_T(b) \ \& \ a = (b)_{last}$,

(3)  $bwb_T \, a :\Leftrightarrow \exists b \ bew_T(b, a)$.

Since $bwb_T$ is a unary predicate that will be met several times in the sequel, we dropped the argument parentheses in writing $bwb_T \, a$. Easily

---

[2] The language $\mathcal{L}_\in$ of ZFC is obviously simpler than $\mathcal{L}_{ar}$. It contains no composed terms and hence only the simplest possible equations, which of course simplifies encoding.

obtained from (1), (2), and (3) are

(4) $\vdash_T \alpha \Leftrightarrow bwb_T \dot\alpha \ (\Leftrightarrow \exists b \, bew_T(b, \dot\alpha))$,

(5) $bew_T(c, a) \& bew_T(d, a \mathbin{\dot\to} b) \Rightarrow bew_T(c * d * \langle b \rangle, b)$, for all $a, b, c, d$,

(6) $bwb_T a \ \& \ bwb_T(a \mathbin{\dot\to} b) \Rightarrow bwb_T b$, for all $a, b$,

(7) $bwb_T \dot\alpha \ \& \ bwb_T(\alpha \to \beta)^{\textbf{·}} \Rightarrow bwb_T \dot\beta$, for all $\alpha, \beta \in \mathcal{L}$.

The equivalence (4) is clear, for if $\vdash_T \alpha$ then there is a proof $\Phi$ for $\alpha$, hence $\exists n \, bew_T(n, \dot\alpha)$ with $n = \dot\Phi$, and conversely. (5) tells us in arithmetical terms the familiar story that concatenating proofs for $\alpha, \alpha \to \beta$ and tacking on $\beta$ yields a proof for $\beta$. (5) immediately yields (6) by particularization, and (6) implies (7) since $(\alpha \to \beta)^{\textbf{·}} = \dot\alpha \mathbin{\dot\to} \dot\beta$.

**Remark 2.** We will not need (5)–(7) until **7.1**. But it is very instructive for our later transfer of proofs to PA to verify (5) first naively. This is simple when we refer to the following facts: $\ell(a * b) = \ell a + \ell b$, $(\forall i < \ell a)(a * b)_i = (a)_i$ and $(\forall i < \ell b)(a * b)_{\ell a + i} = (b)_i$, for all $a, b \in GN$. Note also that $(\langle c \rangle)_0 = c$ for all $c$. Since it would impede the proof of (5), $(\forall k < b)(b)_k \in \dot{\mathcal{L}}$ was not added to the right-hand side of (1). This is in fact dispensable, for induction on the length $\ell b$ of the Gödel number $b$ readily shows that $proof_T(b)$ implies $(\forall k < \ell b)(b)_k \in \dot{\mathcal{L}}$. Here we need $a, a \mathbin{\dot\to} b \in \dot{\mathcal{L}} \Rightarrow b \in \dot{\mathcal{L}}$, for all $a, b \in \mathbb{N}$ (Exercise 3 in **2.3**).

Now we really get down to work and show that the syntactic basic notions up to the predicate $bew_T$ are p.r. In **6.5** only their recursiveness is important; not until Chapter **7** do we make essential use of their p.r. character, which ensures that all involved functions are definable in PA. We shall return to our example $\mathcal{L} = \mathcal{L}_{ar}$ only at the end of **6.3**, because the proofs of the following lemmas are not entirely independent of the language's syntax and the selected encoding, though they can be proved for other arithmetizable languages in nearly the same way.

In addition to the already-defined functions $\dot\neg$, $\dot\wedge$, and $\dot\to$, we define $a \mathbin{\dot=} b := a * \dot= * b \, (= a * 2^2 * b)$ and $\dot\forall(i, a) := \dot\forall * i * a$. $\dot\exists$ is defined similarly. Finally, for the operations $\mathsf{S}, +, \cdot$ define $\tilde{\mathsf{S}} a := \dot{\mathsf{S}} * a$, $a \mathbin{\dot+} b := (\dot* a * \dot+ * b *)$, and similarly for $\cdot$. Then, for example, $(s = t)^{\textbf{·}} = \dot s \mathbin{\dot=} \dot t$ and $(\mathsf{S}t)^{\textbf{·}} = \tilde{\mathsf{S}} \dot t$ for terms $s, t$, as well as $(\forall x \alpha)^{\textbf{·}} = \dot\forall \dot x \dot\alpha \ (= \dot\forall(\dot x, \dot\alpha))$. All these functions are obviously primitive recursive.

For arbitrary strings $\xi, \eta$ let $\xi \leqslant \eta$ mean $\dot\xi \leqslant \dot\eta$ (correspondingly for $<$). For example, $\xi \leqslant \eta$ holds if $\xi$ is a substring of $\eta$, in particular if $\xi$ is a subformula of the formula $\eta$. This follows immediately from the property $a, b \leqslant a * b$ for Gödel numbers $a, b$, mentioned already on page 224.

**Lemma 2.1.** *The set $\mathcal{T}$ of all terms is primitive recursive.*

**Proof.** The set $\mathcal{V}$ of terms $v_i$ is p.r. since $n \in \dot{\mathcal{V}} \Leftrightarrow (\exists k \leqslant n)\, n = 2^{22+2k}$. Thus $\mathcal{T}_{prim} := \mathcal{V} \cup \{0\}$, the *set of all prime terms*, is p.r. as well. By the recursive definition of $\mathcal{T}$, $t \in \mathcal{T}$ if and only if

$$t \in \mathcal{T}_{prim} \vee (\exists t_1, t_2 < t)[t_1, t_2 \in \mathcal{T} \ \& \ (t = \mathsf{St}_1 \vee t = (t_1 + t_2) \vee t = (t_1 \cdot t_2))].$$

Therefore the corresponding arithmetical equivalence holds as well:

$$(*) \quad n \in \dot{\mathcal{T}} \Leftrightarrow n \in \dot{\mathcal{T}}_{prim} \vee (\exists i, k < n)[i, k \in \dot{\mathcal{T}} \ \& \ Q(n, i, k)],$$

where $Q(n, i, k) \Leftrightarrow (n = \tilde{\mathsf{S}}i \vee n = i\tilde{+}k \vee n = i\tilde{\cdot}k)$. We now show how to convert this "informal definition" of $\dot{\mathcal{T}}$, which on the right-hand side makes use of elements of $\dot{\mathcal{T}}$ smaller than $n$ only, into a course-of-values recursion for the characteristic function $\chi_{\dot{\mathcal{T}}}$, whence $\chi_{\dot{\mathcal{T}}}$, and so $\mathcal{T}$ would turn out to be p.r. Consider the p.r. predicate $P$ defined by

$$P(a, n) \Leftrightarrow n \in \dot{\mathcal{T}}_{prim} \vee (\exists i, k < n)[(a)_i = (a)_k = 1 \ \& \ Q(n, i, k)].$$

We claim that $f := \chi_{\dot{\mathcal{T}}}$ satisfies $\boldsymbol{Oq} : fn = \chi_P(\bar{f}n, n)$, where $\bar{f}n$ equals $\langle f(0), \dots, f(n-1) \rangle$, and hence $f$ is p.r. by Theorem 1.1. Indeed, since $fi = fk = 1 \Leftrightarrow i, k \in \dot{\mathcal{T}}$, we obtain in view of $(*)$

$$
\begin{aligned}
n \in \dot{\mathcal{T}} \quad &\Leftrightarrow \quad n \in \dot{\mathcal{T}}_{prim} \vee (\exists i, k < n)[fi = fk = 1 \ \& \ Q(n, i, k)] \\
&\Leftrightarrow \quad P(\bar{f}n, n) \qquad \text{(because } (\bar{f}n)_i = fi \text{ and } (\bar{f}n)_k = fk\text{)}.
\end{aligned}
$$

From this it clearly follows that $fn = 1 \Leftrightarrow \chi_P(\bar{f}n, n) = 1$, which in turn implies $\boldsymbol{Oq}$, since both $f$ and $\chi_P$ take values from $\{0, 1\}$ only. $\blacksquare$

**Lemma 2.2.** *The set $\mathcal{L} (= \mathcal{L}_{ar})$ of all formulas is primitive recursive.*

**Proof.** $\mathcal{L}_{prim}$ is p.r. simply because

$$n \in \dot{\mathcal{L}}_{prim} \Leftrightarrow (\exists i, k < n)[i, k \in \dot{\mathcal{T}} \ \& \ n = i\,\tilde{=}\,k].$$

If we consider $\dot{x} < \dot{\xi}$ for every $\xi \in \mathcal{S}_{\mathcal{L}}$ with $x \in \text{var}\xi$ (because then $\xi = \eta x \theta$ for some strings $\eta, \theta \in \mathcal{S}_{\mathcal{L}}$), then the predicate '$\varphi \in \mathcal{L}$' clearly satisfies

$$
\begin{aligned}
\varphi \in \mathcal{L}_{prim} \vee (\exists \alpha, \beta, x < \varphi)[&\alpha, \beta \in \mathcal{L} \ \& \ x \in \mathcal{V} \\
&\& \ (\varphi = \neg\alpha \vee \varphi = (\alpha \wedge \beta) \vee \varphi = \forall x\alpha)].
\end{aligned}
$$

This "informal definition" can then be transformed just as in Lemma 2.1 into a course-of-values recursion of the characteristic function of $\dot{\mathcal{L}}$ using the characteristic function of the certainly p.r. predicate $P$ given by

$$P(a, n) \Leftrightarrow n \in \dot{\mathcal{L}}_{prim} \vee (\exists i, k, j < n)[(a)_i = (a)_k = 1 \,\&\, j \in \dot{V}$$
$$\&\, (n = \tilde{\neg}i \vee n = i \,\tilde{\wedge}\, k \vee n = \tilde{\forall}jk)]. \quad \blacksquare$$

We now define a ternary p.r. function $(m, i, k) \mapsto [m]_i^k$ such that

$(*)$ $\quad [\dot{\xi}]_x^t = (\xi \frac{t}{x})^{\cdot}$ for all $\xi \in \mathcal{L} \cup \mathcal{T}$, $x \in V$, and $t \in \mathcal{T}$.

$[m]_i^k$ will be constructed essentially by course-of-value recursion on $m$ in two steps, first for $m \in \dot{\mathcal{T}}$ and then for $m \in \dot{\mathcal{L}}$. **Step 1**: Put $[m]_i^k = 0$, if $m \notin \dot{\mathcal{T}}$ or $i \notin \dot{V}$ or $k \notin \dot{\mathcal{T}}$. Otherwise let first $m \in \dot{\mathcal{T}}_{prim}$. In case $m = i$ set $[m]_i^k = k$ (remember that $x \frac{t}{x} = t$), and if $m \neq i$ set $[m]_i^k = m$. Now let $m \in \dot{\mathcal{T}} \setminus \dot{\mathcal{T}}_{prim}$. According to the case distinction in Exercise 2, let first $(\exists m_0 < m)(m = \tilde{S}m_0 \,\&\, m_0 \in \dot{\mathcal{T}})$. Put $[m]_i^k = \tilde{S}[m_0]_i^k$ ($m_0$ is unique). If $(\exists m_1, m_2 < m)(m = m_1 \tilde{+} m_2 \,\&\, m_1, m_2 \in \dot{\mathcal{T}})$, set $[m]_i^k = [m_1]_i^k \tilde{+} [m_2]_i^k$, and similarly for $\tilde{\cdot}$. Thus, $[m]_i^k$ is now well defined and p.r., but $(*)$ holds currently only for $\xi \in \mathcal{T}$, since $[m]_i^k = 0$ for $m \in \dot{\mathcal{L}}$. **Step 2**: We modify the step 1 definition of $[m]_i^k$ for the case $m \in \dot{\mathcal{L}}$, $i \in \dot{V}$, and $k \in \dot{\mathcal{T}}$ in a p.r. way. For $m = (t_1 \tilde{=} t_2)^{\cdot}$ put $[m]_i^k = [t_1]_i^k \tilde{=} [t_2]_i^k$ ($t_1, t_2 \in \mathcal{T}$, prime formula case). Otherwise, $m = \tilde{\neg} m_0$ or $m = m_0 \tilde{\wedge} m_1$ or $m = \tilde{\forall}(m_2, m_0)$ for suitable $m_0, m_1 < a$ from $\dot{\mathcal{L}}$ and $m_2 \in \dot{V}$ according to Exercise 3. We then explain $[m]_i^k$ for all $m \in \dot{\mathcal{L}}$ by course-of-value recursion similarly to what we did in Step 1 and according to the definition of a substitution on formulas. This results in a p.r. function that satisfies $(*)$.

As was already noticed, the predicate '$x$ occurs in $\xi$', or '$x \in \text{var}\,\xi$' for short, is p.r., since $x \in \text{var}\,\xi \Leftrightarrow x \in \dot{V} \,\&\, (\exists \eta, \vartheta \leqslant \xi)(\xi = \eta x \vartheta)$. Replacing here $\eta x \vartheta$ by $\eta \forall x \vartheta$ makes it clear that '$x \in \text{bnd}\,\alpha$' is p.r. as well. The binary predicate '$x \in \text{free}\,\alpha$' is also p.r. because $x \in \text{free}\,\alpha$ if and only if $x \in V \,\&\, \alpha \frac{0}{x} \neq \alpha$ $(\Leftrightarrow x \in \dot{V} \,\&\, [\dot{\alpha}]_x^0 \neq \dot{\alpha})$. Consequently $\mathcal{L}^0$ is p.r. With these preparations we now prove

**Lemma 2.3.** *The set $\Lambda$ of logical axioms is primitive recursive.*

**Proof.** $\Lambda 1$ is p.r. because $\varphi \in \Lambda 1$ if and only if

$$(\exists \alpha, \beta, \gamma < \varphi)[\alpha, \beta, \gamma \in \mathcal{L} \,\&\, \varphi = (\alpha \to \beta \to \gamma) \to (\alpha \to \beta) \to (\alpha \to \gamma)].$$

To characterize the corresponding arithmetical predicate we use the p.r. function $\tilde{\to}$. One reasons similarly for $\Lambda 2$–$\Lambda 4$. For a p.r. characterization of $\Lambda 5$ use the fact that the ternary predicate '$\alpha, \frac{t}{x}$ collision-free' is p.r.

For '$\alpha, \frac{t}{x}$ collision-free' holds iff $(\forall y < \alpha)(y \in \text{bnd}\,\alpha \ \& \ y \in \text{var}\,t \Rightarrow y = x)$. Further, the predicate '$\varphi = \forall x \alpha \to \alpha \frac{t}{x}$', which depends on $\varphi, \alpha, x, t$, is p.r., as can be seen by applying $(m, i, k) \mapsto [m]_i^k$. Hence, $\Lambda 5$ is p.r. as well, because $\varphi \in \Lambda 5$ if and only if

$$(\exists\,\alpha, x, t < \varphi)(\alpha \in \mathcal{L} \ \& \ x \in \mathcal{V} \ \& \ t \in \mathcal{T}$$
$$\& \ \varphi = \forall x \alpha \to \alpha \tfrac{t}{x} \ \& \ \alpha, \tfrac{t}{x} \text{ collision-free}).$$

Similarly it is shown that $\Lambda 6$–$\Lambda 10$ are p.r. Thus, each of the schemata $\Lambda i$ is p.r. and therefore so is $\Lambda_0 := \Lambda 1 \cup \cdots \cup \Lambda 10$. But then the same holds for $\Lambda$ itself, because $k \mapsto \sharp v_k$ is surely p.r. and every $\alpha \in \Lambda$ can be written $\alpha = \forall \vec{x} \alpha_0$ with some (possibly empty) prefix $\forall \vec{x}$ and for some $\alpha_0 \in \Lambda_0$, and then it must hold that

$$n \in \dot{\Lambda} \ \Leftrightarrow \ n \in \dot{\mathcal{L}} \ \& \ (\exists\,m, k < n)(n = m * k \ \& \ 2 | \ell m \ \& \ k \in \dot{\Lambda}_0$$
$$\& \ (\forall i < \ell m)[2 | i \ \& \ (m)_i = \sharp \forall \ \lor \ 2 \nmid i \ \& \ (\exists k \leqslant n)(m)_i = \sharp v_k].$$

The second line of this formula tells us that $m$ is the Gödel number of a prefix $\forall x_1 \cdots \forall x_l$. This is a string of length $m = 2l$. ❑

All of the above holds completely analogously for every arithmetizable language. Hence, given a p.r. or recursive axiom system $X$, $X \cup \Lambda$ is p.r. (resp. recursive) as well. This applies in particular to the axiom systems of **PA** and **ZFC**. These are p.r. like every other common axiom system, despite the difference in their strengths. The proof is carried out in a manner fairly similar to that of Lemma 2.3.

The main result of this section, which now follows, is completely independent of the strength of an axiomatic theory $T$. The strength of a theory $T$ first comes into the picture when we want to prove something about $bew_T$ and $bwb_T$ *within $T$ itself*.

**Theorem 2.4.** *Let $X$ be a p.r. axiom system for a theory $T$ of an arithmetizable language. Then the predicate $bew_T$ is p.r. The same holds if we substitute here "recursive" for "primitive recursive." $T$ is in either case recursively enumerable.*

**Proof.** Definition (2) on page 229 shows that $bew_T$ is p.r. Because of (3) on the same page, $\dot{T} = \{a \in \dot{\mathcal{L}}^0 \mid bwb_T\,a\}$ is the range of a (primitive) recursive relation and thus is r.e. Clearly, the last part of the theorem is proved in the same manner. ❑

Theorem 2.4 can be strengthened only under particular circumstances, for example if $T$ is complete. Although $bew_T$ is a (primitive) recursive predicate for each axiomatic arithmetizable theory $T$, $bwb_T$ need not be recursive, as, for example, in the case $T = \mathsf{Q}$. This is a famous finitely axiomatizable theory presented in the next section, whose particular role for applied recursion theory was revealed in [TMR].

### Exercises

1. Prove that if a theory $T$ has a recursively enumerable axiom system $X$, then $T$ also possesses a recursive axiom system (W. Craig).

2. Let $a \in \dot{T} \backslash \dot{T}_{prim}$. Show that there are $b, c \in \dot{T}$ with $a = \tilde{\mathsf{S}}b$ or $a = b \dot{+} c$ or $a = b \dot{\cdot} c$. Moreover, $b, c$ are unique and $< a$ in each case.

3. Let $a \in \dot{\mathcal{L}} \backslash \dot{\mathcal{L}}_{prim}$. Show that there are $b, c \in \dot{\mathcal{L}}$, $d \in \dot{\mathcal{V}}$ with $a = \tilde{\neg} b$ or $a = b \tilde{\wedge} c$ or $a = \tilde{\forall}(d, b, c)$. $b, c, d$ are unique and $< a$ in each case.

4. Let $T$ be axiomatizable and $\alpha \in \mathcal{L}^0_{ar}$. (a) Define a binary p.r. $f$ such that $bew_{T+\alpha}(\dot{\Phi}, \dot{\varphi}) \Rightarrow bew_T(f(\dot{\Phi}, \dot{\alpha}), (\alpha \to \varphi)^{\cdot})$ (the arithmetized deduction theorem). (b) Show that $bwb_{T+\alpha} \dot{\varphi} \Leftrightarrow bwb_T(\alpha \to \varphi)^{\cdot}$.

## 6.3   Representability of Arithmetical Predicates

First of all we consider the finitely axiomatized theory $\mathsf{Q}$ with the axioms

Q1: $\forall x\ \mathsf{S}x \neq 0$,      Q2: $\forall xy(\mathsf{S}x = \mathsf{S}y \to x = y)$,    Q3: $(\forall x \neq 0)\exists y\ x = \mathsf{S}y$,

Q4: $\forall x\ x + 0 = x$,   Q5: $\forall xy\ x + \mathsf{S}y = \mathsf{S}(x + y)$,

Q6: $\forall x\ x \cdot 0 = 0$,    Q7: $\forall xy\ x \cdot \mathsf{S}y = x \cdot y + x$.

The axioms characterize $\mathsf{Q}$, also called *Robinson's arithmetic*, as a modest subtheory of $\mathsf{PA}$. Both theories are formalized in $\mathcal{L}_{ar}$ and are subtheories of $Th\mathcal{N}$, where $\mathcal{N}$ as always denotes the standard model $(\mathbb{N}, 0, \mathsf{S}, +, \cdot)$. In $\mathsf{Q}$, $\mathsf{PA}$ and related theories in $\mathcal{L}_{ar}$, $\leqslant$, and $<$ are defined by the formulas $x \leqslant y \leftrightarrow \exists z\ z + x = y$ and $x < y \leftrightarrow x \leqslant y \wedge x \neq y$ according to our convention on page 105. The term $\mathsf{S}^n 0$ is denoted by $\underline{n}$.

From the results of this and the next section, not only will the recursive undecidability of $\mathsf{Q}$ be derived, but also that of every subtheory and every consistent extension of $\mathsf{Q}$; see **6.5**. If we were interested only in undecidability results, we could simplify the proof of Theorem 4.2 by noting that all recursive functions can already be obtained with $\mathbf{Oc}$ and $\mathbf{O\mu}$ from the somewhat larger set of initial functions $0, \mathsf{S}, I_\nu^n, +, \cdot, \dot{-}$. But even ignoring the considerable effort required to prove the eliminability of the schema $\mathbf{Op}$ at the price of additional initial functions, such an approach would blur the distinction between primitive recursive and $\mu$-recursive functions, relevant for some details in Chapter **7**.

$\forall x\, x \neq \mathsf{S}x$ is easily provable in $\mathsf{PA}$ by induction, but $\mathsf{Q}$ is too weak to allow a proof of this sentence. Its unprovability follows from the fact that $(\mathbb{N} \cup \{\infty\}, 0, \mathsf{S}, +, \cdot)$ satisfies all axioms of $\mathsf{Q}$, but not $\forall x\, x \neq \mathsf{S}x$. Here $\infty$ is a new object and the operations $\mathsf{S}, +, \cdot$ are extended to $\mathbb{N} \cup \{\infty\}$ by putting $\mathsf{S}\infty = \infty$, $\infty \cdot 0 = 0$, and for all $n$ and all $m \neq 0$,

$$\infty + n = n + \infty = \infty + \infty = n \cdot \infty = \infty \cdot m = \infty.$$

This model shows the unprovability in $\mathsf{Q}$ of many familiar laws of arithmetic, which tell us that $\mathcal{N}$ is the nonnegative part of a discretely ordered commutative ring with unit element $1 := \mathsf{S}0$. These laws are collected in the following axiom system of a finitely axiomatizable theory $\mathsf{N} \subseteq \mathcal{L}_{ar}$, with the order defined as in $\mathsf{Q}$ above:

| | |
|---|---|
| N0: $\quad x + 0 = x$ | N1: $\quad x + y = y + x$ |
| N2: $\quad (x + y) + z = x + (y + z)$ | N3: $\quad x \cdot 1 = x$ |
| N4: $\quad x \cdot y = y \cdot x$ | N5: $\quad (x \cdot y) \cdot z = x \cdot (y \cdot z)$ |
| N6: $\quad x \cdot (y + z) = x \cdot y + x \cdot z$ | N7: $\quad \mathsf{S}x = x + 1$ |
| N8: $\quad x + z = y + z \rightarrow x = y$ | N9: $\quad x \leqslant y \lor y \leqslant x$ |
| N10: $x \leqslant 0 \rightarrow x = 0$ | N11: $\quad x < y \leftrightarrow \mathsf{S}x \leqslant y$ |

$\forall$-quantifiers in the axioms are omitted. $\mathsf{N}$ is also denoted by $\mathsf{PA}^-$ in the literature, and like $\mathsf{Q}$, a subtheory of $\mathsf{PA}$. All $\mathsf{Q}$-axioms are derivable in $\mathsf{N}$ (a recommendable exercise), so that $\mathsf{Q} \subseteq \mathsf{N} \subseteq \mathsf{PA}$. Reflexivity, transitivity, and antisymmetry of $\leqslant$ are provable in $\mathsf{N}$, as are the strong and weak monotonicity laws for $+$ and $\cdot$.

In this section we mostly write $\vdash \alpha$ for $\vdash_\mathsf{Q} \alpha$ and $\alpha \vdash \beta$ for $\alpha \vdash_\mathsf{Q} \beta$ etc. We also write occasionally $\alpha \vdash \beta \vdash \gamma$ for $\alpha \vdash \beta \,\&\, \beta \vdash \gamma$, and apply further self-explanatory abbreviations such as $\vdash t_1 = t_2 = t_3$ for $\vdash t_1 = t_2 \land t_2 = t_3$, and $\vdash \alpha \equiv \beta$ for '$\vdash_\mathsf{Q} \alpha$ and $\alpha$ is equivalent to $\beta$'. The use of $\vdash$ in the subtle derivations carried out below helps one see what is going on and makes the metainduction used there more vivid. Some of the proofs can be seen as "transplanting inductions from PA into the metatheory." For instance, $\forall x\, x \neq Sx$ is provable in PA, but not in Q, as was just shown. Nonetheless, we still can prove $\vdash \underline{n} \neq S\underline{n}$ for all $n$, as is seen by metainduction on $n$. Indeed, $\vdash \underline{0} \neq S\underline{0}$ is clear by Q1. The induction step $\vdash \underline{n} \neq S\underline{n} \Rightarrow \vdash S\underline{n} \neq SS\underline{n}$ derives from $\underline{n} \neq S\underline{n} \vdash S\underline{n} \neq SS\underline{n}$. This in turn easily follows with MP from $S\underline{n} = SS\underline{n} \vdash \underline{n} = S\underline{n}$, an application of Q2. We now prove

C0: $\vdash S x + \underline{n} = x + S\underline{n}$,

C1: $\vdash \underline{m} + \underline{n} = \underline{m+n},\ \underline{m} \cdot \underline{n} = \underline{m \cdot n}$,    C2: $\vdash \underline{n} \neq \underline{m}$   for $n \neq m$,

C3: $\vdash \underline{m} \leqslant \underline{n}$   for $m \leqslant n$,        C4: $\vdash \underline{m} \not\leqslant \underline{n}$   for $m \not\leqslant n$,

C5: $x \leqslant \underline{n} \vdash x = \underline{0} \vee \cdots \vee x = \underline{n}$,    C6: $\vdash x \leqslant \underline{n} \vee \underline{n} \leqslant x$.

From C5 follows $x < \underline{n} \vdash x = \underline{0} \vee \cdots \vee x = \underline{n-1}\ (= \bigvee_{i<n} x = \underline{i})$, which is $\perp$ for $n = 0$. The proofs of C0–C6 will be carried out by induction (more precisely, metainduction) on $n$.

C0: Clear for $n = 0$, because $\vdash S x + \underline{0} = S x = S(x + \underline{0}) = x + S\underline{0}$ by Q4 and Q5. Our induction hypothesis is $\vdash S x + \underline{n} = x + S\underline{n}$. It yields, again by Q5, the induction claim $\vdash S x + S\underline{n} = S(S x + \underline{n}) = S(x + S\underline{n}) = x + SS\underline{n}$.

C1: Clear for $n = 0$, because $\vdash \underline{m} + \underline{0} = \underline{m} = \underline{m+0}$ by Q4. The induction hypothesis $\vdash \underline{m} + \underline{n} = \underline{m+n}$ yields $\vdash \underline{m} + S\underline{n} = S(\underline{m} + \underline{n}) = S\underline{m+n}$, by Q5, and the last term is the same as $\underline{m + Sn}$. This proves the induction step. Analogously we derive $\vdash \underline{m} \cdot \underline{n} = \underline{m \cdot n}$ with Q6, Q7, and what was shown already.

C2: Clear for $n = 0$, since then $m = Sk$ for some $k$, and so $\vdash \underline{0} \neq \underline{m}$ by Q1. Let $Sn \neq m$. By Q1, $\vdash S\underline{n} \neq \underline{m}$ in case $m = 0$. Otherwise, $m = Sk$ for some $k$, so that $n \neq k$; hence $\vdash \underline{n} \neq \underline{k}$ by the induction hypothesis. Thus, $\vdash S\underline{n} \neq S\underline{k} = \underline{m}$ by Q2.

C3: $m \leqslant n$ implies $k + m = n$ for some $k$, hence $\underline{k + m} = \underline{n}$. Thus, $\vdash \underline{k} + \underline{m} = \underline{n}$ by C1. Therefore $\vdash \exists z\, z + \underline{m} = \underline{n}$, i.e., $\vdash \underline{m} \leqslant \underline{n}$.

C4: $m \not\leqslant n \Rightarrow m \neq 0$, hence $m = Sk$ for some $k$. Let $m \not\leqslant 0$. Then $\vdash \underline{m} \not\leqslant \underline{0}$ because $\underline{m} \leqslant \underline{0} \vdash S\underline{k} \leqslant \underline{0} \vdash \exists v\, S(v + \underline{k}) = \underline{0} \vdash \perp$ by Q5, Q1. Let $m \not\leqslant Sn$. Then $k \not\leqslant n$, and so $\vdash \underline{k} \not\leqslant \underline{n}$ by the induction hypothesis. Hence $\vdash \underline{m} \not\leqslant S\underline{n}$, for $x \leqslant y \equiv_Q S x \leqslant S y$ (the latter needs only Q5 and Q2).

C5: Clear for $n = 0$, because $x \neq \underline{0}, x \leqslant \underline{0} \vdash \exists v S v = \underline{0} \vdash \perp$ by Q3, Q5, Q1. The induction claim is equivalent to $x \neq \underline{0}, x \leqslant S\underline{n} \vdash \bigvee_{i=1}^{n+1} x = \underline{i}$. It is derived as follows:

$$x \neq 0, x \leqslant \underline{Sn} \quad \vdash \exists y (x = Sy \wedge y \leqslant \underline{n}) \qquad \text{(Q3, Q5, and Q2)}$$
$$\vdash \exists y (x = Sy \wedge \bigvee_{i \leqslant n} y = \underline{i}) \qquad \text{(induction hypothesis)}$$
$$\vdash \exists y (x = Sy \wedge \bigvee_{i=1}^{n+1} Sy = \underline{i}) \vdash \bigvee_{i=1}^{n+1} x = \underline{i}.$$

C6: Clear for $n = 0$. Further, $\underline{n} < x \vdash \exists y Sy + \underline{n} = x \vdash \exists y y + S\underline{n} = x$, by Q3, C0, and $\vdash 0 + \underline{n} = \underline{n}$ by C1. Thus, $\underline{n} < x \vdash \underline{Sn} \leqslant x$. Now, C5 and C3 easily lead to $x \leqslant \underline{n} \vdash x \leqslant \underline{Sn}$. This and the former yield the inductive step, because $x \leqslant \underline{n} \vee \underline{n} \leqslant x \vdash x \leqslant \underline{n} \vee \underline{n} < x \vdash x \leqslant \underline{Sn} \vee \underline{Sn} \leqslant x$.

With these preparations we now give the following crucial definition, in which $T \supseteq \mathsf{Q}$ is supposed. This will cover all our applications.

**Definition.** Call a predicate $P \subseteq \mathbb{N}^n$ *numeralwise representable* or simply *representable* in $T \, (\supseteq \mathsf{Q})$[3] if there is some $\alpha = \alpha(\vec{x})$, called a *representing formula,* such that

$$\text{R}^+ \colon \ P\vec{a} \ \Rightarrow \ \vdash_T \alpha(\underline{\vec{a}}) \ ; \quad \text{R}^- \colon \ \neg P\vec{a} \ \Rightarrow \ \vdash_T \neg \alpha(\underline{\vec{a}}).$$

**Examples.** The identity relation $\{(a,a) \mid a \in \mathbb{N}\}$ is represented by $x = y$, because $\vdash_\mathsf{Q} \underline{a} = \underline{b}$ is trivial if $a = b$, and $\vdash_\mathsf{Q} \underline{a} \neq \underline{b}$ is derivable for $a \neq b$ by C2. By C3 and C4 the formula $x \leqslant y$ represents the $\leqslant$-predicate. $x \neq x$ represents the empty set, represented as well by any $\alpha$ with $\neg \alpha \in \mathsf{Q}$.

For consistent $T \supseteq \mathsf{Q}$, whenever $\text{R}^+$, $\text{R}^-$ are valid then so too are their converses, so that in fact $P\vec{a} \Leftrightarrow \vdash_T \alpha(\underline{\vec{a}})$ and $\neg P\vec{a} \Leftrightarrow \vdash_T \neg \alpha(\underline{\vec{a}})$. Note also that a representable $P \subseteq \mathbb{N}^n$ is recursive by Church's thesis. For let $P$ be represented by $\alpha(\vec{x})$. Simply turn on the enumeration machine for $\mathsf{Q}$ and wait until $\alpha(\underline{\vec{a}})$ or $\neg \alpha(\underline{\vec{a}})$ appears. The set of $n$-ary representable predicates is closed under union, intersection, and complement, as well as swapping, equating, and adjoining fictional arguments. If $P, Q$ are represented respectively by $\alpha(\vec{x}), \beta(\vec{x})$, then so too are $P \cap Q$ by $\alpha(\vec{x}) \wedge \beta(\vec{x})$ and $\neg P$ by $\neg \alpha(\vec{x})$. Consequently, $P \cup Q$ by $\alpha(\vec{x}) \vee \beta(\vec{x})$, etc.

A predicate $P$ represented in $\mathsf{Q}$ by $\alpha$ is clearly representable by the same $\alpha$ in any consistent extension of $\mathsf{Q}$, in particular in $Th\mathcal{N}$. But this just means definability of $P$ in $\mathcal{N}$ by $\alpha$ in the sense of **2.3**, because $\mathcal{N} \vDash \alpha \, [\vec{a}]$ is equivalent to $\mathcal{N} \vDash \alpha(\underline{\vec{a}})$. In short, definability of $P$ in $\mathcal{N}$ and representability of $P$ in $Th\mathcal{N}$ coincide. In the main, however, we consider

---

[3] 'in $T$' will often be omitted; we then always mean 'in $T = \mathsf{Q}$'. Representable predicates are called *entscheidungsdefinit* in [Gö2] (translated as *decidable* in [Hei]), in [HB] *vertretbar*, in [Kl1] *numeralwise expressible*, and in [TMR] *definable*.

representability in $Q$ to obtain some strong results needed in **6.5**. We always have to look carefully at the representing formulas.

One could define $f \in \mathbf{F}_n$ to be representable if graph $f$ is representable. However, it turns out that this definition is equivalent to a stronger notion of representability for functions that will be introduced after some additional preparation.

Predicates and functions definable in $\mathcal{N}$, that is, by $0, \mathsf{S}. +, \cdot$, are called *arithmetical* after [Gö2]. From now on this word will always have this meaning. The arithmetical predicates encompass the representable ones. In order to discover more about these objects we consider their defining formulas more closely. Prime formulas in $\mathcal{L}_{ar}$ are equations, also called *Diophantine equations* for traditional reasons. If $\delta(\vec{x}, \vec{y})$ is such an equation and $P\vec{a} \Leftrightarrow \mathcal{N} \vDash \exists \vec{y} \delta(\vec{a}, \vec{y})$, then $P$ is called *Diophantine*. A simple example is $\leqslant$, because $a \leqslant b \Leftrightarrow \exists y \, y + a = b$.[4] In fact, all predicates definable in $\mathcal{N}$ by $\exists$-formulas $\exists \vec{y} \varphi$ from $\mathcal{L}_{ar}$ with kernel $\varphi$ are Diophantine. The proof is not difficult: Think of $\varphi$ as being constructed from literals by means of $\wedge, \vee$ (cf. Exercise 4 in **2.4**), and use the following easily provable equivalences in an inductive proof on $\varphi$ of what has been claimed:

$$s \neq t \quad \equiv_{\mathcal{N}} \quad \exists z (\mathsf{S}z + s = t \vee \mathsf{S}z + t = s),$$
$$s_1 = t_1 \vee s_2 = t_2 \quad \equiv_{\mathcal{N}} \quad s_1 s_2 + t_1 t_2 = s_1 t_2 + s_2 t_1,$$
$$s_1 = t_1 \wedge s_2 = t_2 \quad \equiv_{\mathcal{N}} \quad s_1^2 + t_1^2 + s_2^2 + t_2^2 = \underline{2}(s_1 t_1 + s_2 t_2).$$

A classification of arithmetical formulas and predicates helpful not only for the sake of representability is given by the following definition, to be generalized in **6.7**:

**Definition.** A formula is called $\Delta_0$ or a $\Delta_0$-*formula* if it is generated from prime formulas of $\mathcal{L}_{ar}$ by $\wedge, \neg$, and *bounded quantification*, i.e., if $\alpha$ is $\Delta_0$ then so is $(\forall x \leqslant t)\alpha$; here $t$ is any $\mathcal{L}_{ar}$-term with $x \notin \mathrm{var}\, t$. (It is not important that $x \leqslant t$ is not a prime formula of $\mathcal{L}_{ar}$). Let $\varphi$ be $\Delta_0$ and $\vec{x}$ arbitrary. Then $\exists \vec{x} \varphi$ is called a $\Sigma_1$-*formula*, and $\forall \vec{x} \varphi$ a $\Pi_1$-*formula*. Further, $P \subseteq \mathbb{N}^n$ is said to be $\Delta_0$, $\Sigma_1$, or $\Pi_1$ whenever $P$ is defined in $\mathcal{N}$ by a $\Delta_0$-, $\Sigma_1$-, or $\Pi_1$-formula, respectively. $\Delta_0$, $\Sigma_1$, and $\Pi_1$ denote the sets of $\Delta_0$-, $\Sigma_1$-, and $\Pi_1$-predicates. In addition, $\Delta_1 := \Sigma_1 \cap \Pi_1$. There are no $\Delta_1$-formulas, for there is no meaningful definition of such

---

[4] The right side of this equivalence is an informal and more easily readable substitute for the somewhat lengthy notation $\mathcal{N} \vDash \exists y \, y + \underline{a} = \underline{b}$.

formulas. $\varphi$ is called $\Delta_0$, $\Sigma_1$, or $\Pi_1$ also if it is equivalent to an original $\Delta_0$-, $\Sigma_1$-, or $\Pi_1$-formula, respectively. In this sense, if $\alpha$ is $\Delta_0$ then so too are $(\exists x{\leqslant}t)\,\alpha\ \big(\equiv \neg(\forall x{\leqslant}t)\neg\alpha\big)$ and $(\forall x{<}t)\alpha\ \big(\equiv (\forall x{\leqslant}t)(x{=}t \vee \alpha)\big)$.

Clearly, $\Pi_1$ consists of the complements of the $P \in \Sigma_1$. The $P \in \Delta_1$ are both $\Sigma_1$- and $\Pi_1$-definable, with possibly distinct formulas. By Exercise 3 in **2.4**, $\Sigma_1$ and $\Pi_1$ are closed under union and intersection of predicates of the same arity, and $\Delta_1$ like $\Delta_0$ moreover under complements. If $P \in \mathbb{N}^m$ and $g_1, \ldots, g_m \in \mathbf{F}_n$ are $\Sigma_1$, so too is $Q = P[g_1, \ldots, g_m]$, simply because $Q\vec{a} \Leftrightarrow \exists \vec{y}(\bigwedge_{i=1}^{n} y_i{=}g_i\vec{a}\ \&\ P\vec{y})$. Note also that if graph $f$ is $\Sigma_1$ then it is automatically $\Delta_1$, for $f\vec{a}{\neq}b \Leftrightarrow \exists y(f\vec{a}{=}y\ \&\ y{\neq}b)$, so that the complement of graph $f$ is again $\Sigma_1$. Here some examples of $\Delta_0$- and $\Sigma_1$-formulas and sentences. Interesting $\Pi_1$-sentences are found at the end of **6.5**.

**Examples.** Diophantine equations are the simplest $\Delta_0$-formulas. To these belong the formulas $y{=}t(\vec{x})$ with $y \notin \mathrm{var}\,t$, which define the term functions $\vec{a} \mapsto t^{\mathcal{N}}(\vec{a})$. Since $a|b \Leftrightarrow (\exists c{\leqslant}b)(a \cdot c = b)$, divisibility and thus also the predicate $\mathsf{prim}$ are $\Delta_0$. Because $\wp(a, b) = c \Leftrightarrow 2c = (a + b)^2 + 3a + b$, $\mathrm{graph}_\wp$ is $\Delta_0$. The same holds for the relation of being *coprime*, denoted by $\perp$ and defined by $a{\perp}b :\Leftrightarrow (\forall c \leqslant a + b)(c|a, b \Rightarrow c = 1)$. Diophantine predicates are trivially $\Sigma_1$. Surprisingly, by Theorem 5.6 the converse holds as well, although it had originally been conjectured that, for instance, the set $\{a \in \mathbb{N} \mid (\forall p{\leqslant}a)(\mathsf{prim}\,p\ \&\ p|a \Rightarrow p = 2)\}$ of all powers of 2 was not Diophantine. This set is $\Delta_0$. Even the graph of $n \mapsto 2^n$ is $\Delta_0$.

**Remark 1.** More generally, the predicate '$a^b = c$' is $\Delta_0$, though it is difficult to prove this fact. Indeed, even the proof in **6.4** that this predicate is arithmetical requires effort. Earlier results from Bennet, Paris, Pudlak, among others, are generalized in [BA] as follows: if $f \in \mathbf{F}_{n+1}$ (more precisely, graph $f$) is $\Delta_0$ then so is $g: (\vec{a}, n) \mapsto \prod_{i{\leqslant}n} f(\vec{a}, i)$, and the recursion equation $g(\vec{x}, \mathsf{S}y) = g(\vec{x}, y) \cdot f(\vec{x}, y)$ is provable in $I\Delta_0$. This theory is an important weakening of PA. It results from Q by adjoining the induction schema restricted to $\Delta_0$-formulas. $I\Delta_0$ plays a role in various questions, e.g., in complexity theory ([Kra] or [HP]). Induction on the $\Delta_0$-formulas readily shows that all $\Delta_0$-predicates are p.r. The converse does not hold; an example is the graph of the very rapidly growing *hyperexponentiation*, recursively defined by $\mathrm{hex}(a, 0) = 1$ and $\mathrm{hex}(a, \mathsf{S}b) = a^{\mathrm{hex}(a,b)}$. Stated more suggestively, $\mathrm{hex}(a, n) = \underbrace{a^{a^{\cdot^{\cdot^{\cdot^{a}}}}}}_{n}$.

According to Theorem 3.1 below, already the weak theory Q is $\Sigma_1$-complete, i.e., each $\Sigma_1$-sentence true in $\mathcal{N}$ is provable in Q. This can be confirmed in various ways. For instance, one may use that by C1 and C2, $\mathcal{N}$ is a prime model of $Q$ in the sense of **5.1**; in addition, each $\mathcal{A} \vDash Q$ is an end extension of $\mathcal{N}$ as defined in **3.3**. But we choose here a constructive approach, which provides some additional information.

**Theorem 3.1 (on the $\Sigma_1$-completeness of Q).** *Every $\Sigma_1$-sentence true in $\mathcal{N}$ is already provable in $Q$ and hence in each extension $T \supseteq Q$.*

**Proof.** We claim that it suffices to prove

$(*)$     Either $\vdash_Q \alpha$ or $\vdash_Q \neg\alpha$, for each $\Delta_0$-sentence $\alpha$.

Indeed, let $\mathcal{N} \vDash \exists \vec{x}\varphi(\vec{x})$ with the $\Delta_0$-formula $\varphi(\vec{x})$, say $\mathcal{N} \vDash \alpha := \varphi(\vec{a})$. Then $\vdash_Q \alpha$ by $(*)$, for $\vdash_Q \neg\alpha$ is impossible. Hence $\vdash_Q \exists \vec{x}\varphi(\vec{x})$. We verify $(*)$ first for prime sentences. If $t$ is a variable-free term then C1 readily yields $\vdash_Q t = \underline{t}^{\mathcal{N}}$. For example, $\vdash_Q (\underline{3}+\underline{4})\cdot\underline{5} = \underline{35}$. Thus, if $\alpha$ is the prime sentence $t_1 = t_2$, then $\vdash_Q t_1^{\mathcal{N}} = t_2^{\mathcal{N}}$ or $\vdash_Q t_1^{\mathcal{N}} \neq t_2^{\mathcal{N}}$ by C2, which confirms $(*)$ for $\alpha$. The induction steps over $\wedge, \neg$ are simple. For instance, $\vdash_Q \alpha \wedge \beta$ if $\vdash_Q \alpha, \beta$, and $\vdash_Q \neg\alpha \vee \neg\beta \equiv \neg(\alpha \wedge \beta)$ otherwise, i.e. if $\vdash_Q \neg\alpha$ or $\vdash_Q \neg\beta$. These steps suffice already to prove $(*)$, because bounded quantifiers are eliminable from a $\Delta_0$-*sentence* $\alpha$ modulo $Q$. Indeed, let $(\forall x \leqslant t)$ be the first bounded quantifier in $\alpha$ from the left, with the scope $\beta$. Then $\mathrm{var}\, t = \emptyset$, because $x \notin \mathrm{var}\, t$ and any $y \in \mathrm{var}\, t$ must have been bounded further to the left. Moreover, $(\forall x \leqslant t)\beta(x) \equiv_Q (\forall x \leqslant \underline{n})\beta(x)$ with $n := t^{\mathcal{N}}$, since $\vdash_Q t = \underline{n}$. We then easily get $(\forall x \leqslant \underline{n})\beta(x) \equiv_Q \beta(\underline{0}) \wedge \cdots \wedge \beta(\underline{n})$ with C3 and C5. Thus, $(\forall x \leqslant t)$ can be eliminated from $\alpha$ and this process can be repeated if necessary. $\square$

If $\varphi(\vec{x})$ is $\Delta_0$ then $\mathcal{N} \vDash \varphi(\vec{a}) \Rightarrow \vdash_Q \varphi(\vec{a})$ and $\mathcal{N} \vDash \neg\varphi(\vec{a}) \Rightarrow \vdash_Q \neg\varphi(\vec{a})$ by the theorem, because both $\varphi(\vec{a})$ and $\neg\varphi(\vec{a})$ are trivially $\Sigma_1$. Thus, we obtain a first important result on representing formulas:

**Corollary 3.2.** *A $\Delta_0$-formula represents in $Q$ the predicate that it defines in $\mathcal{N}$.*

**Lemma 3.3.** *Let $\alpha(\vec{x}, y)$ represent $P \subseteq \mathbb{N}^{n+1}$. Then $(\exists z < y)\alpha(\vec{x}, z)$ and $(\forall z < y)\alpha(\vec{x}, z)$ represent the predicates $Q$ and $R$, respectively, where*

$$Q(\vec{a}, b) :\Leftrightarrow (\exists c < b)P(\vec{a}, c) \text{ and } R(\vec{a}, b) :\Leftrightarrow (\forall c < b)P(\vec{a}, c).$$

*The same is true if $<$ is replaced by $\leqslant$ in this lemma.*

**Proof.** $R^+$: Suppose $Q(\vec{a}, b)$, that is, $P(\vec{a}, c)$ for some $c < b$. Then $\vdash \underline{c} < \underline{b} \wedge \alpha(\vec{\underline{a}}, \underline{c})$. Consequently, $\vdash (\exists z < \underline{b}) \alpha(\vec{\underline{a}}, z)$. To prove $R^-$ suppose $\neg Q(\vec{a}, b)$, hence $\neg P(\vec{a}, i)$, i.e., $\vdash \neg \varphi(\vec{\underline{a}}, \underline{i})$ for all $i < b$. We thus obtain $\bigvee_{i < b} z = \underline{i} \vdash \neg \alpha(\vec{\underline{a}}, z)$. By C5, $z < \underline{b} \vdash \bigvee_{i < b} z = \underline{i}$ and so $z < \underline{b} \vdash \neg \alpha(\underline{a}, z)$. Therefore, $\vdash (\forall z < \underline{b}) \neg \alpha(\vec{\underline{a}}, z) \equiv \neg (\exists z < \underline{b}) \alpha(\vec{\underline{a}}, z)$. This proves $R^-$. For handling the predicate $R$ simply notice that $R(\vec{a}, b) \Leftrightarrow \neg (\exists c < b) \neg P(\vec{a}, c)$. This proof is literally the same if $<$ is replaced by $\leqslant$ in the lemma. $\quad\blacksquare$

Following [Gö2] and [TMR], we now define the notion of a representable function. Although representability of $f$ is much stronger a notion than representability of graph $f$, Lemma 3.4(b) will show that both properties coincide, provided the axioms of $Q$ are available.

**Definition.** $f \in \mathbf{F}_n$ is *representable* in $T$ if there is a formula $\varphi(\vec{x}, y) \in \mathcal{L}_{ar}$ such that for all $\vec{a} \in \mathbb{N}^n$,

$$R^+ : \quad \vdash_T \varphi(\vec{\underline{a}}, \underline{f\vec{a}}), \qquad R^= : \quad \varphi(\vec{\underline{a}}, y) \vdash_T y = \underline{f\vec{a}}.$$

If $\varphi$ is $\Delta_0$ (respectively $\Sigma_1$ or $\Pi_1$) then $f$ is said to be $\Delta_0$-*representable* (respectively $\Sigma_1$- or $\Pi_1$-*representable*).

For some purposes it is useful to refine this definition: $f \subseteq \mathbf{F}_n$ is said to be $\Delta_1$-representable if $f$ is both $\Sigma_1$- *and* $\Pi_1$-representable, with usually distinct formulas. Corresponding phrases will be used for predicates $P$ instead of functions $f$.

Since $R^+$ is equivalent to $y = \underline{f\vec{a}} \vdash_T \varphi(\vec{\underline{a}}, y)$, it is obvious that $R^+$ and $R^=$ together are replaceable by the single condition $y = \underline{f\vec{a}} \equiv_T \varphi(\vec{\underline{a}}, y)$ for all $\vec{a}$. If $f$ is represented by $\varphi(\vec{x}, y)$ then graph $f$ is represented by the same formula, because if $b \neq f\vec{a}$ and hence $\vdash \underline{b} \neq \underline{f\vec{a}}$ by C2, then $\vdash \neg \varphi(\vec{\underline{a}}, \underline{b})$ by $R^=$, so that the condition $R^-$ holds. The following lemma will show in particular that $f$ is representable provided graph $f$ is representable.

**Lemma 3.4.** (a) *Let $P \subseteq \mathbb{N}^{n+1}$ be represented by $\alpha(\vec{x}, y)$ and suppose that $\forall \vec{a} \exists b P(\vec{a}, b)$. Then $\varphi(\vec{x}, y) := \alpha(\vec{x}, y) \wedge (\forall z < y) \neg \alpha(\vec{x}, z)$ represents the function $f : \vec{a} \mapsto \mu b[P(\vec{a}, b)]$. If $P$ is $\Delta_0$-representable then so too is $f$. If $P$ is $\Delta_1$-representable then so is $f$.*
(b) *$f$ is representable provided graph $f$ is representable.*
(c) *If $f$ is $\Sigma_1$-representable then $f$ is $\Pi_1$-representable as well.*
(d) *If $\chi_P$ is $\Sigma_1$-representable then $P$ is $\Delta_1$-representable.*

**Proof.** By Lemma 3.3, $\varphi(\vec{x}, y)$ represents the predicate defined by $\varphi(\vec{x}, y)$ and this is clearly graph $f$. Hence, R$^+$ holds. We verify R$^=$ by proving

$$(*) \quad \alpha(\vec{\underline{a}}, y) \wedge (\forall z{<}y)\neg\alpha(\vec{\underline{a}}, z) \vdash y = \underline{f\vec{a}}.$$

Suppose $b := f\vec{a}$. Then $\underline{b} < y \vdash (\exists z{<}y)\alpha(\vec{\underline{a}}, z)$, because $\vdash \alpha(\vec{\underline{a}}, \underline{b})$. Contraposition yields $(\forall z{<}y)\neg\alpha(\vec{\underline{a}}, z) \vdash \underline{b} \not< y$. By C5 and R$^-$ we have $y < \underline{b} \vdash \bigvee_{i<b} y = \underline{i} \vdash \neg\alpha(\vec{\underline{a}}, y)$. Hence $\alpha(\vec{\underline{a}}, y) \vdash y \not< \underline{b}$ and so, by C6,

$$\alpha(\vec{\underline{a}}, y) \wedge (\forall z{<}y)\neg\alpha(\vec{\underline{a}}, z) \vdash y \not< \underline{b} \wedge \underline{b} \not< y \vdash y = \underline{b}.$$

This confirms $(*)$. Clearly, $\varphi$ in (a) is $\Delta_0$ if $\alpha$ is $\Delta_0$. Let $P$ be represented at the same time by the $\Pi_1$-formula $\beta$. Repeating the above with $\alpha(\vec{x}, y) \wedge (\forall v{<}y)\neg\beta(\vec{x}, v)$ (a $\Sigma_1$-formula by Exercise 2) in place of $\varphi$ shows that $f$ is $\Sigma_1$-representable. It is then also $\Delta_1$-representable by item (c). (b) follows from applying (a) to $P = \text{graph} f$ while noting that $f\vec{a} = \mu b[P(\vec{a}, b)]$. (c): Let the $\Sigma_1$-formula $\varphi(\vec{x}, y)$ represent $f$ and $z \notin \text{var}\,\varphi$. Then $\varphi'(\vec{x}, y) := \forall z(\varphi(\vec{x}, z) \to z = y)$ is a $\Pi_1$-formula that represents $f$ as well: Application of R$^=$ results in $\vdash \varphi'(\vec{\underline{a}}, \underline{f\vec{a}})$, which confirms R$^+$ for $\varphi'$, and because of $\vdash \varphi(\vec{\underline{a}}, \underline{f\vec{a}})$, we obtain R$^=$ for $\varphi'$ from

$$\varphi'(\vec{\underline{a}}, y) = \forall z(\varphi(\vec{\underline{a}}, z) \to y = z) \vdash \varphi(\vec{\underline{a}}, \underline{f\vec{a}}) \to y = \underline{f\vec{a}} \vdash y = \underline{f\vec{a}}.$$

(d): Let $\chi_P$ be $\Sigma_1$-represented by $\varphi(\vec{x}, y)$. Then $P$ is $\Sigma_1$-represented by $\varphi(\vec{x}, \underline{1})$ and $\Pi_1$-represented by $\neg\varphi(\vec{x}, 0)$, as is easily confirmed. $\qquad\square$

**Remark 2.** $\alpha(x, y, z) := z \cdot \underline{2} = (x + y) \cdot \mathsf{S}(x + y) + x \cdot \underline{2}$ represents graph $\wp$ in Q. Thus, the $\Delta_0$-formula $\alpha(x, y, z) \wedge (\forall u{<}z)\neg\alpha(x, y, u)$ represents $\wp$ according to Lemma 3.4(a). We mention that in PA (but not in Q) the function $\wp$ is represented even by the open formula $\alpha$.

**Lemma 3.5.** *Let $P \subseteq \mathbb{N}^k$ be represented by $\alpha(\vec{y})$, and $g_i \in \mathbf{F}_n$ represented by $\gamma_i$ for $i = 1, \ldots, k$. Then $\beta(\vec{x}) := \exists\vec{y}[\bigwedge_i \gamma_i(\vec{x}, y_i) \wedge \alpha(\vec{y})]$ represents the predicate $Q := P[g_1, \ldots, g_k]$. If the $\gamma_i$ are $\Sigma_1$ and $P$ is $\Sigma_1$-representable or $\Delta_1$-representable, then the corresponding holds for $Q$.*

**Proof.** Let $b_i := g_i\vec{a}$, so that $\vdash \gamma_i(\vec{\underline{a}}, \underline{b_i})$, and $\vec{b} = (b_1, \ldots, b_k)$. If $Q\vec{a}$ holds, hence $P\vec{b}$, then $\vdash \alpha(\vec{\underline{b}})$, whence $\vdash \bigwedge_i \gamma_i(\vec{\underline{a}}, \underline{b_i}) \wedge \alpha(\vec{\underline{b}})$, and so $\vdash \beta(\vec{\underline{a}})$. But if $\neg Q\vec{a}$ and thus $\neg P\vec{b}$, then clearly $\vdash \neg\alpha(\vec{\underline{b}})$. Using R$^=$ for the $\gamma_i$, this then yields $\bigwedge_i \gamma_i(\vec{\underline{a}}, y_i) \vdash \bigwedge_i y_i = \underline{b_i} \vdash \neg\alpha(\vec{y})$. Hence $\vdash \forall\vec{y}[\bigwedge_i \gamma_i(\vec{\underline{a}}, y_i) \to \neg\alpha(\vec{y})] \equiv \neg\beta(\vec{\underline{a}})$. If the $\gamma_i$ and also $\alpha$ are $\Sigma_1$, then so too is $\beta$. If $P$ is represented by the $\Pi_1$-formula $\alpha'(\vec{x})$ at the same time, then the $\Pi_1$-formula $\forall\vec{y}[\bigwedge_i \gamma_i(\vec{x}, y_i) \to \alpha'(\vec{y})]$ represents $Q$ as well. $\qquad\square$

From this lemma, applied to graph $h$, follows without difficulty

**Corollary 3.6.** *If $h \in \mathbf{F}_m$ is representable by $\beta$ and the $g_i \in \mathbf{F}_n$ by $\gamma_i$, then $\varphi(\vec{x}, z) := \exists \vec{y} \, [\bigwedge_i \gamma_i(\vec{x}, y_i) \wedge \beta(\vec{y}, z)]$ represents $f = h[g_1, \dots, g_m]$.*

**Exercises**

1. Let $\exists \vec{x} \alpha$ be $\Sigma_1$ and $\forall \vec{x} \alpha$ be $\Pi_1$. Construct $\Delta_0$-formulas $\beta$ and $\gamma$ such that $\exists \vec{x} \alpha \equiv_{\mathcal{N}} \exists x \beta$ and $\forall \vec{x} \alpha \equiv_{\mathcal{N}} \forall x \gamma$ (*quantifier compression*). Since a $\Delta_0$-predicate is p.r. (Remark 1), each $\Sigma_1$-predicate is r.e. and w.l.o.g. of the form $(\exists b {\in} \mathbb{N}) Q(\vec{a}, b)$ with $Q \in \Delta_0$.

2. Show using (c) from Exercise 4 in **3.3** that $\Sigma_1$ is closed under bounded quantification, that is, if $\alpha = \alpha(\vec{x}, \vec{y}, y)$ and $\exists \vec{x} \alpha$ is $\Sigma_1$ (defines an $(m+1)$-ary $\Sigma_1$-predicate), then also $(\forall z {<} y) \exists \vec{x} \alpha \, {}_{\vec{y}}^{z}$ and $(\exists z {<} y) \exists \vec{x} \alpha \, {}_{\vec{y}}^{z}$ are $\Sigma_1$. Derive the corresponding for $\Pi_1$ and $\Delta_1$.

3. Prove that $\alpha(\vec{x}) \wedge y = \underline{1} \; \vee \; \neg \alpha(\vec{x}) \wedge y = \underline{0}$ represents $\chi_P$ provided $\alpha$ represents $P$.

4. Show that every $\Delta_0$-formula is equivalent to a $\Delta_0$-formula built up from literals by means of $\wedge, \vee$, and the bounded quantifiers $(\forall x {\leqslant} t)$ and $(\exists x {\leqslant} t)$, a correspondence to Exercise 4 in **2.4**.

# 6.4 The Representability Theorem

For the representability of all recursive or just all p.r. functions, it is helpful to have a representable $g \in \mathbf{F}_2$ that satisfies the following: for every $n$ and every sequence $c_0, \dots, c_n$ there exists a number $c$ such that $g(c, i) = c_i$ for all $i \leqslant n$. In short, $c$ can be chosen such that the values $g(c, 0), g(c, 1), \dots, g(c, n)$ are the given ones. Now, there are many p.r. functions $g$ that can do this, for example $g \colon (c, i) \mapsto (c)_i$ for choosing $c = p_0^{1+c_0} \cdots p_n^{1+c_n}$. Initially there is no obvious way to show the representability of such a function $g$ in $\mathbf{Q}$ or in some extension of $\mathbf{Q}$ *within* the language $\mathcal{L}_{ar}$. Therefore, K. Gödel, who around 1930 was working on this and related problems, in the words of A. Mostowski "phoned with God." Although nowadays several possibilities are known, we follow the original, which has not lost any of its attraction.

Let $\alpha(a, b, i) := \text{rem}(a, (1 + (1 + i)b))$, where $\text{rem}(a, d)$ denotes the remainder of $a$ divided by $d$ ($\neq 0$). In addition, $\text{rem}(a, 0) := 0$. Note that $r = \text{rem}(a, d)$ is well defined, since for any $a$ and $d \neq 0$ there are unique $q, r \in \mathbb{N}$ with $a = qd + r$ and $r < d$ (readily shown by induction on $a$). Clearly, graph $\alpha$ has the $\Delta_0$-definition

$$\alpha(a, b, i) = k \iff (\exists q \leqslant a)[a = q(1 + (1 + i)b) + k \ \& \ k < 1 + (1 + i)b].$$

Hence, the function $\alpha$ is $\Delta_0$-representable by Lemma 3.4(b). The same holds for the pairing function $\wp$. Because $\wp$ is bijective there are unary functions $\varkappa_1, \varkappa_2$ such that $\wp(\varkappa_1 k, \varkappa_2 k) = k$ for all $k$. Their explicit form is insignificant; we just require the obvious property $\varkappa_1 k, \varkappa_2 k \leqslant k$. The function $\beta: (c, i) \mapsto \alpha(\varkappa_1 c, \varkappa_2 c, i)$ is called the $\beta$-*function*. Since

$$\beta(c, i) = k \iff (\exists a \leqslant c)(\exists b \leqslant c)[\wp(a, b) = c \ \& \ \alpha(a, b, i) = k],$$

graph $\beta$ is $\Delta_0$. Hence, according to Lemma 3.4, $\beta$ is represented by a $\Delta_0$-formula, which will be denoted by **beta**. Omitting the argument parentheses in **beta**, this means that

(1)    $\vdash_Q \textbf{beta} \underline{c} \underline{i} y \leftrightarrow y = \underline{\beta(c, i)}$, for all $c, i \in \mathbb{N}$.

Clearly, **beta** also defines the $\beta$-function in $\mathcal{N}$. The following simple number-theoretic facts known for ages will be applied in proving the main property of the $\beta$-function stated in Lemma 4.1 below.

**Euclid's lemma.** *Let* $a, b$ *be positive and coprime* ($a \perp b$). *Then there exist* $x, y \in \mathbb{N}$ *such that* $xa + 1 = yb$. (The converse of this claim is obvious: $c | a, b$ implies $c | yb - xa = 1$ and hence $c = 1$.)

**Proof** by $<$-induction on $s = a + b$. Trivial for $s \leqslant 2$, i.e., $a = b = 1$. Let $s > 2$. Then $a \neq b$, say $a > b$, and clearly $a - b \perp b$ as well. Since $(a - b) + b < s$, there are $x, y \in \mathbb{N}$ with $x(a - b) + 1 = yb$ by the induction hypothesis. Hence, $xa + 1 = y'b$ with $y' = x + y$. In the case $a < b$ consider $a \perp b - a$, so that $xa + 1 = y(b - a)$ for some $x, y$ by the induction hypothesis. Hence $(x + y)a + 1 = yb$. ◻

**Chinese remainder theorem.** *Let* $c_i < d_i$ *for* $i = 0, \ldots, k$ *and let* $d_0, \ldots, d_k$ *be pairwise coprime. Then there exists some* $a \in \mathbb{N}$ *such that* $\text{rem}(a, d_i) = c_i$ *for* $i = 0, \ldots, k$.

**Proof** by induction on $k$. For $k = 0$ this is clear with $a = c_0$. Let the assumptions hold for $k > 0$. By the induction hypothesis, $\text{rem}(a, d_i) = c_i$

for some $a$ and all $i < k$. Since $d_0, \ldots, d_k$ are coprime, $m := \operatorname{lcm}\{d_\kappa \mid \nu < k\}$ and $d_k$ are coprime (Exercise 1c). Thus, by Euclid's lemma, there are $x, y \in \mathbb{N}$ such that $xm + 1 = yd_k$. Multiplying both sides by $c_k(m-1) + a$, we obtain $x'm + c_k(m-1) + a = y'd_k$ with new values $x', y' \in \mathbb{N}$. Put $a' := (x' + c_k)m + a = y'd_k + c_k$. Then $\operatorname{rem}(a', d_i) = \operatorname{rem}(a, d_i) = c_i$ for all $i < k$, since $d_i \mid m$. But also $\operatorname{rem}(a', d_k) = c_k$, because $c_k < d_k$. $\quad\blacksquare$

Unlike those in most textbooks of number theory, the proof above is constructive and easily transferable to **PA**, as will be shown in **7.1**. In logic it is occasionally not just important what you prove, but how you prove it. The claim of Euclid's lemma can also be shown by means of the Euclidean algorithm for determining $\operatorname{lcm}(a, b)$.

**Lemma 4.1 (on the $\beta$-function).** *For every $n$ and every sequence $c_0, \ldots, c_n$ there exists some $c$ such that $\beta(c, i) = c_i$ for $i = 0, \ldots, n$.*

**Proof.** It suffices to provide numbers $a$ and $b$ such that $\alpha(a, b, i) = c_i$ for all $i \leqslant n$. Because of $\beta(\wp(a, b), i) = \alpha(a, b, i)$ the claim is then satisfied with $c = \wp(a, b)$. Let $m := \max\{n, c_0, \ldots, c_n\}$ and $b := \operatorname{lcm}\{1, \ldots, m\}$. We claim that the numbers $d_i := 1 + (1 + i) \cdot b > c_i$ ($i \leqslant n$) are pairwise coprime. For otherwise let $p$ be a common prime factor of $d_i, d_j$ with $i < j \leqslant n$. Then $p \mid d_j - d_i = (j - i)b$; hence $p \mid j - i$ or $p \mid b$. But since $j - i \mid b$ in view of $j - i \leqslant n \leqslant m$, it follows that $p \mid b$ in any case. Since $b \mid d_i - 1$ this yields $p \mid d_i - 1$, contradicting $p \mid d_i$. Hence, $d_0, \ldots, d_n$ are indeed pairwise coprime. By the Chinese remainder theorem there is an $a$ such that $\operatorname{rem}(a, d_i) = c_i$, that is, $\alpha(a, b, i) = c_i$ for $i = 0, \ldots, n$. $\quad\blacksquare$

**Remark 1.** Already at this stage we gain the interesting insight that the exponential function $(a, b) \mapsto a^b$ is explicitly definable in $\mathcal{N}$, namely by

$$\delta_{exp}(x, y, z) := \exists u [\beta(u, 0) = \mathsf{S}0 \wedge (\forall v < y) \, \beta(u, \mathsf{S}v) = \beta(u, v) \cdot x \wedge \beta(u, y) = z].$$

This is a $\Sigma_1$-formula, more precisely, the *description* of a $\Sigma_1$-formula arising after the elimination of the occurring $\beta$-terms by means of (1), using some further $\exists$-quantifiers instead. By induction on $b$ one sees that $\mathcal{N} \vDash \delta_{exp}(\underline{a}, \underline{b}, \underline{c})$ implies $a^b = c$. Suppose conversely that $a^b = c$. Then Lemma 4.1 guarantees a suitable $u$ such that $\mathcal{N} \vDash \delta_{exp}(\underline{a}, \underline{b}, \underline{c})$: choose $u$ such that $\beta(u, i) = a^i$ for all $i \leqslant b$. This argument is generalized in Theorem 4.2 below. It tells us in particular that each recursive function is explicitly definable in $\mathcal{N}$.

For simplicity, we assume $T \supseteq \mathsf{Q}$ in Theorem 4.2 below, though it holds as well if $\mathsf{Q}$ is merely interpretable in $T$ in the sense of **6.6**. For the derivation of undecidability results or a simplified version of the first incompleteness theorem, the theorem's "Moreover" part is not needed.

**Theorem 4.2 (Representability theorem).** *Each recursive function* $f$—*and hence every recursive predicate—is representable in an arbitrary consistent axiomatic extension* $T \supseteq \mathsf{Q}$. *Moreover,* $f$ *is* $\Sigma_1$-*representable.*

**Proof.** It suffices to construct a $\Sigma_1$-formula that represents $f$ in $\mathsf{Q}$. For the initial functions $0$, $\mathsf{S}$, $I_\nu^n$ we may choose the formulas $v_0 = 0$, $v_1 = \mathsf{S}v_0$, and $v_n = v_\nu$. As regards $\boldsymbol{Oc}$, let $f = h[g_1, \ldots, g_m]$ and suppose $\beta(\vec{y}, z)$ and $\gamma_i(\vec{x}, y_i)$ are $\Sigma_1$-formulas representing $h$ and the $g_i$. Then by Corollary 3.6, $\varphi(\vec{x}, z) := \exists \vec{y}[\bigwedge_i \gamma_i(\vec{x}, y_i) \wedge \beta(\vec{y}, z)]$ is such a formula for $f$. Next let $f = \boldsymbol{Op}(g, h)$, with $g, h$ both being $\Sigma_1$-representable. Define

$$P(\vec{a}, b, c) :\Leftrightarrow \beta(c, 0) = g\vec{a} \,\&\, (\forall v < b)\beta(c, \mathsf{S}v) = h(\vec{a}, v, \beta(c, v)).$$

According to Lemmas 3.5 and 3.3, $P$ is $\Delta_1$-representable (use composition, instantiation of $\Sigma_1$-representable functions, bounded quantification, and conjunction). Clearly $P(\vec{a}, b, c)$ is equivalent to $(*)$: $\beta(c, i) = f(\vec{a}, i)$ for all $i \leqslant b$. By Lemma 4.1, for given $\vec{a}, b$ there is some $c$ satisfying $(*)$; hence we know that $\forall \vec{a}, b \exists c P(\vec{a}, b, c)$. Thus, $\tilde{f} \colon (\vec{a}, b) \mapsto \mu c[P(\vec{a}, b, c)]$ is $\Sigma_1$-representable; Lemma 3.4(a). Since $P(\vec{a}, b, \tilde{f}(\vec{a}, b))$, $(*)$ holds with $c = \tilde{f}(\vec{a}, b)$, which yields $f(\vec{a}, b) = \beta(\tilde{f}(\vec{a}, b), b)$ for $i = b$. Thus, as a composition of $\Sigma_1$-representable functions, $f$ is $\Sigma_1$-representable. Finally, let $f$ result from $g$ by $\boldsymbol{O\mu}$, $f\vec{a} = \mu b[P(\vec{a}, b)]$, where $P(\vec{a}, b) \Leftrightarrow g(\vec{a}, b) = 0$ and $g$ is $\Sigma_1$-representable. By Lemma 3.4(c), $g$ is $\Pi_1$-representable, too. This clearly implies that $P$ is $\Delta_1$-representable. Hence, $f$ is $\Sigma_1$-representable by Lemma 3.4(a). $\blacksquare$

Let $T \supseteq \mathsf{Q}$ be a theory in $\mathcal{L}_{ar}$. To $\varphi \in \mathcal{L}_{ar}$ corresponds *within* $T$ the term $\underline{n}$ with $n := \dot{\varphi}$, which will be denoted by $\ulcorner \varphi \urcorner$ (or $\dot{\varphi}$) and called the *Gödel term* of $\varphi$. For example, $\ulcorner v_0 = 0 \urcorner$ is $\dot{v}_0 \doteq \dot{0}$ $(= 2^{2^2} \cdot 3^2 \cdot 5^{14})$. Analogously $\ulcorner t \urcorner$ is defined for terms $t$. For instance, $\ulcorner \underline{1} \urcorner = \ulcorner \mathsf{S}0 \urcorner = 2^{16} \cdot 3^{14}$. If $T$ is axiomatized, also $\ulcorner \Phi \urcorner = \dot{\Phi}$ for proofs $\Phi$ in $T$ is well defined. For instance, $(v_0 = v_0)$ is for such a $T$ a trivial proof of length 1 by axiom $\Lambda 9$ in **3.6**. Its Gödel term is $2^{\dot{v}_0 \doteq \dot{v}_0 + 1}$. The predicate $\mathsf{bew}_T$ is p.r. (Theorem 2.4), hence $\Sigma_1$-representable (Theorem 4.2), by the formula $\mathsf{bew}_T(y, x)$, say. Define $\mathsf{bwb}_T(x) := \exists y\, \mathsf{bew}_T(y, x)$. Then Theorem 4.2 and (4) from page 230 obviously yield the following important

**Corollary 4.3.** *If $T \supseteq Q$ is axiomatizable then $\vdash_T \varphi \Rightarrow \vdash_T \mathsf{bew}_T(\underline{n}, \ulcorner \varphi \urcorner)$ for some $n$, and $\nvdash_T \varphi \Rightarrow \vdash_T \neg \mathsf{bew}_T(\underline{n}, \ulcorner \varphi \urcorner)$ for all $n$. Hence, $\vdash_T \varphi$ implies $\vdash_T \mathsf{bwb}_T(\ulcorner \varphi \urcorner)$ in any case.*

The converse $\vdash_T \mathsf{bwb}_T(\ulcorner \varphi \urcorner) \Rightarrow \vdash_T \varphi$ need not hold; see **7.1**. Theorem 4.2 has several important consequences, for example Theorem 4.5 below. Before stating it we will acquaint ourselves with a method of eliminating Church's thesis from certain intuitively clear arguments that demand justification when 'decidable' is identified with 'recursive'. Of course, such an elimination must in principle always be possible if the thesis is to retain its legitimacy. For instance, Church's thesis was essentially used in the proof of Theorem 3.5.2. We reformulate the theorem and will give a rigorous proof.

**Theorem 4.4.** *A complete axiomatizable theory $T$ is recursive.*

**Proof.** Because of completeness, the function

$$f : a \mapsto \mu b[a \in \dot{\mathcal{L}}^0 \Rightarrow bew_T(b, a) \vee bew_T(b, \tilde{\neg} a)]$$

is well defined. Indeed, let $P(a, b)$ denote the recursive predicate in square brackets. Then $\forall a \exists b P(a, b)$ (note that $P(a, 0)$ if $a \notin \dot{\mathcal{L}}^0$). By $O\mu$, then, $f$ is recursive. We claim $(*)$: $a \in \dot{T} \Leftrightarrow a \in \dot{\mathcal{L}}^0$ & $bew_T(fa, a)$. This clearly implies the recursiveness of $T$. In order to prove $(*)$ let $a \in \dot{T}$, so certainly $a \in \dot{\mathcal{L}}^0$. Then for $b = fa$, the smallest $b$ such that $bew_T(b, a) \vee bew_T(b, \tilde{\neg} a)$, the first disjunct must hold, because due to the consistency of $T$, no $c \in \mathbb{N}$ with $bew_T(c, \tilde{\neg} a)$ can exist at all. Hence, $bew_T(fa, a)$. The $\Leftarrow$-direction in $(*)$ is obvious. ∎

This proof illustrates sufficiently well the distinction between a primitive recursive and a recursive decision procedure. Even when $X$ and thus the predicate $P$ in the proof above are primitive recursive, the defined recursive function $f$ need not be so, because the completeness of $T$ may have been established in a nonconstructive way. The use of Church's thesis in the proofs of (i)⇒(ii) and (iii)⇒(ii) of the following theorem can be eliminated in almost exactly the same manner as above, although then the proof would lose much of its transparency.

**Theorem 4.5.** *For a predicate $P \subseteq \mathbb{N}^n$ and any consistent axiomatizable theory $T \supseteq Q$ the following are equivalent:*

(i) *$P$ is representable in $T$,* (ii) *$P$ is recursive,* (iii) *$P$ is $\Delta_1$.*

**Proof.** (i)⇒(ii): Suppose $P$ is represented in $T$ by $\alpha(\vec{x})$. Given $\vec{a}$ we set going the enumeration machine of $T$ and wait until $\alpha(\vec{a})$ or $\neg\alpha(\vec{a})$ appears. Thus, $P$ is decidable and hence recursive by Church's thesis. (ii)⇒(i),(iii): By Theorem 4.2, $\chi_P$ is representable in $T$ by a $\Sigma_1$-formula; hence $P$ is $\Delta_1$-representable by Lemma 3.4(d) and of course by the corresponding formulas also defined in $\mathcal{N}$. Thus, $P \in \Delta_1$. (iii)⇒(ii): Let $P$ be defined by the $\Sigma_1$-formula $\alpha(\vec{x})$ and the $\Pi_1$-formula $\beta(\vec{x})$. Given $\vec{a}$ we start the enumeration machine for $\mathbf{Q}$ and wait until the $\Sigma_1$-sentence $\alpha(\vec{a})$ or $\neg\beta(\vec{a})$ appears. In the first case $P\vec{a}$ holds; in the second it does not. This procedure terminates because $\mathbf{Q}$ is $\Sigma_1$-complete by Theorem 3.1. ◻

This theorem tells us that in all consistent axiomatic extensions of $\mathbf{Q}$ exactly the same predicates are representable, namely the recursive ones. Moreover, $\Delta_1$ contains precisely the recursive predicates, from which it easily follows that $\Sigma_1$ consists just of all r.e. predicates. Theorem 4.5 clarifies fairly well the close relationship between logic and recursion theory. It is independent of Church's thesis. Even if the thesis for certain theoretical or practical reasons had to be revised, the distinguished role of the $\mu$-recursive functions would not be affected.

**Remark 2.** The above results allow us to define recursive or decidable predicates directly as follows: $P \subseteq N^n$ is recursive iff there is some finitely axiomatizable theory in which $P$ is representable. We need only to notice that a predicate representable in any finitely axiomatizable theory in which representability makes sense is recursive by Church's thesis. In this and the previous section we met several formulas or classes of those that represent predicates in $\mathbf{Q}$ and hence are recursive. It would of course be nice to provide a somewhat more uniform system of formulas that represent the recursive predicates, or at least that define them in $\mathcal{N}$. Unfortunately, such a system of formulas cannot be recursively enumerated. Indeed, suppose there is such an enumeration. Let $\alpha_0, \alpha_1, \ldots$ be the resulting subenumeration of its members in $\mathcal{L}^1_{ar}$. These define in $\mathcal{N}$ the recursive sets. Then also $\{n \in \mathbb{N} \mid n \notin \alpha_n^{\mathcal{N}}\}$ is recursive, hence is defined in $\mathcal{N}$ by $\alpha_m$, say, so that $n \in \alpha_m^{\mathcal{N}} \Leftrightarrow n \notin \alpha_n^{\mathcal{N}}$. However, this equivalence yields for $n = m$ the contradiction $m \in \alpha_m^{\mathcal{N}} \Leftrightarrow m \notin \alpha_m^{\mathcal{N}}$.

In **6.5** we need a p.r. "substitution" function and in **7.1** a generalization. Let $\mathrm{cf}\, n := \dot{\underline{n}}\ (= (\underline{n})^{\cdot})$ denote the Gödel number of the "cipher term" $\underline{n}\ (= \mathsf{S}^n 0)$. Then $n \mapsto \mathrm{cf}\, n$ is p.r., since $\mathrm{cf}\, 0 = \dot{0}$ and $\mathrm{cf}\, \mathsf{S}n = \dot{\mathsf{S}} * \mathrm{cf}\, n$. Let $\mathrm{sb}_x(m, n) = [m]_{\dot{x}}^{\mathrm{cf}\, n}$ and define $\mathrm{sb}_{\vec{x}} \in \mathbf{F}_{n+1}$ inductively on the length $n$ of $\vec{x} \in \mathrm{Var}^n$ by $\mathrm{sb}_\emptyset(m) = m$ and $\mathrm{sb}_{\vec{x}x}(m, \vec{a}, a) = \mathrm{sb}_x(\mathrm{sb}_{\vec{x}}(m, \vec{a}), a)$. Here $x_1, \ldots, x_n, x$ denote distinct variables. Clearly, the $\mathrm{sb}_{\vec{x}}$ are all p.r.

Let $\dot{\varphi}_{\vec{x}}(\vec{a})$ denote the Gödel number of the formula $\varphi_{\vec{x}}(\vec{a})$ that arises from $\varphi$ by stepwise substituting $\underline{a_i}$ at the free occurrences of $x_i$ in $\varphi$ for $i = 1, \ldots, n$ (cf. also page 60). Then the main property of the p.r. functions $\mathrm{sb}_{\vec{x}}$ is expressed by

**Theorem 4.6.** $\mathrm{sb}_{\vec{x}}(\dot{\varphi}, \vec{a}) = \dot{\varphi}_{\vec{x}}(\vec{a})$, *for arbitrary* $\varphi \in \mathcal{L}$ *and all* $\vec{a} \in \mathbb{N}^n$.

**Proof.** Since $\varphi_{\vec{x}}(\vec{a})$ results from applying simple substitutions stepwise, we need only show that $\mathrm{sb}_x(\dot{\varphi}, a) = \dot{\varphi}_x(\underline{a})$ for all $\varphi \in \mathcal{L}$, $x \in \mathrm{Var}$, and $a \in \mathbb{N}$. This holds, since $\mathrm{sb}_x(\dot{\alpha}, a) = [\dot{\alpha}]_{\vec{x}}^{\mathrm{cf}\,a} = [\dot{\alpha}]_{\vec{x}}^{\dot{a}} = (\alpha_x(\underline{a}))^{\cdot} = \dot{\alpha}_x(\underline{a})$ in view of $(*)$ from page 232. ∎

**Example.** Let $\alpha$ be $\mathsf{S}x = y$. Then $\mathrm{sb}_{xy}(\dot{\alpha}, a, b) = (\mathsf{S}\underline{a} = \mathsf{S}\underline{b})^{\cdot}$ for $a, b \in \mathbb{N}$. Further, $\mathrm{sb}_{xy}(\dot{\alpha}, a, \mathsf{S}a) = (\mathsf{S}\underline{a} = \mathsf{S}\underline{a})^{\cdot} = (\mathsf{S}\underline{a} = \mathsf{S}\underline{a})^{\cdot} = \mathrm{sb}_x(\dot{\alpha}\frac{\mathsf{S}x}{y}, a)$. This equation will be generalized in Exercise 3c.

### Exercises

1. Let $a, b, a_0, \ldots, a_n$ ($n > 0$) be positive natural numbers and $p$ a prime. Prove (a) $p|ab \Rightarrow p|a \vee p|b$, (b) $p|\operatorname{lcm}\{a_\nu \mid \nu \leqslant n\} \Rightarrow p|a_\nu$ for some $\nu \leqslant n$, and (c) $\operatorname{lcm}\{a_\nu \mid \nu < n\}$ and $a_n$ are coprime provided $a_0, \ldots, a_n$ are pairwise coprime.

2. Provide a defining $\Sigma_1$-formula for the prime enumeration $n \mapsto p_n$.

3. Expand the $\mathcal{L}_{ar}$-structure $\mathcal{N}$ by the functions $\tilde{\wedge}, \tilde{\neg}, \tilde{\rightarrow}, \tilde{\forall}$, and all $\mathrm{sb}_{\vec{x}}$, and if necessary, further p.r. base functions to a structure $\mathcal{N}^*$. Then the terms $\mathrm{sb}_{\vec{x}}(\ulcorner\varphi\urcorner, \vec{x})$ for $\varphi \in \mathcal{L}_{ar}$ are well defined in the theory of $\mathcal{N}^*$.[5] Verify for arbitrary $\alpha, \beta, \varphi \in \mathcal{L}_{ar}$ the following equations in $\mathcal{N}^*$:

   (a) $\mathrm{sb}_{\vec{x}}((\alpha \tilde{\wedge} \beta)^{\cdot}, \vec{x}) = \mathrm{sb}_{\vec{x}}(\dot{\alpha}, \vec{x}) \tilde{\wedge} \mathrm{sb}_{\vec{x}}(\dot{\beta}, \vec{x})$, and analogously for $\neg$, $\rightarrow$, and $\forall$.
   (b) $\mathrm{sb}_{\vec{x}}(\dot{\varphi}, \vec{x}) = \mathrm{sb}_{\vec{x}'}(\dot{\varphi}, \vec{x}')$, where $\vec{x}'$ covers all $x \in \mathrm{free}\,\varphi$ such that $x \in \mathrm{var}\,\vec{x}$.
   (c) Let $y \notin \mathrm{bnd}\,\varphi$. Then $\mathrm{sb}_{\vec{x},x}(\dot{\varphi}, \vec{x}, t) = \mathrm{sb}_{\vec{x},y}((\varphi\frac{t}{x})^{\cdot}, \vec{x}, y)$ for $t \in \{0, y, \mathsf{S}y\}$ in case $x \in \mathrm{free}\,\varphi$ and $y \notin \mathrm{var}\,\vec{x}$; otherwise $\mathrm{sb}_{\vec{x},x}(\dot{\varphi}, \vec{x}, t) = \mathrm{sb}_{\vec{x}}((\varphi\frac{t}{x})^{\cdot}, \vec{x})$.

---

[5] The expansion of $\mathcal{N}$ alleviates the later transfer of this exercise to PA. In **7.1** it will be shown that the additional functions of $\mathcal{N}^*$ are explicitly definable already in PA.

## 6.5   The Theorems of Gödel, Tarski, Church

Call a theory $T \subseteq \mathcal{L}$ *arithmetizable* if $\mathcal{L}$ is arithmetizable and a sequence $(\underline{n})_{n \in \mathbb{N}}$ of constant terms is available such that $\vdash_T \underline{n} \neq \underline{m}$ for $n \neq m$ and $\mathrm{cf} : n \mapsto (\underline{n})^{\cdot}$ is p.r. These are minimal requirements that representability of arithmetical predicates in $T$ make sense. They are trivially satisfied for $T \supseteq \mathsf{Q}$, but also for $\mathsf{ZFC}$ with respect to $\omega$-terms (page 115). Terms and formulas are coded within $T$, similarly to what is done in theories in $\mathcal{L}_{ar}$. In particular, $\ulcorner \alpha \urcorner$ $(= (\dot{\alpha}))$ denotes the already defined Gödel term of a formula $\alpha$. In order to evoke a concrete picture of the following two fairly general lemmas, take $\mathcal{L} = \mathcal{L}_{ar}$ and $T = \mathsf{PA}$ as standard examples.

A sentence $\gamma$ is called a *fixed point of* $\alpha = \alpha(x)$ *in* $T$ if $\gamma \equiv_T \alpha(\ulcorner \gamma \urcorner)$; equivalently, $\vdash_T \gamma \leftrightarrow \alpha(\ulcorner \gamma \urcorner)$. In intuitive terms, $\gamma$ then says "$\alpha$ applies to me." The p.r. function $\mathrm{sb}_x$ from **6.4** is representable in $T$ under relatively weak assumptions by Theorem 4.2. Hence, the lemmas below have a large spectrum of application.

**Fixed point lemma.** *Let $T$ be an arithmetizable theory and suppose that $\mathrm{sb}_x$ is representable in $T$. Then for each $\alpha = \alpha(x) \in \mathcal{L}$ there is some $\gamma \in \mathcal{L}^0$ such that*

(1)   $\gamma \equiv_T \alpha(\ulcorner \gamma \urcorner)$.

**Proof.** Let $x_1, x_2, y \neq x$ and $\mathsf{sb}(x_1, x_2, y)$ be a formula representing $\mathrm{sb}_x$ in $T$. Then $\mathsf{sb}(\ulcorner \varphi \urcorner, \underline{n}, y) \equiv_T y = \ulcorner \varphi(\underline{n}) \urcorner$ for all $\varphi = \varphi(x)$ and all $n$. With $\underline{n} = \ulcorner \varphi \urcorner$ we then get

(2)   $\mathsf{sb}(\ulcorner \varphi \urcorner, \ulcorner \varphi \urcorner, y) \equiv_T y = \ulcorner \varphi(\ulcorner \varphi \urcorner) \urcorner$.

Let $\beta(x) := \forall y (\mathsf{sb}(x, x, y) \to \alpha \frac{y}{x})$. Then $\gamma := \beta(\ulcorner \beta \urcorner)$ yields what we require. Indeed,

$$
\begin{aligned}
\gamma \;&=\; \forall y (\mathsf{sb}(\ulcorner \beta \urcorner, \ulcorner \beta \urcorner, y) \to \alpha \tfrac{y}{x}) \\
&\equiv_T\; \forall y (y = \ulcorner \beta(\ulcorner \beta \urcorner) \urcorner \to \alpha \tfrac{y}{x}) && ((2) \text{ with } \varphi := \beta(x)) \\
&=\; \forall y (y = \ulcorner \gamma \urcorner \to \alpha \tfrac{y}{x}) && (\text{because } \gamma = \beta(\ulcorner \beta \urcorner)) \\
&\equiv\; \alpha(\ulcorner \gamma \urcorner). && \qquad \blacksquare
\end{aligned}
$$

A fixed point can in the most interesting cases of $\alpha$ be constructed fairly easily; see **7.5**. The following lemma also formulates a frequently appearing argument.

**Nonrepresentability lemma.** Let $T$ be a theory as in the fixed point lemma. Then $T$ (more precisely $\dot{T}$) is not representable in $T$ itself.

**Proof.** Let $T$ be represented by the formula $\tau(x)$. We show that even the weaker assumption (a): $(\forall \alpha \in \mathcal{L}^0)\ \vdash_T \alpha \Leftrightarrow \vdash_T \neg \tau(\ulcorner \alpha \urcorner)$ leads to a contradiction. Indeed, let $\gamma$ be a fixed point of $\neg \tau(x)$ according to (1), so that (b): $\vdash_T \gamma \Leftrightarrow \vdash_T \neg \tau(\ulcorner \gamma \urcorner)$. Choosing $\alpha = \gamma$ in (a) clearly yields with (b) the contradiction $\nvdash_T \gamma \Leftrightarrow \vdash_T \gamma$. $\blacksquare$

We now formulate Gödel's first incompleteness theorem, giving three versions, of which the second corresponds essentially to the original. For simplicity, let henceforth $\mathcal{L} \supseteq \mathcal{L}_{ar}$ and $T \supseteq \mathsf{Q}$, ensuring the applicability of the two lemmas above. However, all of the following holds for theories $T$, such as ZFC, in which Q is just interpretable in the sense of **6.6**.

**Theorem 5.1 (the popular version).** *Every consistent (recursively) axiomatizable theory $T \supseteq \mathsf{Q}$ is incomplete.*

**Proof.** If $T$ is complete then it is recursive by Theorem 4.4, hence representable in $T$ by Theorem 4.2, which is impossible by the nonrepresentability lemma. $\blacksquare$

Unlike the proofs of Theorems 5.1$'$ and 5.1$''$, the above proof is nonconstructive, for it does not explicitly provide a sentence $\alpha$ such that $\nvdash_T \alpha$ and $\nvdash_T \neg \alpha$.

Stronger than the consistency of $T$ is the so-called *$\omega$-consistency* of $T$ ($\subseteq \mathcal{L}_{ar}$), i.e., for all $\varphi = \varphi(x)$ such that $\vdash_T \exists x \varphi(x)$ we have $\nvdash_T \neg \varphi(\underline{n})$ for at least one $n$, or equivalently, if $\vdash_T \neg \varphi(\underline{n})$ for all $n$, then $\nvdash_T \exists x \varphi(x)$. Clearly, if $\mathcal{N} \vDash T$ then $T$ is surely $\omega$-consistent, because the supposition $\vdash_T \exists x \alpha$ and $\vdash_T \neg \alpha(\underline{n})$ for all $n$ implies the contradiction $\mathcal{N} \vDash \exists x \alpha, \forall x \neg \alpha$. Thus, from a semantic perspective the theories Q and PA are certainly $\omega$-consistent, hence also consistent.[6]

**Theorem 5.1$'$ (the original version).** *For every $\omega$-consistent theory $T \supseteq \mathsf{Q}$ axiomatized by a p.r. axiom system $X$, there is a $\Pi_1$-sentence $\alpha$ such that neither $\vdash_T \alpha$ nor $\vdash_T \neg \alpha$, i.e., $\alpha$ is independent in $T$. There is a p.r. function that assigns such an $\alpha$ to a formula representing $X$.*

---

[6] There are famous (relative) consistency proofs for PA that presuppose considerably less than the full semantic approach; cf. e.g. [Tak].

**Proof.** Let $bew_T$ be represented in $T$ by the $\Sigma_1$-formula $\mathtt{bew}(y,x)$, see page 247. For $\mathtt{bwb}(x) = \exists y\,\mathtt{bew}(y,x)$ from Corollary 4.3 we obtain (a): $\vdash_T \varphi \Rightarrow \vdash_T \mathtt{bwb}(\ulcorner\varphi\urcorner)$, for all $\varphi$. Let $\gamma$ be a fixed point of $\neg\,\mathtt{bwb}(x)$ according to the fixed point lemma, so that (b): $\gamma \equiv_T \neg\,\mathtt{bwb}(\ulcorner\gamma\urcorner)$. The assumption $\vdash_T \gamma$ yields $\vdash_T \mathtt{bwb}(\ulcorner\gamma\urcorner)$ by (a), but $\vdash_T \neg\,\mathtt{bwb}(\ulcorner\gamma\urcorner)$ by (b), contradicting the consistency of $T$. Thus, $\nvdash_T \gamma$. Now assume $\vdash_T \neg\gamma$, so that $\vdash_T \mathtt{bwb}(\ulcorner\gamma\urcorner)$ by (b); hence (c): $\vdash_T \exists y\,\mathtt{bew}(y,\ulcorner\gamma\urcorner))$. Obviously $\nvdash_T \gamma$, because $T$ is consistent. Applying Corollary 4.3 once again, we infer that $\vdash_T \neg\,\mathtt{bew}(\underline{n},\ulcorner\gamma\urcorner)$ for all $n$. However, this and (c) contradict the $\omega$-consistency of $T$. Consequently $\vdash_T \neg\gamma$ is impossible as well. Thus, $\gamma$ is independent in $T$. But then too is the $\Pi_1$-sentence $\alpha := \neg\,\mathtt{bwb}(\ulcorner\gamma\urcorner)$, which is equivalent to $\gamma$ in $T$. The claim of the p.r. assignment follows evidently from the construction of $\gamma$ in the proof of (1). ☐

This theorem remains valid without restriction if the axiom system $X$ is just r.e. In this case $X$ can be replaced by some recursive $X'$ (Exercise 1 in **6.2**), so that $bew_T$ is still recursive according to Theorem 2.4.

**Theorem 5.1″ (Rosser's strengthening of Theorem 5.1′).** *The assumption of $\omega$-consistency in Theorem 5.1′ can be weakened to the consistency of $T$.*

**Proof.** Instead of $\mathtt{bew}(y,x)$ we consider the arithmetical predicate

$$\mathtt{prov}(x) := \exists y[\mathtt{bew}(y,x) \wedge (\forall z{<}y)\neg\,\mathtt{bew}(z,\dot\neg x)],$$

where $\mathtt{bew}$ is $bew_T$ and $T$ is consistent. We think here of the p.r. function $\dot\neg$ as having been eliminated in the usual way by a formula representing it. Because of the consistency of $T$, $\mathtt{prov}(x)$ says essentially the same as $\mathtt{bwb}(x)$ and has the following fundamental properties:

(a) $\vdash_T \alpha \Rightarrow \vdash_T \mathtt{prov}(\ulcorner\alpha\urcorner)$,     (b) $\vdash_T \neg\alpha \Rightarrow \vdash_T \neg\,\mathtt{prov}(\ulcorner\alpha\urcorner)$.[7]

Indeed, suppose $\vdash_T \alpha$, so that $\vdash_T \mathtt{bew}(\underline{n},\ulcorner\alpha\urcorner)$ for some $n$ (Corollary 4.3). Since $\nvdash_T \neg\alpha$, it follows that $\vdash_T \neg\,\mathtt{bew}(\underline{k},\ulcorner\neg\alpha\urcorner)$ for all $k$. Therefore, C5 in **6.3** gives $\vdash_T (\forall z{<}\underline{n})\neg\,\mathtt{bew}(z,\ulcorner\neg\alpha\urcorner)$, and so

$$\vdash_T \mathtt{bew}(\underline{n},\ulcorner\alpha\urcorner) \wedge (\forall z{<}\underline{n})\neg\,\mathtt{bew}(z,\ulcorner\neg\alpha\urcorner),$$

---

[7] In particular $\vdash_T \neg\,\mathtt{prov}(\ulcorner\bot\urcorner)$. That the latter is not the case if we write $\mathtt{bwb}$ instead of $\mathtt{prov}$ is the import of Gödel's second incompleteness theorem, Theorem 7.3.2. Thus, $\mathtt{bwb}$ and $\mathtt{prov}$ behave *within* $T$ very differently, although $\mathtt{bew}(y,x) \equiv_{\mathcal{N}} \mathtt{prov}(y,x)$.

whence particularization yields the claim $\vdash_T \mathsf{prov}(\ulcorner\alpha\urcorner)$. Proof of (b): Suppose $\vdash_T \neg\alpha$, say $\vdash_T \mathsf{bew}(\underline{m},\ulcorner\neg\alpha\urcorner)$. Then $\vdash_T (\forall y{\leqslant}\underline{m})\neg\,\mathsf{bew}(y,\ulcorner\alpha\urcorner)$ by C5, since $\nvdash_T \alpha$. This gives $\mathsf{bew}(y,\ulcorner\alpha\urcorner) \vdash_T y > \underline{m}$ by C6. Because of $y > \underline{m} \vdash_T (\exists z{<}y)\,\mathsf{bew}(z,\ulcorner\neg\alpha\urcorner)$ (choose $\underline{m}$ for $z$) we clearly obtain that $\vdash_T \forall y[\mathsf{bew}(y,\ulcorner\alpha\urcorner) \to (\exists z{<}y)\,\mathsf{bew}(z,\ulcorner\neg\alpha\urcorner)] \equiv \neg\,\mathsf{prov}(\ulcorner\alpha\urcorner)$. This confirms (b). Now let (c): $\gamma \equiv_T \neg\,\mathsf{prov}(\ulcorner\gamma\urcorner)$ by (1). The assumption $\vdash_T \gamma$ yields with (a) and (c) the contradiction $\vdash_T \mathsf{prov}(\ulcorner\gamma\urcorner), \neg\,\mathsf{prov}(\ulcorner\gamma\urcorner)$, and the assumption $\vdash_T \neg\gamma$ yields with (b) and (c) the same contradiction. Thus, neither $\vdash_T \gamma$ nor $\vdash_T \neg\gamma$. ◻

$T \subseteq \mathcal{L}^0_{ar}$ is called $\omega$-*incomplete* if there is some $\varphi = \varphi(x)$ such that $\vdash_T \varphi(\underline{n})$ for all $n$ and yet $\nvdash_T \forall x\varphi$. We claim that PA is not only incomplete but $\omega$-incomplete. Let $\gamma \equiv_{\mathsf{PA}} \neg\,\mathsf{bwb}_{\mathsf{PA}}(\ulcorner\gamma\urcorner)$ and $\varphi(x) := \neg\,\mathsf{bew}_{\mathsf{PA}}(x,\ulcorner\gamma\urcorner)$. By Theorem 5.1′, $\nvdash_{\mathsf{PA}} \gamma \equiv_{\mathsf{PA}} \neg\,\mathsf{bwb}_{\mathsf{PA}}(\ulcorner\gamma\urcorner) \equiv \forall x\varphi$, that is, $\nvdash_{\mathsf{PA}} \forall x\varphi$. On the other hand, since $\nvdash_{\mathsf{PA}} \gamma$, we know that $\vdash_{\mathsf{PA}} \varphi(\underline{n})\ (= \neg\,\mathsf{bew}_{\mathsf{PA}}(\underline{n},\ulcorner\gamma\urcorner))$ for all $n$ (Corollary 4.3) which confirms our claim. Note that $\varphi(x)$ is even a $\Pi_1$-formula, which is particularly interesting.

$\alpha \in \mathcal{L}^0$ is said to be *true in* $\mathcal{A}$ if $\mathcal{A} \vDash \alpha$. In particular, $\alpha \in \mathcal{L}^0_{ar}$ is called *true* (more precisely, true in $\mathcal{N}$ or true in reality, as some people like to say) if $\mathcal{N} \vDash \alpha$. If there is some $\tau(x) \in \mathcal{L}$ with a single free variable such that $\mathcal{A} \vDash \alpha \Leftrightarrow \mathcal{A} \vDash \tau(\ulcorner\alpha\urcorner)$, for all $\alpha \in \mathcal{L}^0$, it is said that *truth of* $\mathcal{A}$ *is definable in* $\mathcal{A}$. Clearly, this is equivalent to the representability of $Th\,\mathcal{A}$ in $Th\,\mathcal{A}$. For $\mathcal{A} = \mathcal{N}$, however, such a possibility is excluded by the nonrepresentability lemma. We therefore obtain

**Theorem 5.2 (Tarski's nondefinability theorem).** *The notion of truth in* $\mathcal{N}$ *is not definable in* $\mathcal{N}$; *in other words,* $Th\,\mathcal{N}$ *is not arithmetical.*

In this theorem lies the origin of a highly developed theory of definability in $\mathcal{N}$ (see also **6.7**). The theorem holds correspondingly for every domain of objects $\mathcal{A}$ whose language is arithmetizable and in which the function $\mathsf{sb}_x$ is representable for some variable $x$.

We now turn to undecidability results. First of all we prove the claim in Exercise 1 in **3.5** in a somewhat stronger framework: 'decidable' will now have the precise meaning of 'recursive'.

**Lemma 5.3.** *Every finite extension* $T'$ *of a decidable theory* $T$ *of one and the same (arithmetizable) language* $\mathcal{L}$ *is decidable.*

**Proof.** Suppose $T'$ extends $T$ by $\alpha_0, \ldots, \alpha_n$. Put $\alpha := \bigwedge_{i \leqslant n} \alpha_i$, so that $T' = T + \alpha$. Since $\beta \in T' \Leftrightarrow \alpha \to \beta \in T$, we obtain

$$n \in \dot{T}' \;\Leftrightarrow\; n \in \dot{\mathcal{L}}^0 \;\&\; \dot{\alpha} \stackrel{\sim}{\to} n \in \dot{T}.$$

Now, $\dot{T}$, $\dot{\mathcal{L}}^0$, and $\stackrel{\sim}{\to}$ are recursive. Hence the same applies to $\dot{T}'$. ◻

   That $T'$ belongs to the same language as $T$ is important. A decidable theory $T$ axiomatized by $X \subseteq \mathcal{L}^0$ but considered as a theory in $\mathcal{L}' \supset \mathcal{L}$ *with the same axiom system X* may well be undecidable, due to a higher complexity of the additional tautologies of $\mathcal{L}'$.

   $T_0 \subseteq \mathcal{L}_0$ is called *strongly undecidable* if $T_0$ is consistent and each theory $T \subseteq \mathcal{L}_0$ compatible with $T_0$ (i.e., $T + T_0$ is consistent) is undecidable. Then each $T$ compatible with $T_0$ in a language $\mathcal{L} \supseteq \mathcal{L}_0$ is also undecidable, for otherwise $T \cap \mathcal{L}_0$ would clearly be decidable. If $T_0$ is strongly undecidable then so is every consistent $T_1 \supseteq T_0$, for if $T$ is compatible with $T_1$ then it is also compatible with $T_0$. Moreover, each subtheory of $T_0$ in $\mathcal{L}_0$ is then undecidable, or $T_0$ is *hereditarily undecidable* in the terminology of [TMR]. The weaker a strongly undecidable theory, the wider the scope of applications. This will become plain by means of examples in the next section. The following theorem is the main result from [TMR].

**Theorem 5.4.** Q *is strongly undecidable.*

**Proof.** Let $T \cup Q$ be consistent. Assume $T$ is decidable. Then the same does hold for the finite extension $T' = T + Q$ of $T$; Lemma 5.3. Thus, by Theorem 4.2, $T'$ is representable in itself, which is impossible by the nonrepresentability lemma. ◻

**Theorem 5.5 (Church's undecidability theorem).** *The set* $\mathsf{Taut}_\mathcal{L}$ *of all tautological sentences is undecidable for* $\mathcal{L} \supseteq \mathcal{L}_{ar}$.

**Proof.** $\mathsf{Taut}_\mathcal{L}$ is surely compatible with Q and hence is undecidable by Theorem 5.4. ◻

   This result readily carries over to the language with a single binary relation, as will be shown in the next section, and hence to all expansions of this language. Indeed, it carries over to all languages with the exception of those containing unary predicate symbols only and at most one unary function symbol. For the tautologies of these languages there exist various decision procedures; see [ML, vol. I].

By Theorem 5.4, in particular $Th\mathcal{N}$ is undecidable; likewise is every subtheory of $Th\mathcal{N}$, for instance Peano arithmetic PA and each of its subtheories, as well as all consistent extensions of PA, because these are all compatible with Q. $Th\mathcal{N}$ is not even axiomatizable, since an axiomatizable complete theory is decidable. Further conclusions concerning undecidable theories will be drawn in **6.6**.

Alongside undecidability results concerning formalized theories, numerous special results can also be obtained in a similar manner; for instance negative solutions to word problems of all kinds, and halting problems (see e.g. [Rog] or [Bar, C2]). Of these perhaps the most spectacular was the solution to Hilbert's tenth problem: Does an algorithm exist that for every polynomial $p(\vec{x})$ with integer coefficients decides whether the Diophantine equation $p(\vec{x}) = 0$ has a solution in $\mathbb{Z}$? The answer is no, as Matiyasevich proved in 1970.

We briefly sketch the proof. Note first that it suffices to show that no algorithm exists for the solvability of all Diophantine equations in $\mathbb{N}$. Indeed, by a famous theorem from Lagrange, every natural number is the sum of four squares of integers. Consequently, $p(\vec{x}) = 0$ is solvable in $\mathbb{N}$ iff $p(u_1^2 + v_1^2 + w_1^2 + z_1^2, \ldots, u_n^2 + v_n^2 + w_n^2 + z_n^2) = 0$ is solvable in $\mathbb{Z}$. Thus, if we could decide the solvability of Diophantine equations in $\mathbb{Z}$, then we could solve as well the corresponding problem in $\mathbb{N}$. For the latter notice first of all that the question of solvability of $p(\vec{x}) = 0$ in natural numbers is equivalent to the solvability of a Diophantine equation of $\mathcal{L}_{ar}$ (i.e., an equation $s(\vec{x}) = t(\vec{x})$), by simply bringing all terms of $p(\vec{x})$ preceded by a minus sign "to the other side." Thus, Hilbert's problem is reduced to the question of a decision procedure for the problem $\mathcal{N} \vDash \exists \vec{x} \delta(\vec{x})$, where $\delta(\vec{x})$ runs through all Diophantine equations $s(\vec{x}) = t(\vec{x})$ in $\mathcal{L}_{ar}$.

The negative solution to the last question follows easily from the much further-reaching Theorem 5.6, which establishes a surprising connection between number theory and recursion theory, proved in detail for instance in [Mat]. This theorem is a paradigm of the experience that the solution of certain mathematical questions lead to results whose significance extends way beyond that of an answer to the original question.

**Theorem 5.6.** *An arithmetical predicate $P$ is Diophantine if and only if $P$ is recursively enumerable.*

To give at least an indication of the proof, let the Diophantine predicate $P \subseteq \mathbb{N}^m$ be defined by $P\vec{a} \Leftrightarrow \mathcal{N} \vDash \exists \vec{x} \delta^{\mathcal{N}}(\vec{x}, \vec{a})$, with the equation $\delta(\vec{x}, \vec{y})$, $\vec{y} = (y_1, \ldots, y_m)$. The defining formula for $P$ is $\Sigma_1$, and since $\delta^{\mathcal{N}}(\vec{x}, \vec{a})$ is recursive by Theorem 4.5, $P$ is r.e. This is the trivial direction of the claim. The converse—every r.e. predicate is Diophantine—is too large in scope to be given here. Much tricky inventiveness is used in order to show that certain arithmetical predicates and functions are Diophantine, among them the ternary predicate '$a^b = c$', which for a long time resisted the proof of being Diophantine. Theorem 5.6 yields

**Corollary 5.7.** (a) *Hilbert's tenth problem has a negative answer.*
(b) *For every axiomatizable theory $T \supseteq \mathsf{Q}$, in particular $T = \mathsf{PA}$, there is an unsolvable Diophantine equation whose unsolvability is provable in $T$.*

**Proof.** $bwb_{\mathsf{Q}}$ is r.e. by **6.2**. Hence, by Theorem 5.6, there is a Diophantine equation $\delta(\vec{x}, y)$ such that $bwb_{\mathsf{Q}}(n) \Leftrightarrow \mathcal{N} \vDash \exists \vec{x} \delta(\vec{x}, \underline{n})$. We claim that even for the set $\{\delta(\vec{x}, \underline{n}) \mid n \in \mathbb{N}\}$ of equations it is undecidable whether $\mathcal{N} \vDash \exists \vec{x} \delta(\vec{x}, \underline{n})$. Otherwise, $\{n \in \mathbb{N} \mid \mathcal{N} \vDash \exists \vec{x} \delta(\vec{x}, \underline{n})\}$ and hence also $bwb_{\mathsf{Q}}$ would be recursive. This is a contradiction to Theorem 5.4 and proves (a).
(b): If the unsolvability of every unsolvable Diophantine equation $\delta(\vec{x})$ were provable in $T$, then either $\vdash_T \neg \exists \vec{x} \delta(\vec{x})$ (provided $\delta(\vec{x})$ is unsolvable) or else $\vdash_T \exists \vec{x} \delta(\vec{x})$, for $T$ is $\Sigma_1$-complete. Since the theorems of $T$ are r.e., one would then have a decision procedure for the solvability of Diophantine equations, which contradicts part (a). ∎

Theorem 5.6 can be yet further strengthened; namely, it can be proved within $\mathsf{PA}$. Thus, one obtains the following theorem, whose name stems from **M**atiyasevich, **R**obinson, **D**avis, and **P**utnam, all of whom made significant contributions to the solution of Hilbert's tenth problem. Because of its lengthy proof, we shall not use this theorem, though in fact many things would thereby be simplified.

**MRDP theorem.** *For every $\Sigma_1$-formula $\alpha$ there exists an $\exists$-formula $\varphi$ in $\mathcal{L}_{ar}$ such that $\alpha \equiv_{\mathsf{PA}} \varphi$. Here $\varphi$ is without loss of generality of the form $\exists \vec{x}\, s = t$ with certain $\mathcal{L}_{ar}$-terms $s, t$.*

$\Pi_1$-formulas and -sentences have a corresponding simple representation. A famous example of a $\Pi_1$-sentence is *Goldbach's conjecture*

$$(*) \quad \forall x(2 < x \wedge 2 \mid x \rightarrow (\exists p, q < x)(\mathsf{prim}\, p, q \wedge x = p + q)).$$

(each even number $>2$ is a sum of two primes). $(*)$ represents an example of a $\Pi_1$-sentence in $\mathcal{L}_{ar}^1$ whose truth in $\mathcal{N}$ is still unknown, hence may be independent in PA. Clearly, if $(*)$ is false then its negation is provable, even in Q. But $(*)$ may be true and nevertheless unprovable.

Fermat's conjecture, which has a still longer history and was finally proved at the end of the twentieth century, is the statement

$$(\dagger) \quad (\forall\, x,\, y,\, z \in \mathbb{N}_+)(\forall n{>}2)\, x^n + y^n \neq z^n.$$

This is equivalent to a $\Pi_1$-sentence, because $(a,b) \mapsto a^b$ is not only $\Sigma_1$- but even $\Delta_0$-definable in $\mathcal{N}$, as was noticed in Remark 1 in **6.3**. Hence, $(\dagger)$ is a candidate for a sentence that may be independent in PA.

**Remark.** It would be interesting to discover whether the proof of Fermat's conjecture or a suitable modification of this proof can be carried out in PA. A demonstration that this is not the case would hardly be less spectacular than the solution of the problem itself. However, it seems that the proof can be carried out in a suitable conservative extension of PA (communicated by G. Kreisel). Note also the following: Since PA is $\omega$-incomplete already for $\Pi_1$-formulas (see page 253), it may even be the case that $\vdash_{\mathsf{PA}} (\forall x \forall y \forall z{\neq}0)\, x^{\underline{n}} + y^{\underline{n}}{\neq}z^{\underline{n}}$ for every single $n > 2$, although $(*)$ is not provable in PA. Similarly, it may well be that $\exists p,q\,(\mathsf{prim}\,p, q \wedge 2\underline{n} = p + q)$ is true for each $n > 1$ but $(\dagger)$ is still unprovable. This would be no less sensational than a proof of Goldbach's conjecture itself.

## Exercises

1. Show that an $\omega$-incomplete theory in $\mathcal{L}_{ar}$ has a consistent but $\omega$-inconsistent extension.

2. Suppose $T$ is complete; prove the equivalence of

    (i) $T$ is strongly undecidable,     (ii) $T$ is hereditarily undecidable.

3. A consistent theory $T_0 \subseteq \mathcal{L}_0$ is called *essentially undecidable* if each consistent $T \supseteq T_0$ is undecidable. Show that a finitely axiomatizable theory $T$ is essentially decidable iff $T$ is strongly undecidable.

4. Let $\Delta$ be a finite list containing explicit definitions of new symbols in terms of those occurring in $\mathcal{L}$. Show that if $T$ is decidable then so is $T + \Delta$ (independent of whether all definitions in $\Delta$ are legitimate in $T$; in the worst case $T + \Delta$ is inconsistent).

5. Construct a primitive recursive function $f : \mathbb{N} \to \mathbb{N}$ such that $\mathrm{ran}\, f$ is not recursive (although it is surely recursively enumerable).

## 6.6   Transfer by Interpretation

Interpretability is a powerful method to transfer model-theoretic and other properties, such as undecidability, from one theory to another. Roughly speaking, interpreting a theory $T_0 \subseteq \mathcal{L}_0$ into a theory $T_1 \subseteq \mathcal{L}_1$ means to make the basic notions of $T_0$ understandable in $T_1$ via explicit definitions. 'for all $x$' from $T_0$ is replaced in $T_1$ by 'for all $x \in$ P ', where P is a new unary predicate symbol for the domains of $T_0$-models, i.e., the $T_0$-quantifiers run over the subdomains $\mathsf{P}^\mathcal{A}$ of the domains of the $T_1$-models $\mathcal{A}$. We consider the most important concepts, interpretability from Tarski (also called *relative interpretability*) and interpretability from Rabin, called *model interpretability*. All theories considered in this section are supposed to be consistent.

Let P be a unary predicate symbol not occurring in $T_1$. The formula $\varphi^\mathsf{P}$, the P-*relativized* of a formula $\varphi$, results from $\varphi$ by replacing all subformulas of the form $\forall x\alpha$ by $\forall x(\mathsf{P}\,x \to \alpha)$. A precise definition of $\varphi^\mathsf{P}$ runs by induction: $\varphi^\mathsf{P} = \varphi$ if $\varphi$ is a prime formula, $(\neg\varphi)^\mathsf{P} = \neg\varphi^\mathsf{P}$, $(\varphi\wedge\psi)^\mathsf{P} = \varphi^\mathsf{P}\wedge\psi^\mathsf{P}$, and $(\forall x\varphi)^\mathsf{P} = \forall x(\mathsf{P}\,x \to \varphi^\mathsf{P})$, so that $\varphi^\mathsf{P} = \varphi$ for open $\varphi$. One readily confirms $(\exists x\varphi)^\mathsf{P} \equiv \exists x(\mathsf{P}\,x \wedge \varphi^\mathsf{P})$. The right-side formula shows clearly what relativation is intending to mean. Set $X^\mathsf{P} := \{\alpha^\mathsf{P} \mid \alpha \in X\}$ for $X \subseteq \mathcal{L}_0$.

**Example.** $(\forall x\exists y\, y = \mathsf{S}x)^\mathsf{P} \equiv \forall x(\mathsf{P}\,x \to \exists y(\mathsf{P}\,y \wedge y = \mathsf{S}x)) \equiv \forall x(\mathsf{P}\,x \to \mathsf{P}\,\mathsf{S}x)$. The last equivalence results from $\exists y(\mathsf{P}\,y \wedge y = \mathsf{S}x) \equiv \mathsf{P}\,\mathsf{S}x$, cf. (12) in **2.4**.

**Definition.** $T_0 \subseteq \mathcal{L}_0$ is called *interpretable* in $T_1 \subseteq \mathcal{L}_1$ (where for simplicity we assume that $T_0$ has finite signature) if there is a list $\Delta$ of explicit definitions legitimate in $T_1$ of the symbols of $T_0$ not occurring in $T_1$ and of a new unary predicate symbol P such that $T_0^\mathsf{P} \subseteq T_1 + \Delta$, *the definitorial extension of $T_1$ by $\Delta$.*

This definition expresses only that all notions of $T_0$ "are understood" in $T_1$, and what is provable in $T_0$ is also provable in $T_1$. Examples will be given later. The theory $T + \Delta$ ($T \subseteq \mathcal{L}_1$) will henceforth be denoted by $T^\Delta$, and its language by $\mathcal{L}_1^\Delta$. Interpretability generalizes the notion of a subtheory: If $T_0 \subseteq T_1$ then $T_0$ is trivially interpretable in $T_1$, i.e., only the *trivial relativation* $\mathsf{P}\,x \leftrightarrow x = x$ belongs to $\Delta$. In this case $\alpha^\mathsf{P} \equiv \alpha$.

Let $CA$ denote the set of the so-called *closure axioms*

$$\exists x\,\mathsf{P}\,x,\quad \mathsf{P}\,c,\quad \forall\vec{x}\,(\textstyle\bigwedge_{i=i}^{n}\mathsf{P}\,x_i \to \mathsf{P}\,f\vec{x}) \qquad (c, f \in L_0).$$

These are equivalent to $(\exists x\, x = x)^{\mathrm{P}}$, $(\exists x\, x = c)^{\mathrm{P}}$, and $(\forall \vec{x}\, \exists y\, y = f\vec{x})^{\mathrm{P}}$, respectively. Thus, $CA$ is up to equivalence a set of the form $F^{\mathrm{P}}$ for some finite set $F$ of $\mathcal{L}_0$-tautologies, so that $CA \subseteq T_0^{\mathrm{P}}$ for *each* theory $T_0 \subseteq \mathcal{L}_0$. The sentences of $CA$ guarantee that for a given $\mathcal{L}_0^{\Delta}$-structure $\mathcal{B} \vDash \Delta$ there is a well-defined $\mathcal{L}_0$-structure $\mathcal{A}$ whose domain is $A = \mathsf{P}^{\mathcal{B}}$. The relations and operations of $\mathcal{A}$ are the ones defined by $\Delta$ but restricted to $A$. This structure $\mathcal{A}$ will be denoted by $\mathcal{B}_\Delta$. It is a substructure of the $\mathcal{L}_0$-reduct of $\mathcal{B}$, whose role will become clear in the next lemma.

**Lemma 6.1.** *Let $\mathcal{B} \vDash CA$. Then $\mathcal{B}_\Delta \vDash \alpha \Leftrightarrow \mathcal{B} \vDash \alpha^{\mathrm{P}}$, for all sentences $\alpha$ of the language $\mathcal{L}_0$.*

**Proof.** $\mathcal{A} := \mathcal{B}_\Delta$ is an $\mathcal{L}_0$-structure. Claim: $(\mathcal{A}, w) \vDash \varphi \Leftrightarrow (\mathcal{B}, w) \vDash \varphi^{\mathrm{P}}$, for any $w \colon Var \to A$. This proves the lemma, since $\alpha$ is a sentence. We prove the claim by induction on $\varphi \in \mathcal{L}_0$. It is clear for prime formulas $\pi$ since $\alpha^{\mathrm{P}} = \pi$. The induction steps for $\wedge, \neg$ proceed without difficulty, and the one for $\forall$ is obtained as follows:

$$(\mathcal{A}, w) \vDash \forall x \varphi \Leftrightarrow (\mathcal{A}, w_x^a) \vDash \varphi \text{ for all } a \in A$$
$$\Leftrightarrow (\mathcal{B}, w_x^a) \vDash \varphi^{\mathrm{P}} \text{ for all } a \in A \quad \text{(induction hypothesis)}$$
$$\Leftrightarrow (\mathcal{B}, w_x^a) \vDash \mathsf{P}\, x \to \varphi^{\mathrm{P}}, \text{ for all } a \in B \quad \text{(because } \mathsf{P}^{\mathcal{B}} = A\text{)}$$
$$\Leftrightarrow (\mathcal{B}, w) \vDash \forall x (\mathsf{P}\, x \to \varphi^{\mathrm{P}}) = (\forall x \varphi)^{\mathrm{P}}. \quad \blacksquare$$

**Remark 1.** If $T_0$ is axiomatized by $X_0$ then in the definition of interpretability it suffices to require just $X_0^{\mathrm{P}} \cup CA \subseteq T_1^{\Delta}$ instead of $T_0^{\mathrm{P}} \subseteq T_1^{\Delta}$. That is, we have only to check $\alpha^{\mathrm{P}} \in T_1^{\Delta}$ for the axioms $\alpha$ of $T_0$, and $CA \subseteq T_1^{\Delta}$ (cf. the example below). This fact is highly important. It follows immediately from
$$(*) \qquad S \vdash \alpha \ \Rightarrow \ S^{\mathrm{P}} \cup CA \vdash \alpha^{\mathrm{P}} \quad (S \cup \{\alpha\} \subseteq \mathcal{L}_0^0).$$
For proving $(*)$ let $S \vdash \alpha$ and $\mathcal{B} \vDash S^{\mathrm{P}} \cup CA$. Then $\mathcal{B}_\Delta \vDash S$ by the lemma. Thus, $\mathcal{B}_\Delta \vDash \alpha$ because $S \vdash \alpha$, and so $\mathcal{B} \vDash \alpha^{\mathrm{P}}$. Since $\mathcal{B} \vDash S^{\mathrm{P}} \cup CA$ was arbitrary, we get $S^{\mathrm{P}} \cup CA \vdash \alpha^{\mathrm{P}}$.

**Theorem 6.2.** *Let $T_0$ be interpretable in $T_1$. If $T_0$ is strongly undecidable so is $T_1$.*

**Proof.** Let $T \subseteq \mathcal{L}_1$ be compatible with $T_1$. Then $T + T_1$ is consistent and so is $(T + T_1)^{\Delta}$. Now, $S := \{ \alpha \in \mathcal{L}_0^0 \mid \alpha^{\mathrm{P}} \in T^{\Delta} + CA \}$ is a theory, for $S^{\mathrm{P}} \subseteq T^{\Delta} + CA$ and $(*)$ yield $S \vdash \alpha \Rightarrow T^{\Delta} \cup CA \vdash \alpha^{\mathrm{P}} \Rightarrow \alpha \in S$. Let $\mathcal{B} \vDash (T + T_1)^{\Delta} \supseteq T_0^{\mathrm{P}} \cup CA \cup S^{\mathrm{P}}$. Thus, $\mathcal{B}_\Delta \vDash T_0, S$ by Lemma 6.1; hence $S$ is compatible with $T_0$ and so undecidable. If $T$ were decidable, then so would be $T^{\Delta}$ (Exercise 3 in **6.5**). Hence also $T^{\Delta} + CA$ (Lemma 5.3), and so clearly $S$. This is a contradiction. $\blacksquare$

**Example.** Q is interpretable in the theory $T_d$ of discretely ordered rings $\mathcal{R} = (R, 0, +, \times, <)$. These have a smallest positive element $e$, defined by $x = e \leftrightarrow 0 < x \wedge \forall y(0 < y \rightarrow x \leqslant y)$, that need not be a unit element of $\mathcal{R}$. Ring multiplication is denoted by $\times$, to distinguish it from multiplication in Q. Here are the definitions for P, S, $\cdot$ $(0, +$ remain unaltered):

$$P\,x \leftrightarrow x \geqslant 0 \wedge x \times e = e \times x \wedge \forall y \exists z\, z \times e = y \times x,$$

$$y = S\,x \leftrightarrow y = x + e, \quad z = x \cdot y \leftrightarrow z \times e = x \times y \vee \forall u(u \times e \neq x \times y \wedge z = x).$$

With some patience, all P-relativized Q-axioms can be proved in $T_d^\Delta$, with the list $\Delta$ of the above definitions. Thus, by Remark 1, Q is interpretable in $T_d$, and $T_d$ is strongly undecidable according to Theorem 6.2.

While Q is not directly interpretable in the theory $T_F$ of fields, it is in a certain finite extension of $T_F$, whereby $T_F$ is shown to be undecidable (Julia Robinson). The same also holds for the theory of groups $T_G$ [TMR]. However, none of these theories is strongly undecidable.

Q and also PA are interpretable in ZFC, as is nearly every other theory. Let $P\,x \leftrightarrow x \in \omega$, and define S, $+$, $\cdot$ within ZFC such that their restrictions to $\omega$ coincide with the usual operations. In particular, S is defined by $y = S\,x \leftrightarrow y = x \cup \{x\}$. This immediately yields the incompleteness and the undecidability of ZFC, assuming of course its consistency. Q is also interpretable in weak subtheories of ZFC, e.g., in the *Tarski fragment* TF, by which we mean the theory in $\mathcal{L}_\in$ with the following three axioms.[8] Hence, like Q, the theory TF is strongly undecidable.

$$\exists x \forall y\, y \notin x \qquad\qquad (\emptyset \text{ exists}),$$
$$\forall x \forall y(\forall z(z \in x \leftrightarrow z \in y) \rightarrow x = y) \quad (\text{extensionality}),$$
$$\forall x \forall y \exists z \forall u(u \in z \leftrightarrow u \in x \vee u = y) \quad (x \cup \{y\} \text{ exists}).$$

In particular, the set of tautologies in a binary relation is undecidable, even without identity in the language; for $=$ can conservatively be eliminated from $T_\in$ by means of $x = y \leftrightarrow \forall z(z \in x \leftrightarrow z \in y)$. Q is surely interpretable in $Th\mathcal{N}$, and $Th\mathcal{N}$ in turn in $Th\mathcal{Z}$ with $\mathcal{Z} = (\mathbb{Z}, 0, 1, +, \cdot)$. This is a consequence of Lagrange's theorem. Hence, $Th\mathcal{Z}$ is strongly undecidable, and thus every subtheory is undecidable, e.g., the theory of commutative rings. $Th\mathcal{N}$ and $Th\mathcal{Z}$ have the same degree of complexity, because $Th\mathcal{Z}$ is (in various ways) interpretable in $Th\mathcal{N}$; Exercise 3.

---

[8] Claimed in [TMR, p. 34]. The lengthy proof is presented in [Mo, pp. 283–290].

**Remark 2.** Remarkable is the mutual interpretability of PA and $\mathsf{ZFC_{fin}}$, the theory of (hereditarily) finite sets. It arises from ZFC by replacing AI by the schema Sfin: $\varphi(\emptyset) \wedge \forall xy(\varphi(x) \to \varphi(x \cup \{y\})) \to \forall x \varphi(x))$ and AF by the schema of foundation Sfnd: $\exists x\varphi \to \exists(\varphi \wedge (\forall y \in x)\neg\varphi\frac{y}{x})$; see [De] or [Ra4]. A surprisingly simple interpretation of $\mathsf{ZFC_{fin}}$ in PA was given by Ackermann in [Ac]: Each natural number represents a finite set in PA (relativization is trivial), and the $\in$-relation is defined by $i \in a :\Leftrightarrow \mathsf{b}_i a = 1$.[9] For instance, 0 represents $\emptyset$, since $\mathsf{b}_i 0 = 0$ for all $i$. 1, 2, and 3 (more precisely, $\underline{1}$, $\underline{2}$, and $\underline{3}$) represent $\{\emptyset\}$, $\{1\}$, and $\{0,1\}$ ($= \{\emptyset, \{\emptyset\}\}$), respectively. To see that 3 represents $\{0,1\}$, notice that $\mathsf{b}_0 3 = \mathsf{b}_1 3 = 1$ and $\mathsf{b}_i 3 = 0$ for $i > 1$. See also [Fi] for more details.

We now describe a related notion of interpretability. For simplicity, we omit some details. Let $\boldsymbol{K}_0$ and $\boldsymbol{K}$ be nonempty classes of $\mathcal{L}_0$- and $\mathcal{L}$-structures, respectively. Further, let $\Delta$ be a list of definitions of the $\mathcal{L}_0$-symbols and a predicate symbol P, and let $\mathcal{L}^\Delta$, $CA$, and $\mathcal{B}_\Delta$ for $\mathcal{B} \vDash CA$ be defined as above. $\mathcal{A}^\Delta$ denotes the expansion of $\mathcal{A} \in \boldsymbol{K}$ in $\mathcal{L}^\Delta$ according to $\Delta$ (the $\Delta$-*expansion* of $\mathcal{A}$). If $\gamma \in \mathcal{L}^\Delta$ is a sentence, let $\boldsymbol{K}_\gamma$ denote the class of all $\mathcal{A}^\Delta$ for $\mathcal{A} \in \boldsymbol{K}$ such that $\mathcal{A}^\Delta \vDash \gamma$.

**Definition.** $\boldsymbol{K}_0$ (or $Th\,\boldsymbol{K}_0$) is called *model interpretable* in $\boldsymbol{K}$ (or in $Th\,\boldsymbol{K}$ respectively) if for suitable $\Delta$ and a suitable sentence $\gamma \in \mathcal{L}^\Delta$,

(1)  $\boldsymbol{K}_\gamma \vDash CA$ and $\mathcal{B}_\Delta \in \boldsymbol{K}_0$ for each $\mathcal{B} \in \boldsymbol{K}_\gamma$,

(2)  For every $\mathcal{A} \in \boldsymbol{K}_0$ there is some $\mathcal{B} \in \boldsymbol{K}_\gamma$ such that $\mathcal{A} \simeq \mathcal{B}_\Delta$.

Clearly, we can construct as in **2.6** for each sentence $\alpha \in \mathcal{L}^\Delta$ a reduced sentence $\alpha^{rd} \in \mathcal{L}$ such that

(3)  $\mathcal{A}^\Delta \vDash \alpha \Leftrightarrow \mathcal{A} \vDash \alpha^{rd}$, for all $\mathcal{A} \in \boldsymbol{K}$.

**Theorem 6.3.** *Let $\boldsymbol{K}_0$ be model interpretable in $\boldsymbol{K}$. If $Th\,\boldsymbol{K}_0$ is undecidable then so too is $Th\,\boldsymbol{K}$.*

**Proof.** Put $\hat{\alpha} := (\gamma \to \alpha^\mathsf{P})^{rd} = \gamma^{rd} \to (\alpha^\mathsf{P})^{rd}$ for $\alpha \in \mathcal{L}_0^0$. It suffices to prove $(*): \boldsymbol{K}_0 \vDash \alpha \Leftrightarrow \boldsymbol{K} \vDash \hat{\alpha}$, because a decision procedure for $Th\,\boldsymbol{K}$ then clearly extends to $Th\,\boldsymbol{K}_0$. $\Rightarrow$: Let $\boldsymbol{K}_0 \vDash \alpha$, $\mathcal{A} \in \boldsymbol{K}$, $\mathcal{A}^\Delta \vDash \gamma$ so that $\mathcal{A} \vDash \gamma^{rd}$ by (3), and $\mathcal{B} := \mathcal{A}^\Delta \in \boldsymbol{K}_\gamma$. By (1), $\mathcal{B}_\Delta \in \boldsymbol{K}_0$. Thus, $\mathcal{B}_\Delta \vDash \alpha$. Hence $\mathcal{B} \vDash \alpha^\mathsf{P}$ by Lemma 6.1, and so $\mathcal{A} \vDash (\alpha^\mathsf{P})^{rd}$. This confirms $\mathcal{A} \vDash \hat{\alpha}$ for all $\mathcal{A} \in \boldsymbol{K}$, in other words, $\boldsymbol{K} \vDash \hat{\alpha}$. $\Leftarrow$: Assume that $\boldsymbol{K}_0 \nvDash \alpha$, say $\mathcal{A} \nvDash \alpha$. Choose some $\mathcal{B} \in \boldsymbol{K}_\gamma$ according to (2), so that $\mathcal{B}_\Delta \nvDash \alpha$. Then $\mathcal{B} \nvDash \gamma \to \alpha^\mathsf{P}$; hence $\boldsymbol{K} \nvDash \hat{\alpha}$. This confirms $(*)$. $\quad\square$

---

[9]$a \mapsto \mathsf{b}_i a$ (the $i$th binary digit of $a$; see **6.1**) is a p.r. function and hence explicitly definable in PA, as is every p.r. function according to Theorem 7.1.1.

**Example.** The class $K_0$ of all graphs $(A, R)$ is model interpretable in the class $K$ of simple graphs $(B, S)$, i.e. $S$ is irreflexive and symmetric. The figure below shows some $A \in K_0$ with $aRa$, $aRb$, $bRa$, $bRc$, and on the right some $B \in K$, called the *encoding structure of* $A$ because it completely describes $A$ in order to satisfy (2). Roughly put, a set $N$ of new points is adjoined to $A$ so that $B = A \cup N$. Since the edges in $B$ are undirected, we need new points for coding the edge directions of $A$.

The "old" points in $B$, the larger dots in the figure, are the ones from $A = \mathsf{P}^B$. Such a point neighbors three or two endpoints (i.e., points in which only one $S$-edge ends), depending on whether the point is reflexive in $A$ or not (only $a$ is reflexive). Informally, the definition for $R$ in $B$ reads as follows: "$xRy$ iff $x, y \in \mathsf{P}^B$ and either $x = y$ and $x$ neighbors three endpoints, or there exists exactly one new point $z$ such that $xSzSy$, or there are exactly two new points $u, v$ such that $xSuSvSy$ and $uSy$." $\gamma$ is rendered informally into "$\exists x\, \mathsf{P}\,x$ and all new points are either endpoints or neighbor precisely two old points or one old and one new point."

In the example, $Th\,K_0$ is the logical theory of a binary relation, already established as undecidable. Accordingly the theory of all simple graphs is undecidable. The latter can be used to show, for instance, that the theory SL of semilattices is undecidable. By Theorem 5.4 the same then follows for the theory SG of semigroups, for SL is a finite extension of SG. In order to show that $Th\,K_0$ is model interpretable in SG,

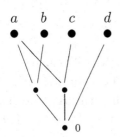

it suffices to provide, similarly to the last example, for a simple graph $(A, S)$ the encoding semilattice $(B, \circ)$. The figure on the left shows the ordering diagram of $B$ for $A = \{a, b, c, d\}$ and $S = \{\{a, b\}, \{a, c\}\}$; here $S$ is understood as a set of edges. The old points are precisely the maximal points of $B$. By construction, $B$ has a smallest element 0 and is of depth 3, that is, there are at most three consecutive points in $B$ with respect to $<$. This must now be expressed by the sentence $\gamma$ required in the definition of model interpretability.

The theory of *finite* simple graphs with or without some additional feature (for instance planarity) is undecidable; see e.g. [RZ]. The above construction shows that the undecidable theory of finite simple graphs is model interpretable in the theory of finite semilattices, which hence is undecidable. This clearly implies the undecidability of the theory FSG of finite semigroups. Setting an element on top of the maximal elements in the last figure results in the order diagram of a finite lattice, so that the theory of finite lattices turns out to be undecidable. The same holds for the theory FPO of finite partial orders because for the description of $(A, S)$ only the partial order of $B$ is relevant.

**Remark 3.** Somewhat more mathematics is required to prove the undecidability of the theory FDL of all *finite distributive lattices*. The previous figure illustrates that also FPO is undecidable. But FPO is model interpretable in FDL, in that one identifies the elements of $g$ with the $\cap$-irreducible elements of the lattice, $\mathcal{A}$ say. Here we need to know that $\mathcal{A}$'s structure is completely determined by the partial order of its irreducible elements and that this order can be given completely arbitrarily.

Positive results are also transferable. For instance, the (logical) theory of a unary function is interpretable in the first-order theory of (undirected) trees [KR], and with the latter the former is also decidable. The decidability of the theory of a single unary function was first proved by Ehrenfeucht with a different method. We mention that the theory of two or more unary functions is undecidable. Decidability of the theory of simple trees also follows from the decidability of the second-order monadic theory of binary trees [Bar, C3], a very strong result with an immense scope of applications. One of these applications is a simple proof of decidability of all modal systems considered in Chapter **7** (see e.g. [Ga]).

### Exercises

1. Show that if a theory $T_0$ is essentially undecidable and interpretable in $T_1 \subseteq \mathcal{L}_1$ then $T_1$ is essentially undecidable as well.

2. Show (informally) that PA is interpretable not only in ZFC but also in $\mathsf{ZFC}_{\mathrm{fin}}$. (*Attention*: $\omega$ is no longer a set in $\mathsf{ZFC}_{\mathrm{fin}}$.)

3. Show in detail that $Th(\mathbb{Z}, 0, 1, +, \cdot, \leqslant)$ is interpretable in $Th\mathcal{N}$.

4. Prove that all axioms of $\mathsf{ZFC}_{\mathrm{fin}}$ derive from TF + Sfin + Sfnd. This makes interpretability of $\mathsf{ZFC}_{\mathrm{fin}}$ in PA an easy task.

## 6.7   The Arithmetical Hierarchy

We now add a little more on the complexity of predicates of $\mathbb{N}$ including subsets of $\mathbb{N}$. The set of the Gödel numbers of all sentences valid in $\mathcal{N}$ is an example of a rather simply defined nonarithmetical subset of $\mathbb{N}$; by Theorem 5.2 it has no definition in $\mathcal{L}_{ar}$.[10] However, relatively simply defined arithmetical sets and predicates may be recursion-theoretically highly complicated. It is useful to classify these according to the complexity of the defining formulas. The result is the *arithmetical hierarchy*, also called the *first-order Kleene–Mostowski hierarchy*. The following definition builds upon the one in **6.3** of the $\Sigma_1$- and $\Pi_1$-formulas and the $\Sigma_1$-, $\Pi_1$-, and $\Delta_1$-predicates defined by these.

**Definition.** A $\Sigma_{n+1}$-formula is a formula of the form $\exists \vec{x}\alpha(\vec{x},\vec{y})$, where $\alpha$ is a $\Pi_n$-formula from $\mathcal{L}_{ar}$; analogously, we call $\forall \vec{x}\beta(\vec{x},\vec{y})$ a $\Pi_{n+1}$-formula if $\beta$ is a $\Sigma_n$-formula. Here $\vec{x},\vec{y}$ are arbitrary tuples of variables. A $\Sigma_n$-*predicate* (resp. $\Pi_n$-*predicate*) is an arithmetical predicate $P$ defined in $\mathcal{N}$ by a $\Sigma_n$-formula (resp. $\Pi_n$-formula). If $P$ is both $\Sigma_n$ and $\Pi_n$ (i.e., a $\Sigma_n$- and $\Pi_n$-predicate) then we say that $P$ is a $\Delta_n$-predicate, or $P$ *is* $\Delta_n$ for short. We denote by $\Sigma_n$, $\Pi_n$, and $\Delta_n$ the sets of the $\Sigma_n$-, $\Pi_n$- and $\Delta_n$-predicates, respectively. In addition, $\Sigma_0 := \Pi_0 := \Delta_0$.

According to this definition, a $\Sigma_n$-formula is a prenex formula $\varphi$ with $n$ alternating blocks of quantifiers, the first of which is an $\exists$-block, which may also be empty. $\varphi$'s kernel is $\Delta_0$. Clearly, each $\varphi \in \mathcal{L}_{ar}$ is equivalent to a $\Sigma_n$- or $\Pi_n$-formula for a suitable $n$, for $\varphi$ can be brought into prenex normal form and the quantifiers can be grouped into blocks of the same quantifiers. Obviously, $\Delta_n \subseteq \Sigma_n, \Pi_n$. When considering the hierarchy it is convenient to have $\Sigma_n$- and $\Pi_n$-formulas closed under equivalence in $\mathcal{N}$. Hence, we say that $\alpha$ *is* $\Sigma_n$ or $\Pi_n$ to indicate that $\alpha$ is equivalent to an *original* $\Sigma_n$- or $\Pi_n$-formula, respectively. Note that since $\exists \vec{x}\varphi \equiv \forall \vec{x}\varphi \equiv \varphi$ in case $\mathrm{var}\,\vec{x} \cap \mathrm{var}\,\varphi = \emptyset$, every $\Sigma_n$- or $\Pi_n$-formula is also both $\Sigma_{n+1}$ and $\Pi_{n+1}$. Therefore $\Sigma_n, \Pi_n \subseteq \Delta_{n+1}$. This yields the following inclusion diagram, where all the inclusions, indicated by lines, are proper:

---

[10] *Th*$\mathcal{N}$ is definable only in second-order arithmetic, which along with variables for numbers has variables for sets of natural numbers. However, the set of sentences $\alpha$ from *Th*$\mathcal{N}$ with bounded quantifier rank, i.e. qr $\alpha \leqslant n$, *is* definable for each $n$. In this sense, *Th*$\mathcal{N}$ has an "approximate" elementary definition.

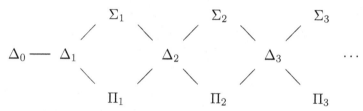

We have already come across $\Sigma_1$-, $\Pi_1$-, and $\Delta_1$-predicates; for instance, the solvability claims of Diophantine equations are $\Sigma_1$, and the unsolvability claims are $\Pi_1$. Below we provide an example of a $\Pi_2$-predicate. It is also convenient to say that $\Sigma_n$- and $\Pi_n$-sentences define $0$-*ary* $\Sigma_n$- and $\Pi_n$-predicates, respectively. In this sense the consistency of PA (in arithmetical terms $\neg\, bwb(\ulcorner\emptyset\neq\emptyset\urcorner))$ is $\Pi_1$, the incompleteness of PA is $\Sigma_2$, and the $\omega$-consistency is $\Pi_3$ (Exercise 3). The hierarchy serves various purposes. More recent investigations have considered also $\Delta_0$- or $\Sigma_n$- or $\Pi_n$-*induction*. Here the schema IS is restricted to the corresponding class of formulas. An example is $I\Delta_0$ mentioned on page 239.

As already shown in **6.4**, the $\Sigma_1$-predicates are the recursively enumerable ones, the $\Pi_1$-predicates their complements, and the $\Delta_1$-predicates are exactly the recursive predicates, which are the ones whose complements are r.e. as well. Thus, we are provided with a purely recursion-theoretic way of regarding $\Sigma_1$, $\Pi_1$, and $\Delta_1$. This underscores the importance of the arithmetical hierarchy, which is fairly stable with respect to minor changes in the definition of $\Delta_0$.

In view of Theorem 5.6 one could begin, for instance, with a $\Delta_0$ consisting of all polynomially (or equivalently, quantifier-free) definable relations. In some presentations, a system of formulas is effectively enumerated (and denoted by $\Delta_0$) that define exactly the p.r. predicates in $\mathcal{N}$. Section **7.1** will indicate how such a system can be defined. Between these and the $\Delta_0$-formulas (which themselves may still be classified) lie many r.e. sets of formulas that define computable functions significant in both the theory and practice of computability, e.g., the elementary functions mentioned in the introduction to this chapter.

However, by Remark 2 in **6.4** we know that there exists no effectively enumerable system of formulas in $\mathcal{L}_{ar}$ through which all recursive, or equivalently all $\Delta_1$-predicates, are defined, so that the definition of the arithmetical hierarchy cannot start in a feasible manner with a representative "set of $\Delta_1$-formulas."

**Remark.** We mention that the first-order arithmetical hierarchy considered so far extends in a natural way to the second-order arithmetic. Also this extended hierarchy is closely related to recursion theory (see e.g. [Shoe]). A treatment lies outside the scope of this book.

Similarly to the case $n = 1$, one readily shows that a conjunction or disjunction of $\Sigma_n$-formulas is equivalent to some other $\Sigma_n$-formula; likewise for $\Pi_n$-formulas. The negation of a $\Sigma_n$-formula is equivalent to a $\Pi_n$-formula, and vice versa; this is certainly correct for $n = 1$, which initiates an easy induction on $n$. The complement of a $\Sigma_n$-predicate is therefore a $\Pi_n$-predicate, and vice versa. From this it easily follows that $\Delta_n$ is closed under all the mentioned operations, including complementation.

By "compression of quantifiers," the idea of which was illustrated in Exercise 1 in **6.3**, one obtains a somewhat simpler presentation of the quantifier blocks. The $\exists$- and $\forall$-blocks can each be collapsed into one quantifier. This procedure is fairly easy, provided we are dealing with equivalence in $\mathcal{N}$ as is the case here, and not in a possibly too weakly axiomatized theory over $\mathcal{N}$ (in fact, PA would suffice for the proof):

**Theorem 7.1.** *A (proper) $\Sigma_n$-predicate is definable by a prenex formula* $\exists x_1 \forall x_2 \cdots Q_n x_n \alpha$ *with a $\Delta_0$-formula $\alpha$, where $Q_n$ is either the $\forall$- or $\exists$-quantifier, depending on whether $n$ is even or odd. Similarly, a $\Pi_n$-predicate is definable by a formula* $\forall x_1 \exists x_2 \cdots Q_n x_n \alpha$.

**Proof** by (simultaneous) induction on $n$. Exercise 1 in **6.3** formulates the case for $\Sigma_1$- and for $\Pi_1$-predicates. Assume that this is the case for $n$ and let $\exists \vec{x} \alpha$ be the defining formula of a $\Sigma_{n+1}$-predicate, where $\alpha$ defines a $\Pi_n$-predicate and $\exists \vec{x}$ is a block of length $m \geqslant 1$. By using the (defining $\Delta_0$-formula of the) pairing function, $\exists \vec{x}$ can be compressed stepwise to a single $\exists$-quantifier $\exists x$. The arising bounded quantifiers commute with the following $\forall$-block (Exercise 1). The case $m = 0$ can also be included in the argument, using a "vacuous quantifier" $\exists x$ (i.e., $x \notin \mathrm{var}\,\alpha$). The $\Pi_{n+1}$-formulas are treated completely analogously. $\blacksquare$

It is quite often a nontrivial task to determine a well-defined predicate's exact position in the arithmetical hierarchy, or better, like every fastidious game, it requires sufficient training. In the example below, we consider a set that is neither recursive nor r.e. For the sake of simplicity, we apply Church's thesis in one place, although it can be eliminated using a little recursion theory, as was demonstrated previously in the proof of Theorem 4.4. The example is also a good preparation for **7.6**.

**Example.** Let $\mathcal{L}_r$ denote the set of the $\alpha \in \mathcal{L}^1_{ar}$ that represent in Q recursive subsets of $\mathbb{N}$. Thus, the $\alpha \in \mathcal{L}_r$ have at most one free variable, namely the first one. For instance, all $\Delta_0$-formulas in $\mathcal{L}^1_{ar}$ belong to this $\mathcal{L}_r$. Since $\mathbb{N}$ and $\emptyset$ are recursive and $\mathcal{L}^0_{ar} \subseteq \mathcal{L}^1_{ar}$, all members of the set $Q^* := Q \cup \{\alpha \mid \neg\alpha \in Q\}$ also belong to $\mathcal{L}_r$, because each $\alpha \in Q$ trivially represents $\mathbb{N}$, and each $\alpha$ with $\neg\alpha \in Q$ represents $\emptyset$. Conversely, each closed formula of $\mathcal{L}_r$ belongs to $Q^*$. Obviously then, $Q^* = \mathcal{L}_r \cap \mathcal{L}^0_{ar}$.

We now show that $\mathcal{L}_r$ is arithmetical; more precisely, it is $\Pi_2$, and indeed properly $\Pi_2$, that is, neither $\Sigma_1$ nor $\Pi_1$. By definition,

$$\alpha \in \mathcal{L}_r \Leftrightarrow \alpha \in \mathcal{L}^1_{ar} \ \& \ \forall n \exists \Phi [\Phi \text{ is a proof for } \alpha(\underline{n}) \text{ or for } \neg\alpha(\underline{n})].$$

This equivalence readily yields a definition of $\mathcal{L}_r$ (more exactly, of $\dot{\mathcal{L}}_r$) by a $\Pi_2$-formula $\varphi(x)$ (that is, a $\forall$-formula $\forall x\alpha$ such that $\alpha$ is $\Sigma_1$). Let the p.r. predicate '$a \in \mathcal{L}^1_{ar}$' be $\Sigma_1$-defined by the formula $\lambda_1(x)$. With $\text{sb} = \text{sb}_{v_0}$, we then set

$$\varphi(x) := \lambda_1(x) \wedge \forall y \exists u [\text{bew}_Q(u, \text{sb}(x, y)) \vee \text{bew}_Q(u, \dot{\neg}\, \text{sb}(u, y))].$$

More precisely, $\varphi$ should be the reduced in $\mathcal{L}_{ar}$ after eliminating the occurring p.r. function terms using more $\exists$-quantifiers inside the brackets. Thus, $\varphi$ describes a $\Pi_2$-formula, that is, $\mathcal{L}_r$ is $\Pi_2$. It is not $\Sigma_1$, because $\mathcal{L}_r$ is not r.e. by Remark 2 in **6.4**, nor is it $\Pi_1$. Indeed, assume this were the case; then $Q^* = \mathcal{L}_r \cap \mathcal{L}^0_{ar}$ would also be $\Pi_1$, for $\mathcal{L}^0_{ar}$ is $\Delta_1$. Now $Q^*$ is certainly r.e. and thus $\Sigma_1$, and so by Theorem 4.5, $Q^*$ would be recursive. But then we obtain a decision procedure for Q (hence a contradiction) as follows: Let $\alpha \in \mathcal{L}^0_{ar}$ be given. If $\alpha \notin Q^*$ then also $\alpha \notin Q$; if $\alpha \in Q^*$, we turn on the enumeration machine for Q and wait until either $\alpha$ or $\neg\alpha$ appears. This obviously is a decision procedure for Q.

**Special $\Sigma_1$-formulas.** We end this chapter with a result useful for proving the $\Sigma_1$-completeness of PA *inside* PA in **7.2**, the so-called *provable $\Sigma_1$-completeness*. It will be shown that the $\Sigma_1$-predicates are definable without reference to $\Delta_0$, using special $\Sigma_1$-formulas. To this end somewhat stronger axioms are considered than those of Q, namely the axioms of the theory N presented in **6.3**. All axioms of N are derivable in PA, as was pointed out already on page 235.

**Definition.** *Special* $\Sigma_1$-*formulas are defined as follows:*

(a) $Sx = y$, $x + y = z$, and $x \cdot y = z$ are special $\Sigma_1$-formulas, where $x, y, z$ denote distinct variables (the special prime formula condition);

(b) if $\alpha, \beta$ are special $\Sigma_1$-formulas then so too are $\alpha \wedge \beta$, $\alpha \vee \beta$, $\alpha \frac{0}{x}$, and $\alpha \frac{y}{x}$, where $x, y$ are distinct and not in $bnd\,\alpha$ (*prime-term substitution*), as well as $\exists x \alpha$ and $(\forall x < y)\alpha$ for $y \notin var\,\alpha$.

**Theorem 7.2.** *Every original* $\Sigma_1$-*formula is equivalent to a special* $\Sigma_1$-*formula in the theory* N, *thus in* PA *and a fortiori in* $\mathcal{N}$.

**Proof.** It suffices to verify the claim for all $\Delta_0$-formulas, since the set of special $\Sigma_1$-formulas is closed under $\exists$-quantification. Since

$$s = t \equiv \exists x (x = s \wedge x = t) \text{ with } x \notin var\,s, t,$$

it is enough to consider prime formulas of the form $x = t$. For prime terms $t$ this clearly follows from $x = 0 \equiv (x = y)\frac{0}{y}$ and $x = y \equiv_N (x + z = y)\frac{0}{z}$. The induction steps on the operations $S, +$ are obtained as follows:

$$x = St \equiv \exists y (x = Sy \wedge y = t), \quad x = s + t \equiv \exists y \exists z (x = y + z \wedge y = s \wedge z = t),$$

and similarly for $\cdot$. The claim holds for all literals because

$$s \neq t \equiv \exists y \exists z (x \neq y \wedge x = s \wedge y = t),$$
$$x \neq y \equiv_N \exists u \exists z (Su = z \wedge (x + z = y \vee y + z = x)).$$

By Exercise 4 in **6.3** we need only carry out induction on $\wedge, \vee, (\forall x \leqslant t)$ and $(\exists x \leqslant t)$. For $\wedge, \vee$ this is clear. For the remainder note that $(\forall x \leqslant t)\alpha$ and $(\exists x \leqslant t)\alpha$ are N-equivalent respectively to $\exists y (y = t \wedge (\forall x < y)\alpha \wedge \alpha \frac{y}{x})$ and $\exists x \exists y \exists z (x + y = z \wedge z = t \wedge \alpha)$. ◻

## Exercises

1. Show that $\Sigma_n$ and $\Pi_n$ and hence $\Delta_n$ are closed under bounded quantification (bounded quantifiers "commute" with the next $\exists$- or $\forall$-block, Exercise 2 in **6.3**).

2. Confirm that $\Delta_0 \subset \Delta_1 \subset \Sigma_1, \Pi_1$, which therefore shows that these four classes of arithmetical predicates are distinct.

3. Prove that $\omega$-inconsistency is (at most) $\Sigma_3$. Theorem 7.6.2 will show that $\omega$-inconsistency is properly $\Sigma_3$. Hence, $\omega$-consistency is $\Pi_3$.

# Chapter 7

# On the Theory of Self-Reference

By self-reference we basically mean the possibility of talking inside a theory $T$ about $T$ itself or related theories. Here we can give merely a glimpse into this recently much advanced area of research; see e.g. [Bu]. We will prove Gödel's second incompleteness theorem, Löb's theorem, and many other results related to self-reference, while further results are discussed only briefly and elucidated by means of applications. All this is of great interest both for epistemology and the foundations of mathematics.

The mountain we first have to climb is the proof of the derivability conditions for PA and related theories in **7.1**, and the derivable $\Sigma_1$-completeness in **7.2**. But anyone contented with leafing through these sections can begin straight away in **7.3**; from then on we will just be reaping the fruits of our labor. However, one would forgo a real adventure in doing so, namely the fusion of logic and number theory in the analysis of PA. For a comprehensive understanding of self-reference, the material of **7.1** and **7.2** (partly prepared in Chapter **6**) should be studied anyway.

Gödel himself tried to interpret the notion "provable" using a modal operator in the framework of the modal system S4. This attempt reflects some of his own results, though not adequately. Only after 1970, when modal logic was sufficiently advanced, could such a program be successfully carried out. A suitable instrument turned out to be the modal logic denoted by G (or GL). The Kripke semantics for G introduced in **7.4** is an excellent tool for confirming or refuting self-referential statements. Solovay's completeness theorem and the completeness theorem of Kripke semantics for G in **7.5** are fortunately of the kind that allows application without knowing the completeness proof itself, which in both cases are not quite easy and use several technical tricks.

W. Rautenberg, *A Concise Introduction to Mathematical Logic*,
Universitext, DOI 10.1007/978-1-4419-1221-3_7,
© Springer Science+Business Media, LLC 2010

There are several extensions of **G**, for example, the bimodal logic **GD** in **7.6**. This logic is related to Hilbert's famous $\omega$-rule. A weakening of it can expressed by the modal operator $\boxdot$ of **GD**. A comprehensive survey can be found in [Bu, Chapter VII]; see also [Vi2]. In **7.7** we discuss some questions regarding self-reference in axiomatic set theory.

# 7.1 The Derivability Conditions

Put somewhat simply, Gödel's second incompleteness theorem states that $\vdash_T \mathsf{Con}_T$ cannot hold for a sufficiently strong and consistent axiomatizable theory $T$. Here $\mathsf{Con}_T$ is a sentence reflecting the metatheoretic statement of consistency of $T$ *inside* $T$, more precisely, inside the (first-order) language $\mathcal{L}$ of $T$. In a popular formulation: *If $T$ is consistent, then this consistency is unprovable in $T$*. As was outlined by Gödel and will be verified in this chapter, the italicized sentence is not only true but also formalizable in $\mathcal{L}$ and even provable in the framework of $T$.

The easiest way to obtain Gödel's theorem is first to prove the *derivability conditions* stated below. Their formulation supposes the arithmetizability of $T$, which includes the distinguishing of a sequence $\underline{0}, \underline{1}, \ldots$ of ground terms; see page 250. Let $\mathsf{bew}_T(y, x)$ be a formula that represents the recursive predicate $bew_T$ in $T$ as in **6.4**. For $\mathsf{bwb}_T(x) = \exists y\, \mathsf{bew}_T(y, x)$ we write $\Box(x)$, and $\Box\alpha$ is to mean $\mathsf{bwb}_T \frac{\ulcorner \alpha \urcorner}{x}$. We may read $\Box\alpha$ as "box $\alpha$" or more suggestively "$\alpha$ is provable in $T$," because $\Box\alpha$ reflects the metatheoretic property $\vdash_T \alpha$ *in $T$*. If $\Box$ refers to some theory $T' \neq T$ then $\Box$ has to be indexed correspondingly. For instance, $\Box_{\mathsf{ZFC}}\alpha$ for $\alpha \in \mathcal{L}_\in$ can easily expressed also in $\mathcal{L}_{ar}$. Note that $\Box\alpha$ is always a sentence, even if $\alpha$ contains free variables.

Further, set $\Diamond\alpha := \neg\Box\neg\alpha$ for $\alpha \in \mathcal{L}$. If $\alpha$ is a sentence, $\Diamond\alpha$ may be read as $\alpha$ *is compatible with $T$*, because it formalizes '$\nvdash_T \neg\alpha$', which is, as we know, equivalent to the consistency of $T + \alpha$. First of all, we define $\mathsf{Con}_T$ in a natural way by

$$\mathsf{Con}_T := \neg\Box\bot \left( = \neg\, \mathsf{bwb}_T(\ulcorner \bot \urcorner) \right),$$

where $\bot$ is a contradiction, $\underline{0} \neq \underline{0}$, for instance. We shall see in a moment that $\mathsf{Con}_T$ is independent modulo $T$ of the choice of $\bot$. The mentioned derivability conditions then read as follows:

$D1$: $\vdash_T \alpha \Rightarrow \vdash_T \Box\alpha$,

$D2$: $\Box\alpha \wedge \Box(\alpha \to \beta) \vdash_T \Box\beta$,

$D3$: $\vdash_T \Box\alpha \to \Box\Box\alpha$.

Here $\alpha, \beta$ run through all sentences of $\mathcal{L}$. These conditions are due to Löb, but they were considered in a slightly different setting already in [HB]. Sometimes $D2$ is written in the equivalent form $\Box(\alpha \to \beta) \vdash_T \Box\alpha \to \Box\beta$, and $D3$ as $\Box\alpha \vdash_T \Box\Box\alpha$.

A consequence of $D1$ and $D2$ is $D0$: $\alpha \vdash_T \beta \Rightarrow \Box\alpha \vdash_T \Box\beta$. This results from the following chain of implications:

$$\alpha \vdash_T \beta \Rightarrow \vdash_T \alpha \to \beta \Rightarrow \vdash_T \Box(\alpha \to \beta) \Rightarrow \vdash_T \Box\alpha \to \Box\beta \Rightarrow \Box\alpha \vdash_T \Box\beta.$$

From $D0$ it clearly follows that $\alpha \equiv_T \beta \Rightarrow \Box\alpha \equiv_T \Box\beta$. In particular, the choice of $\bot$ in $\mathsf{Con}_T$ is arbitrary as long as $\bot \equiv_T 0 \neq 0$.

**Remark 1.** Any operator $\partial\colon \mathcal{L} \to \mathcal{L}$ satisfying the conditions $d1$: $\vdash_T \alpha \Rightarrow \vdash_T \partial\alpha$ and $d2$: $\partial(\alpha \to \beta) \vdash_T \partial\alpha \to \partial\beta$ thus satisfies also $d0$: $\alpha \vdash_T \beta \Rightarrow \partial\alpha \vdash_T \partial\beta$, and hence $d00$: $\alpha \equiv_T \beta \Rightarrow \partial\alpha \equiv_T \partial\beta$, for all $\alpha, \beta \in \mathcal{L}$. It likewise satisfies $d\wedge$: $\partial(\alpha \wedge \beta) \equiv_T \partial\alpha \wedge \partial\beta$, for $\alpha \wedge \beta \vdash_T \alpha, \beta$, hence $\partial(\alpha \wedge \beta) \vdash_T \partial\alpha, \partial\beta \vdash_T \partial\alpha \wedge \partial\beta$ in view of $d0$. The converse direction $\partial\alpha \wedge \partial\beta \vdash_T \partial(\alpha \wedge \beta)$ readily follows from $\alpha \vdash_T \beta \to \alpha \wedge \beta$ by first applying $d0$ and then $d2$.

Whereas $D2$ and $D3$ represent sentence schemata in $T$, condition $D1$ is of metatheoretic nature and follows obviously from the representability of $\mathsf{bew}_T$ in $T$. Thus, $D1$ holds even for weak theories such as $T = \mathsf{Q}$. On the other hand, the converse of $D1$,

$D1^*$:    $\vdash_T \Box\alpha \Rightarrow \vdash_T \alpha$, for all $\alpha \in \mathcal{L}^0$,

may fail. Fortunately, it holds for all $\omega$-consistent axiomatic extensions $T \supseteq \mathsf{Q}$ such as $T = \mathsf{PA}$. Indeed, $\nvdash_T \alpha$ implies $\vdash_T \neg\,\mathsf{bew}_T(\underline{n}, \ulcorner\alpha\urcorner)$ for all $n$ (Corollary 6.4.3). Hence, $\nvdash_T \exists y\,\mathsf{bew}_T(y, \ulcorner\alpha\urcorner)$ in view of the $\omega$-consistency of $T$, that is, $\nvdash_T \Box\alpha$.

Unlike $D1$, the properties $D2$ and $D3$ are not so easily obtained. The theory $T$ must be able not only to *speak* about provability in $T$ (perhaps via arithmetization), but also to *prove* basic properties about provability. $D3$ is nothing else than *condition $D1$ formalized within $T$*, while $D2$ formalizes (7) from page 230, the closure under MP in arithmetical terms. Let us first realize that $D2$ holds, provided it has been shown that

$D2^*$:    $\mathsf{bew}_T(u, x) \wedge \mathsf{bew}_T(v, x \mathbin{\dot\to} y) \vdash_T \mathsf{bew}_T(u * v * \langle y \rangle, y)$, where the p.r. functions $\tilde{\rightarrow}$, $*$, and $y \mapsto \langle y \rangle$ appearing in $D2^*$ must either be present or definable in $T$. Generally speaking, $f \in \mathbf{F}_n$ is called *definable* in an arithmetizable theory $T \subseteq \mathcal{L}$ (with respect to the sequence of terms $(\underline{n})_{n \in \mathbb{N}}$ in $T$) if there is a formula $\delta(\vec{x}, y) \in \mathcal{L}$ such that

(1)    (a) $\vdash_T \delta(\underline{\vec{a}}, \underline{f\vec{a}})$ for all $\vec{a}$,    (b) $\vdash_T \forall\vec{x}\,\exists!y\delta(\vec{x}, y)$.

Clearly, $f$ is then also represented by $\delta(\vec{x}, y)$. For $T = \mathsf{PA}$ and related theories, (1) means that $f$ is explicitly definable in $T$ in the sense of **2.6** and may be introduced in $T$ (using a corresponding symbol). From now on we will no longer distinguish between $T$ and its definitorial extensions and apply $\vdash_T y = f\vec{x} \leftrightarrow \delta(\vec{x}, y)$ without comment. This and (1) easily imply $\vdash_T \underline{f\vec{a}} = \underline{f\vec{a}}$, e.g. $\vdash_T \underline{a} \tilde{\rightarrow} \underline{b} = \underline{a} \tilde{\rightarrow} \underline{b}$. With $\ulcorner\alpha\urcorner$, $\ulcorner\beta\urcorner$ for $x$, $y$, we thus obtain from $D2^*$ in view of $\ulcorner\alpha \rightarrow \beta\urcorner = \dot{\alpha} \tilde{\rightarrow} \dot{\beta} = \underline{\dot{\alpha}} \tilde{\rightarrow} \underline{\dot{\beta}} = \ulcorner\alpha\urcorner \tilde{\rightarrow} \ulcorner\beta\urcorner$,

$$\mathsf{bew}_T(u, \ulcorner\alpha\urcorner) \wedge \mathsf{bew}_T(v, \ulcorner\alpha \rightarrow \beta\urcorner) \vdash_T \mathsf{bew}_T(u * v * \langle\ulcorner\beta\urcorner\rangle, \ulcorner\beta\urcorner).$$

Particularization yields $D2$. But the real work, the definability of the functions appearing in $D2^*$ in theories like $T = \mathsf{PA}$, still lies ahead.

In order to better keep track of things, we restrict our considerations to the theories $\mathsf{ZFC}$ and $\mathsf{PA}$, which are of central interest in nearly all foundational questions. $\mathsf{ZFC}$ is only briefly discussed. Here the proofs of $D2$ and $D3$ (with $\Box = \Box_{\mathsf{ZFC}}$) are much easier than in $\mathsf{PA}$ and need only a few lines as follows: $D2^*$ and hence $D2$ are clear, because the naive proof of $D2^*$ above with $\mathsf{bew}_T = \mathsf{bew}_{\mathsf{ZFC}}$ can easily be formalized *inside* $\mathsf{ZFC}$. This includes the definability of all functions occurring in $D2^*$, for we *did* define them; for instance, the operation $*$ on page 224 may be defined by setting $a * b = \emptyset$ if $a \notin \omega$ or $b \notin \omega$. We arithmetize $\mathcal{L}_{\in}$ according to the pattern in **6.2**, encoding formulas with Gödel numbers,[1] so that $\mathcal{L}_{\in}$-formulas are encoded within $\mathsf{ZFC}$ by certain $\omega$-terms, defined in **3.4**. Formulas from $\mathcal{L}_{ar}$ are identified with their $\omega$-relativized in $\mathcal{L}_{\in}$, called the *arithmetical formulas* of $\mathcal{L}_{\in}$. Moreover, the arithmetical predicate $\mathsf{bew}_{\mathsf{ZFC}}$ is certainly representable in $\mathsf{ZFC}$ by Theorem 6.4.2, since this theorem can be viewed, just like every theorem in this book, as a theorem *within* $\mathsf{ZFC}$. Thus, the naive proof of $D1$ based on this theorem (up to Corollary 6.4.3) can as a whole be carried out in $\mathsf{ZFC}$, and so $D3$ is proved.

---

[1] This is not actually necessary, since in $\mathsf{ZFC}$ one can talk directly about finite sequences and hence about $\mathcal{L}_{\in}$-formulas (Remark 2 in **6.6**), but we do so in order to maintain coherence with the exposition in **6.2**.

Roughly speaking, $D2$ and $D3$ hold for ZFC because ordinary mathematics, in particular the material in Chapter **6**, is formalizable in ZFC. In all of the above, no special set-theoretic constructions such as transfinite recursion are needed. Only relatively simple combinatorial facts are required. Hence there is some hope that the proofs of $D2$ and $D3$ can also be carried out in sufficiently strong arithmetical theories like PA. This is indeed so. The proof of $D3$ for PA will need the most effort and will be completed only in **7.2**. Our first goal will be to show that the p.r. functions occurring in $D2^*$, and in fact *all* p.r. functions, are explicitly definable in PA.[2] They turn out to be definable even in a sense stronger than required by (1) from the previous page.

**Definition.** An $n$-ary recursive function $f$ is called *provably recursive* or $\Sigma_1$-*definable in* PA if there is a $\Sigma_1$-formula $\delta_f(\vec{x}, y)$ in $\mathcal{L}_{ar}$ such that

(2)   (a) $\vdash_{PA} \delta_f(\vec{a}, f\vec{a})$ for all $\vec{a} \in \mathbb{N}^n$;   (b) $\vdash_{PA} \forall \vec{x} \exists! y \delta_f(\vec{x}, y)$.

Since PA is $\Sigma_1$-complete, 2(a) is equivalent to $\mathcal{N} \vDash \delta_f(\vec{a}, f\vec{a})$ for all $\vec{a}$, which is often more easily verified than 2(a) and could replace 2(a). We will show that *all* p.r. functions are $\Sigma_1$-definable in PA, which strengthens their explicit definability in PA. Thereafter we may treat all occurring p.r. functions in PA as if they had been available in the language right from the outset. Essentially this fundamental fact allows a treatment of elementary number theory and combinatorics within the boundaries of PA and hence is particularly interesting for a critical foundation of mathematics.

If $\delta_f(\vec{x}, y)$ in (2) is $\Delta_0$ then $f$ is called $\Delta_0$-definable. An example is the $\beta$-function (Exercise 1), which from now on may be supposed to be present in PA. Basic for the $\Sigma_1$-definability of all p.r. functions is $\beta$'s main property, Lemma 6.4.1, of which we need, of course, some provable version in PA. Since Euclid's lemma and the Chinese remainder theorem are involved here, these should be derived first. Clearly, the basic arithmetical laws applied in their proofs in **6.3** should be at our disposal, including those on the order relation and on $a - b$ for $a \geqslant b$, all provable in N.

The proof of Euclid's lemma is straightforward, Exercise 2. As for the Chinese remainder theorem, we avoid the quantification over finite

---

[2] In [Gö2], Gödel presented a list of 45 definable p.r. functions; the last was $\chi_{bew}$. Following [WR], Gödel considered a higher-order arithmetical theory. That Gödel's theorems also hold in first-order arithmetic was probably first noticed in [HB].

sequences for the time being, by stating the theorem as a scheme. Let $c, d$ denote unary provably recursive functions, which may depend on further parameters. Each such $c$ determines for given $n$ the sequence $c_0, \ldots, c_n$, with $c_\nu = c(\nu)$ for $\nu \leqslant n$. For suggestive reasons from now on also letters such as $n, \nu, \ldots$ may denote variables in $\mathcal{L}_{ar}$. With the $\Delta_0$-definable relation $\perp$ of coprimeness, the Chinese remainder theorem can provisionally be stated as follows: for arbitrary $c, d$ as arranged above, we get

$$(3) \quad \vdash_{\mathsf{PA}} \forall n[(\forall i, j \leqslant n)(c_i < d_i \wedge (i \neq j \rightarrow d_i \perp d_j))$$
$$\rightarrow \exists a (\forall \nu \leqslant n) \operatorname{rem}(a, d_\nu) = c_\nu].$$

To convert the original proof of the remainder theorem to one for (3) we require, for given provably recursive $d$, the term $\operatorname{lcm}\{d_\nu \mid \nu \leqslant n\}$, the least common multiple of $d_0, \ldots, d_n$. **Claim:** $f : n \mapsto \operatorname{lcm}\{d_\nu \mid \nu \leqslant n\}$ is defined in $\mathsf{PA}$ by the $\Sigma_1$-formula

$$\delta_f(x, y) := (\forall \nu \leqslant x) d_\nu \mid y \wedge (\forall z < y)(\exists \nu \leqslant x) d_\nu \nmid z.$$

More precisely, $\delta_f(x, y)$ *describes* a $\Sigma_1$-formula in $\mathcal{L}_{ar}$ that is even $\Delta_0$, provided $d$ is $\Delta_0$-definable. Clearly $\mathcal{N} \models \delta(\underline{n}, \operatorname{lcm}\{d_\nu \mid \nu \leqslant n\})$ for all $n$. Thus, 2(a) holds. With the minimum schema (Exercise 4 in **3.3**) applied to $\beta(x, y) := (\forall \nu \leqslant x) d_\nu \mid y$, we obtain $\vdash_{\mathsf{PA}} \exists! y \delta_f(x, y)$, provided it has been shown that $\vdash_{\mathsf{PA}} \exists y \beta(x, y)$ ('$c_0, \ldots, c_x$ have a common multiple'), which is easily derived by induction on $x$; see Example 1 in **2.5**. This proves the claim. After having derived Euclid's lemma in $\mathsf{PA}$ (Exercise 2) we confirm (3) by following the proof of the remainder theorem in **6.2**, and, writing $\beta st$ for $\beta(s, t)$, a suitable version of Lemma 6.4.1 as follows:

$$(4) \quad \vdash_{\mathsf{PA}} \forall n \exists u (\forall \nu \leqslant n) c_\nu = \beta u\nu, \text{ for any given provably recursive } c.$$

**Theorem 1.1.** *Each p.r. function $f$ is provably recursive. Moreover, the recursion equations for $f$ are provable in $\mathsf{PA}$ whenever $f = \boldsymbol{Op}(g, h)$.*

**Proof.** For the initial functions and $+, \cdot$ the formulas $v_0 = 0$, $v_1 = Sv_0$, $v_n = v_\nu$ along with $v_2 = v_0 + v_1$ and $v_2 = v_0 \cdot v_1$ are obviously defining $\Sigma_1$-formulas. For the composition $f = h[g_1, \ldots, g_m]$, let $\delta_f(\vec{x}, y)$ be the formula $y = h(g_1 \vec{x}, \ldots, g_m \vec{x})$. In this case (2) is clear, because we might think of $h, g_1, \ldots, g_m$ as being already introduced in $\mathsf{PA}$, so that $\delta_f(\vec{x}, y)$ belongs to the expanded language. Only the construction of $\delta_f$ for the case $f = \boldsymbol{Op}(g, h)$ requires some skill. We may assume that besides $\beta$ also $g, h$ have already been introduced in the language. Consider

$$(5)\ \delta_f(\vec{x}, y, z) := \exists u[\underbrace{\beta u0 = g\vec{x} \wedge (\forall v{<}y)\beta uSv = h(\vec{x}, v, \beta uv) \wedge \beta uy = z}_{\gamma(u,\vec{x},y,z)}].$$

$\delta_f$ is similar to $\delta_{\exp}$ from Remark 1 in **6.4**. It is $\Sigma_1$, because $\beta$, $g$, $h$ are $\Sigma_1$-definable. Lemma 6.4.1 applied with $c_i = f(\vec{a}, i)$ for $i \leqslant b$ shows that $\mathcal{N} \vDash \delta_f(\vec{a}, \underline{b}, f\vec{a})$, equivalently 2(a). Uniqueness in 2(b), that is,

$$\delta_f(\vec{x}, y, z) \wedge \delta_f(\vec{x}, y, z') \vdash_{\mathsf{PA}} z = z',$$

derives easily from $\gamma(u, \vec{x}, y, z) \wedge \gamma(u', \vec{x}, y, z') \vdash_{\mathsf{PA}} z = z'$, which clearly follows from $\gamma(u, \vec{x}, y, z) \wedge \gamma(u', \vec{x}, y, z') \vdash_{\mathsf{PA}} (\forall v{\leqslant}y)\beta uv = \beta u'v$. This is easily shown by induction on $y$. Also, $\vdash_{\mathsf{PA}} \exists z \delta_f(\vec{x}, y, z)$ will be shown inductively on $y$. We get $\vdash_{\mathsf{PA}} \exists u \beta u0 = g\vec{x}$ (hence $\vdash_{\mathsf{PA}} \exists z \delta_f(\vec{x}, 0, z)$) from (4), choosing $\mathsf{c}$ therein such that $\mathsf{c}_0 = g\vec{x}$ and $\mathsf{c}_\nu = 0$ for $\nu \neq 0$. $\mathsf{c}$ is provably recursive, for the term $g\vec{x}$ is $\Sigma_1$-definable. The inductive step will be verified informally, that is, we shall prove

$$(*)\qquad \exists z \delta_f(\vec{x}, y, z) \vdash_{\mathsf{PA}} \exists z' \delta_f(\vec{x}, Sy, z').$$

Suppose $\gamma(u, \vec{x}, y, z)$. Consider the provably recursive $\mathsf{c} \colon \nu \mapsto \mathsf{c}_\nu$ defined by $\mathsf{c}_\nu = \beta u\nu$ for $\nu \leqslant Sy$ and $\mathsf{c}_{Sy} = h(\vec{x}, y, \beta uy)$. Here $u, \vec{x}, y$ are parameters in the defining $\Sigma_1$-formula for $\mathsf{c}$. So by (4) (taking $Sy$ for $n$) there is some $u'$ with $\beta u'\nu = \mathsf{c}_\nu = \beta u\nu$ for all $\nu \leqslant y$ and $\beta u'Sy = \mathsf{c}_{Sy} = h(\vec{x}, y, \beta uy)$. With this $u'$ and $z' = \beta u'Sy$ we obtain $\gamma(u', \vec{x}, Sy, z')$, and so $\exists z' \delta_f(\vec{x}, Sy, z')$. This confirms $(*)$ and hence 2(b). Thus, $f$ is provably recursive and may now be introduced in $\mathsf{PA}$. We finally sketch a proof of the recursion equations for $f$ in $\mathsf{PA}$, which also in $\mathsf{PA}$ may be written as usual, i.e.,

$$(A)\ \vdash_{\mathsf{PA}} f(\vec{x}, 0) = g\vec{x}, \quad (B)\ \vdash_{\mathsf{PA}} f(\vec{x}, Sy) = h(\vec{x}, y, f(\vec{x}, y)).$$

(A) holds because $\vdash_{\mathsf{PA}} \delta_f(\vec{x}, 0, f(\vec{x}, 0)) \equiv_{\mathsf{PA}} \exists u(\beta u0 = g\vec{x} \wedge \beta u0 = f(\vec{x}, 0))$ and clearly $\exists u(\beta u0 = g\vec{x} \wedge \beta u0 = f(\vec{x}, 0)) \vdash f(\vec{x}, 0) = g\vec{x}$. (B) follows by $<$-induction on $y$ applied to $\alpha = \alpha(\vec{x}, y) := f(\vec{x}, Sy) = h(\vec{x}, y, f(\vec{x}, y))$. Assume that $(\forall v{<}y)\alpha \frac{v}{y}$. Choosing $u$ in (5) such that $\gamma(u, \vec{x}, Sy, f(\vec{x}, Sy))$, we readily obtain $(\forall v{\leqslant}y)f(\vec{x}, v) = \beta uv$, so that

$$f(\vec{x}, Sy) = \beta uSy = h(\vec{x}, y, \beta uy) = h(\vec{x}, y, f(\vec{x}, y)).$$

This confirms $\forall y((\forall v{<}y)\alpha \frac{v}{y} \rightarrow \alpha)$, hence $\vdash_{\mathsf{PA}} \forall y\alpha$ by $<$-induction. $\qquad\square$

We thus have achieved our first goal. Next observe that the properties of $*, \ell, \ldots$ from the remark on page 230 along with the basic property (5) stated there are also readily proved *within* $\mathsf{PA}$. This is a little extra program that includes the proof of unique prime factorization, see Exercise 4.

Thus, $D2^*$ and hence $D2$ are indeed provable for $T = \mathsf{PA}$. In particular, the property (6) from page 230 carries over to $\mathsf{PA}$, so that

$\quad$ (6) $\quad \Box(x \overset{\cdot}{\to} y) \vdash_{\mathsf{PA}} \Box(x) \to \Box(y)$.

We mention that $\Box$ in (3) may even denote the formula $\mathsf{bwb}_T$ for any axiomatizable (and arithmetizable) theory $T$. $D3$ will be proved in the next section in a somewhat broader context.

**Remark 2.** The formalized equations of Exercise 3 in **6.4** are now also provable in $\mathsf{PA}$. For instance, item (b) reads $\vdash_{\mathsf{PA}} \mathrm{sb}_{\vec{x}}(\ulcorner \varphi \urcorner, \vec{x}) = \mathrm{sb}_{\vec{x}'}(\ulcorner \varphi \urcorner, \vec{x}')$ for $\varphi = \varphi(\vec{x})$, where $\vec{x}'(\subseteq \vec{x})$ enumerates the free variables of $\varphi$. As regards (c), consider first a special case. Let $\varphi$ be $\mathsf{S}x = y$. Then $\mathrm{sb}_{xy}(\dot{\varphi}, x, \mathsf{S}x) = \mathrm{sb}_x((\varphi \frac{\mathsf{S}x}{y})', x)$, formalized $\mathrm{sb}_{xy}(\ulcorner \varphi \urcorner, x, y) \frac{\mathsf{S}x}{y} = \mathrm{sb}_x(\ulcorner \varphi \frac{\mathsf{S}x}{y} \urcorner, x)$. For the proof of this equation in $\mathsf{PA}$, just $\vdash_{\mathsf{PA}} \mathrm{cf}\, \mathsf{S}x = \tilde{\mathsf{S}}\, \mathrm{cf}\, x$ is required, which holds by Theorem 1.1. Whoever wants to write down a detailed proof should follow the example on page 249.

## Exercises

1. Prove in $\mathsf{PA}$ the $\Delta_0$-definability of the remainder function rem, the pairing function, and the $\beta$-function; see **6.4**. In particular, rem is defined by $\delta_{\mathrm{rem}}(a, b, r) := (\exists q \leqslant a)(a = b \cdot q + r \wedge r < b) \vee b = r = 0$. The laws of arithmetic as given by $\mathsf{N}$ (page 235) may be used.

2. Prove *in* $\mathsf{PA}$ (a) $(\forall a, b > 0) \exists x \exists y (a \perp b \to ax + \underline{1} = by)$, that is, Euclid's lemma. (b) $(\forall a > 1) \exists p (\mathrm{prim}\, p \wedge p | a)$ ('each number $\geqslant 2$ has a prime divisor'), (c) $\vdash_{\mathsf{PA}} (\forall a, b > 0) \forall p (\mathrm{prim}\, p \wedge p | ab \to p | a \vee p | b)$.

3. Show that $\vdash_{\mathsf{PA}} \mathrm{prim}\, p \wedge p | \mathrm{lcm}\{d_\nu \mid \nu \leqslant n\} \to (\exists i \leqslant n) p | d_i$, required for carrying out the proof of the Chinese remainder theorem in $\mathsf{PA}$.

4. One of several possibilities of formalizing the prime factorization in $\mathsf{PA}$ is $(\forall n \geqslant 2)(\exists m \geqslant 2) n = \prod_{i \leqslant \ell m} p_i^{(m)_i}$, where $m$ serves as a variable for the sequence of prime exponents.[3] Prove this in $\mathsf{PA}$, as well as its uniqueness, which is essentially based on Exercise 2.

5. Let $T' = T + \alpha$ and $T$ satisfy $D1$–$D3$. Show that

$\quad$ (a) $\vdash_T \Box_{T'} \varphi \leftrightarrow \Box_T(\alpha \to \varphi)$ (the formalized deduction theorem),

$\quad$ (b) $D1$–$D3$ hold also for $T'$.

---

[3] An equivalent formalization of the prime factorization in $\mathsf{PA}$ using the $\beta$-function is $(\forall k \geqslant 2) \exists u \exists n (k = \prod_{i \leqslant n} p_i^{\beta u i} \wedge \beta u n \neq 0)$.

## 7.2 The Provable $\Sigma_1$-Completeness

D3 is a special case of the *provable* $\Sigma_1$-completeness. This is essentially the statement $\vdash_{\mathsf{PA}} \alpha \to \Box\alpha$ for $\Sigma_1$-sentences $\alpha$. The proof demands still additional preparation, and even good textbooks do not carry out all proof steps. All steps described in this section and not handled in detail can easily be completed in full by the sufficiently assiduous reader. Life could be made easier through the mutual interpretability of PA and $\mathsf{ZFC}_{\mathrm{fin}}$ mentioned in **6.6**. Let $\Box = \Box(x)$ denote the formula $\mathsf{bwb}_{\mathsf{PA}}(x)$ till the end of this section. We first introduce an additional notation. Let $\varphi = \varphi(\vec{x})$.

**Definition.** $\Box[\varphi] := \Box(\mathrm{sb}_{\vec{x}}(\ulcorner\varphi\urcorner, \vec{x}))$ $(= \mathsf{bwb}_{\mathsf{PA}} \frac{\mathrm{sb}_{\vec{x}}(\ulcorner\varphi\urcorner, \vec{x})}{x})$.

By Remark 2 in **7.1**, $\vdash_{\mathsf{PA}} \mathrm{sb}_{\vec{x}}(\ulcorner\varphi\urcorner, \vec{x}) = \mathrm{sb}_{\vec{x}'}(\ulcorner\varphi\urcorner, \vec{x}')$, where $\vec{x}'$ enumerates $free\,\varphi$. Hence, we may assume w.l.o.g. that $free\,\Box[\varphi] = free\,\varphi$. Moreover, for $\alpha \in \mathcal{L}^0_{ar}$ we have $\vdash_{\mathsf{PA}} \mathrm{sb}_{\vec{x}}(\ulcorner\alpha\urcorner, \vec{x}) = \mathrm{sb}_{\emptyset}(\ulcorner\alpha\urcorner) = \ulcorner\alpha\urcorner$, hence $\Box[\alpha]$ and $\Box\alpha$ may be identified. '$\vdash_{\mathsf{PA}} \varphi(\vec{a})$ for all $\vec{a} \in \mathbb{N}^n$' is reflected in PA by '$\vdash_{\mathsf{PA}} \forall\vec{x}\,\Box[\varphi]$'. The latter thus reflects in PA the existence of a *collection of proofs* which, due to the $\omega$-incompleteness of PA, may be less than $\vdash_{\mathsf{PA}} \Box\forall\vec{x}\varphi$, or what amounts to the same, $\vdash_{\mathsf{PA}} \Box\varphi$.

**Example.** Let $\varphi = \varphi(x, y)$ be $\mathsf{S}x = y$. We prove $\varphi \vdash_{\mathsf{PA}} \Box[\varphi]$, or equivalently, $\vdash_{\mathsf{PA}} \Box[\varphi] \frac{\mathsf{S}x}{y}$, where w.l.o.g. $x, y$ do not occur bound in $\Box(x)$. In order to prove $\vdash_{\mathsf{PA}} \Box[\varphi] \frac{\mathsf{S}x}{y}$ observe that in view of Remark 2 in **7.1**,

$$\Box[\varphi] \tfrac{\mathsf{S}x}{y} = \Box(\mathrm{sb}_{xy}(\ulcorner\varphi\urcorner, x, \mathsf{S}x)) \equiv_{\mathsf{PA}} \Box(\mathrm{sb}_x(\ulcorner\varphi \tfrac{\mathsf{S}x}{y}\urcorner, x)) = \Box[\alpha(x)]$$

with $\alpha(x) := \mathsf{S}x = \mathsf{S}x$. Thus, it suffices to verify $\vdash_{\mathsf{PA}} \Box[\alpha(x)]$ (equivalently, $\vdash_{\mathsf{PA}} \forall x\Box[\alpha(x)]$). This reflects in PA 'for arbitrary $n$, $\vdash_{\mathsf{PA}} \mathsf{S}\underline{n} = \mathsf{S}\underline{n}$'. We verify $\vdash_{\mathsf{PA}} \Box[\alpha(x)]$ in detail. Consider the p.r. function $\tilde{\alpha} \colon n \mapsto \mathrm{sb}_x(\dot{\alpha}, n)$ (the Gödel number of $\alpha(\underline{n})$). By axiom $\Lambda 9$, $\langle\tilde{\alpha}(n)\rangle$ is for each $n$ a trivial arithmetized proof of length 1. Stated within PA, $\vdash_{\mathsf{PA}} \mathsf{bew}_{\mathsf{PA}}(\langle\tilde{\alpha}(x)\rangle, \tilde{\alpha}(x))$. This clearly yields $\vdash_{\mathsf{PA}} \exists y\,\mathsf{bew}_{\mathsf{PA}}(y, \tilde{\alpha}(x)) = \Box(\tilde{\alpha}(x)) = \Box[\alpha]$.

Next we prove some modifications D1, D2 for $\alpha = \alpha(\vec{x})$ and $\beta = \beta(\vec{x})$:

(1)  (a) $\vdash_{\mathsf{PA}} \alpha \Rightarrow \vdash_{\mathsf{PA}} \Box[\alpha]$;  (b) $\Box[\alpha \to \beta] \vdash_{\mathsf{PA}} \Box[\alpha] \to \Box[\beta]$.

To see (a) let $\vdash_{\mathsf{PA}} \alpha$, hence also $\vdash_{\mathsf{PA}} \forall\vec{x}\alpha$ and so $\vdash_{\mathsf{PA}} \Box\forall\vec{x}\alpha$. Just as in the above example, a proof for $\forall\vec{x}\alpha$ provides one for $\alpha_{\vec{x}}(\vec{a})$ in a p.r. way, or stated *within* PA: $\Box\forall\vec{x} \vdash_{\mathsf{PA}} \Box(\mathrm{sb}_{\vec{x}}(\ulcorner\alpha\urcorner, \vec{x}))$ $(= \Box[\alpha]$; thus, $\vdash_{\mathsf{PA}} \Box[\alpha])$.

(b) follows from (6) in **7.1** with $\mathrm{sb}_{\vec{x}}(\ulcorner\alpha\urcorner, \vec{x}), \mathrm{sb}_{\vec{x}}(\ulcorner\beta\urcorner, \vec{x})$ for $x, y$, observing that $\vdash_{\mathsf{PA}} \mathrm{sb}_{\vec{x}}(\ulcorner\alpha \to \beta\urcorner, \vec{x}) = \mathrm{sb}_{\vec{x}}(\ulcorner\alpha\urcorner, \vec{x}) \stackrel{\sim}{\to} \mathrm{sb}_{\vec{x}}(\ulcorner\beta\urcorner, \vec{x})$, see Exercise 3 in **6.4**. (c) of this exercise yields for all not necessarily distinct $x, y$

(2)    $\Box[\alpha] \frac{t}{x} \equiv_{\mathsf{PA}} \Box[\alpha \frac{t}{x}]$    $(t \in \{0, y, \mathsf{S}y\}$ and $y \notin \mathrm{bnd}\,\alpha)$.

Now, $D3$ is only a special case of the *provable $\Sigma_1$-completeness* of PA, stated not only for sentences, but for arbitrary formulas as follows:

(3)    $\varphi \vdash_{\mathsf{PA}} \Box[\varphi]$ (equivalently, $\vdash_{\mathsf{PA}} \varphi \to \Box[\varphi]$), for all $\Sigma_1$-formulas $\varphi$.

Indeed, choose in (3) for $\varphi$ the $\Sigma_1$-sentence $\Box\alpha$ for any $\alpha \in \mathcal{L}_{ar}^0$. Then $\Box\alpha \vdash_{\mathsf{PA}} \Box[\Box\alpha] \equiv \Box\Box\alpha$, and $D3$ is proved. We obtain (3) from Theorem 2.1 below, since by (1), (2), and since w.l.o.g. $\mathrm{free}\,\alpha = \mathrm{free}\,\Box[\alpha]$, the operator $\partial\colon \alpha \mapsto \Box[\alpha]$ satisfies the conditions of the theorem.

**Theorem 2.1.** *Let $\partial\colon \mathcal{L}_{ar} \to \mathcal{L}_{ar}$ be any operator with $\mathrm{free}\,\partial\alpha \subseteq \mathrm{free}\,\alpha$ satisfying*

$d1$:    $\vdash_{\mathsf{PA}} \alpha \Rightarrow \vdash_{\mathsf{PA}} \partial\alpha$,

$d2$:    $\partial(\alpha \to \beta) \vdash_{\mathsf{PA}} \partial\alpha \to \partial\beta$,

$ds$:    $\partial\alpha \frac{t}{x} \equiv_{\mathsf{PA}} \partial(\alpha\frac{t}{x})$    $(t \in \{0, y, \mathsf{S}y\}, \; y \notin \mathrm{bnd}\,\alpha)$.

*Then $\vdash_{\mathsf{PA}} \varphi \to \partial\varphi$ holds for all $\Sigma_1$-formulas $\varphi \in \mathcal{L}_{ar}$.*

**Proof.** $\partial$ satisfies also $d0$, $d00$, and $d\wedge$ (see Remark 1 in **7.1**). Hence, by Theorem 6.7.2 and $d00$ we need to carry out the proof only for special $\Sigma_1$-formulas. First let $\varphi$ be $\mathsf{S}x = y$. Clearly, $\vdash_{\mathsf{PA}} \varphi \to \partial\varphi$ is equivalent to $\vdash_{\mathsf{PA}} \partial\varphi \frac{\mathsf{S}x}{y}$, and this to $\vdash_{\mathsf{PA}} \partial \mathsf{S}x = \mathsf{S}x$ by $ds$, which is obvious from $d1$. Now let $\varphi$ be $x + y = z$. We shall prove $\vdash_{\mathsf{PA}} \forall yz(\varphi \to \partial\varphi)$ by induction on $x$. Observing that $y = z \vdash_{\mathsf{PA}} \partial y = z$ (equivalently $\vdash_{\mathsf{PA}} \partial z = z$), we obtain $\varphi \frac{0}{x} \vdash_{\mathsf{PA}} y = z \vdash_{\mathsf{PA}} \partial y = z \equiv_{\mathsf{PA}} \partial(\varphi\frac{0}{x}) \equiv_{\mathsf{PA}} \partial\varphi \frac{0}{x}$. Thus, $\vdash_{\mathsf{PA}} \forall yz(\varphi \to \partial\varphi) \frac{0}{x}$. Now $\varphi \frac{\mathsf{S}y}{y} \equiv_{\mathsf{PA}} \varphi \frac{\mathsf{S}x}{x}$; hence $\partial\varphi \frac{\mathsf{S}y}{y} \equiv_{\mathsf{PA}} \partial\varphi \frac{\mathsf{S}x}{x}$, by $d00$, $ds$. The induction step $\forall yz(\varphi \to \partial\varphi) \vdash_{\mathsf{PA}} \forall yz(\varphi \to \partial\varphi) \frac{\mathsf{S}x}{x}$ follows then from

$$\forall yz(\varphi \to \partial\varphi) \vdash \varphi \frac{\mathsf{S}y}{y} \to \partial\varphi \frac{\mathsf{S}y}{y} \vdash_{\mathsf{PA}} \varphi \frac{\mathsf{S}x}{x} \to \partial\varphi \frac{\mathsf{S}x}{x} = (\varphi \to \partial\varphi)\frac{\mathsf{S}x}{x}.$$

The formula $x \cdot y = z$ is left to the reader, who should observe $d\wedge$, $d2$, the induction steps for $\wedge$ and $\exists$, and $\mathsf{S}x \cdot y = z \equiv_{\mathsf{PA}} \exists u(x \cdot y = u \wedge u + y = z)$.

We now treat the logical connectives. The induction steps for $\wedge, \vee, \exists$ are simple. Indeed, from $d\wedge$ we obtain

$$\alpha \wedge \beta \vdash \alpha, \beta \vdash_{\mathsf{PA}} \partial\alpha \wedge \partial\beta \vdash_{\mathsf{PA}} \partial(\alpha \wedge \beta).$$

For $\vee$ note that $\alpha \vdash_{\mathsf{PA}} \partial\alpha \vdash_{\mathsf{PA}} \partial(\alpha \vee \beta)$, and similarly for $\beta$. Further, since $\varphi \vdash \exists x\varphi$ we get $\varphi \vdash_{\mathsf{PA}} \partial\varphi \vdash_{\mathsf{PA}} \partial\exists x\varphi$ by $d0$, and from $x \notin \mathrm{free}\,\partial\exists x\varphi$

follows $\exists x\varphi \vdash_{\mathsf{PA}} \partial \exists x\varphi$. The prime-term substitution step ($t$ is prime in $\frac{t}{x}$) also runs smoothly: $\varphi \vdash_{\mathsf{PA}} \partial\varphi$ yields $\varphi\frac{t}{x} \vdash_{\mathsf{PA}} \partial\varphi\frac{t}{x} \vdash_{\mathsf{PA}} \partial(\varphi\frac{t}{x})$ by $ds$.

It remains to verify the step for bounded quantification. Suppose that $\alpha \vdash_{\mathsf{PA}} \partial\alpha$ and $y \notin \operatorname{var}\alpha$. We prove $\varphi := (\forall x{<}y)\,\alpha \vdash_{\mathsf{PA}} \partial\varphi$ by induction on $y$. The initial step is obvious: $\vdash_{\mathsf{PA}} \varphi\frac{0}{y}$, and therefore

$$\vdash_{\mathsf{PA}} \partial(\varphi\tfrac{0}{y}) \vdash_{\mathsf{PA}} \partial\varphi\tfrac{0}{y} \vdash_{\mathsf{PA}} \varphi\tfrac{0}{y} \to \partial\varphi\tfrac{0}{y}.$$

Clearly, $\varphi\frac{Sy}{y} \equiv_{\mathsf{PA}} \varphi \wedge \alpha\frac{y}{x}$. Hence $\alpha\frac{y}{x} \vdash_{\mathsf{PA}} \partial\alpha\frac{y}{x} \vdash_{\mathsf{PA}} \partial(\alpha\frac{y}{x})$ because of $\alpha \vdash_{\mathsf{PA}} \partial\alpha$. That leads to

$$\varphi\tfrac{Sy}{y} \wedge (\varphi \to \partial\varphi) \vdash_{\mathsf{PA}} \varphi \wedge \alpha\tfrac{y}{x} \wedge (\varphi \to \partial\varphi) \vdash_{\mathsf{PA}} \partial\varphi \wedge \partial(\alpha\tfrac{y}{x})$$
$$\vdash_{\mathsf{PA}} \partial(\varphi \wedge \alpha\tfrac{y}{x}) \vdash_{\mathsf{PA}} \partial(\varphi\tfrac{Sy}{y}).$$

Thus, $\varphi \to \partial\varphi \vdash_{\mathsf{PA}} \varphi\frac{Sy}{y} \to \partial(\varphi\frac{Sy}{y})$, which is obviously equivalent to the inductive step. □

**Remark 3.** $D1$–$D3$ are also provable for much weaker theories than PA, e.g., for the so-called *elementary arithmetic* $\mathsf{EA} = I\Delta_0 + \forall xy\exists z\delta_{exp}(x,y,z)$. Here $I\Delta_0$ is defined in Remark 1 in **6.3** and $\delta_{exp}$ is a defining $\Delta_0$-formula for exp, see also [FS]. Also Theorem 1.1 can essentially be strengthened and has many variants. For instance, the provably recursive functions of $I\Sigma_1$ (like PA but IS restricted to $\Sigma_1$-formulas) are precisely the p.r. ones, [Tak]. The same provably recursive functions has EA augmented by the $\Pi_2$-induction schema without parameters, [Be4]. It is noteworthy that the provable recursive functions of EA itself are precisely the elementary ones, [Si]. For more material on the metatheory of PA and related theories see [Bar, Part D], and in particular [HP].

## 7.3  The Theorems of Gödel and Löb

We are now in a position to harvest the yields of our efforts. As long as not stated otherwise, let $T$ denote any arithmetizable axiomatic theory in $\mathcal{L}$, that satisfies the derivability conditions $D1$–$D3$ of **7.1** along with the fixed point lemma of **6.5**. We direct attention straight away to the uniqueness statement of Lemma 3.1(b) below. According to this claim, up to equivalence in $T$ at most $\Box\alpha \to \alpha$ can be the fixed point of the formula $\Box(x) \to \alpha$. The proof of Theorem 3.2 will show that $\neg\Box(x)$ too has only one fixed point modulo $T$. Beneath all this lies, as we shall see from Corollary 5.6, a completely general result.

**Lemma 3.1.** *Let $T$ be as arranged above, and let $\alpha, \gamma \in \mathcal{L}^0$ be such that* $\gamma \equiv_T \Box\gamma \to \alpha$. *Then* (a) $\Box\gamma \equiv_T \Box\alpha$ *and* (b) $\gamma \equiv_T \Box\alpha \to \alpha$.

**Proof.** The supposition yields $\Box\gamma \vdash_T \Box(\Box\gamma \to \alpha) \vdash_T \Box\Box\gamma \to \Box\alpha$, by $D0$ and $D2$. Now by $D3$, we clearly obtain $\Box\gamma \vdash_T \Box\Box\gamma$, hence $\Box\gamma \vdash_T \Box\alpha$. Since $\alpha \vdash_T \Box\gamma \to \alpha \equiv_T \gamma$ and so $\alpha \vdash_T \gamma$, it follows that $\Box\alpha \vdash_T \Box\gamma$ by $D0$. Together with the already verified $\Box\gamma \vdash_T \Box\alpha$ we get (a). Using (a) we may replace $\Box\gamma$ with $\Box\alpha$ in $\gamma \equiv_T \Box\gamma \to \alpha$, which results in (b). ∎

**Theorem 3.2 (Second incompleteness theorem).** PA *satisfies alongside the fixed point lemma also $D1$–$D3$. Every theory $T$ with these properties satisfies the conditions*

    (1) $\nvdash_T \mathrm{Con}_T$ *provided $T$ is consistent,*    (2) $\vdash_T \mathrm{Con}_T \to \neg\Box\,\mathrm{Con}_T$.

**Proof.** $D1$–$D3$ were proved for PA in **7.1**. (1) follows from (2). Assume $\vdash_T \mathrm{Con}_T$. Then $\vdash_T \Box\,\mathrm{Con}_T$ by $D1$, as well as $\vdash_T \neg\Box\,\mathrm{Con}_T$ by (2). Thus, $T$ is inconsistent. To verify (2), let $\gamma$ be a fixed point of $\neg\Box(x)$, i.e.,

    (∗)   $\gamma \equiv_T \neg\Box\gamma \ (\equiv \Box\gamma \to \bot)$.

By Lemma 3.1(b) with $\alpha = \bot$, we obtain $\gamma \equiv_T \Box\bot \to \bot \equiv \neg\Box\bot = \mathrm{Con}_T$. Replacing $\gamma$ in (∗) with $\mathrm{Con}_T$ gives $\mathrm{Con}_T \equiv_T \neg\Box\,\mathrm{Con}_T$. Half of this is the claim (2). ∎

Thus, by (1), no sufficiently strong consistent theory can prove its own consistency. In particular, $\nvdash_{\mathsf{PA}} \mathrm{Con}_{\mathsf{PA}}$ as long as PA is consistent which is assumed throughout this book and is a minimal assumption for a far-reaching metamathematics. The above proof shows that $\mathrm{Con}_T$ is the only fixed point of $\neg\,\mathbf{bwb}_T$ modulo $T$. Actually, it shows a bit more, namely

    (3)   $\mathrm{Con}_T \equiv_T \neg\Box\,\mathrm{Con}_T$.

This strengthens (2), but only by a little: $\neg\Box\,\mathrm{Con}_T \vdash_T \mathrm{Con}_T$ is just a special case of

    (4)   $\neg\Box\alpha \vdash_T \mathrm{Con}_T$ (equivalently, $\neg\,\mathrm{Con}_T \vdash_T \Box\alpha$), for every $\alpha \in \mathcal{L}$.

This follows from $\bot \vdash_T \alpha$, since $\neg\,\mathrm{Con}_T \equiv \Box\bot \vdash_T \Box\alpha$ by $D0$. (4) reflects in $T$ 'If $T$ is inconsistent then every formula is provable'. From (1) and (3) we get in particular $\nvdash_{\mathsf{PA}} \neg\Box_{\mathsf{PA}}\,\mathrm{Con}_{\mathsf{PA}}$, although '$\mathrm{Con}_{\mathsf{PA}}$ is unprovable in PA' is true according to (1) (again we tacitly use the consistence of PA). $\neg\Box_{\mathsf{PA}}\,\mathrm{Con}_{\mathsf{PA}}$ reflects '$\mathrm{Con}_{\mathsf{PA}}$ is unprovable in PA'; hence $\nvdash_{\mathsf{PA}} \neg\Box_{\mathsf{PA}}\,\mathrm{Con}_{\mathsf{PA}}$ is just another formulation of the second incompleteness theorem.

The above claims hold independently of the "truth content" of the sentences provable in $T$. Namely, a consequence of the second incompleteness theorem is the existence of consistent theories $T \supseteq \mathsf{PA}$ in which along with claims true in $\mathcal{N}$ also false ones are provable, i.e., in which truth and untruth live in peaceful coexistence with each other. Such "dream theories" are highly rich in content, for all of them include ordinary number theory. An example is $\mathsf{PA}^{\perp} := \mathsf{PA} + \neg\, \mathsf{Con}_{\mathsf{PA}}$. This theory is consistent because *the consistency of* $\mathsf{PA}^{\perp}$ *is equivalent to the unprovability of* $\mathsf{Con}_{\mathsf{PA}}$ *in* $\mathsf{PA}$. The italicized sentence is even provable in $\mathsf{PA}$, as (5) below will show. By the formalized deduction theorem (Exercise 5 in **7.1**), $\Box_{T+\alpha}\bot \equiv_T \Box(\alpha \to \bot) \equiv \Box\neg\alpha$; hence $\neg\Box_{T+\alpha}\bot \equiv_T \neg\Box\neg\alpha\ (\equiv \Diamond\alpha)$, and consequently,

(5)  $\mathsf{Con}_{T+\alpha} \equiv_T \neg\Box\neg\alpha$  (in particular, $\mathsf{Con}_{\mathsf{PA}^{\perp}} \equiv_{\mathsf{PA}} \neg\Box_{\mathsf{PA}}\mathsf{Con}_{\mathsf{PA}}$).

The special cases under (5) and (3) for $T = \mathsf{PA}$ now clearly yield

(6)  $\mathsf{Con}_{\mathsf{PA}} \equiv_{\mathsf{PA}} \mathsf{Con}_{\mathsf{PA}^{\perp}}$  (hence also $\mathsf{Con}_{\mathsf{PA}} \equiv_{\mathsf{PA}^{\perp}} \mathsf{Con}_{\mathsf{PA}^{\perp}}$).

Put together, $\mathsf{PA}^{\perp}$ contains ordinary number theory as known to us, but also proves the indubitably false sentence $\mathsf{bwb}_{\mathsf{PA}}(\ulcorner 0 \neq 0 \urcorner)$. Moreover, because of $\vdash_{\mathsf{PA}^{\perp}} \neg\, \mathsf{Con}_{\mathsf{PA}}$ and hence $\vdash_{\mathsf{PA}^{\perp}} \neg\, \mathsf{Con}_{\mathsf{PA}^{\perp}}$ by (6), $\mathsf{PA}^{\perp}$ proves (the reflection of) its own inconsistency, although along with $\mathsf{PA}$ also $\mathsf{PA}^{\perp}$ is consistent. It claims to have a mysterious proof of $\bot$. Thus, consistency of $T$ can have a different meaning within $T$ and seen from outside, just as the meanings of *countable* diverge, depending on whether one is situated in $\mathsf{ZFC}$ or is looking at it from outside. One may even say that $\mathsf{PA}^{\perp}$ is lying to us with the claim $\neg\, \mathsf{Con}_{\mathsf{PA}^{\perp}}$.

We learn from the preceding that the extension $T + \mathsf{Con}_T$ of a consistent theory $T$ need not be consistent. $T = \mathsf{PA}^{\perp}$ is a concrete example, and in fact only one of arbitrarily many others. More on the meaning of $\neg\, \mathsf{Con}_T$ will be said in Theorem 3.4.

We now discuss what is, along with (3), the most famous example of a self-referential sentence. Clearly, a fixed point $\alpha$ of $\Box(x)$ claims just its own provability, that is, $\alpha \equiv_T \Box\alpha$. A trivial example is $\alpha = \top$, because $\vdash_T \Box\top \to \top$, and since $\vdash_T \top$, clearly $\vdash_T \Box\top$, so that $\top \equiv_T \Box\top$. What is surprising here is that $\top$ turns out to be the only fixed point of $\Box(x)$ modulo $T$. By $D4°$ below, $\vdash_T \Box\alpha \to \alpha$ implies $\vdash_T \alpha$ and so $\alpha \equiv_T \top$ (which confirms the uniqueness), although one might perhaps expect $\vdash_T \Box\alpha \to \alpha$ for all $\alpha \in \mathcal{L}^0$ because $\Box\alpha \to \alpha$ is intuitively true.

**Theorem 3.3 (Löb's theorem).** *Take $T$ to satisfy D1–D3 and the fixed point lemma. Then $T$ has the properties*

$$D4: \vdash_T \Box(\Box\alpha \to \alpha) \to \Box\alpha, \quad D4^\circ: \vdash_T \Box\alpha \to \alpha \Rightarrow \vdash_T \alpha \quad (\alpha \in \mathcal{L}^0).$$

**Proof.** Let $\gamma$ be a fixed point of $\Box(x) \to \alpha$, i.e., $\gamma \equiv_T \Box\gamma \to \alpha$. Then $\gamma \equiv_T \Box\alpha \to \alpha$ by Lemma 3.1(b). This and $D0$ imply $\Box\gamma \equiv_T \Box(\Box\alpha \to \alpha)$. Lemma 3.1(a) states $\Box\gamma \equiv_T \Box\alpha$, hence $\Box\alpha \equiv_T \Box(\Box\alpha \to \alpha)$. Half of this is $D4$. Now suppose $\vdash_T \Box\alpha \to \alpha$. Then by $D1$, $\vdash_T \Box(\Box\alpha \to \alpha)$. Using $D4$ results in $\vdash_T \Box\alpha$, and $\vdash_T \Box\alpha \to \alpha$ yields $\vdash_T \alpha$, thus proving $D4^\circ$. ∎

$D4$ reflects just $D4^\circ$ in $T$. One application of Löb's theorem is an extremely easy proof of $\nvdash_{\mathsf{PA}} \mathsf{Con_{PA}}$. Indeed, $\vdash_{\mathsf{PA}} \mathsf{Con_{PA}}$ ($\equiv \Box\bot \to \bot$) implies $\vdash_{\mathsf{PA}} \bot$ by $D4^\circ$. That's all. Similarly, $D4$ implies (2) for $\alpha = \bot$ by contraposition. Thus, Löb's theorem is stronger than Gödel's second incompleteness theorem, which is not obvious at first glance.

Unlike $\mathsf{PA}^\perp$, $\mathsf{PA} + \mathsf{Con_{PA}}$ conforms to truth (in $\mathcal{N}$). Unfortunately it is not quite clear what $\mathsf{Con_{PA}}$ means in number-theoretic terms. This is clear, however, for an arithmetical statement discovered by Paris and Harrington (see [Bar]) that implies $\mathsf{Con_{PA}}$; this statement is provable in ZFC but not in PA. Since then, many such sentences have been found, mostly of a combinatorial nature. A popular example is

**Goodstein's theorem.** *Every Goodstein sequence ends in 0.*

A *Goodstein sequence* is a number sequence $(a_n)_{n\in\mathbb{N}}$, with arbitrary $a_0$ given in advance, such that $a_{n+1}$ is obtained from $a_n$ as follows: Let $b_n = n + 2$, so that $b_0 = 2$, $b_1 = 3$, etc. Expand $a_n$ in $b$-adic base for $b := b_n$, so that for suitable $k$,

$$(*) \quad a_n = \textstyle\sum_{i\leqslant k} b^{k-i} c_i, \text{ with } 0 \leqslant c_i < b.$$

Also the powers $k - i$ are represented in $b$-adic form, so too the powers of powers, and so on. Now replace $b$ everywhere with $b + 1$ ($= b_{n+1}$) and subtract 1 from the output. The result is $a_{n+1}$. The table below gives an example beginning with $a_0 = 11$; already $a_6$ has the value $134\,217\,727$.

| | | |
|---|---|---|
| $a_0 = 11 = 2^{2+1} + 2 + 1$ | $2 \rightsquigarrow 3$ | $3^{3+1} + 3 + 1 = 85$ |
| $a_1 = 84 = 3^{3+1} + 3$ | $3 \rightsquigarrow 4$ | $4^{4+1} + 4 = 1028$ |
| $a_2 = 1027 = 4^{4+1} + 3$ | $4 \rightsquigarrow 5$ | $5^{5+1} + 3 = 15\,628$ |
| $a_3 = 15\,627 = 5^{5+1} + 2$ | $5 \rightsquigarrow 6$ | $6^{6+1} + 2 = 279\,938$ |
| $a_4 = 279\,937 = 6^{6+1} + 1$ | $6 \rightsquigarrow 7$ | $7^{7+1} + 1 = 5\,764\,802$ |

As one sees from this example, $a_n$ initially increases enormously, and it is hardly believable that the sequence ever starts to decrease and ends in 0. But the proof of the theorem is not particularly difficult; one estimates $a_n$ from above by the ordinal number $\lambda_n$, which, crudely put, results from $a_n$ on replacing the basis $b$ in $(*)$ by $\omega$. With some ordinal arithmetic it can readily be shown that $\lambda_{n+1} < \lambda_n$ as long as $\lambda_n \neq 0$. Since there is no properly decreasing infinite sequence of ordinal numbers (these are well-ordered), the sequence $(a_n)_{n\in\mathbb{N}}$ must eventually end in 0. For more detailed information see for instance [HP].

Many metatheoretic properties can be expressed using the provability operator $\Box$ in $T$, often using sentence schemata. The following ones turn out to be equivalent and facilitate a better understanding of the meaning of $\neg\mathsf{Con}_T$ within $T$. None of these properties hold for a consistent $T$ from the outside (Theorem 6.5.1$'$), but all of them are provable in $T = \mathsf{PA}^\perp$.

(i)     $\neg\mathsf{Con}_T:$   $\Box\perp$                          (provable inconsistency),
(ii)    $\mathsf{SyComp}:$   $\Box\alpha \vee \Box\neg\alpha$                    (syntactic completeness),
(iii)   $\mathsf{SeComp}:$   $\alpha \to \Box\alpha$                      (semantic completeness),
(iv)    $\omega\text{-}\mathsf{Comp}:$   $\forall x\Box[\varphi(x)] \to \Box\forall x\varphi(x)$   ($\omega$-completeness).

**Theorem 3.4.** *The properties* (i)–(iv) *are all equivalent in a theory $T$ satisfying the properties named at the beginning of this section.*

**Proof.** By (4) (i)$\Rightarrow$(ii),(iii),(iv) are clear. (ii)$\Rightarrow$(i): By Rosser's theorem formulated in $T$ (see **7.5**), $\mathsf{Con}_T \vdash_T \neg\Box\alpha \wedge \neg\Box\neg\alpha$ for some $\alpha$. Thus, $\Box\alpha \vee \Box\neg\alpha \vdash_T \neg\mathsf{Con}_T$. (iii)$\Rightarrow$(i): For $\alpha := \mathsf{Con}_T$, SeComp and (2) yield $\alpha \vdash_T \Box\alpha$, $\neg\Box\alpha$ and so $\vdash_T \neg\alpha$. (iv)$\Rightarrow$(i): By (3) in **7.2**, we obtain $\neg\,\mathsf{bew}_T(x,\ulcorner\perp\urcorner) \vdash_T \Box[\neg\,\mathsf{bew}_T(x,\ulcorner\perp\urcorner)]$, for $\neg\,\mathsf{bew}_T(x,\ulcorner\perp\urcorner)$ is $\Sigma_1$. Hence,

$$\mathsf{Con}_T = \forall x\neg\,\mathsf{bew}_T(x,\ulcorner\perp\urcorner) \vdash_T \forall x\Box[\neg\,\mathsf{bew}_T(x,\ulcorner\perp\urcorner)].$$

$\omega$-Comp and (2) yield $\mathsf{Con}_T \vdash_T \Box\forall x\neg\,\mathsf{bew}(x,\ulcorner\perp\urcorner) = \Box\,\mathsf{Con}_T \vdash_T \neg\mathsf{Con}_T$. Therefore, $\vdash_T \neg\mathsf{Con}_T$. $\blacksquare$

**Remark.** $\mathsf{Con}_T$ is also equivalent in $T$ to other properties, for example to the schema $\Box\alpha \to \alpha$ for $\Pi_1$-formulas $\alpha$ (the *local* $\Pi_1$-reflection principle) as well as the *uniform* $\Pi_1$-reflection principle $\forall x\Box[\alpha(x)] \to \forall x\alpha(x)$ for $\Pi_1$-formulas $\alpha$. Both the theorems of Paris–Harrington and of Goodstein are equivalent in PA to the uniform $\Sigma_1$-reflection, or equivalently, to the consistency of PA plus all true $\Pi_1$-sentences; see e.g. [Bar, D8].

Define inductively $T^0 = T$ and $T^{n+1} = T^n + \mathsf{Con}_{T^n}$. This *n-times-iterated consistency extension* $T^n$ can be written as $T^n = T + \neg\square^n\bot$ with $\square = \mathsf{bwb}_T$, $\square^0\alpha = \alpha$ and $\square^{n+1}\alpha = \square\square^n\alpha$ (Exercise 3). Thus, the consistency of $T^n$ can be expressed by an iterated consistency statement on $T$. Let $T^\omega := \bigcup_{n\in\omega} T^n$. Since $T^n \subseteq T^{n+1}$ and $T^n = T + \neg\square^n\bot$ (hence $T^\omega = T \cup \{\neg\square^n\bot \mid n \in \omega\}$), the following three items are equivalent:

(i) $T^\omega$ is consistent, (ii) $T^n$ is consistent for all $n$, (iii) $\nvdash_T \square^n\bot$ for all $n$.

Like $\mathsf{PA}^1 = \mathsf{PA} + \mathsf{Con}_{\mathsf{PA}}$, also $\mathsf{PA}^\omega$ conforms to truth looking at $\mathsf{PA}$ from outside. When considered more closely, this means only that $\mathsf{PA}^\omega$ is relatively consistent with respect to $\mathsf{ZFC}$. In other terms, $\vdash_{\mathsf{ZFC}} \mathsf{Con}_{\mathsf{PA}^\omega}$. The argument (to be formalized in $\mathsf{ZFC}$) runs as follows: $\vdash_{\mathsf{PA}^\omega} \bot$ implies $\vdash_{\mathsf{PA}^n} \bot$ for some $n$, as was noticed above, hence $\vdash_{\mathsf{PA}} \square^n\bot$. But this is impossible, as is seen by a repeated application of $D1^*$ (p. 271) on $\mathsf{PA}$.

### Exercises

1. Prove $D4^\circ$ for $T$ by applying Theorem 3.2 to $T' = T + \neg\alpha$.

2. Show by means of Löb's theorem that $\mathsf{Con}_{\mathsf{PA}} \to \neg\square\neg\mathsf{Con}_{\mathsf{PA}}$ is unprovable in $\mathsf{PA}$, although this formula is true if seen from outside.

3. Let $T^n$ recursively be defined as in the text above. Prove that $T^n = T + \neg\square^n\bot$ and $\mathsf{Con}_{T^n} \equiv_T \neg\square^{n+1}\bot$, where $\square$ is $\mathsf{bwb}_T$.

4. Show that $\vdash_{\mathsf{ZFC}} \square_{\mathsf{PA}}\alpha \to \alpha$ for all arithmetical sentences $\alpha$ from $\mathcal{L}_\in$ (the $\mathcal{L}_\in$-sentences relativized to $\omega$).

## 7.4   The Provability Logic G

In **7.3** first-order logic was hardly required. It comes then as no surprise that many of the results there can be obtained propositionally, more precisely, in a certain modal propositional calculus. This calculus contains alongside $\wedge, \neg$ the falsum symbol $\bot$, and a further unary connective $\square$ to be interpreted as the proof operator in $\mathcal{L}_{ar}$, denoted by $\square$ as well. First we define a propositional language $\mathcal{F}_\square$, whose formulas are denoted by $H, G, F$: (a) the variables $p_1, p_2, \ldots$ from $PV$ (page 4) and $\bot$ belong to $\mathcal{F}_\square$; (b) if $H, G$ belong to $\mathcal{F}_\square$ then so too $(H \wedge G)$, $\neg H$, and $\square H$.

No other strings belong to $\mathcal{F}_\square$ in this context. $H \vee G$, $H \to G$, and $H \leftrightarrow G$ are defined as in **1.4**, $\top := \neg\bot$. Further, set $\Diamond H := \neg\square\neg H$ and define recursively $\square^0 H = H$, $\square^{n+1} H = \square\square^n H$. Let G be the set of those formulas in $\mathcal{F}_\square$ derivable using substitution in $\mathcal{F}_\square$, modus ponens MP, and the rule MN: $H/\square H$ from the tautologies of two-valued propositional logic, augmented by the axioms (called also the G-axioms)

$$\square(p \to q) \to \square p \to \square q, \quad \square p \to \square\square p,[4] \quad \square(\square p \to p) \to \square p.$$

For $H \in$ G we mostly write $\vdash_G H$ (read "$H$ is derivable in G"). Rule MN corresponds to $D1$. The first G-axiom reflects $D2$, the middle $D3$, and the last (called *Löb's formula*) $D4$, hence the name *provability logic*. The connection between G and PA is described in **7.5**. Here we are concerned with the modal logic G and its *Kripke semantics*. For simplicity, we restrict ourselves to finite *Kripke frames*, which are just finite directed graphs. We can do so, since all modal logics considered here have the finite model property. We begin without further ado with the following

**Definition.** A G-*frame* or *Kripke frame for* G is a finite poset $(g, <)$. A *valuation* is a mapping $w$ that assigns to every variable $p$ a subset $wp$ of $g$. The relation $P \Vdash H$, dependent on $w$, between points $P \in g$ and formulas $H \in \mathcal{F}_\square$ (read "$P$ accepts $H$") is defined inductively by

$$P \Vdash p \text{ iff } P \in wp, \quad P \not\Vdash \bot, \quad P \Vdash H \wedge G \text{ iff } P \Vdash H \ \& \ P \Vdash G,$$
$$P \Vdash \neg H \text{ iff } P \not\Vdash H, \quad P \Vdash \square H \text{ iff } P' \Vdash H \text{ for all } P' > P.$$

These conditions easily imply $P \Vdash \Diamond H$ iff $P' \Vdash H$ for some $P' > P$, and $P \Vdash H \to G$ iff $P \Vdash H \Rightarrow P \Vdash G$. If $P \Vdash H$ for all $w$ and all $P \in g$, we write $g \vDash H$ and say $H$ *holds in* $g$. If $g \vDash H$ for all G-frames $g$, we write $\vDash_G H$ and say $H$ *is* G-*valid*. The G-frame on the right, consisting of two points $P, P'$ with $P < P'$, shows that $\nvDash_G p \to \square p$. $\overset{P}{\bullet} \longrightarrow \overset{P'}{\bullet}$ Indeed, let $wp = \{P\}$. Then $P \Vdash p$, but $P \not\Vdash \square p$ because $P' \not\Vdash p$. Note also $\nvDash_G \square p \to p$, for $P' \not\Vdash p$ but $P' \Vdash \square p$ because there is no $P'' > P'$.

We may tacitly assume that G-frames are *initial* (have a smallest point), for $g \vDash H$ is verified pointwise. We write $H \equiv_G H'$ for $\vDash_G H \leftrightarrow H'$. It is readily seen that $\equiv_G$ is a congruence in $\mathcal{F}_\square$ that extends the usual logical equivalence conservatively. For instance, $\neg\square H \equiv_G \neg\square\neg\neg H \equiv_G \Diamond\neg H$. Many more equivalences are presented in the following examples. These will later be translated into statements about self-reference.

---

[4] This axiom is dispensable; it is provable from the remaining, see e.g. [Boo] or [Ra1].

**Examples.** (a) Let $g$ be an arbitrary G-frame. Although always $P \nVdash \bot$, we have $P \Vdash \Box\bot$, provided $P$ is maximal in $g$, that is, no $Q > P$ exists. Likewise, $\Box\neg\Box\bot$ is accepted precisely at the maximal points of $g$. Thus, $\Box\bot \equiv_G \Box\neg\Box\bot$, or equivalently, $\neg\Box\bot \equiv_G \Diamond\Box\bot$ $(= \neg\Box\neg\Box\bot)$. This reflects in G the second incompleteness theorem, as will be seen in **7.5**.

(b) Let $\{P_0, \ldots, P_n\}$ be the ordered G-frame with $P_n < \cdots < P_0$. Clearly, $P_0 \Vdash \Box^m\bot$ for each $m > 0$. Induction on $n$ shows that $P_n \Vdash \Box^m\bot$ for all $m > n$, but $P_n \nVdash \Box^n\bot$, and therefore $P_n \nVdash \Box^{n+1}\bot \to \Box^n\bot$. Hence, $\nvDash_G \Box^{n+1}\bot \to \Box^n\bot$, and a fortiori $\nvDash_G \Box^n\bot$ and $\nvDash_G \neg\Box^{n+1}\bot$, for all $n$.

(c) $\vDash_G \Box(\Box p \to p) \to \Box p$. For take an arbitrary $g$ and $P \in g$. If $P \nVdash \Box p$ then there is, since $g$ is finite, some $Q > P$ with $Q \Vdash \neg p$ and $Q' \Vdash p$ for all $Q' > Q$. Thus $Q \Vdash \Box p$; hence $Q \nVdash \Box p \to p$ and so $P \nVdash \Box(\Box p \to p)$. Consequently, $P \Vdash \Box(\Box p \to p) \to \Box p$, which proves our claim. Note also that $\vDash_G \Box p \to \Box\Box p$. Only the transitivity of $<$ is relevant for the proof.

(d) $\vDash_G \neg\Box^{n+1}\bot \to \Diamond R_n$, where $R_n := \bigwedge_{i=1}^{n}(\Box p_i \to p_i)$. For let $P \in g$, $P \Vdash \neg\Box^{n+1}\bot$. Then there must be a chain $P = P_0 < P_1 < \cdots < P_{n+1}$ in $g$. Now, it is a nice separate exercise to verify that each conjunct of $R_n$ fails to be accepted by at most one of the $n + 1$ points $P_1, \ldots, P_{n+1}$. Thus, at least one of these accepts all conjuncts. In other words, $P_i \Vdash R_n$ for some $i > 0$; hence $P \Vdash \Diamond R_n$. This nontrivial example will essentially be employed in the proof of Theorem 7.1.

By induction on $\vdash_G H$ one easily proves $\vdash_G H \Rightarrow \vDash_G H$ (soundness of Kripke semantics for $\vdash_G$). Example (c) is a part of the initial step. The induction steps over the rules are easy. For instance, $g \vDash H$ clearly implies $g \vDash \Box H$. The converse, $\vDash_G H \Rightarrow \vdash_G H$, holds as well. Thus, $\vdash_G H$ can be confirmed by proving $\vDash_G H$, and vice versa. This is the content of

**Theorem 4.1 (Completeness of Kripke semantics for G).** *For each formula $H$ from $\mathcal{F}_\Box$ it holds that $\vdash_G H \Leftrightarrow \vDash_G H$.*

The nontrivial direction $\Leftarrow$ follows directly from the finite model property of G, i.e., each $H \notin G$ is falsified or refuted by some finite G-frame, proved, for example, in [Boo], [Ra1], and [CZ]. For the relatively simple formulas considered here, $\vDash_G H$ is in general more easily checked than $\vdash_G H$.

Both the formulas provable in G and those refutable are clearly recursively enumerable, thanks to the finite model property of G. Thus, in analogy to Exercise 2 in **3.6**, we obtain

**Theorem 4.2.** $G$ *is decidable.*

**Remark.** The finite model property, decidability, and some other properties such as interpolation can all be proved in one move, see e.g. [Ra2]. An important fragment of G is $G^0 := G \cap \mathcal{F}_\square^0$, where $\mathcal{F}_\square^0$ denotes the set of variable-free formulas of $\mathcal{F}_\square$. The formulas $\neg \square^n \bot$ ($\equiv_G \Diamond^n \top$) form a Boolean base in $G^0$. One proves this most easily by showing that $G^0$ is complete with respect to all (totally) ordered G-frames, including the infinite ones, and applying Theorem 5.2.3 accordingly.

### Exercises

1. Let $g$ be *any* finite Kripke frame (a graph) that satisfies the axioms of $G$. Show that $g$ is necessarily a poset. Only this fact justifies the identification of G-frames with posets.

2. Prove $\vdash_G \square p \to \square(\square p \to p)$, the inverse of Löb's formula. (Only the first of the three G-axioms is needed in the proof.)

## 7.5 The Modal Treatment of Self-Reference

Let $T$ be a theory as in **7.3**. A mapping $\imath$ from $PV$ to $\mathcal{L}^0$ with $p_i^\imath = \alpha_i$ is called an *insertion*. $\imath$ can be extended to the whole of $\mathcal{F}_\square$ by the clauses $\bot^\imath = \bot$, $(\neg H)^\imath = \neg H^\imath$, $(H \wedge G)^\imath = H^\imath \wedge G^\imath$, and $(\square H)^\imath = \square H^\imath$ ($= \mathsf{bwb}_T(\ulcorner H^\imath \urcorner)$). Briefly speaking, $H^\imath$ results from $H(p_1, \ldots, p_n)$ by replacing the $p_\nu$ by the sentences $\alpha_\nu$ from $\mathcal{L}$. For instance, if $p^\imath = \alpha$ then $(\square p \wedge \neg \square \bot)^\imath = \square \alpha \wedge \neg \square \bot$, and $(\neg \square \bot)^\imath = \neg \square \bot = \mathsf{Con}_T$. The following lemma shows that $\vdash_G$ is "sound" for $\vdash_T$. Already this simple fact considerably simplifies proofs about self-referential statements.

**Lemma 5.1.** *For each $H$ with $\vdash_G H$ and each insertion $\imath$, $\vdash_T H^\imath$.*

**Proof** by induction on $\vdash_G H$. If $H$ is a propositional tautology then $H^\imath \in \mathsf{Taut}_\mathcal{L} \subseteq T$. If $H$ is one of the modal axioms of G, then $\vdash_T H^\imath$ by D2, D3, or D4. If $\vdash_G H$ and $\sigma : \mathcal{F}_\square \to \mathcal{F}_\square$ is a substitution, then $\vdash_T H^{\sigma \imath}$, since $H^{\sigma \imath} = H^{\imath'}$ with $\imath' : p \mapsto p^{\sigma \imath}$, and $\vdash_T H^{\imath'}$ holds by the induction hypothesis. As regards the induction step over MP, consider $(F \to G)^\imath = F^\imath \to G^\imath$. Finally, if MN is applied, and $\vdash_T H^\imath$ by the induction hypothesis, then $\vdash_T \square H^\imath = (\square H)^\imath$, due to D1. $\quad\square$

**Example 1.** We prove (3) of Theorem 3.2 with the calculus $\vdash_G$. By Lemma 5.1 and Theorem 4.1 it suffices to show that $\vDash_G \neg \square \bot \leftrightarrow \neg \square \neg \square \bot$.

This holds by Example (a) in **7.4**. Next example: $\vDash_G \Box(p \leftrightarrow \Diamond p) \rightarrow \neg \Diamond p$ is easily confirmed. Thus, $\vdash_T \Box(\alpha \leftrightarrow \Diamond \alpha) \rightarrow \neg \Diamond \alpha$. This tells us (if everything is related to $T = \mathsf{PA}$) that a sentence claiming its own consistency with $\mathsf{PA}$ is incompatible with $\mathsf{PA}$, which hardly seems plausible. Even the converse is provable in $\mathsf{PA}$ since $\vDash_G \neg \Diamond p \rightarrow \Box(p \leftrightarrow \Diamond p)$.

We now explain certain facts that expand upon the reasoning of above. For $\mathsf{PA}$ and related theories, the converse of Lemma 5.1 holds as well. That is to say, the derivability conditions and Löb's theorem already contain everything worth knowing about self-referential formulas or schemes. This is essentially the content of Theorem 5.2. For the subtle proofs of Theorems 5.2, 5.4, and 5.5, the reader is referred to [Boo].

**Theorem 5.2 (Solovay's completeness theorem).** *For all $H \in \mathcal{F}_\Box$: $\vdash_G H$ (equivalently $\vDash_G H$) if and only if $\vdash_{\mathsf{PA}} H^\iota$ for all insertions $\iota$.*

**Example 2 (applications).** (a) $\nvdash_{\mathsf{PA}} \Box^{n+1} \bot \rightarrow \Box^n \bot$ because by Example (b) in **7.4**, $\nvDash_G \Box^{n+1} \bot \rightarrow \Box^n \bot$. In particular, $\nvdash_{\mathsf{PA}} \mathsf{Con}_{\mathsf{PA}}$ ($\equiv \Box\bot \rightarrow \bot$). (b) $\nvdash_{\mathsf{PA}} \neg\Box^{n+1}\bot$, since $\nvDash_G \neg\Box^{n+1}\bot$. (c) It is easily verified with the 2-point frame on page 285 that $\nvDash_G \neg\Box p \rightarrow \Box\neg\Box p$, in particular $\nvDash_G \neg\Box\bot \rightarrow \Box\neg\Box\bot$. Therefore, $\nvdash_{\mathsf{PA}} \mathsf{Con}_{\mathsf{PA}} \rightarrow \Box\mathsf{Con}_{\mathsf{PA}}$. (d) $\mathsf{PA}_n := \mathsf{PA} + \Box^n\bot$ is consistent for $n > 0$ by (b), but is $\omega$-inconsistent. Otherwise, by $D1^*$ (page 271), $\vdash_{\mathsf{PA}_n} \Box^n\bot \Rightarrow \vdash_{\mathsf{PA}_n} \Box^{n-1}\bot \Rightarrow \cdots \Rightarrow \vdash_{\mathsf{PA}_n} \bot$, contradicting $\nvdash_{\mathsf{PA}_n} \bot$. Since $\vdash_{\mathsf{PA}} \Box^n\bot \rightarrow \Box^{n+1}\bot$ by $D3$, we get $\mathsf{PA}_n \supseteq \mathsf{PA}_{n+1}$, and since $\mathsf{PA}_n \neq \mathsf{PA}_{n+1}$ by (a), we have $\mathsf{PA}_0 \supset \mathsf{PA}_1 \supset \cdots \supset \mathsf{PA}$. Observe that $\mathsf{PA}_1$ is just $\mathsf{PA}^\bot$.

Note also the following: Since $\nvDash_G \Box p \rightarrow p$, there must be some $\alpha \in \mathcal{L}^0_{ar}$ such that $\nvdash_{\mathsf{PA}} \Box\alpha \rightarrow \alpha$. Indeed, choose $\alpha = \bot$. The above examples point out that Theorem 5.2 and the decidability of $G$ are very efficient tools in deciding the provability of self-referential statements.

Many other theories have the same provability logic as $\mathsf{PA}$, where in general a modal propositional logic $H$ is the *provability logic for $T$* when the analogue of Theorem 5.2 holds with respect to $T$ and $H$. For some theories, the provability logic may be a proper extension of $G$. For example, the $\omega$-inconsistent theory $\mathsf{PA}_n$ from Example 2(d) has the provability logic $G_n := G + \Box^n\bot$, the smallest extension of $G$ closed under all rules of $G$ with the additional axiom $\Box^n\bot$ (Exercise 1; note that $G_0$ is inconsistent). By the following theorem, which will be proved in **7.7**, other extensions of $G$ to be considered as provability logics are out of the question.

**Theorem 5.3 ([Vi1]).** *Let $T$ be at least as strong as* PA. *Then*

(a) *If $T^\omega$ (page 284) is consistent, then* G *is the provability logic of $T$;*

(b) *if $\vdash_{T^\omega} \perp$ and $n$ is minimal such that $\vdash_{T^n} \perp$, then $T$'s provability logic is* $G_n$.

The formulas $H \in \mathcal{F}_\square$ such that $\mathcal{N} \models H^\imath$ for all insertions $\imath$ in $\mathcal{L}_{ar}$ can also be surprisingly easily characterized. All $H \in$ G are obviously included; but in addition also $\square p \to p$ belongs to this sort of formula, because $\mathcal{N} \models \square\alpha \to \alpha$ for $\alpha \in \mathcal{L}^0_{ar}$. Indeed, if $\mathcal{N} \models \square\alpha$ then *there is* some $n$ that codes a proof of $\alpha$ in PA, hence $\mathcal{N} \models \alpha$.

Let GS ($\supseteq$ G) be the set of all formulas in $\mathcal{F}_\square$ that can be obtained from those in $G \cup \{\square p \to p\}$ using substitution and modus ponens only. Induction in GS readily yields $H \in$ GS $\Rightarrow \mathcal{N} \models H^\imath$ for all $\imath$. Again, the converse holds as well:

**Theorem 5.4 ([So]).** $H \in$ GS *if and only if* $\mathcal{N} \models H^\imath$ *for all insertions* $\imath$.

GS is decidable as well, because it can be shown that $H \in$ GS $\Leftrightarrow H^* \in$ G, where $H^* := [\bigwedge_{\square G \in Sf^\square H}(\square G \to G)] \to H$. Here $Sf^\square H$ is the set of sub-formulas of $H$ of the form $\square G$. Thus, Theorem 5.4 reduces the decidability of GS to that of G. Using this theorem, many questions concerning the relations between *provable* and *true* are effectively decidable. For instance,

$$H(p) := \neg\square(\neg\square\perp \to \neg\square p \wedge \neg\square\neg p) \notin \text{GS}$$

is readily verified. Hence $\mathcal{N} \models \neg H(\alpha) \equiv \square(\neg\square\perp \to \neg\square\alpha \wedge \neg\square\neg\alpha)$ for some $\alpha \in \mathcal{L}^0_{ar}$ by Theorem 5.4. Translated into English: *It is provable in* PA *that the consistency of* PA *implies the independence of $\alpha$ for some sentence $\alpha$.* This is exactly Rosser's theorem, which in this way turns out to be provable in PA. As was shown in [Bel], the box in the formulas $H \in$ GS in Theorem 5.4 may denote $bwb_T$ for *any* axiomatizable $T \supseteq$ PA, provided $T \subseteq Th\mathcal{N}$. However, if $T$ proves false sentences (as does e.g. $PA^\perp$) then GS has to be redefined in a feasible manner and is always decidable.

A variable $p$ in $H$ is called *modalized in $H$* if every occurrence of $p$ is contained within the scope of a $\square$, as is the case in $\neg\square p$, $\neg\square\neg p$, and $\square(p \to q)$. By contrast, $p$ is not modalized in $\square p \to p$. Another particularly interesting theorem is

**Theorem 5.5 (DeJongh–Sambin fixed point theorem).** *Let $p$ be modalized in $H(p, q_1, \ldots, q_n)$, $n \geqslant 0$. Then a formula $F = F(\vec{q})$ from $\mathcal{F}_\square$ can effectively be constructed such that*

   (a) $F \equiv_\mathsf{G} H(F, \vec{q})$,
   (b) $\vdash_\mathsf{G} \bigwedge_{i=1}^{2} [(p_i \leftrightarrow H(p_i, \vec{q})) \wedge \square(p_i \leftrightarrow H(p_i, \vec{q}))] \rightarrow (p_1 \leftrightarrow p_2)$.

This theorem easily yields a corresponding result for theories $T$:

**Corollary 5.6.** *Let $p$ be modalized in $H = H(p, \vec{q})$ and suppose $T$ satisfies D1–D4. Then there is an $F = F(\vec{q}) \in \mathcal{F}_\square$ with $F(\vec{\alpha}) \equiv_T H(F(\vec{\alpha}), \vec{\alpha})$ for all $\vec{\alpha} = (\alpha_1, \ldots, \alpha_n)$, $\alpha_i \in \mathcal{L}^0$. For each $\vec{\alpha}$ there is only one $\beta \in \mathcal{L}^0$ modulo $T$ such that $\beta \equiv_T H(\beta, \vec{\alpha})$.*

**Proof.** Choose $F$ as in (a) of the theorem. Then $F(\vec{\alpha}) \equiv_T H(F(\vec{\alpha}), \vec{\alpha})$ by Lemma 5.1 ($\vec{q}^{\,\iota} = \vec{\alpha}$). To prove uniqueness let $\beta_i \equiv_T H(\beta_i, \vec{\alpha})$ for $i = 1, 2$. By D1, $\vdash_T (\beta_i \leftrightarrow H(\beta_i, \vec{\alpha})) \wedge \square(\beta_i \leftrightarrow H(\beta_i, \vec{\alpha}))$. Inserting $\beta_i$ for $p_i$ and $\alpha_i$ for $q_i$ in the formula under (b) in the theorem then yields $\vdash_T \beta_1 \leftrightarrow \beta_2$ by Lemma 5.1. ∎

**Example 3.** For $H = \neg\square p$ ($n = 0$), $F = \neg\square\bot$ is a "solution" of (a) in Theorem 5.5 because $\neg\square\bot \equiv_\mathsf{G} \neg\square(\neg\square\bot)$. According to Corollary 5.6, $\mathrm{Con}_T$ ($= \neg\square\bot$) is modulo $T$ the only fixed point of $\neg\,\mathbf{bwb}_T$. This is just the claim of (3) from **7.3**.

Many special cases of the corollary represent older self-reference results from Gödel, Löb, Rogers, Jeroslow, and Kreisel, which, stated in terms of modal logic, concern fixed points of $\neg\square p$, $\square p$, $\neg\square\neg p$, $\square\neg p$, and $\square(p \rightarrow q)$ in PA. Incidentally, one gets the fixed points of these formulas—namely $\neg\square\bot$, $\top$, $\bot$, $\square\bot$, and $\square q$—according to a simple recipe. All first listed formulas are of the form $H = G \frac{\square H'}{p}$, where $p$ is not modalized in $G(p, \vec{q})$ and $H'(p, \vec{q})$ is chosen appropriately. In this case, $F = H \frac{G(\top, \vec{q})}{p}$ is the fixed point of $H$, as is seen after some calculation. For $H = \neg\square p$ from Example 3 is $G = \neg p$. Thus, according to the recipe the fixed point is

$$F = \neg\square p \, \tfrac{\neg\top}{p} = \neg\square\neg\top \equiv_\mathsf{G} \neg\square\bot.$$

For Kreisel's formula $\square(p \rightarrow q)$ is $G = p$. Hence, it has the fixed point

$$F = \square(p \rightarrow q) \, \tfrac{\top}{p} = \square(\top \rightarrow q) \equiv_\mathsf{G} \square q.$$

The recipe also works for $H = \square p \rightarrow q$, by choosing $G = p \rightarrow q$. Hence $F = (\square p \rightarrow q) \frac{\top \rightarrow q}{p} = \square(\top \rightarrow q) \rightarrow q \equiv_\mathsf{G} \square q \rightarrow q$ is the only fixed point of

$H$ modulo $T$. Exactly this is the claim of Lemma 3.1(b), used in Gödel's second incompleteness theorem.

### Exercises

1. Prove that the theory $\mathsf{PA}_n$ from Example 2(d) has the provability logic $\mathsf{G}_n$.

2. Show that $\mathsf{PA}^n_\perp := \mathsf{PA}^n + \neg\,\mathsf{Con}_{\mathsf{PA}^n}$ equals $\mathsf{PA} + \Box^{n+1}\perp \wedge \neg\Box^n\perp$ and that it has the provability logic $\mathsf{G}_1 = \mathsf{G} + \Box\perp$. Here $\Box$ means $\Box_{\mathsf{PA}}$.

3. Prove that $\top$, $\perp$, and $\Box\perp$ are the fixed points of $\Box p$, $\neg\Box\neg p$, and $\Box\neg p$.

4. (Mostowski). Let $T \supseteq \mathsf{PA}$ be axiomatizable and suppose $\mathcal{N} \vDash T$. Show that there are two mutually independent $\Sigma_1$-sentences $\alpha, \beta$ in $T$, that is, $\alpha \to \beta$, $\alpha \to \neg\beta$, $\beta \to \alpha$, $\beta \to \neg\alpha$ (hence also $\alpha$, $\beta$, $\neg\alpha$, and $\neg\beta$) are unprovable in $T$.

## 7.6   A Bimodal Provability Logic for PA

Hilbert remarked jokingly that the incompleteness phenomenon can be forcefully removed from the world by use of the so-called $\omega$-*rule*

$$\rho_\omega : \quad \frac{X \vdash \varphi(\underline{n}) \text{ for all } n}{X \vdash \forall x\varphi}.$$

$\rho_\omega$ has infinitely many premises. It is an easy exercise to derive with the aid of $\rho_\omega$ every sentence $\alpha$ valid in $\mathcal{N}$ from the axioms of $\mathsf{PA}$, even from those of $\mathsf{Q}$. Indeed, all sentences can (up to equivalence) be obtained from variable-free literals with $\wedge, \vee, \forall, \exists$, bypassing formulas with free variables. Due to the $\Sigma_1$-completeness of $\mathsf{Q}$, all valid variable-free literals are derivable. The inductive steps for $\wedge, \vee, \exists$ are simple, applying $\Sigma_1$-completeness in the $\exists$-step once again. Only in the $\forall$-step is $\rho_\omega$ used.

Clearly, an unrestricted use of the infinitistic rule $\rho_\omega$ (in spite of its relevance for higher order arithmetic) contradicts Hilbert's own intention of giving mathematics a finitistic foundation. However, things look different if we restrict $\rho_\omega$ each time to a *single* application. In view of Remark 1 in **6.2**, we no longer distinguish between $\varphi$ and $\dot\varphi$, so that $\varphi$ itself is a number and $\ulcorner\varphi\urcorner = \underline\varphi$ is the corresponding Gödel term. Let us define

$$1bwb_{PA}(\alpha) := (\exists \varphi \in \mathcal{L}^1_{ar})[bwb_{PA}(\forall x \varphi \rightarrow \alpha) \ \& \ \forall n \ bwb_{PA}(\varphi(\underline{n}))].$$

$1bwb_{PA}$ is arithmetical, in fact it is $\Sigma_3$, for $bwb_{PA}$ is $\Sigma_1$ and $\forall n \ bwb_{PA}(\varphi(\underline{n}))$ is $\Pi_1$. We read $1bwb_{PA}(\alpha)$ as "$\alpha$ is 1-provable." Let $\mathtt{1bwb}(x)$ be the $\Sigma_3$-formula in $\mathcal{L}_{ar}$ defining $1bwb_{PA}$. Here let $x$ be $\boldsymbol{v}_0$. Write $\boxed{1}\alpha$ for $\mathtt{1bwb}(\ulcorner\alpha\urcorner)$ and $\diamondsuit\alpha$ for $\neg\boxed{1}\neg\alpha$. Clearly, $\Box\alpha$ for $\alpha \in \mathcal{L}^0_{ar}$ ($\Box = \Box_{PA}$) can be read 'PA $+ \neg\alpha$ is inconsistent', while $\boxed{1}\alpha$, by Lemma 6.1, formalizes 'PA $+ \neg\alpha$ is $\omega$-inconsistent'. Thus, $\diamondsuit\top$ ($\equiv \neg\boxed{1}\bot$) means 'PA ($=$ PA $+ \neg\bot$) is $\omega$-consistent'. This explains the interest in the operator $\boxed{1}$.

If $bwb_{PA}(\alpha)$ *then certainly* $1bwb_{PA}(\alpha)$ (choose $\alpha$ for $\varphi$). The italicized statement is reflected in PA as '$\vdash_{PA} \Box\alpha \rightarrow \boxed{1}\alpha$ for every $\alpha \in \mathcal{L}^0_{ar}$'. The converse fails, since $\nvdash_{PA} \mathtt{Con}_{PA}$, while $\mathtt{Con}_{PA}$ is easily 1-provable: $\vdash_{PA} \varphi(\underline{n})$ for all $n$, with $\varphi(x) := \neg \ \mathtt{bew}_{PA}(x, \bot)$, and trivially $\vdash_{PA} \forall x \varphi(x) \rightarrow \mathtt{Con}_{PA}$. In what follows, some claims will not be proved in detail.

Define $\Omega := \{\varphi \in \mathcal{L}^1_{ar} \mid \ \vdash_{PA} \varphi(\underline{n}) \text{ for all } n\}$. By its definition, $\Omega$ and hence also $\mathsf{PA}^\Omega := \mathsf{PA} + \Omega$ are formally $\Sigma_3$. As Theorem 6.2 will show, $\mathsf{PA}^\Omega$ is properly $\Sigma_3$ and hence is no longer recursively axiomatizable.

**Lemma 6.1.** *The following properties are equivalent for* $\alpha \in \mathcal{L}^0_{ar}$:

    (i) $1bwb_{PA}(\alpha)$,   (ii) $\vdash_{\mathsf{PA}^\Omega} \alpha$,   (iii) PA $+ \neg\alpha$ *is* $\omega$-*inconsistent*.

**Proof.** (i)$\Rightarrow$(ii) follows with a glance at the definitions (read (i) naively). (ii)$\Rightarrow$(iii): Let $\vdash_{\mathsf{PA}^\Omega} \alpha$. Since $\Omega$ is closed under conjunctions, there is some $\forall x \varphi(x) \in \Omega$ with $\forall x \varphi \vdash_{PA} \alpha$, hence $\vdash_{PA} \neg\alpha \rightarrow \exists x \neg\varphi$ and so $\vdash_{PA+\neg\alpha} \exists x \neg\varphi$. Now, $\forall x \varphi \in \Omega$, therefore $\vdash_{PA} \varphi(\underline{n})$ and a fortiori $\vdash_{PA+\neg\alpha} \varphi(\underline{n})$, for all $n$. Thus, PA $+ \neg\alpha$ is $\omega$-inconsistent. (iii)$\Rightarrow$(i): Let $\vdash_{PA+\neg\alpha} \beta(\underline{n})$ for all $n$, but $\vdash_{PA+\neg\alpha} \exists x \neg\beta$. Then $\vdash_{PA} \forall x \beta \rightarrow \alpha$. With $\varphi(x) := \neg\alpha \rightarrow \beta(x)$ clearly $\vdash_{PA} \varphi(\underline{n})$ for all $n$. Now, $\forall x \varphi \equiv \alpha \vee \forall x \beta \vdash_{PA} \alpha$. Hence $\vdash_{PA} \forall x \varphi \rightarrow \alpha$. Thus, altogether $1bwb_{PA}(\alpha)$. $\blacksquare$

**Theorem 6.2 (the 1-provable $\Sigma_3$-completeness of PA).** *All true* $\Sigma_3$-*sentences are 1-provable. Moreover, for every* $\beta$ *of this kind,* $\vdash_{PA} \beta \rightarrow \boxed{1}\beta$.

**Proof.** Let $\mathcal{N} \vDash \beta := \exists y \forall x \gamma(y, x)$ where $\gamma(y, x)$ is $\Sigma_1$. Then there is some $m$ such that $\mathcal{N} \vDash \gamma(\underline{m}, \underline{n})$ for all $n$. Therefore, $\vdash_{PA} \gamma(\underline{m}, \underline{n})$ for all $n$, because PA is $\Sigma_1$-complete. Hence, $\forall x \gamma(\underline{m}, x) \in \Omega$ and so $\vdash_{\mathsf{PA}^\Omega} \exists z \forall x \gamma$, or equivalently, $1bwb_{PA}(\beta)$ by Lemma 6.1. Because of the provable $\Sigma_1$-completeness of PA, this argumentation is comprehensible in PA, so that also $\vdash_{PA} \beta \rightarrow \boxed{1}\beta$. $\blacksquare$

$D1$–$D4$ are also valid for the operator $\boxdot\colon \mathcal{L}_{ar}^0 \to \mathcal{L}_{ar}^0$. Indeed, $D1$ holds because $\vdash_{\mathsf{PA}} \alpha \Rightarrow\ \vdash_{\mathsf{PA}} \Box\alpha \Rightarrow\ \vdash_{\mathsf{PA}} \boxdot\alpha$, and $D2$ formalizes (or reflects) '$\vdash_{\mathsf{PA}\Omega} \alpha, \alpha \to \beta \Rightarrow\ \vdash_{\mathsf{PA}\Omega} \beta$' in $\mathsf{PA}$ (observe Lemma 6.1). $D3$ follows from Theorem 6.2 with $\beta = \boxdot\alpha$. The proof of $D4$ in **7.3** uses, along with the fixed point lemma, only $D1$–$D3$; so $D4$ holds as well. Therefore, nearly everything said in **7.3** on $\Box$ applies also to $\boxdot$, including Theorem 3.2, which now reads $\nvdash_{\mathsf{PA}} \neg\boxdot\bot$ ($\equiv \Diamond\top$). To put it more concisely, although the consistency of $\mathsf{PA}$ is provable with the extended means, $\omega$-consistency is not. Hence, this property, which is $\Pi_3$-definable according to Exercise 3 in **6.7**, cannot be $\Sigma_3$ by Theorem 6.2, and must therefore be properly $\Pi_3$. Equivalently, $\omega$-inconsistency is properly $\Sigma_3$.

Alongside $\Box\alpha \to \boxdot\alpha$, there are other noteworthy interactions between $\Box$ and $\boxdot$, in particular $\vdash_{\mathsf{PA}} \neg\Box\alpha \to \boxdot\neg\Box\alpha$. This formalizes 'If $\nvdash_{\mathsf{PA}} \alpha$ then $\neg\Box\alpha$ is 1-provable'. To verify the latter notice that $\nvdash_{\mathsf{PA}} \alpha$ implies $\vdash_{\mathsf{PA}} \varphi(\underline{n})$ for all $n$, where $\varphi(x)$ is $\neg\,\mathsf{bew}_{\mathsf{PA}}(x, \ulcorner\alpha\urcorner)$, and since $\vdash_{\mathsf{PA}} \forall x\varphi \to \neg\Box\alpha$, we get $\vdash_{\mathsf{PA}} \boxdot\neg\Box\alpha$. On the other hand, $\vdash_{\mathsf{PA}} \neg\Box\alpha \to \Box\neg\Box\alpha$ fails in general; Example 2(c) in **7.5** yields a counterexample.

The language of the bimodal propositional logic $\mathsf{GD}$ now to be defined results from $\mathcal{F}_\Box$ by adding a further connective $\boxdot$ to $\mathcal{F}_\Box$, which is treated syntactically just as $\Box$. The axioms of $\mathsf{GD}$ are those of $\mathsf{G}$ stated both for $\Box$ and $\boxdot$, augmented by the axioms

$$\Box p \to \boxdot p \quad \text{and} \quad \neg\Box p \to \boxdot\neg\Box p.$$

The rules of $\mathsf{GD}$ are the same as those for $\mathsf{G}$. Insertions $\imath$ to $\mathcal{L}_{ar}^0$ are defined as in **7.5**, but with the additional clause $(\boxdot H)^\imath = \boxdot H^\imath$, that is, $(\boxdot H)^\imath = \mathtt{1bwb}(\ulcorner H^\imath \urcorner)$. By the reasoning above, all axioms and rules of $\mathsf{GD}$ are sound. This proves (the easier) half of the following remarkable theorem from Dzhaparidze (1985):

**Theorem 6.3.** $\vdash_{\mathsf{GD}} H \Leftrightarrow\ \vdash_{\mathsf{PA}} H^\imath$ *for all insertions $\imath$ as defined above. Furthermore, $\mathsf{GD}$ is decidable.*

Thus, the modal system $\mathsf{GD}$ completely captures the interaction between $bwb_{\mathsf{PA}}$ and $1bwb_{\mathsf{PA}}$; also Theorem 5.5 carries over. However, $\mathsf{GD}$ no longer has an adequate Kripke semantics, which complicates the decision procedure. For further references see [Boo] or [Be3].

As an exercise, the reader should derive $\boxdot(\Box p \to p)$ from the axioms of $\mathsf{GD}$. Thus, $\vdash_{\mathsf{PA}} \boxdot(\Box\alpha \to \alpha)$ for *every* $\alpha \in \mathcal{L}_{ar}^0$, while $\vdash_{\mathsf{PA}} \Box(\Box\alpha \to \alpha)$ is the case only provided $\vdash_{\mathsf{PA}} \alpha$. In other words, the *local reflection principle* $\{\Box\alpha \to \alpha \mid \alpha \in \mathcal{L}_{ar}^0\}$ is 1-provable in PA. **Be careful:** GD expands G conservatively, so that $\nvdash_{\mathsf{GD}} \Box p \to p$.

## 7.7 Modal Operators in ZFC

Considerations regarding self-reference in ZFC are technically sometimes easier, but from the foundational point of view more involved because there is no superordinate theory. If ZFC is consistent, as we assume it is, then $\mathrm{Con}_{\mathsf{ZFC}}$ is a true arithmetical statement that is unprovable in ZFC. Thus, true arithmetical statements may even be unprovable in ZFC, not only in PA or similarly strong arithmetical theories. It makes sense, therefore, to consider $\mathsf{ZFC}^+ := \mathsf{ZFC} + \mathrm{Con}_{\mathsf{ZFC}}$, because after all, we want set theory to embrace as many facts about numbers and sets as possible from which interesting consequences may result.

As **7.3** shows, the consistency of ZFC alone does not guarantee that $\mathsf{ZFC}^+$ is consistent. The second incompleteness theorem clearly excludes $\vdash_{\mathsf{ZFC}} \mathrm{Con}_{\mathsf{ZFC}}$ but does not preclude $\vdash_{\mathsf{ZFC}} \mathrm{Con}_{\mathsf{ZFC}} \to \mathrm{Con}_{\mathsf{ZFC}^+}$. In this case $\vdash_{\mathsf{ZFC}^+} \mathrm{Con}_{\mathsf{ZFC}^+}$, and so $\vdash_{\mathsf{ZFC}^+} \bot$ by the same theorem. On the other hand, from certain assumptions about the existence of large cardinals, the consistency of $\mathsf{ZFC}^+$ readily follows. These assumptions would have to be jettisoned in case $\vdash_{\mathsf{ZFC}^+} \bot$, i.e. $\vdash_{\mathsf{ZFC}} \neg\mathrm{Con}_{\mathsf{ZFC}}$. Moreover, the consistency of ZFC would then not correctly be reflected in ZFC, and ZFC proves along with true arithmetical facts also false ones. This sounds strange, but there is hardly a convincing argument that this cannot be so.

Even if $\mathsf{ZFC}^+$ is consistent, i.e. $\nvdash_{\mathsf{ZFC}} \neg\mathrm{Con}_{\mathsf{ZFC}}$, it may still be that one of the sentences from the sequence $\Box\neg\mathrm{Con}_{\mathsf{ZFC}}, \Box\Box\neg\mathrm{Con}_{\mathsf{ZFC}}, \ldots$ is provable in ZFC (where $\Box$ denotes $\Box_{\mathsf{ZFC}}$ as long as it is not redefined). The latter is excluded only if we assume that the $\omega$-iterated consistency extension $\mathsf{ZFC}^\omega$ is consistent, hence $\nvdash_{\mathsf{ZFC}} \Box^n\bot$, for all $n$ (see page 284), so that by Theorem 5.3, G would be the provability logic of ZFC.

In fact, the assumption $(\forall n{\in}\mathbb{N})\ \nvdash_{\mathsf{ZFC}} \Box^n\bot$ is equivalent to G's being the provability logic of ZFC, by the general Theorem 7.1 below. Therein $Rf_T := \{\Box\alpha \to \alpha \mid \alpha \in \mathcal{L}^0\}$ denotes the already encountered reflection principle. Also Theorem 5.3 is a corollary of the theorem, simply because $(\forall n{\in}\mathbb{N})\ \nvdash_T \Box^{n+1}\bot$ is equivalent to the consistency of $T^\omega$.

**Theorem 7.1.** *For a sufficiently expressive theory $T$ [5] the following conditions are equivalent:*

   (i)   *$T^\omega$ is consistent,*

  (ii)  *$T + Rf_T$ is consistent,*

 (iii)  *$\mathsf{G}$ is the provability logic of $T$.*

**Proof.** (i)$\Rightarrow$(ii) indirect: Suppose that $T + Rf_T$ is inconsistent. Then there are formulas $\alpha_0, \ldots, \alpha_n$ such that $\vdash_T \neg\varphi$, $\varphi := \bigwedge_{i=1}^{n}(\Box\alpha_i \to \alpha_i)$. Hence $\vdash_T \Box\neg\varphi \equiv_T \neg\Diamond\varphi$. Now, because $\vdash_{T^\omega} \neg\Box^{n+1}\bot$, by Example (d) in **7.4** and Lemma 5.1, we get $\vdash_{T^\omega} \Diamond R_n^i$ ($p_i^i = \alpha_i$). Clearly, $R_n^i = \varphi$ and so $\vdash_{T^\omega} \Diamond\varphi$. Since also $\vdash_{T^\omega} \neg\Diamond\varphi$, $T^\omega$ is inconsistent. (ii)$\Rightarrow$(iii): The proof of Theorem 5.2 for $\mathsf{PA}$, as presented in [Boo], runs nearly in the same way for $T$, because $\mathsf{PA}$ is transgressed in one place only: one uses the fact that $\mathcal{N} \vDash Rf_{\mathsf{PA}}$. However, the existence of a corresponding $T$-model is ensured by (ii). (iii)$\Rightarrow$(i): $\nvdash_{\mathsf{G}} \Box^{n+1}\bot$, hence $\nvdash_T \Box^{n+1}\bot \equiv_T \neg\mathsf{Con}_{T^n}$ for all $n$, and so $T^\omega$ is consistent. $\quad\square$

    The equivalence (i)$\Leftrightarrow$(ii) is a purely proof-theoretic one. It is called *Goryachev's theorem*; see [Gor] or [Be2]. We obtained it using essentially some elementary modal logic. For $T = \mathsf{ZFC}$, perhaps a bit more interesting than (i) or (ii) is the assumption of the $\omega$-*consistency* of $\mathsf{ZFC}$, that is,

    (∗)  $\vdash_{\mathsf{ZFC}} (\exists x \in \omega)\varphi(x) \Rightarrow \nvdash_{\mathsf{ZFC}} \neg\varphi(\underline{n})$ for some $n$  ($\varphi(x) \in \mathcal{L}_\in$).

This assumption implies $D1^*$, which in turn ensures $\nvdash_{\mathsf{ZFC}} \Box^{n+1}\bot$ for all $n$, that is, (i), and hence all other conditions in Theorem 7.1 hold for $T = \mathsf{ZFC}$. It is worthwhile to observe that the consistency of $\mathsf{ZFC} + Rf_{\mathsf{ZFC}}$ and thereby the proof of Solovay's completeness theorem for $\mathsf{ZFC}$ follow directly from (∗), without appealing to Goryachev's theorem. What is needed to see that the latter is the case is the following

**Lemma.** *Suppose that $\mathsf{ZFC}$ is $\omega$-consistent. Then there exists a model $\mathcal{V} \vDash \mathsf{ZFC}$ such that $\mathcal{V} \vDash Rf_{\mathsf{ZFC}}$.*

**Proof.** Let $\Omega := \{(\forall x \in \omega)\alpha \mid \alpha = \alpha(x) \in \mathcal{L}_\in, \vdash_{\mathsf{ZFC}} \alpha(\underline{n})$ for all $n\}$. Then $\mathsf{ZFC} + \Omega$ is consistent. Indeed, otherwise $\vdash_{\mathsf{ZFC}} \neg(\forall x \in \omega)\alpha \equiv (\exists x \in \omega)\neg\alpha$ for some $(\forall x \in \omega)\alpha \in \Omega$ (since $\Omega$ is closed under conjunction), in contradiction to (∗). Any $\mathcal{V} \vDash \mathsf{ZFC} + \Omega$ satisfies the reflection principle $Rf_{\mathsf{ZFC}}$, for if

---

[5] By such a $T$ we mean that the proof steps of Solovay's Theorem 5.2 not transgressing $\mathsf{PA}$ can be carried out in $T$. This does not yet imply the provability of the theorem itself. Which steps are transgressing $\mathsf{PA}$ is described in the following proof.

$V \nvDash \alpha$ then $\nvdash_{\mathsf{ZFC}} \alpha$ and therefore $\vdash_{\mathsf{ZFC}} \neg \mathsf{bew}_{\mathsf{ZFC}}(\underline{n}, \ulcorner \alpha \urcorner)$ for all $n$. Hence $(\forall y \in \omega) \neg \mathsf{bew}_{\mathsf{ZFC}}(y, \ulcorner \alpha \urcorner) \in \Omega$, which clearly implies $V \nvDash \Box \alpha$. ∎

Now we interpret the modal operator $\Box$ no longer as *provable in* ZFC, which is equivalent to *valid in all* ZFC-*models*, but rather as *valid in particular classes of* ZFC-*models*. For undefined notions used in the sequel we refer to [Ku]. A 'model' is to mean throughout a ZFC-model.

Particularly interesting are *transitive* models, i.e. models $V = (V, \in^{V})$, where the set $V$ is *transitive*. This is to mean $a \in b \in V \Rightarrow a \in V$. In these models, $\in^{V}$ coincides with the ordinary $\in$-relation restricted to $V$ (a set in our metatheory that itself is ZFC). We write $V$ for $V$. Let $\rho a$ denote the *ordinal rank* of the set $a$, i.e., the smallest ordinal $\rho$ with $a \in V_{\rho+1}$. To prove the soundness half of Theorem 7.3 we will need

**Lemma 7.2.** ([JK]) *Let $V, W$ be transitive models such that $\rho V < \rho W$ and let $V \vDash \alpha$. Then $W \vDash$ 'there is a transitive model $U$ with $U \vDash \alpha$'.*[6]

Let the modal logic $\mathsf{Gi}$ result from augmenting $\mathsf{G}$ by the axiom

   (i)  $\Box(\Box p \to \Box q) \vee \Box(\Box q \to p)$.

$\mathsf{Gi}$ is complete with respect to all *preference orders* $g$, i.e., $g$ is a finite poset together with some function $\pi : g \to n \ (= \{0, \dots, n-1\})$ such that $P < Q \Leftrightarrow \pi P < \pi Q$, for all $P, Q \in g$. This implies the finite model property of $\mathsf{Gi}$, which, as for $\mathsf{G}$, ensures the decidability of $\mathsf{Gi}$. More suitable for our aims is the characterization of preference orders $g$ by the property

   (p)  $P < P'$ implies $P < Q$ or $Q < P'$, for all $P, P', Q \in g$,

which at once follows from the definition: Let $P < P'$, hence $\pi P < \pi P'$. If $P \nless Q$, i.e. $\pi P \nless \pi Q$, then $\pi Q \leqslant \pi P < \pi P'$, so that $Q < P'$. The proof of the converse is Exercise 1. The figure shows a poset $g$ that is *not* a preference order (for neither $P < Q$ nor $Q < P'$). Axiom (i) is easily refuted in $g$ choosing $wp = \{P'\}$, $wq = \emptyset$, and verifying that $O \Vdash \Diamond(\Box p \wedge \neg \Box q)$ and $O \Vdash \Diamond(\Box q \wedge \neg p)$ (for notice that $P \Vdash \Box p \wedge \neg \Diamond q$ and $Q \Vdash \Box q \wedge \neg p$). Hence, (i) does not belong to $\mathsf{G}$, so that $\mathsf{Gi}$ is a proper extension of $\mathsf{G}$. We mention that in [So] and in [Boo] a somewhat more complex axiomatization of $\mathsf{Gi}$ has been considered.

---

[6] In transitive models $W$ the sentence in ' ' (which with some encoding can be formulated in $\mathcal{L}_{\in}$) is absolute, and therefore equivalent to the existence of a transitive model $U \in W$ with $U \vDash \alpha$.

**Remark on splittings in modal logic.** The completeness of Gi with respect to all preference orders follows also from the fact that Gi is the split logic arising from splitting the lattice of all extensions of G (see e.g. [Kra]) by the subdirect irreducible G-algebra belonging to the frame from the previous page.

We define insertions $\imath\colon \mathcal{F}_\square \to \mathcal{L}_\in^0$ as in **7.5** as usual by $(\square H)^\imath = \square H^\imath$, where $\square\alpha$ for the set-theoretic sentence $\alpha = H^\imath \in \mathcal{L}_\in^0$ is now to mean '$\alpha$ is valid in all transitive models'. Accordingly, $\Diamond\alpha = \neg\square\neg\alpha$ states '$\alpha$ holds in at least one transitive model'.

**Theorem 7.3.** $\vdash_{\mathsf{Gi}} H$ *iff* $\vdash_{\mathsf{ZFC}} H^\imath$ *for all insertions $\imath$ as defined above.*

We prove only the direction $\Rightarrow$, that is, soundness. The converse is much more difficult, see [Boo]. As regards the axioms of Gi, since $\square p \to \square\square p$ is provable from the other axioms of G (see **7.4**), it suffices to prove

(A) $\square(\alpha \to \beta) \wedge \square\alpha \vdash_{\mathsf{ZFC}} \square\beta$,  (B) $\square(\square\alpha \to \alpha) \vdash_{\mathsf{ZFC}} \square\alpha$,

(C) $\vdash_{\mathsf{ZFC}} \square(\square\alpha \to \square\beta) \vee \square(\square\beta \to \alpha)$, for all $\alpha, \beta \in \mathcal{L}_\in^0$.

(A) is trivial, because the sentences valid in any class of models are closed under MP. (B) is equivalent to (B'): $\Diamond\neg\alpha \vdash_{\mathsf{ZFC}} \Diamond(\square\alpha \wedge \neg\alpha)$. Here is the proof: Suppose $\Diamond\neg\alpha$, i.e. there is a transitive model in which $\neg\alpha$ holds. Then there is also one with minimal rank, $V$ say. We claim $V \vDash \square\alpha$. Otherwise $V \vDash \Diamond\neg\alpha$, and hence there would be a transitive model $U \in V$ with $U \vDash \neg\alpha$ and $\rho U < \rho V$, contradicting our choice of $V$. Therefore, $V \vDash \square\alpha \wedge \neg\alpha$. Thus, there is a transitive model in which $\square\alpha \wedge \neg\alpha$ holds, which confirms (B'). Finally, (C) is verified by contraposition: suppose there are transitive models $V, W$ and sentences $\alpha, \beta$ such that

(a) $V \vDash$ '$\alpha$ holds in all transitive models and there is a transitive model in which $\neg\beta$ holds',

(b) $W \vDash$ '$\beta$ holds in all transitive models',  (c) $W \vDash \neg\alpha$.

From these assumptions it follows first of all that $\rho W < \rho V$. Indeed, suppose by (a) that $U \in V$ is a transitive model for $\neg\beta$. If $\rho V \leqslant \rho W$ then $\rho U < \rho W$. Hence, by Lemma 7.2, $W \vDash$ 'there is a transitive model for $\neg\beta$', contradicting (b). Now, since $W \vDash \neg\alpha$ by (c) and because of $\rho W < \rho V$, in $V$ holds 'there is some transitive model for $\neg\alpha$' by Lemma 7.2, in contradiction to (a). This proves (C). Soundness of the substitution rule follows as for G in **7.5**. MN is trivially sound, because if $\alpha$ is provable in ZFC then, of course, $\alpha$ is valid in all transitive models. Also MP is obvious: If $\alpha$ and $\alpha \to \beta$ hold in any class of models, then also $\beta$.

Another interesting model-theoretic interpretation of $\Box\alpha$ is '$\alpha$ is valid in all $V_\kappa$'. Here $\kappa$ runs through all inaccessible cardinal numbers. According to [So], the adequate modal logic for this interpretation of $\Box$ is
$$\mathsf{Gj} := \mathsf{G} + \Box(\Box p \wedge p \rightarrow q) \vee \Box(\Box q \rightarrow p).$$
More precisely, if there are infinitely many inaccessibles then we have

**Theorem 7.4.** $\vdash_{\mathsf{Gj}} H$ *iff* $\vdash_{\mathsf{ZFC}} H^\imath$ *for all insertions* $\imath$, *where* $\Box\alpha$ *is to mean '$\alpha$ is valid in all $V_\kappa$', $\kappa$ running through all inaccessible cardinals.*

$\mathsf{Gj}$ is also denoted by $\mathsf{G.3}$. This logic is complete with respect to all finite strict linear orders. These, of course, are also frames for $\mathsf{Gi}$, so that $\mathsf{Gi} \subseteq \mathsf{Gj}$. The figure shows a $\mathsf{Gi}$-frame, also called "the fork," on which the additional axiom is easily refuted at its initial point $O$ with $wp = \{P\}$ and $wq = \emptyset$. Hence the fork is not a $\mathsf{Gj}$-frame, and so $\mathsf{Gi} \subset \mathsf{Gj}$. The completeness of $\mathsf{Gj}$ with respect to finite orders entails the finite model property of $\mathsf{Gj}$ and hence its decidability.

We recommend that the reader carry out the proof of the soundness part of Theorem 7.4, without consulting the hints to the solutions (Exercise 4). It is easier than the soundness part of Theorem 7.3 proved above. All one needs to know besides Lemma 7.2 is that each $V_\kappa$ is a transitive ZFC-model and that $V_\kappa \in V_\lambda$ or $V_\lambda \in V_\kappa$, for arbitrary inaccessible cardinals $\kappa \neq \lambda$. Maybe the reader can also find a new and lucid proof of the hard direction of Theorem 7.4: If $\vdash_{\mathsf{ZFC}} H^\imath$ for all $\imath$ then $H$ holds in all finite strict linear orders, or equivalently, $\mathsf{Gj} \vdash H$.

## Exercises

1. Let $g$ be a $\mathsf{G}$-frame with property (**p**), page 296. Show by induction on the length of a maximal path in $g$ that $g$ is a preference order.

2. Show (using Exercise 1) that axiom (**i**) for $\mathsf{Gi}$ holds in a $\mathsf{G}$-frame $g$ iff $g$ is a preference order. This is an essential step in proving the completeness of $\mathsf{Gi}$ with respect to all preference orders.

3. This exercise is a crucial step in the completeness proof of $\mathsf{Gj}$. Show that a $\mathsf{G}$-frame $g$ is a frame for $\mathsf{Gj}$, i.e., $\Box(\Box p \wedge p \rightarrow q) \vee \Box(\Box q \rightarrow p))$ holds in $g$ if and only if $g$ is (totally) ordered.

4. Verify the soundness part of Theorem 7.4, i.e., $\vdash_{\mathsf{Gj}} \Rightarrow \vdash_{\mathsf{ZFC}} H^\imath$ for all insertions $\imath$.

# Bibliography

[AGM]   S. ABRAMSKY, D. M. GABBAY, T. S. E. MAIBAUM (editors), *Handbook of Computer Science*, I–IV, Oxford Univ. Press, Vol. I, II 1992, Vol. III 1994, Vol. IV 1995.

[Ac]    W. ACKERMANN, *Die Widerspruchsfreiheit der Allgemeinen Mengenlehre*, Mathematische Annalen 114 (1937), 305–315.

[Bar]   J. BARWISE (editor), *Handbook of Mathematical Logic*, North-Holland 1977.

[BF]    J. BARWISE, S. FEFERMAN (editors), *Model-Theoretic Logics*, Springer 1985.

[Be1]   L. D. BEKLEMISHEV, *On the classification of propositional provability logics*, Math. USSR – Izvestiya 35 (1990), 247–275.

[Be2]   ——, *Iterated local reflection versus iterated consistency*, Ann. Pure Appl. Logic 75 (1995), 25–48.

[Be3]   ——, *Bimodal logics for extensions of arithmetical theories*, J. Symb. Logic 61 (1996), 91–124.

[Be4]   ——, *Parameter free induction and reflection*, in *Computational Logic and Proof Theory*, Lecture Notes in Computer Science 1289, Springer 1997, 103–113.

[BM]    J. BELL, M. MACHOVER, *A Course in Mathematical Logic*, North-Holland 1977.

[Ben]   M. BEN-ARI, *Mathematical Logic for Computer Science*, New York 1993, 2nd ed. Springer 2001.

[BP]    P. BENACERRAF, H. PUTNAM (editors), *Philosophy of Mathematics, Selected Readings*, Englewood Cliffs NJ 1964, 2nd ed. Cambridge Univ. Press 1983, reprint 1997.

[BA]    A. BERARDUCCI, P. D'AQUINO, $\Delta_0$-*complexity of the relation* $y = \prod_{i \leqslant n} F(i)$, Ann. Pure Appl. Logic 75 (1995), 49–56.

[Bi]    G. BIRKHOFF, *On the structure of abstract algebras*, Proceedings of the Cambridge Philosophical Society 31 (1935), 433–454.

[Boo]   G. BOOLOS, *The Logic of Provability*, Cambridge Univ. Press 1993.

[BJ]    G. BOOLOS, R. JEFFREY, *Computability and Logic*, 3$^{rd}$ ed. Cambridge Univ. Press 1989.

[BGG]   E. BÖRGER, E. GRÄDEL, Y. GUREVICH, *The Classical Decision Problem*, Springer 1997.

[Bue]   S. BUECHLER, *Essential Stability Theory*, Springer 1996.

[Bu]    S. R. BUSS (editor), *Handbook of Proof Theory*, Elsevier 1998.

[Ca]    G. CANTOR, *Gesammelte Abhandlungen* (editor E. ZERMELO), Berlin 1932, Springer 1980.

[CZ]    A. CHAGROV, M. ZAKHARYASHEV, *Modal Logic*, Clarendon Press 1997.

[CK]    C. C. CHANG, H. J. KEISLER, *Model Theory*, Amsterdam 1973, 3$^{rd}$ ed. North-Holland 1990.

[Ch]    A. CHURCH, *A note on the Entscheidungsproblem*, J. Symb. Logic 1 (1936), 40–41, also in [Dav, 108–109].

[CM]    W. CLOCKSIN, C. MELLISH, *Programming in PROLOG*, 3$^{rd}$ ed. Springer 1987.

[Da]    D. VAN DALEN, *Logic and Structure*, Berlin 1980, 4$^{th}$ ed. Springer 2004.

[Dav]   M. DAVIS (editor), *The Undecidable*, Raven Press 1965.

[Daw]   J. W. DAWSON, *Logical Dilemmas, The Life and Work of Kurt Gödel*, A. K. Peters 1997.

[De]    O. DEISER, *Axiomatische Mengenlehre*, Springer, to appear 2010.

[Do]    K. DOETS, *From Logic to Logic Programming*, MIT Press 1994.

[EFT]   H.-D. EBBINGHAUS, J. FLUM, W. THOMAS, *Mathematical Logic*, New York 1984, 2$^{nd}$ ed. Springer 1994.

[FF]    A. B. FEFERMAN, S. FEFERMAN, *Alfred Tarski, Live and Logic*, Cambridge Univ. Press 2004.

[Fe1]   S. FEFERMAN, *Arithmetization of metamathematics in a general setting*, Fund. Math. 49 (1960), 35–92.

[Fe2]   ———, *In the Light of Logic*, Oxford Univ. Press 1998.

[Fel1]  W. FELSCHER, *Berechenbarkeit*, Springer 1993.

[Fe2]     ——, *Lectures on Mathematical Logic*, Vol. 1–3, Gordon & Breach 2000.

[Fi]      M. FITTING, *Incompleteness in the Land of Sets*, College Publ. 2007.

[Fr]      T. FRANZÉN, *Gödel's Theorem: An Incomplete Guide to Its Use and Abuse*, A. K. Peters 2005.

[Fre]     G. FREGE, *Begriffsschrift, eine der arithmetischen nachgebildete Formelsprache des reinen Denkens*, Halle 1879, G. Olms Verlag 1971, also in [Hei, 1–82].

[FS]      H. FRIEDMAN, M. SHEARD, *Elementary descent recursion and proof theory*, Ann. Pure Appl. Logic 71 (1995), 1–47.

[Ga]      D. GABBAY, *Decidability results in non-classical logic III*, Israel Journal of Mathematics 10 (1971), 135–146.

[GJ]      M. GAREY, D. JOHNSON, *Computers and Intractability, A Guide to the Theory of NP-Completeness*, Freeman 1979.

[Ge]      G. GENTZEN, *The Collected Papers of Gerhard Gentzen* (editor M. E. SZABO), North-Holland 1969.

[Gö1]     K. GÖDEL, *Die Vollständigkeit der Axiome des logischen Funktionenkalküls*, Monatshefte f. Math. u. Physik 37 (1930), 349–360, also in [Gö3, Vol. I, 102–123], [Hei, 582–591].

[Gö2]     ——, *Über formal unentscheidbare Sätze der Principia Mathematica und verwandter Systeme I*, Monatshefte f. Math. u. Physik 38 (1931), 173–198, also in [Gö3, Vol. I, 144–195], [Hei, 592–617], [Dav, 4–38].

[Gö3]     ——, *Collected Works* (editor S. FEFERMAN), Vol. I–V, Oxford Univ. Press, Vol. I 1986, Vol. II 1990, Vol. III 1995, Vol. IV, V 2003.

[Gor]     S. N. GORYACHEV, *On the interpretability of some extensions of arithmetic*, Mathematical Notes 40 (1986), 561–572.

[Gr]      G. GRÄTZER, *Universal Algebra*, New York 1968, 2$^{nd}$ ed. Springer 1979.

[HP]      P. HÁJEK, P. PUDLÁK, *Metamathematics of First-Order Arithmetic*, Springer 1993.

[Hei]     J. VAN HEIJENOORT (editor), *From Frege to Gödel*, Harvard Univ. Press 1967.

[He]      L. HENKIN, *The completeness of the first-order functional calculus*, J. Symb. Logic 14 (1949), 159–166.

[Her]     J. HERBRAND, *Recherches sur la théorie de la démonstration*, C. R. Soc. Sci. Lett. Varsovie, Cl. III (1930), also in [Hei, 525–581].

[HR]    H. HERRE, W. RAUTENBERG, *Das Basistheorem und einige Anwen-dungen in der Modelltheorie*, Wiss. Z. Humboldt-Univ., Math. Nat. R. 19 (1970), 579–583.

[HeR]   B. HERRMANN, W. RAUTENBERG, *Finite replacement and finite Hilbert-style axiomatizability*, Zeitsch. Math. Logik Grundlagen Math. 38 (1982), 327–344.

[HA]    D. HILBERT, W. ACKERMANN, *Grundzüge der theoretischen Logik*, Berlin 1928, 6$^{th}$ ed. Springer 1972.

[HB]    D. HILBERT, P. BERNAYS, *Grundlagen der Mathematik*, I, II, Berlin 1934, 1939, 2$^{nd}$ ed. Springer, Vol. I 1968, Vol. II 1970.

[Hi]    P. HINMAN, *Fundamentals of Mathematical Logic*, A. K. Peters 2005.

[Ho]    W. HODGES, *Model Theory*, Cambridge Univ. Press 1993.

[Hor]   A. HORN, *On sentences which are true of direct unions of algebras*, J. Symb. Logic 16 (1951), 14–21.

[Hu]    T. W. HUNGERFORD, *Algebra*, Springer 1980.

[Id]    P. IDZIAK, *A characterization of finitely decidable congruence modular varieties*, Trans. Amer. Math. Soc. 349 (1997), 903–934.

[Ig]    K. IGNATIEV, *On strong provability predicates and the associated modal logics*, J. Symb. Logic 58 (1993), 249–290.

[JK]    R. JENSEN, C. KARP, *Primitive recursive set functions*, in *Axiomatic Set Theory, Vol. I* (editor D. SCOTT), Proc. Symp. Pure Math. 13, I AMS 1971, 143–167.

[Ka]    R. KAYE, *Models of Peano Arithmetic*, Clarendon Press 1991.

[Ke]    H. J. KEISLER, *Logic with the quantifier "there exist uncountably many"*, Annals of Mathematical Logic 1 (1970), 1–93.

[Kl1]   S. KLEENE, *Introduction to Metamathematics*, Amsterdam 1952, 2$^{nd}$ ed. Wolters-Noordhoff 1988.

[Kl2]   ———, *Mathematical Logic*, Wiley & Sons 1967.

[KR]    I. KOREC, W. RAUTENBERG, *Model interpretability into trees and applications*, Arch. math. Logik 17 (1976), 97–104.

[Kr]    M. KRACHT, *Tools and Techniques in Modal Logic*, Elsevier 1999.

[Kra]   J. KRAJÍČEK, *Bounded Arithmetic, Propositional Logic, and Complexity Theory*, Cambridge Univ. Press 1995.

[KK]  G. KREISEL, J.-L. KRIVINE, *Elements of Mathematical Logic*, North-Holland 1971.

[Ku]  K. KUNEN, *Set Theory, An Introduction to Independence Proofs*, North-Holland 1980.

[Le]  A. LEVY, *Basic Set Theory*, Springer 1979.

[Li]  P. LINDSTRÖM, *On extensions of elementary logic*, Theoria 35 (1969), 1–11.

[Ll]  J. W. LLOYD, *Foundations of Logic Programming*, Berlin 1984, 2$^{nd}$ ed. Springer 1987.

[Lö]  M. LÖB, *Solution of a problem of Leon Henkin*, J. Symb. Logic 20 (1955), 115–118.

[MS]  A. MACINTYRE, H. SIMMONS, *Gödel's diagonalization technique and related properties of theories*, Colloquium Mathematicum 28 (1973), 165–180.

[Ma]  A. I. MAL'CEV, *The Metamathematics of Algebraic Systems*, North-Holland 1971.

[Mal]  J. MALITZ, *Introduction to Mathematical Logic*, Springer 1979.

[Mar]  D. MARKER, *Model Theory, An Introduction*, Springer 2002.

[Mat]  Y. MATIYASEVICH, *Hilbert's Tenth Problem*, MIT Press 1993.

[MV]  R. MCKENZIE, M. VALERIOTE, *The Structure of Decidable Locally Finite Varieties*, Progress in Mathematics 79, Birkhäuser 1989.

[Me]  E. MENDELSON, *Introduction to Mathematical Logic*, Princeton 1964, 4$^{th}$ ed. Chapman & Hall 1997.

[Mo]  D. MONK, *Mathematical Logic*, Springer 1976.

[Moo]  G. H. MOORE, *The emergence of first-order logic*, in *History and Philosophy of Modern Mathematics* (editors W. ASPRAY, P. KITCHER), University of Minnesota Press 1988, 95–135.

[ML]  G. MÜLLER, W. LENSKI (editors), *The $\Omega$-Bibliography of Mathematical Logic*, Springer 1987.

[Po]  W. POHLERS, *Proof Theory, An Introduction*, Lecture Notes in Mathematics 1407, Springer 1989.

[Pz]  B. POIZAT, *A Course in Model Theory*, Springer 2000.

[Pr]     M. PRESBURGER, *Über die Vollständigkeit eines gewissen Systems der Arithmetik ganzer Zahlen, in welchem die Addition als einzige Operation hervortritt*, Congrès des Mathématiciens des Pays Slaves 1 (1930), 92–101.

[RS]     H. RASIOWA, R. SIKORSKI, *The Mathematics of Metamathematics*, Warschau 1963, $3^{rd}$ ed. Polish Scientific Publ. 1970.

[Ra1]    W. RAUTENBERG, *Klassische und Nichtklassische Aussagenlogik*, Vieweg 1979.

[Ra2]    ———, *Modal tableau calculi and interpolation*, Journ. Phil. Logic 12 (1983), 403–423.

[Ra3]    ———, *A note on completeness and maximality in propositional logic*, Reports on Mathematical Logic 21 (1987), 3–8.

[Ra4]    ———, *Einführung in die mathematische Logik*, Wiesbaden 1996, $3^{rd}$ ed. Vieweg+Teubner 2008.

[Ra5]    ———, *Messen und Zählen, Eine einfache Konstruktion der reellen Zahlen*, Heldermann 2007.

[RZ]     W. RAUTENBERG, M. ZIEGLER, *Recursive inseparability in graph theory*, Notices Amer. Math. Soc. 22 (1975), A–523.

[Ro1]    A. ROBINSON, *Introduction to Model Theory and to the Metamathematics of Algebra*, Amsterdam 1963, $2^{nd}$ ed. North-Holland 1974.

[Ro2]    ———, *Non-Standard Analysis*, Amsterdam 1966, $3^{rd}$ ed. North-Holland 1974.

[Rob]    J. ROBINSON, *A machine-oriented logic based on the resolution principle*, Journal of the ACM 12 (1965), 23–41.

[Rog]    H. ROGERS, *Theory of Recursive Functions and Effective Computability*, New York 1967, $2^{nd}$ ed. MIT Press 1988.

[Ros]    J. B. ROSSER, *Extensions of some theorems of Gödel and Church*, J. Symb. Logic 1 (1936), 87–91, also in [Dav, 230–235].

[Rot]    P. ROTHMALER, *Introduction to Model Theory*, Gordon & Breach 2000.

[Ry]     C. RYLL-NARDZEWSKI, *The role of the axiom of induction in elementary arithmetic*, Fund. Math. 39 (1952), 239–263.

[Sa]     G. SACKS, *Saturated Model Theory*, W. A. Benjamin 1972.

[Sam]    G. SAMBIN, *An effective fixed point theorem in intuitionistic diagonalizable algebras*, Studia Logica 35 (1976), 345–361.

[Sc]     U. SCHÖNING, *Logic for Computer Scientist*, Birkhäuser 1989.

[Se]  A. SELMAN, *Completeness of calculi for axiomatically defined classes of algebras*, Algebra Universalis 2 (1972), 20–32.

[Sh]  S. SHAPIRO (editor), *The Oxford Handbook of Philosophy of Mathematics and Logic*, Oxford Univ. Press 2005.

[She]  S. SHELAH, *Classification Theory and the Number of Nonisomorphic Models*, Amsterdam 1978, 2$^{nd}$ ed. North-Holland 1990.

[Shoe]  J. R. SHOENFIELD, *Mathematical Logic*, Reading Mass. 1967, A. K. Peters 2001.

[Si]  W. SIEG, *Herbrand analyses*, Arch. Math. Logic 30 (1991), 409–441.

[Sm]  P. SMITH, *An Introduction to Gödel's Theorems*, Cambridge Univ. Press 2007.

[Smo]  C. SMORYŃSKI, *Self-reference and Modal Logic*, Springer 1984.

[Smu]  R. SMULLYAN, *Theory of Formal Systems*, Princeton Univ. Press 1961.

[So]  R. SOLOVAY, *Provability interpretation of modal logic*, Israel Journal of Mathematics 25 (1976), 287–304.

[Sz]  W. SZMIELEW, *Elementary properties of abelian groups*, Fund. Math. 41 (1954), 203–271.

[Tak]  G. TAKEUTI, *Proof Theory*, Amsterdam 1975, 2$^{nd}$ ed. Elsevier 1987.

[Ta1]  A. TARSKI, *Der Wahrheitsbegriff in den formalisierten Sprachen*, Studia Philosophica 1 (1936), 261–405 (first edition in Polish, 1933), also in [Ta4, 152-278].

[Ta2]  ――――, *Introduction to Logic and to the Methodology of Deductive Sciences*, Oxford 1941, 3$^{rd}$ ed. Oxford Univ. Press 1965 (first edition in Polish, 1936).

[Ta3]  ――――, *A Decision Method for Elementary Algebra and Geometry*, Santa Monica 1948, Berkeley 1951, Paris 1967.

[Ta4]  ――――, *Logic, Semantics, Metamathematics* (editor J. CORCORAN), Oxford 1956, 2$^{nd}$ ed. Hackett 1983.

[TMR]  A. TARSKI, A. MOSTOWSKI, R. M. ROBINSON, *Undecidable Theories*, North-Holland 1953.

[TV]  A. TARSKI, R. VAUGHT, *Arithmetical extensions and relational systems*, Compositio Mathematica 13 (1957), 81–102.

[Tu]  A. TURING, *On computable numbers, with an application to the Entscheidungsproblem*, Proc. London Math. Soc., 2$^{nd}$ Ser. 42 (1937), 230–265, also in [Dav, 115–154].

[Vi1]   A. VISSER, *Aspects of Diagonalization and Provability*, Dissertation, University of Utrecht 1981.

[Vi2]   _____, *An overview of interpretability logic*, in *Advances in Modal Logic, Vol. 1* (editors M. KRACHT et al.), CSLI Lecture Notes 87 (1998), 307–359.

[Wae]   B. L. VAN DER WAERDEN, *Algebra I*, Berlin 1930, 4$^{th}$ ed. Springer 1955.

[Wag]   F. WAGNER, *Simple Theories*, Kluwer Academic Publ. 2000.

[Wa1]   H. WANG, *From Mathematics to Philosophy*, Routlegde & Kegan Paul 1974.

[Wa2]   _____, *Computer, Logic, Philosophy*, Kluwer Academic Publ. 1990.

[Wa3]   _____, *A Logical Journey, From Gödel to Philosophy*, MIT Press 1997.

[WR]    A. WHITEHEAD, B. RUSSELL, *Principia Mathematica*, I–III, Cambridge 1910, 1912, 1913, 2$^{nd}$ ed. Cambridge Univ. Press, Vol. I 1925, Vol. II, III 1927.

[Wi]    A. WILKIE, *Model completeness results for expansions of the ordered field of real numbers by restricted Pfaffian functions and the exponential function*, Journal Amer. Math. Soc. 9 (1996), 1051–1094.

[WP]    A. WILKIE, J. PARIS, *On the scheme of induction for bounded arithmetic formulas*, Ann. Pure Appl. Logic 35 (1987), 261–302.

[Zi]    M. ZIEGLER, *Model theory of modules*, Ann. Pure Appl. Logic 26 (1984), 149–213.

# Index of Terms and Names

# Index of Symbols

317